YANMAR MARINE DIESEL ENGINE 1GM10, 2GM20, 3GM30, 3HM35

SERVICE MANUAL

YANMAR MARINE DIESEL ENGINE 1GM10, 2GM20, 3GM30, 3HM35

SERVICE MANUAL

ISBN/EAN: 9783954275076
Erscheinungsjahr: 2012
Erscheinungsort: Bremen, Deutschland

© maritimepress in Europäischer Hochschulverlag GmbH & Co. KG, Fahrenheitstr. 1, 28359 Bremen. Alle Rechte beim Verlag und bei den jeweiligen Lizenzgebern.

www.maritimepress.de | office@maritimepress.de

Bei diesem Titel handelt es sich um den Nachdruck eines historischen, lange vergriffenen Buches. Da elektronische Druckvorlagen für diese Titel nicht existieren, musste auf alte Vorlagen zurückgegriffen werden. Hieraus zwangsläufig resultierende Qualitätsverluste bitten wir zu entschuldigen.

YANMAR
SERVICE MANUAL

MARINE DIESEL ENGINE

MODELS

1GM10(C)
2GM20(F)(C)
3GM30(F)(C)
3HM35(F)(C)

FOREWORD

This service manual has been compiled for engineers engaged in sales, service, inspection and maintenance. Accordingly, descriptions of the construction and functions of the engine are emphasized in this manual while items which should already be common knowledge are omitted.

One characteristic of a marine diesel engine is that its performance in a vessel is governed by its applicability to the vessel's hull construction and its steering system.

Engine installation, fitting out and propeller selection have a substantial effect on the performance of the engine and the vessel. Moreover, when the engine runs unevenly or when trouble occurs, it is essential to check a wide range of operating conditions—such as installation on the hull and suitability of the ship's piping and propeller—and not just the engine itself. To get maximum performance from this engine, you should completely understand its functions, construction and capabilities, as well as proper use and servicing.

Use this manual as a handy reference in daily inspection and maintenance, and as a text for engineering guidance.

Models **1GM10(C)**
2GM20(F)(C)
3GM30(F)(C)
3HM35(F)(C)

A. Engine Model Name 0-1
B. Engine Model Name Plate
   and Clutch Model Name Plate 0-1
C. Cylinder Number 0-3

CHAPTER 1 GENERAL
1. Specifications 1-1
2. Principal Construction 1-4
3. Performance Curves 1-5
4. Features 1-9
5. Engine Cross-Sections 1-10
6. Dimensions 1-17
7. Piping Diagrams 1-24

CHAPTER 2 BASIC ENGINE
1. Cylinder Block 2-1
2. Cylinder Head 2-9
3. Piston 2-28
4. Connecting Rod 2-34
5. Crankshaft 2-38
6. Flywheel and Housing 2-49
7. Camshaft 2-53
8. Timing Gear 2-59

CHAPTER 3 FUEL SYSTEM
1. Fuel Injection System 3-1
2. Injection Pump 3-3
3. Injection Nozzle 3-25
4. Fuel Filter 3-29
5. Fuel Feed Pump 3-30
6. Fuel Tank (Option) 3-33

CHAPTER 4 GOVERNOR
1. Governor 4-1
2. Injection Limiter 4-9
3. No-Load Maximum Speed Limiter 4-11
4. Idling Adjuster 4-12
5. Engine Stop Lever 4-13

CHAPTER 5 INTAKE AND EXHAUST SYSTEM
1. Intake and Exhaust System 5-1
2. Intake Silencer 5-3
3. Exhaust System 5-4
4. Breather 5-6

CHAPTER 6 LUBRICATION SYSTEM
1. Lubrication System 6-1
2. Oil Pump 6-5
3. Oil Filter 6-9
4. Oil Pressure Regulator Valve 6-12
5. Oil Pressure Measurement 6-14

CHAPTER 7 DIRECT SEA-WATER COOLING SYSTEM
1. Cooling System 7-1
2. Water Pump 7-5
3. Thermostat 7-11
4. Anticorrosion Zink 7-14
5. Kingston Cock (Option) 7-16
6. Bilge Pump and Bilge Strainer (Option) ... 7-17

CHAPTER 8 FRESH WATER COOLING SYSTEM
1. Cooling System 8-1
2. Sea Water Pump 8-3
3. Fresh Water Pump 8-4
4. Heat Exchanger 8-7
5. Filler Cap and Subtank 8-11
6. Thermostat 8-13
7. Cooling Water Temperature Switch 8-16
8. Precautions 8-17

CHAPTER 9 MODIFYING THE COOLING SYSTEM
1. General 9-1
2. Disassembly of Sea Water-Cooled Engine ... 9-2
3. Assembling modified parts
   to the Fresh Water-Cooled Engine 9-7
4. Cautions When the Engine is Installed Inboard ... 9-12

CHAPTER 10 REDUCTION AND REVERSING GEAR
[A] For Engine Models 1GM10, 2GM20(F) and 3GM30(F)
1. Construction 10-1
2. Shifting Device 10-7
3. Inspection and Servicing 10-14
4. Disassembly 10-19
5. Reassembly 10-24

[B] For Model 3GM35(F)
1. Construction 10-29
2. Installation 10-33
3. Operation and Maintenance 10-34
4. Inspection and Servicing 10-35
5. Disassembly 10-40
6. Reassembly 10-44

[C] Marine Gear Models KM2P, KM3P and KM3V
    for Engine Models 1GM10, 2GM20(F) and 3GM30(F)
1. Construction 10-50
2. Shifting Device 10-56
3. Inspection and Servicing 10-61
4. Disassembly 10-68
5. Reassembly 10-73

[D] V-drive Gear, Model KM3V
1. Construction 10-77
2. Specifications 10-80
3. Power Transmission System 10-81
4. Cooling System (Sea-water Cooling Engine) ... 10-82
5. Piping Diagrams 10-85
6. Inspection and Servicing 10-90
7. Shim Adjustment for V-drive Gear Shaft,
   and Backlash Adjustment for V-drive Gear Shaft and
   Drive Gear 10-92
8. Disassembly 10-94
9. Reassembly 10-97

CHAPTER 11 REMOTE CONTROL SYSTEM
1. Construction 11-1
2. Clutch and Speed Regulator Remote Control 11-3
3. Engine Stop Remote Control 11-7

CHAPTER 12 ELECTRICAL SYSTEM
1. Electrical System 12-1
2. Battery 12-4
3. Starter Motor 12-7
4. Alternator Standard, 12V/55A 12-18
4A. Alternator Option, 12V/35A 12-28
5. Instrument Panel 12-37
6. Tachometer 12-43

CHAPTER 13 OPERATING INSTRUCTIONS
1. Fuel Oil and Lubricating Oil 13-1
2. Engine Operating Instructions 13-8
3. Troubleshooting and Repair 13-13

CHAPTER 14 DISASSEMBLY AND REASSEMBLY
(Direct Sea-Water Cooling Engine)
1. Disassembly and Reassembly Precautions 14-1
2. Disassembly and Reassembly Tools 14-2
3. Others 14-13
4. Disassembly 14-14
5. Reassembly 14-28

CHAPTER 15 DISASSEMBLY AND REASSEMBLY
(Fresh Water Cooling Engine)
1. Disassembly of Fresh Water-Cooled Engine 15-1
2. Reassembly of Fresh Water-Cooled Engine 15-11
3. Tightening Torque 15-21
4. Packing Supplement and Adhesive Application Point 15-24

A. Engine Model Name
B. Engine Model Name Plate and Clutch Model Name Plate

SM/GM(F)(C)-HM(F)(C)

A. Engine Model Name

The nomenclature of the New GM(F)/HM(F) series follows the order shown below.

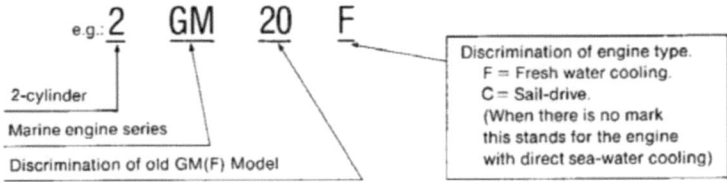

e.g.: 2 GM 20 F

- 2-cylinder
- Marine engine series
- Discrimination of old GM(F) Model
- Discrimination of engine type.
  F = Fresh water cooling.
  C = Sail-drive.
  (When there is no mark this stands for the engine with direct sea-water cooling)

B. Engine Model Name Plate and Clutch Model Name Plate

To every engine model described in this manual, an engine model name plate and clutch model name plate are fitted as shown in the following figures. In addition, the engine serial number is stamped on the cylinder body.
Specifications of the engine and clutch to be shipped are recorded and filed using the numbers marked on the engine model name plate and clutch model name plate.

The specifications or components of the engine or clutch may have been partially altered to improve performance, and the components involved may not necessarily be interchangeable.
Therefore, when parts are ordered, please furnish the item description in the blank spaces shown in the figures, using the descriptions given on these plates.

B-1 Item descriptions on the model name plates and information to be forwarded to us

[Item descriptions on Model name plates]

[Information to be forwarded to us]

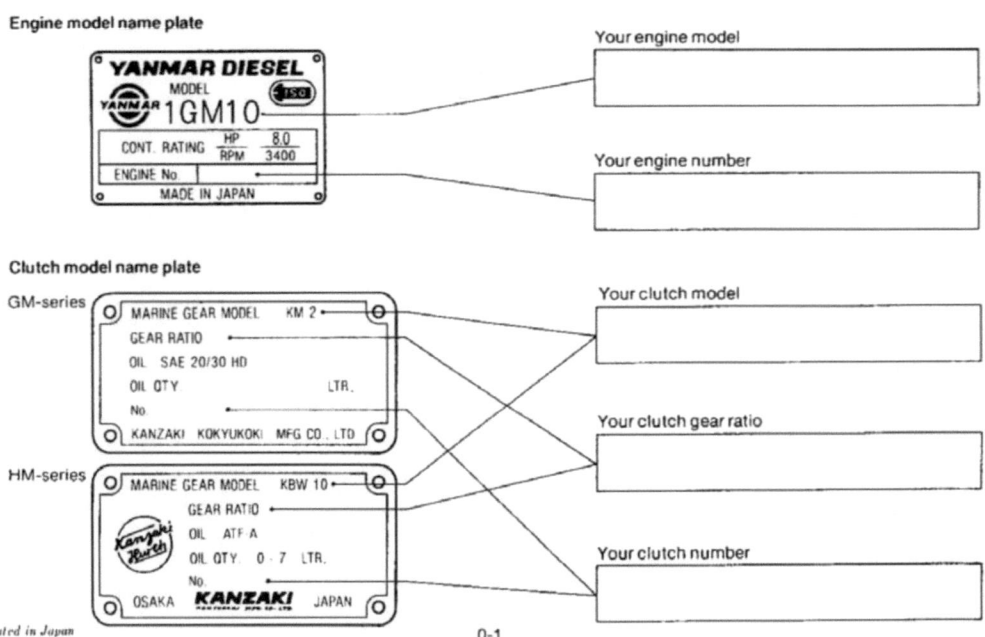

B. Engine Model Name Plate and Clutch Model Name Plate

B-2 Location of engine model name plate and clutch model name plate

B-2.1 1GM10(C)

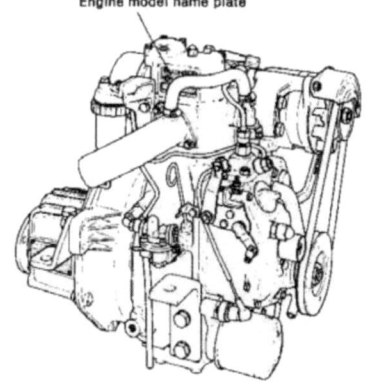

B-2.2 2GM20(F)(C), 3GM30(F)(C)

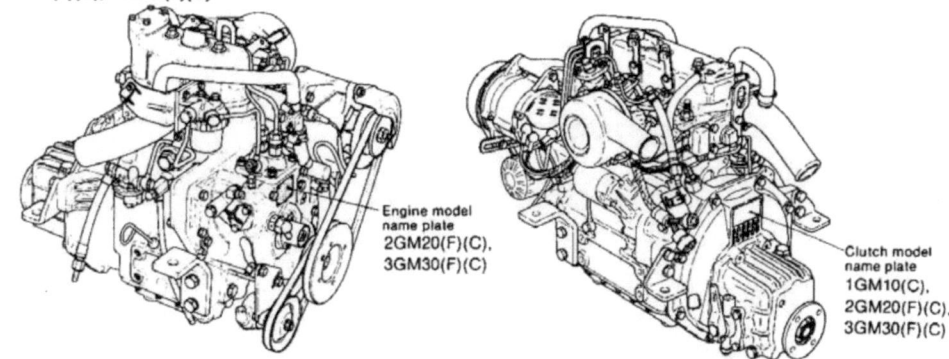

B-2.3 3HM35(F)(C)

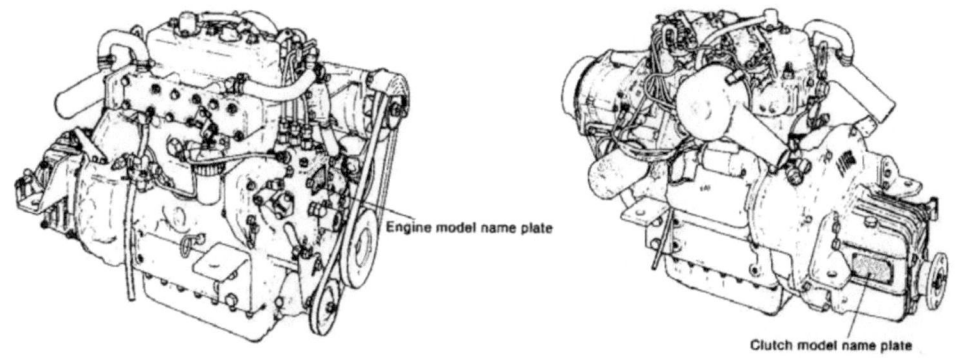

B. *Engine Model Name Plate and Clutch Model Name Plate*
C. *Cylinder Number*

SM/GM(F)(C)·HM(F)(C)

B-3 Location of stamped engine serial number

B-3.1 1GM10(C)

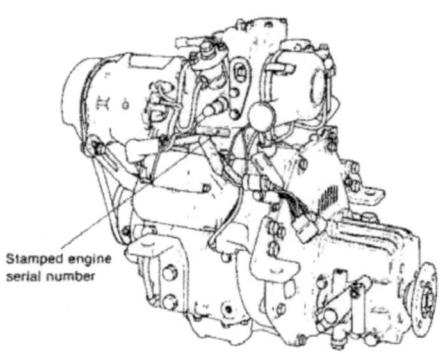

Stamped engine serial number

B-3.2 2GM20(F)(C), 3GM30(F)(C), 3HM35(F)(C)

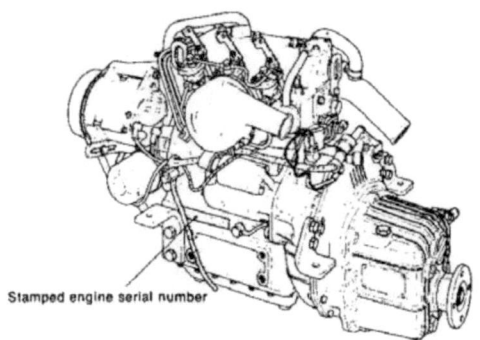

Stamped engine serial number

C. Cylinder Number

The cylinder numbers of the 2 cylinder engine and 3 cylinder engine described in this manual are designated as follows.

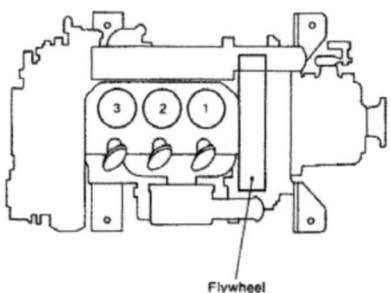

Flywheel

(1) The sequence of cylinder numbers is given as No. 1, No. 2 and No. 3 starting from the flywheel side.
(2) These cylinder numbers are consistently used for devices and parts connected with the cylinder head and valve moving mechanism. However, please note that items related to the fuel injection pump do not correspond to the numbering of the cylinders.

CHAPTER 1
GENERAL

1. Specifications 1-1
2. Principal Construction 1-4
3. Performance Curves 1-5
4. Features 1-9
5. Engine Cross-Sections 1-10
6. Dimensions 1-17
7. Piping Diagrams 1-24

Chapter 1 General
1. Specifications

1-1. Direct Sea Water Cooling Type

Model		1GM10	2GM20	3GM30	3HM35
Type		Vertical 4-cycle water cooled diesel engine			
Combustion chamber		Swirl pre-combustion chamber			
Number of cylinders		1	2	3	3
Bore × stroke	mm (in.)	75 × 72 (2.95 × 2.83)			80 × 85 (3.15 × 3.35)
Displacement	ℓ (in.³)	0.318 (19.40)	0.636 (38.81)	0.954 (58.21)	1.282 (78.23)
Continuous rating output (DIN6270A)	kW/rpm (HP/rpm)	5.9/3400 (8.07/3400)	11.8/3400 (16.0/3400)	17.7/3400 (24.1/3400)	22.4/3200 (30/3200)
	kgf/cm² (lb/in.²)		6.58 (94.71)		6.58 (93.57)
	m/sec (ft/sec.)		8.16 (26.77)		9.07 (29.76)
One hour rating output (DIN6270B)	kW/rpm (HP/rpm)	6.7/3600 (9.1/3600)	13.4/3600 (18.2/3600)	20.1/3600 (27.3/3600)	25.4/3400 (34/3400)
	kgf/cm² (lb/in.²)		7.07 (100.54)		7.02 (99.82)
	m/sec (ft/sec.)		8.64 (28.35)		9.63 (31.59)
Compression ratio			23.0		24.8
Fuel injection timing (F.I.D)	degree	b TDC15±1	b TDC15±1	b TDC18±1	b TDC21±1
Fuel injection pressure	kgf/cm² (lb/in.²)		170±5 (2347−2489)		160±5 (2204−2347)
Main power take off		at Flywheel side			
Front power take off		at Crankshaft V pulley side			
Direction of rotation	Crankshaft	Counter-clockwise viewed from stern			
	Propeller shaft (Ahead)	Clockwise viewed from stern			
Cooling system		Direct sea water cooling (rubber impeller water pump)			
Lubrication system		Complete enclosed forced lubrication			
Starting system		Electric and manual			Electric

Model		KM2C		KM3A		KBW10E	
Type		Mechanical cone clutch with single stage for both ahead and astern				Wet multi-disc mechanical type	
Clutch	Reduction ratio (Ahead/Astern)	2.21/3.06	3.22/3.06	2.36/3.16	3.20/3.16	2.14/2.50	2.83/2.50
	Propeller speed DIN A rating (Ahead/Astern)	1540/1113	1055/1113	1441/1076	1063/1076	1498/1260	1129/1260
	Lubricating oil capacity ℓ (in.³)	0.25 (15.26)		0.3 (18.31)		0.7 (42.72)	
	Clutch weight kg (lb.)	9.5 (20.95)		11.0 (24.26)		17.5 (38.58)	
Dimensions	Overall length mm (in.)	547 (21.53)		638 (25.12)		786 (30.94)	
	Overall width mm (in.)	410 (16.14)		455 (17.91)		485 (19.09)	
	Overall height mm (in.)	485 (19.09)		495 (19.50)		617 (24.29)	
Lubricating oil capacity (rake angle 8°)	Total ℓ (in.³)	1.3 (79.33)		2.6 (158.65)		5.4 (329.51)	
	Effective ℓ (in.³)	0.6 (36.61)		1.3 (79.33)		2.7 (164.75)	
Engine weight with clutch (dry)	Kg (lb.)	76 (168)		106 (234)		130 (287)	158 (348)

Chapter 1 General
1. Specifications

SM/GM(FK)·HM(FK)

1-2. Fresh Water Cooling Type

Model			2GM20F	3GM30F	3HM35F
Type			Vertical 4-cycle water cooled diesel engine		
Combustion chamber			Swirl pre-combustion chamber		
Number of cylinders			2	3	3
Bore x stroke		mm (in.)	75 x 72 (2.95 x 2.83)		80 x 85 (3.15 x 3.35)
Displacement		ℓ (in.³)	0.636 (38.81)	0.954 (58.71)	1.282 (78.22)
Continuous rating output (DIN 6270A)	Output/Crankshaft speed	kW/rpm(HP/rpm)	11.8/3400(16.0/3400)	17.7/3400(24.1/3400)	22.4/3200(30/3200)
	Brake mean effective pressure	kgf/cm² (lb/in.²)	6.66 (94.71)		6.58 (93.57)
	Piston speed	m/sec (ft/sec)	8.16 (26.77)		9.07 (29.76)
One hour rating output (DIN 6270B)	Output/Crankshaft speed	kW/rpm(HP/rpm)	13.4/3600(18.2/3600)	20.1/3600(27.3/3600)	25.4/3400(34/3400)
	Brake mean effective pressure	kgf/cm² (lb/in.²)	7.07 (100.54)		7.02 (99.82)
	Piston speed	m/sec (ft/sec)	8.64 (28.35)		9.63 (31.59)
Compression ratio			23.0		24.8
Fuel injection timing (FID)		degree	b TDC 15±1	b TDC 18±1	b TDC 21±1
Fuel injection pressure		kgf/cm² (lb/in.²)	170±5 (2347−2489)		160±5 (2204−2347)
Main power take off			at Flywheel side		
Front power take off			at Crankshaft V-pulley side		
Direction of rotation	Crankshaft		Counter-clockwise viewed from stern		
	Propeller shaft (Ahead)		Clockwise viewed from stern		
Cooling system			Fresh water cooling with heat exchanger		
Lubrication system			Complete enclosed forced lubrication		
Starting system			Electric		
Clutch	Model		KM2-C	KM3A	KBW10E
	Type		Mechanical cone clutch with single stage for both ahead and astern		Wet multi-disc mechanical type
	Reduction ratio (Ahead/Astern)		2.21/3.06	2.62/3.06	3.22/3.06	2.36/3.16	2.61/3.16	3.20/3.16	2.14/2.50	2.83/2.50
	Propeller speed DIN A rating (Ahead/Astern)	rpm	1540/1113	1298/1113	1055/1113	1441/1076	1303/1076	1062/1076	1496/1280	1129/1280
	Lubricating oil capacity	ℓ (in.³)	0.25 (15.26)	0.30 (18.31)	0.70 (42.72)	
Clutch weight		kg (lb.)	9.5 (20.95)	11.0 (24.26)	17.5 (38.58)	
Dimensions	Overall length	mm (in.)	643 (25.31)	740 (29.13)	791 (31.14)	
	Overall width	mm (in.)	482 (19.00)	455 (17.91)	475 (18.70)	
	Overall height	mm (in.)	545 (21.46)	545 (21.46)	638 (25.12)	
Lubricating oil capacity (rake angle 8°)	Total	ℓ (in.³)	2.0 (122.05)	2.6 (158.66)	5.4 (329.51)	
	Effective	ℓ (in.³)	1.3 (79.33)	1.6 (97.63)	2.7 (164.75)	
Engine weight with clutch (dry)		kg (lb.)	114 (251)	136 (304)	167 (368)	

1-3. Direct Sea Water Cooling Type (Sail-drive)

Model			1GM10C	2GM20C	3GM30C	3HM35C
Type				Vertical, 4-cycle water cooled diesel engine		
Combustion chamber				Swirl pre-combustion chamber		
Number of cylinders			1	2	3	3
Bore x stroke		mm (in.)	75 x 72 (2.95 x 2.83)			80 x 85 (3.15 x 3.35)
Displacement		ℓ (in.³)	0.318 (19.40)	0.636 (38.81)	0.954 (58.21)	1.282 (78.23)
Continuous rating output (DIN 6270A)	Output/Crankshaft speed	kW/rpm(HP/rpm)	5.9/3400(8.02/3400)	11.8/3400(16.0/3400)	17.7/3400(24.1/3400)	22.4/3200(30.0/3200)
	Brake mean effective pressure	kgf/cm² (lb/in.²)		6.66 (94.71)		6.58 (93.57)
	Piston speed	m/sec. (ft/sec.)		8.16 (26.77)		9.07 (29.76)
One hour rating output (DIN 6270B)	Output/Crankshaft speed	kW/rpm(HP/rpm)	6.7/3600(9.1/3600)	13.4/3600(18.2/3600)	20.1/3600(27.3/3600)	25.4/3400(34./3400)
	Brake mean effective pressure	kgf/cm² (lb/in.²)		7.07 (100.54)		7.02 (99.82)
	Piston speed	m/sec. (ft/sec.)		8.64 (28.35)		9.63 (31.59)
Compression ratio				23.0		24.5
Fuel injection timing (FID)		Degree	b TDC 15±1	b TDC 15±1	b TDC 18±1	b TDC 21±1
Fuel injection pressure		kgf/cm² (lb/in.²)		170±5 (2247~2489)		160±5 (2204~2347)
Main power take off				at Flywheel side		
Front power take off				at Crankshaft V-pulley side		
Direction of rotation	Crankshaft			Counter-clockwise viewed from stern		
	Propeller shaft (Sail-drive)			Counter-clockwise viewed from stern		
Cooling system				Direct sea water cooling (rubber impeller water pump)		
Lubrication system				Complete enclosed forced lubrication		
Starting system				Electric and manual		Electric
Sail-drive	Model			SD 20		SD 30
	Reduction system			Constant mesh gear with dog clutch		
	Reduction ratio (Ahead/Astern)			2.64/2.64		
	Propeller speed DIN A rating	rpm	1289			1212
Lubricating capacity		ℓ (in.³)		2.2 (134.24)		
Dry weight		kg (lb.)	30 (66)			32 (70)
Lubricating oil capacity (Engine side)	Total	ℓ (in.³)	1.3 (79.33)	2.0 (122.05)	2.6 (158.66)	5.4 (329.51)
	Effective	ℓ (in.³)	0.6 (36.61)	1.3 (79.33)	1.6 (97.63)	2.7 (164.75)
Engine weight with Sail-drive unit (Dry)		kg (lb.)	104 (229)	134 (295)	153 (337)	180 (397)

Chapter 1 General
2. Principal Construction

2. Principal Construction

Engine model		1GM10	2GM20	3GM30	3HM35
Group	Part	Construction			
Engine block	Cylinder block	Integrally-cast water jacket and crankcase			
	Cylinder liner	Sleeveless			
	Main bearing	Metal housing type			
	Oil sump	Oil pan			
Intake and exhaust systems and valve mechanism	Cylinder head	Integrated type cylinders			
	Intake and exhaust valves	Poppet type, seat angle 90°			
	Exhaust manifold	—		Water-cooled type	Water-cooled type
	Exhaust silencer	Water-cooled mixing elbow type			
	Valve mechanism	Overhead valve push rod, rocker arm system			
	Intake silencer	Round polyurethane sound absorbing type			
Main moving elements	Crankshaft	Stamped forging			
	Flywheel	Attached to crankshaft by flange, with ring gear			
	Piston	Oval type			
	Piston pin	Floating type			
	Piston rings	2 compression rings, 1 oil ring			
Lubrication system	Oil pump	Trochoid pump			
	Oil filter	Full-flow cartridge type, paper element			
	Oil level gauge	Dipstick			
Cooling system	Water pump	Rubber impeller type			
	Thermostat	Wax pellet type			
Fuel system	Fuel injection pump	YPFR-0707-1	YPFR-0707-2	YPFR-0707-3	
	Fuel injection valve	Throttle valve, OSDYD1			
	Fuel feed pump	Mechanical type			
	Fuel strainer	Filter paper			
Governor	Governor	Centrifugal all-speed mechanical type			
Starting system	Electric	Pinion ring gear type starter motor			
	Manual	Camshaft starting			—
Electrical system	Charger	Alternator (with built-in IC regulator)			
Reduction reversing	Reduction gear	Helical gear constant-mesh system			
Clutch system	Clutch	Servo-cone type			Wet multi-disc mechanical type

Fresh-water cooling system (2GM20F, 3GM30F and 3HM35F)

Cooling system	Sea water pump	Rubber impeller type
	Fresh water pump	Centrifugal type
	Thermostat	Wax pellet type
	Heat exchanger	Multi-tube type

3. Performance Curves

3-1. 1GM10(C)

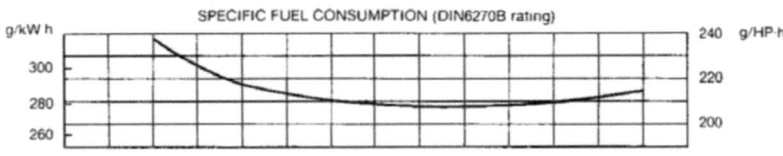

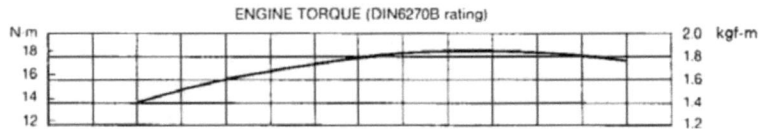

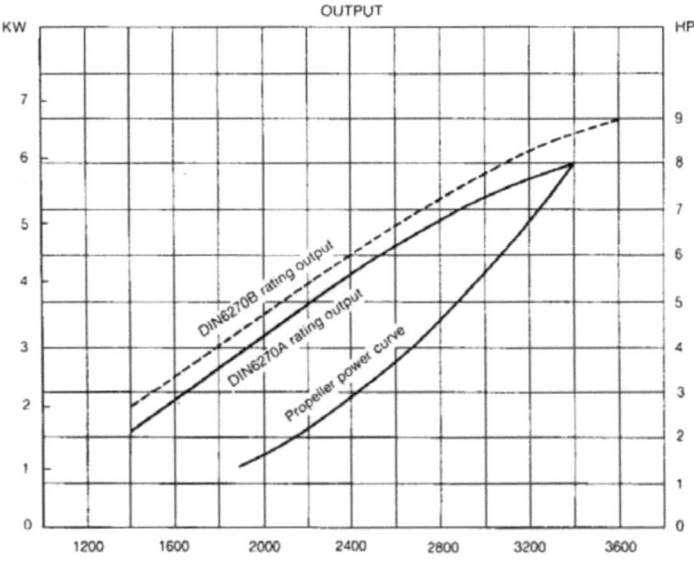

SPEED OF CRANKSHAFT: rpm
THE ENGINE FLYWHEEL OUTPUT IS APPROX 3% HIGHER.

NOTE: These curves show the average performance of respective engine in test operation at our plant.

Chapter 1 General
3. Performance Curves

SM/GM(F)(C)·HM(F)(C)

3-2 2GM20(F)(C)

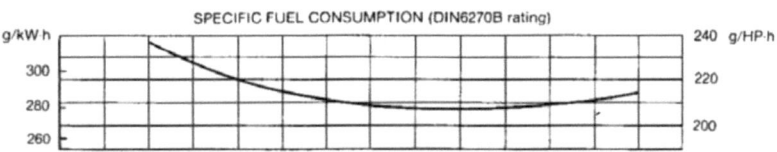

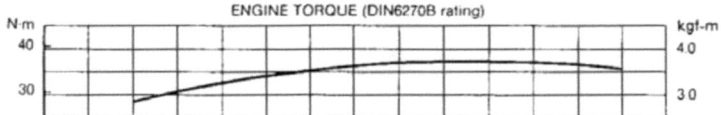

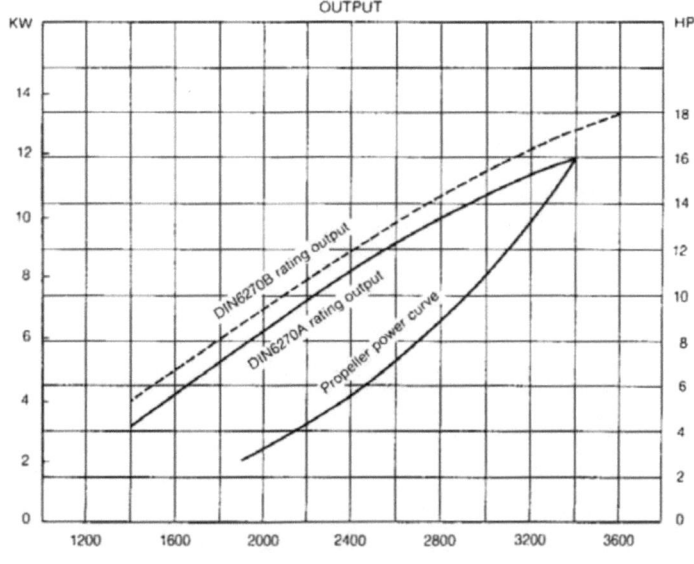

SPEED OF CRANKSHAFT: rpm
THE ENGINE FLYWHEEL OUTPUT IS APPROX 3% HIGHER.

NOTE: These curves show the average performance of respective engine in test operation at our plant.

3-3 3GM30(F)(C)

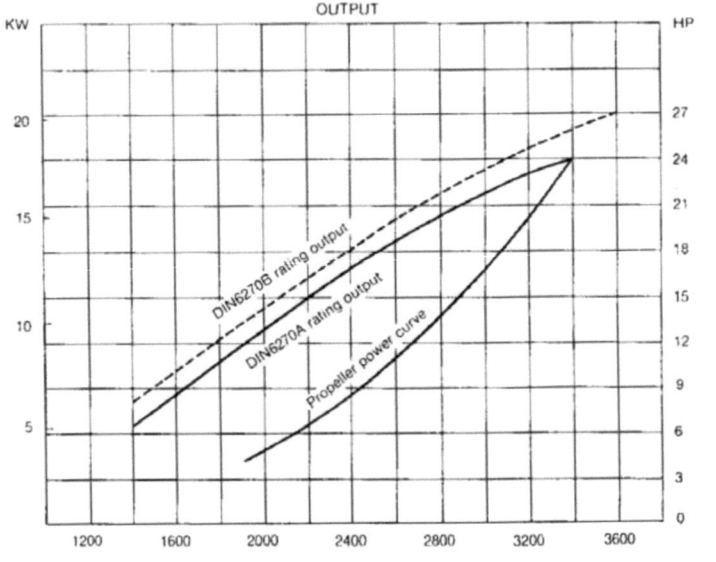

SPEED OF CRANKSHAFT rpm
THE ENGINE FLYWHEEL OUTPUT IS APPROX 3% HIGHER

NOTE: These curves show the average performance of respective engine in test operation at our plant.

3-4 3HM35(F)(C)

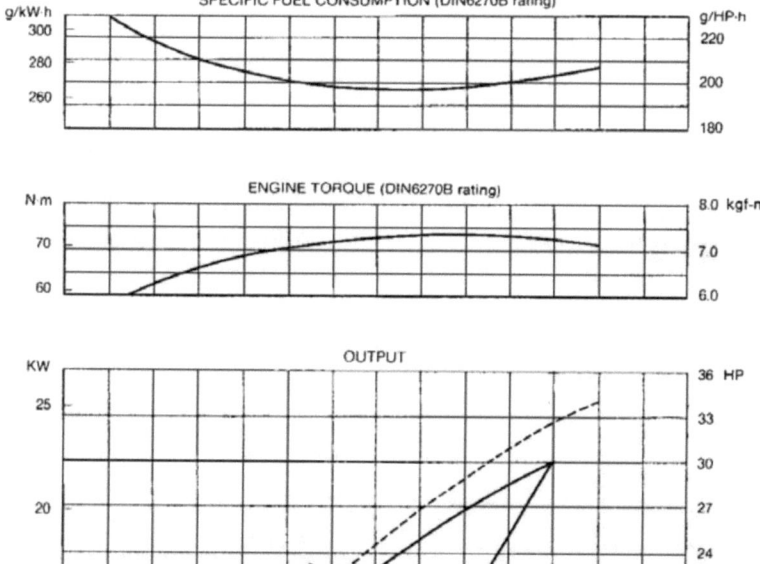

SPEED OF CRANK SHAFT. rpm
THE ENGINE FLYWHEEL OUTPUT IS APPROX 3% HIGHER

NOTE: These curves show the average performance of respective engine in test operation at our plant.

4. Features

4-1 Superior combustion performance

The unique Yanmar swirl precombustion chamber and new cooling system display superior combustion performance in all types of operation. Low-speed, low-load combustion performance, especially demanded for marine applications, is also superb, and stable performance is maintained over a wide range of speeds. Since starting characteristics are also excellent and warm-up is fast, full engine performance can be obtained within a short time.

4-2 Low operating costs

Excellent combustion and low friction reduce fuel costs, while the optimized piston shape ring configuration and improved cooling system reduce oil consumption. continuous operating time has been extended and operating costs reduced through improved durability.

4-3 Compact, lightweight

The cylinder head is the integrally-cast type, and the crankshaft is the housing type. Minimum weight has been pursued for each engine part, and a reduction reversing gear employing a special new mechanism has been incorporated to obtain revolutionary engine lightness.

4-4 Long term continuous operation

Improved durability has been achieved by adopting special construction and materials for main moving parts and the valve mechanism, which are the areas most subject to trouble in high-speed engines. Moreover, a bypass system with a thermostat maintains the cooling water at a stable high temperature, resulting in reduced cylinder liner and piston ring wear, reduced thermal load around the combustion chamber, and substantially improved durability. Long-term continuous operation is possible by correct operation and proper attention to fuel and lubricating oil.

4-5 Low vibration

Vibration has been reduced by minimizing the weights of the pistons, connecting rods, and other sources of vibration, stringent weight management at assembly, and balancing of the flywheel, V-pulley, etc. Vibration has also been suppressed through the adoption of a special cylinder block rib construction and improved rigidity. Rubber shock mounts are available when the engine is to be used under conditions which may lead to severe vibration.

4-6 Quiet operation

Intake and exhaust noises have been lowered by adopting an intake silencer, water-cooled exhaust manifold and water mixing elbow type exhaust system.
The precombustion chamber system and semi-throttle type injection valve suppress combustion noise substantially.
Moreover, gear noise has been reduced by the use of helical gears around the gear train and clutch gear, and by the buffering effect of a damper disc.
In addition, noise prevention measures have also been taken at the control valve mechanism and other parts.

4-7 Superior matching to the hull

(1) Four-point support engine installation feet make installation easy.
(2) Mist intake system prevents contamination of the engine room.
(3) Since the fuel pump is mounted on the engine, the fuel tank can be installed anywhere.
(4) Water-cooled manifold prevents a rise in the engine room temperature.
(5) Independent type instrument panel can be installed wherever it is easiest to see.
(6) Speed, clutch forward and reverse, and engine stop can all be remotely controlled.
(7) The use of rubber and vinyl hoses for ship interior piping not only facilitates piping work, but also eliminates brazing faults caused by vibration.
(8) Electric type bilge pump is available as an option.

4-8 Easy to operate

(1) Cooling water temperature switch and lubricating oil pressure switch are provided, and alarm lamps and buzzer are mounted on the instrument panel.
(2) Manual starting handle permits manual starting.
 (Except model 3HM35(C) and fresh water cooling type)
(3) Positive clutch engagement and disengagement; propeller shaft does not rotate when clutch is placed in neutral position.

5. Engine Cross-Sections

5-1 1GM10

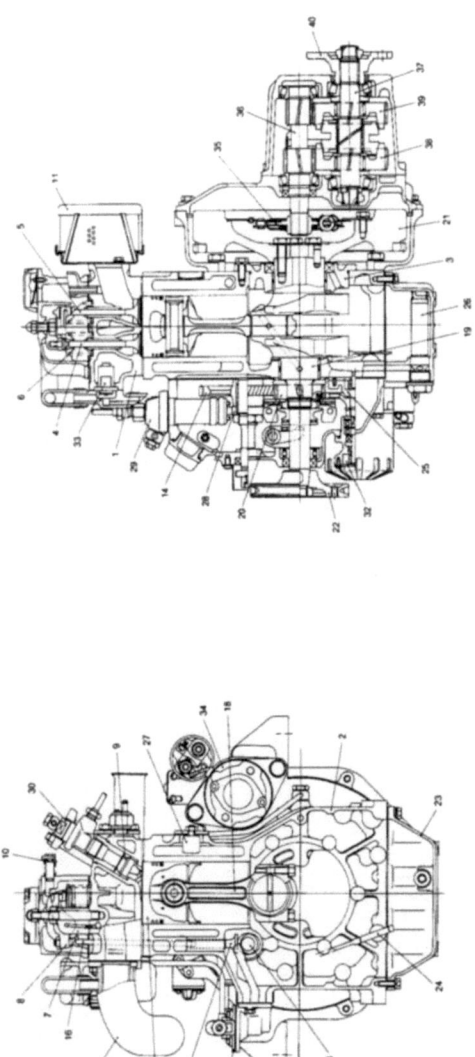

1. Cylinder head
2. Cylinder body
3. Main bearing housing
4. Exhaust valve
5. Intake valve
6. Valve spring
7. Valve rocker arm support
8. Valve rocker arm
9. Precombustion chamber
10. Decompression lever
11. Intake silencer
12. Mixing elbow
13. Camshaft
14. Camshaft gear
15. Tappet
16. Push rod
17. Piston
18. Connecting rod
19. Crankshaft
20. Crankshaft gear
21. Flywheel
22. Crankshaft V-pulley
23. Oil pan
24. Dipstick
25. Lubricating oil pump
26. Lubricating oil inlet pipe
27. Anticorrosion zinc
28. Fuel injection pump cam
29. Fuel injection pump
30. Fuel injection nozzle
31. Fuel feed pump
32. Cooling water pump
33. Thermostat
34. Starter motor
35. Damper disc
36. Input shaft
37. Output shaft
38. Forward large gear
39. Reverse large gear
40. Output shaft coupling

Chapter 1 General
5. Engine Cross-Sections

5-2 2GM20

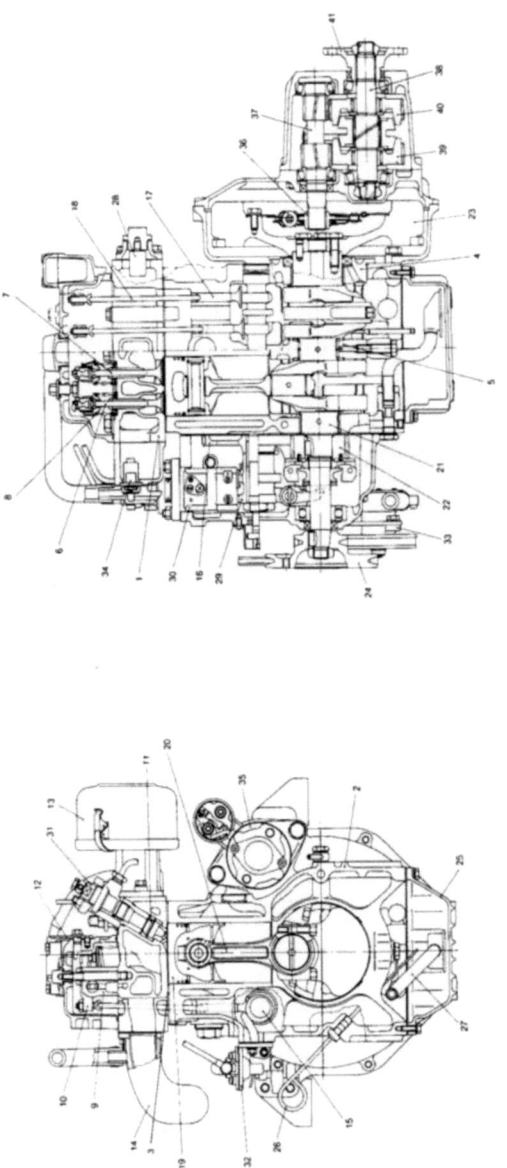

1. Cylinder head
2. Cylinder body
3. Cylinder head gasket
4. Main bearing housing
5. Intermediate main bearing housing
6. Exhaust valve
7. Intake valve
8. Valve spring
9. Valve rocker arm support
10. Valve rocker arm
11. Precombustion chamber
12. Decompression lever
13. Intake silencer
14. Exhaust manifold
15. Camshaft
16. Camshaft gear
17. Tappet
18. Push rod
19. Piston
20. Connecting rod
21. Crankshaft
22. Crankshaft gear
23. Flywheel
24. Crankshaft V-pulley
25. Oil pan
26. Dipstick
27. Lubricating oil inlet pipe
28. Anticorrosion zinc
29. Fuel injection pump cam
30. Fuel injection pump
31. Fuel injection nozzle
32. Fuel feed pump
33. Cooling water pump
34. Thermostat
35. Starter motor
36. Damper disc
37. Input shaft
38. Output shaft
39. Forward large gear
40. Reverse large gear
41. Output shaft coupling

Chapter 1 General
5. Engine Cross-Sections

5-3 3GM30

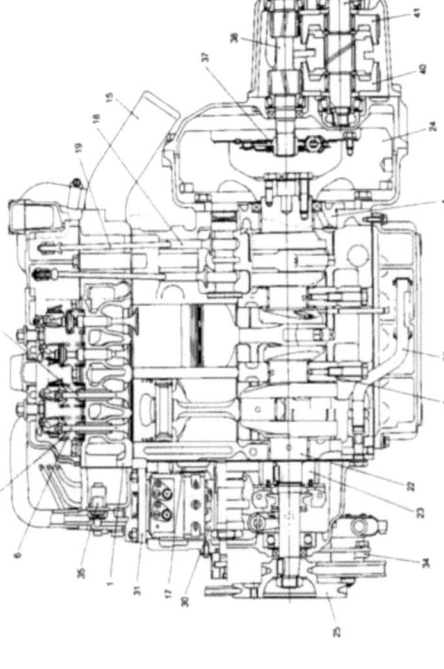

1. Cylinder head
2. Cylinder body
3. Cylinder head gasket
4. Main bearing housing
5. Intermediate main bearing housing
6. Exhaust valve
7. Intake valve
8. Valve spring
9. Valve rocker arm support
10. Valve rocker arm
11. Precombustion chamber
12. Decompression lever
13. Intake silencer
14. Exhaust manifold
15. Mixing elbow
16. Camshaft
17. Camshaft gear
18. Tappet
19. Push rod
20. Piston
21. Connecting rod
22. Crankshaft
23. Crankshaft gear
24. Camshaft V-pulley
25. Flywheel
26. Oil pan
27. Dipstick
28. Lubricating oil inlet pipe
29. Anticorrosion zinc
30. Fuel injection pump cam
31. Fuel injection pump
32. Fuel injection nozzle
33. Fuel feed pump
34. Cooling water pump
35. Thermostat
36. Starter motor
37. Damper disc
38. Input shaft
39. Output shaft
40. Forward large gear
41. Reverse large gear
42. Output shaft coupling

Chapter 1 General
5. Engine Cross-Sections

5-4 3HM35

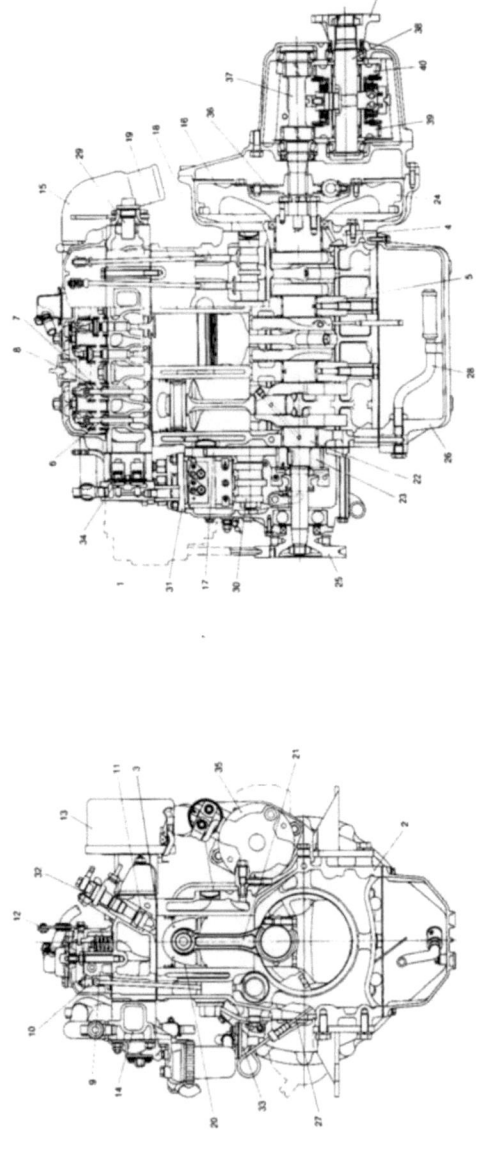

1. Cylinder head
2. Cylinder body
3. Cylinder head gasket
4. Main bearing housing
5. Intermediate main bearing housing
6. Exhaust valve
7. Intake valve
8. Valve spring
9. Valve rocker arm support
10. Valve rocker arm
11. Precombustion chamber
12. Decompression lever
13. Intake silencer
14. Exhaust manifold
15. Mixing elbow
16. Camshaft
17. Camshaft gear
18. Tappet
19. Push rod
20. Piston
21. Connecting rod
22. Crankshaft
23. Crankshaft gear
24. Flywheel
25. Crankshaft V-pulley
26. Oil pan
27. Dipstick
28. Lubricating oil inlet pipe
29. Anticorrosion zinc
30. Fuel injection pump cam
31. Fuel injection pump
32. Fuel injection nozzle
33. Fuel feed pump
34. Thermostat
35. Starter motor
36. Damper disc
37. Input shaft
38. Output shaft
39. Forward large gear
40. Reverse large gear
41. Output shaft coupling

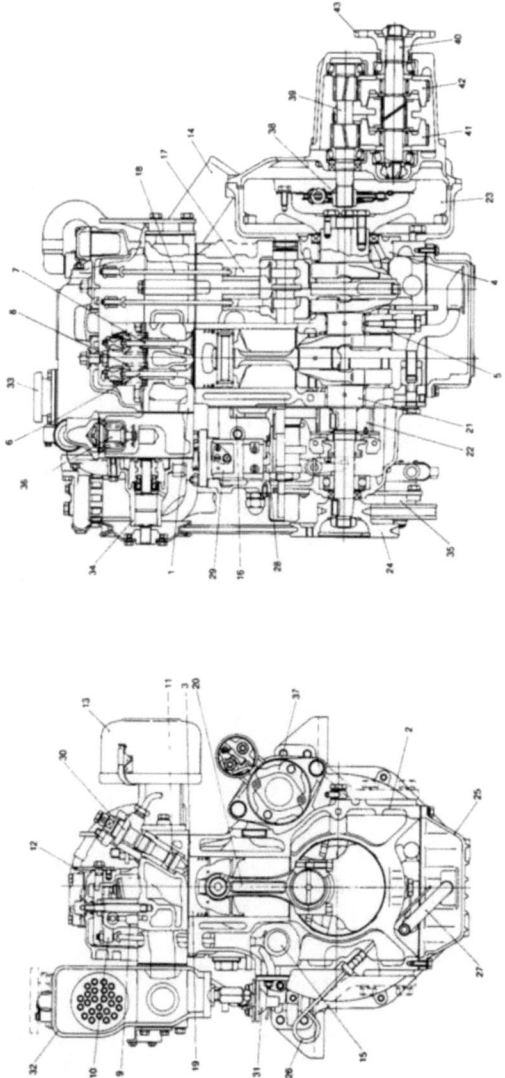

Chapter 1 General
5. Engine Cross-Sections

SM/GM(F)(C)/HM(F)(C)

5-6 3GM30F

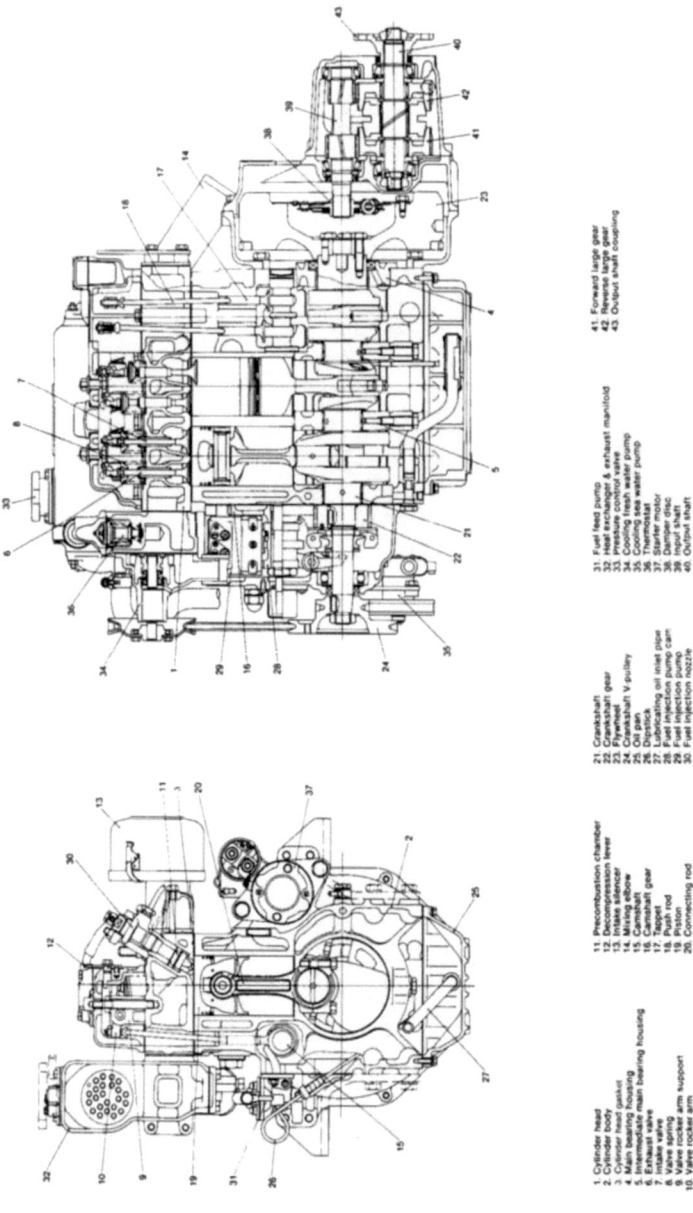

1. Cylinder head
2. Cylinder body
3. Cylinder head gasket
4. Main bearing housing
5. Intermediate main bearing housing
6. Exhaust valve
7. Intake valve
8. Valve spring
9. Valve rocker arm support
10. Valve rocker arm
11. Precombustion chamber
12. Decompression lever
13. Intake silencer
14. Mixing elbow
15. Camshaft
16. Camshaft gear
17. Tappet
18. Push rod
19. Piston
20. Connecting rod
21. Crankshaft
22. Crankshaft gear
23. Flywheel
24. Crankshaft V-pulley
25. Oil pan
26. Dipstick
27. Lubricating oil inlet pipe
28. Fuel injection pump
29. Fuel injection pump cam
30. Fuel injection nozzle
31. Fuel feed pump
32. Heat exchanger & exhaust manifold
33. Pressure control valve
34. Cooling fresh water pump
35. Cooling sea water pump
36. Thermostat
37. Starter motor
38. Damper disc
39. Input shaft
40. Output shaft
41. Forward large gear
42. Reverse large gear
43. Output shaft coupling

Chapter 1 General
5. Engine Cross-Sections

5-7 3HM35F

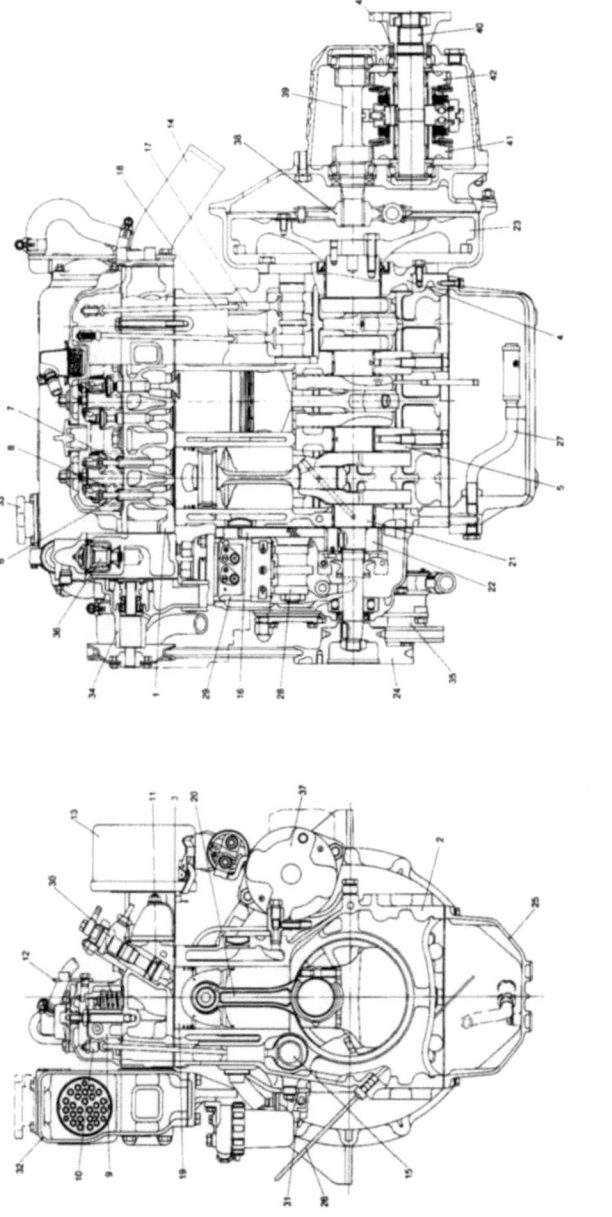

1. Cylinder head
2. Cylinder body
3. Cylinder head gasket
4. Main bearing housing
5. Intermediate main bearing housing
6. Exhaust valve
7. Intake valve
8. Valve spring
9. Valve rocker arm support
10. Valve rocker arm
11. Precombustion chamber
12. Decompression lever
13. Intake silencer
14. Mixing elbow
15. Camshaft
16. Crankshaft gear
17. Tappet
18. Push rod
19. Piston
20. Connecting rod
21. Crankshaft
22. Crankshaft gear
23. Flywheel
24. Crankshaft V-pulley
25. Oil pan
26. Dipstick
27. Lubricating oil inlet pipe
28. Fuel injection pump cam
29. Fuel injection pump
30. Fuel injection nozzle
31. Fuel feed pump
32. Heat exchanger & exhaust manifold
33. Pressure cap
34. Cooling fresh water pump
35. Cooling sea water pump
36. Thermostat
37. Starter motor
38. Damper disc
39. Input shaft
40. Output shaft
41. Forward large gear
42. Reverse large gear
43. Output shaft coupling

1-16

Chapter 1 Gen.
6. Dimensions

SM/GM(F)(C)-HM(F)(C)

6. Dimensions

6-1 1GM10

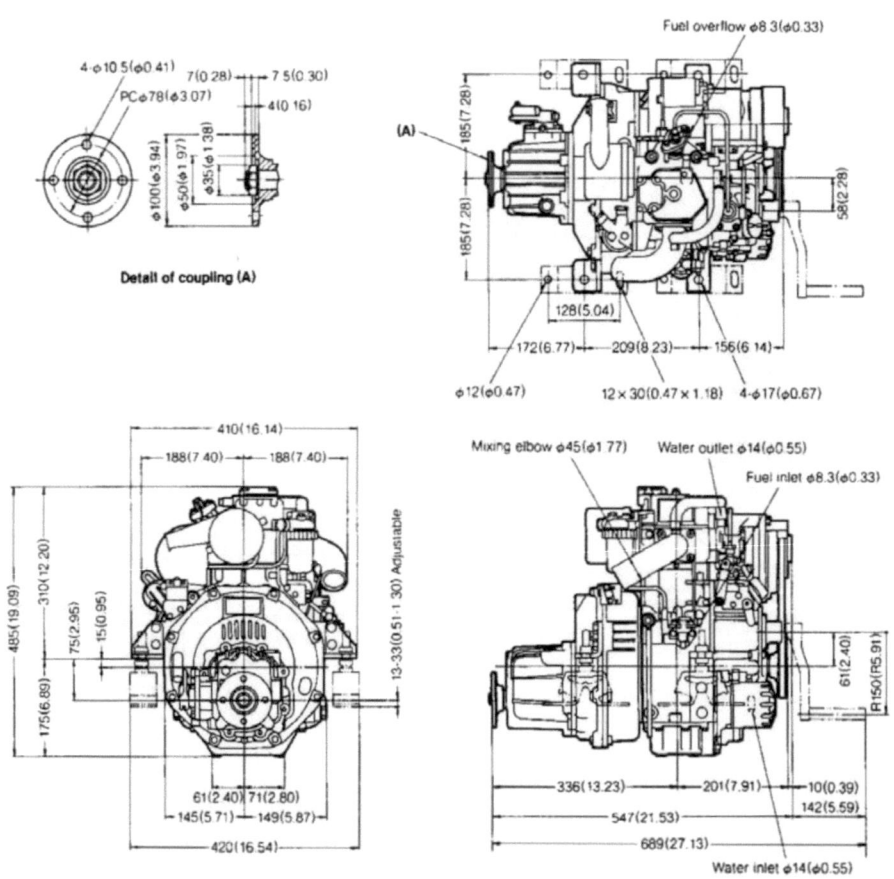

Chapter 1 General
6. Dimensions

6-2 2GM20

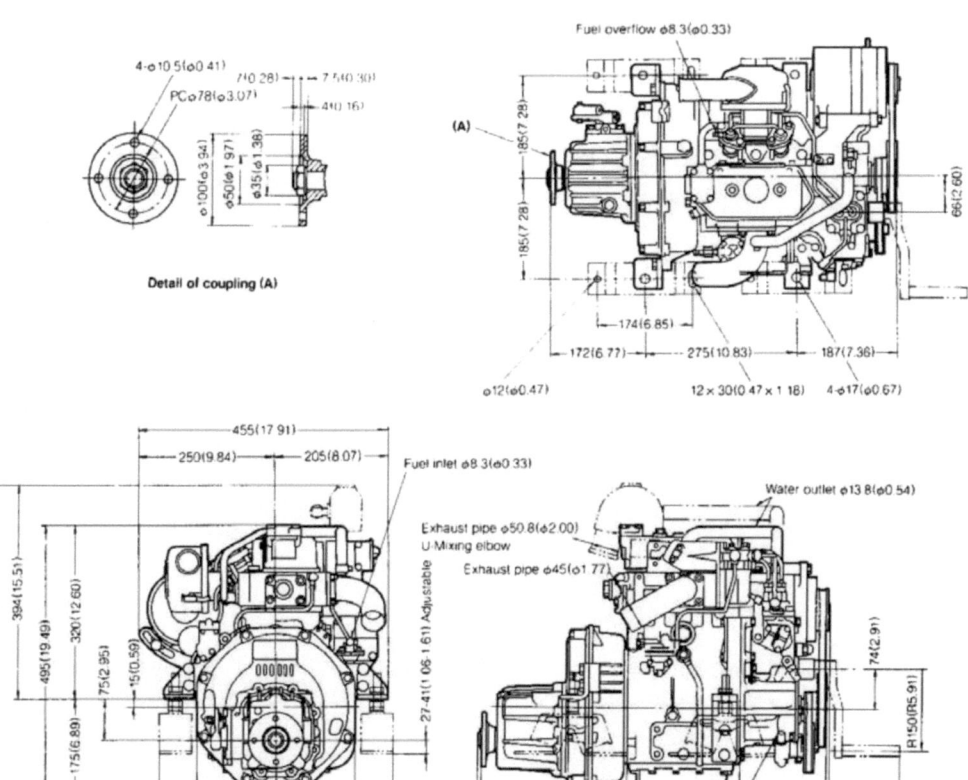

Chapter 1 General
6. Dimensions

SM/GM(F)(C)-HM(F)(C)

6-3 3GM30

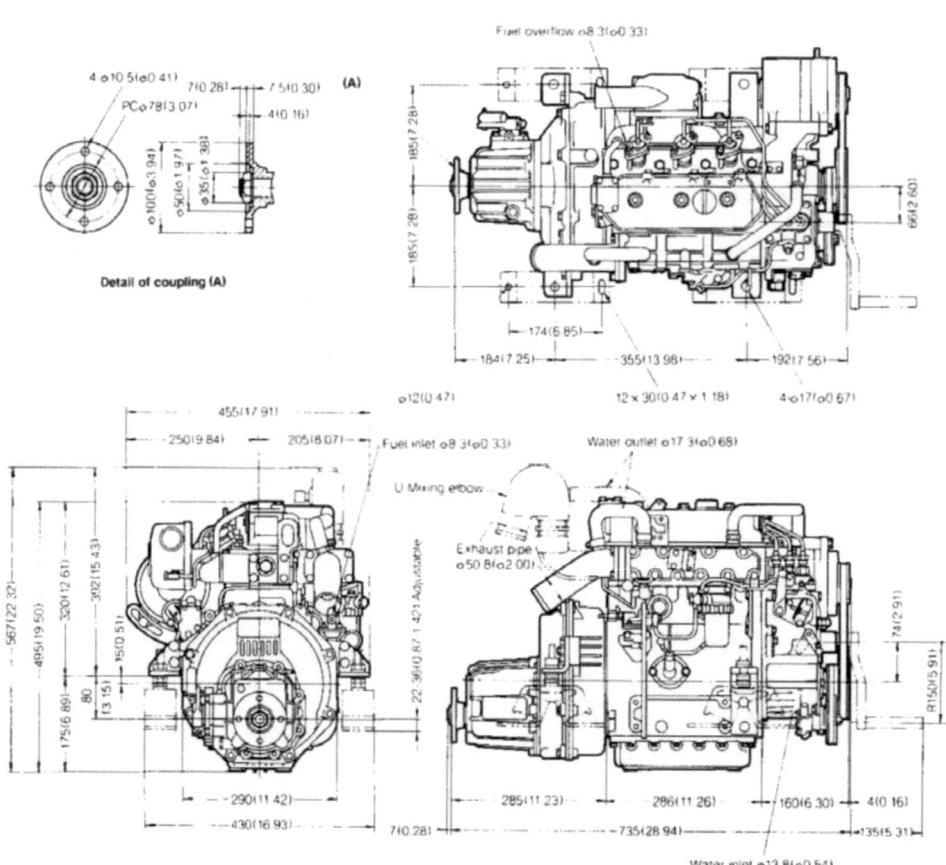

Chapter 1 General
6. Dimensions

6-4 3HM35

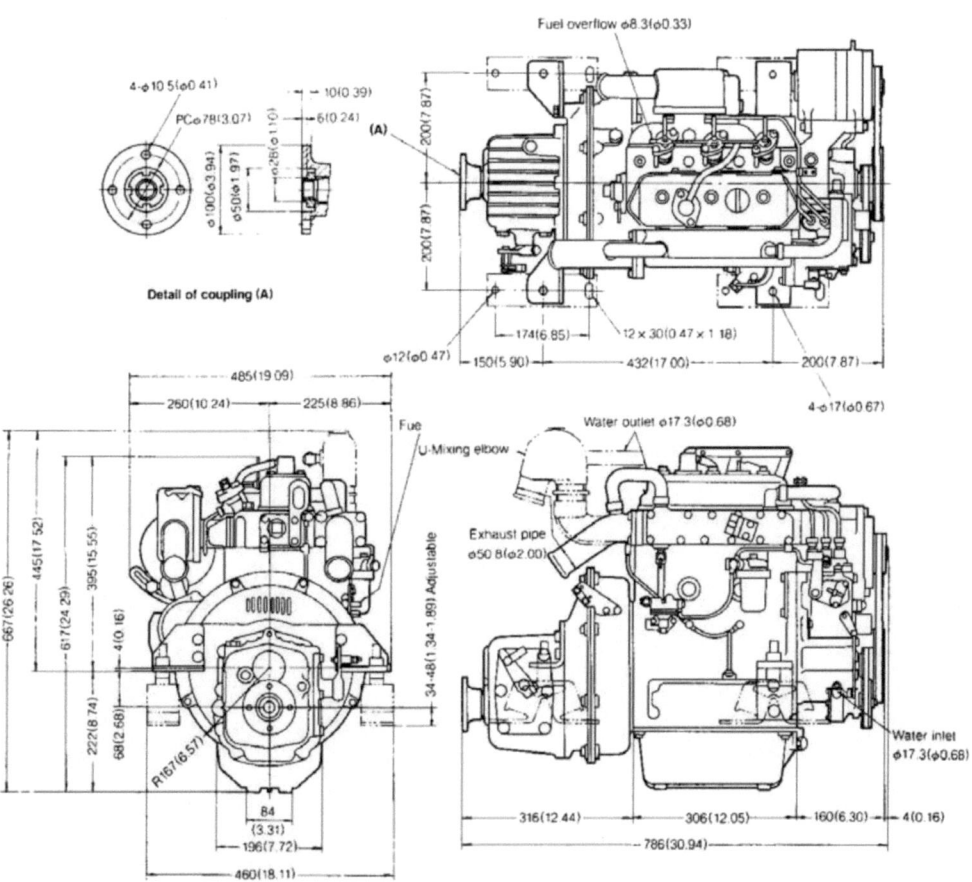

6-5 2GM20F

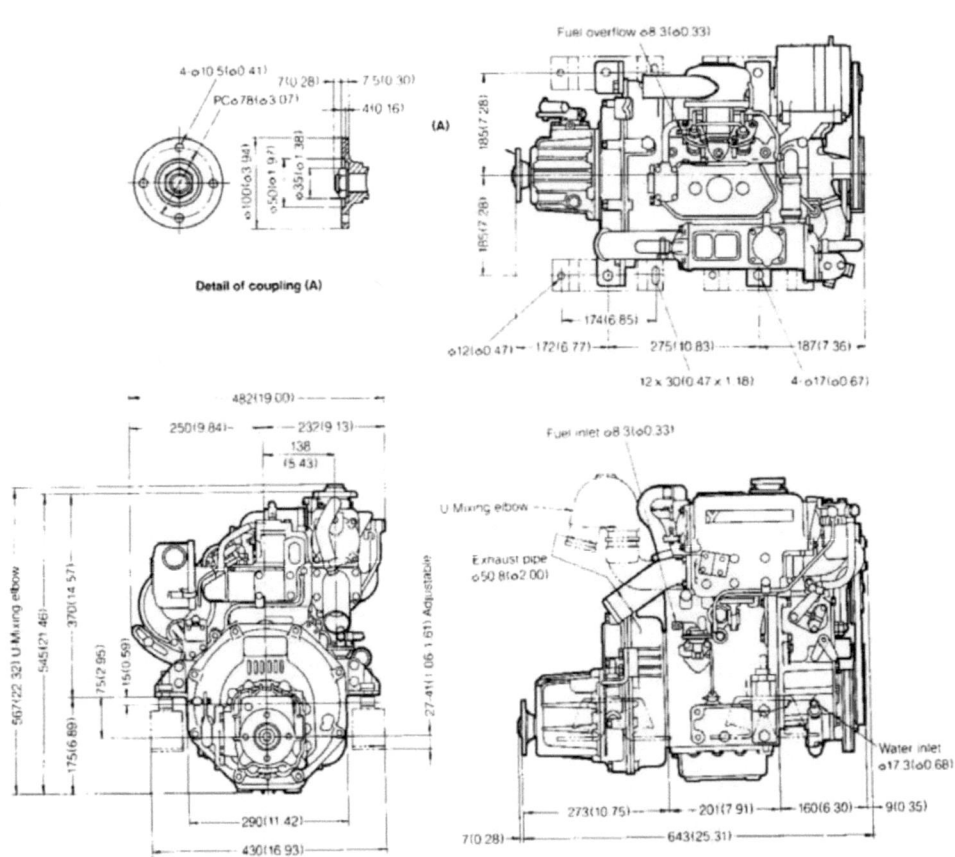

Chapter 1 General
6. Dimensions

6-6 3GM30F

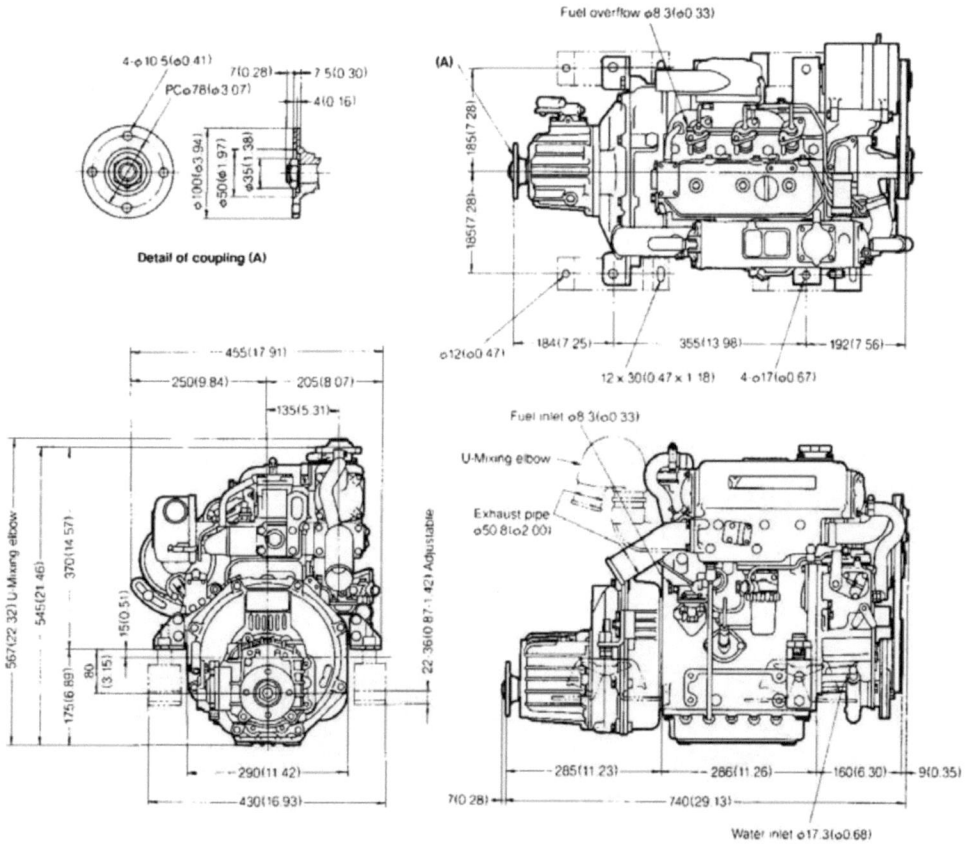

Chapter 1 General
6. Dimensions

SM/GM(F)(C)-HM(F)(C)

6-7 3HM35F

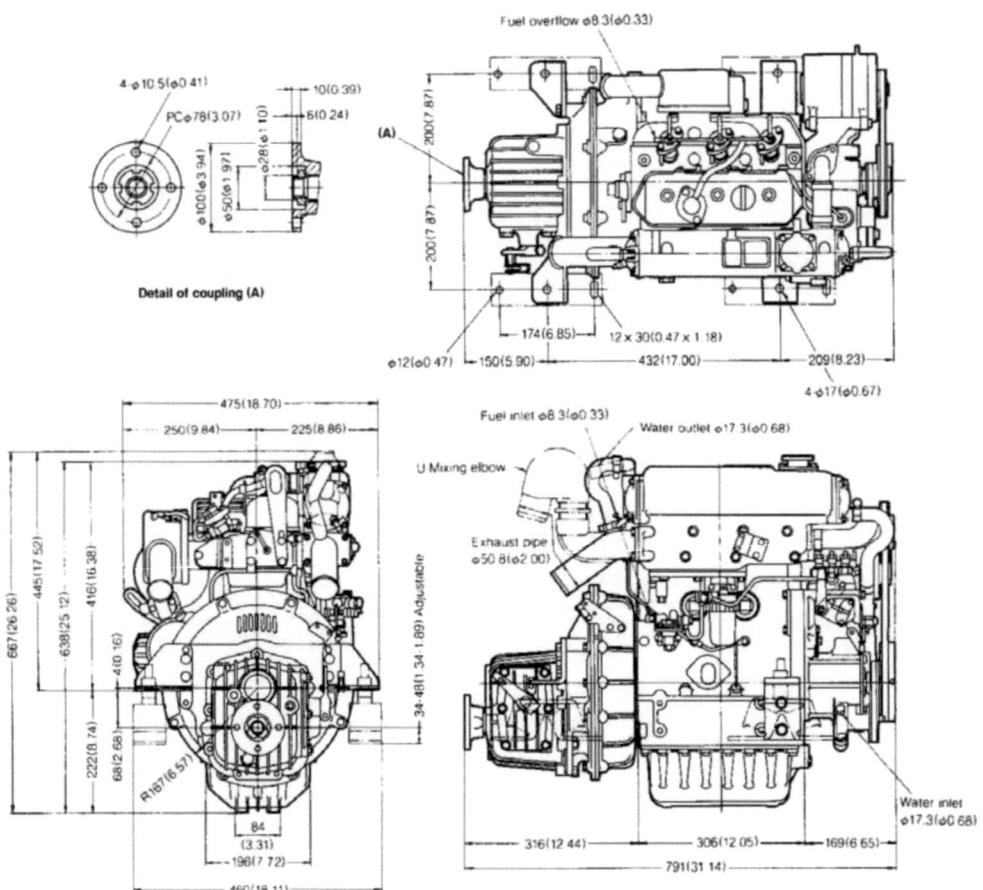

7. Piping Diagrams

7-1 1GM10

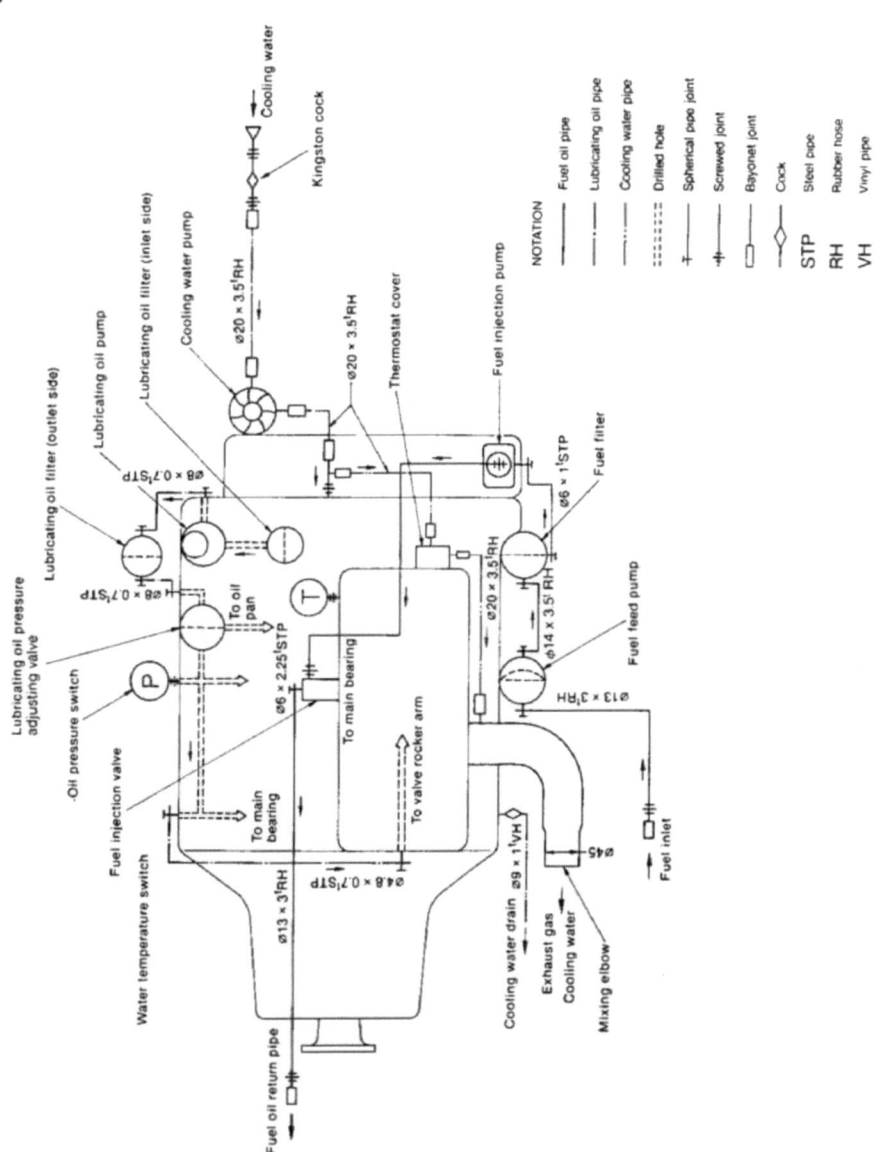

Chapter 1 General
7. Piping Diagrams

7-2 2GM20

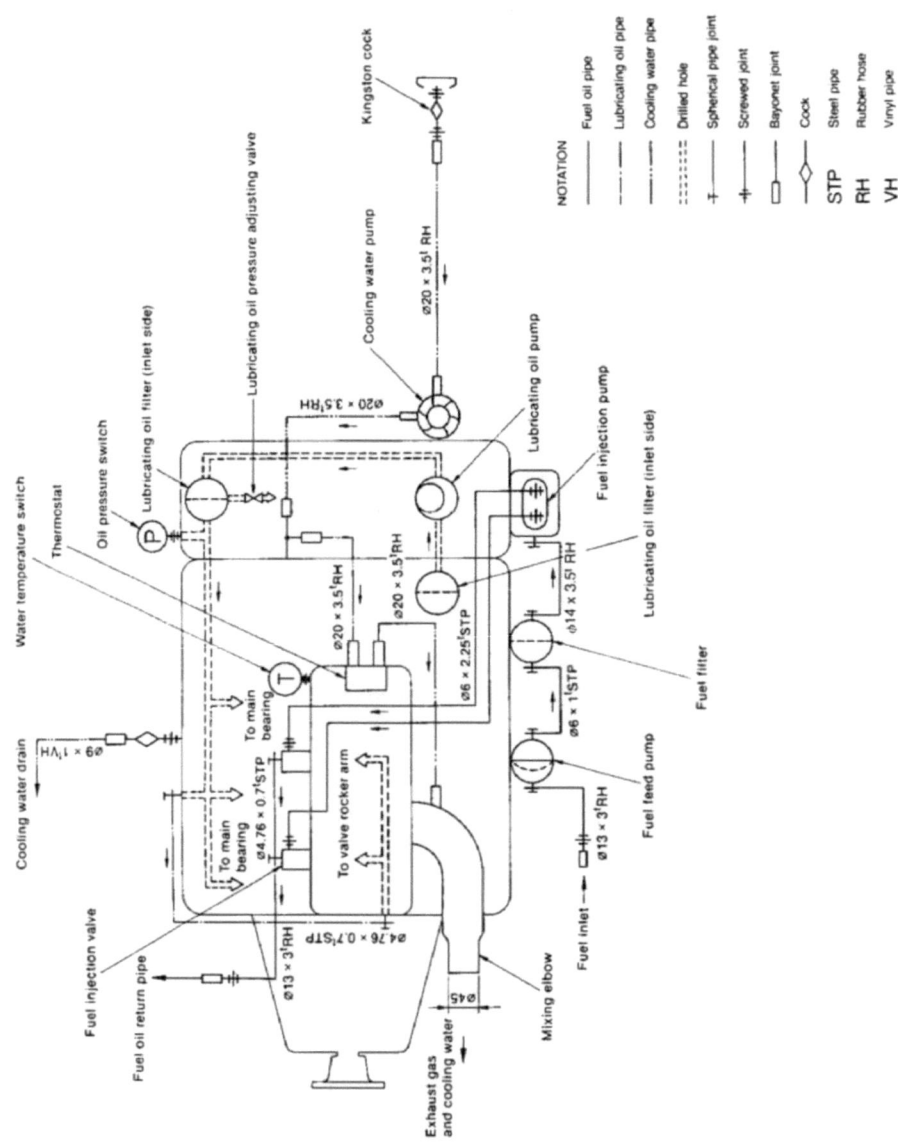

7-3 3GM30 and 3HM35

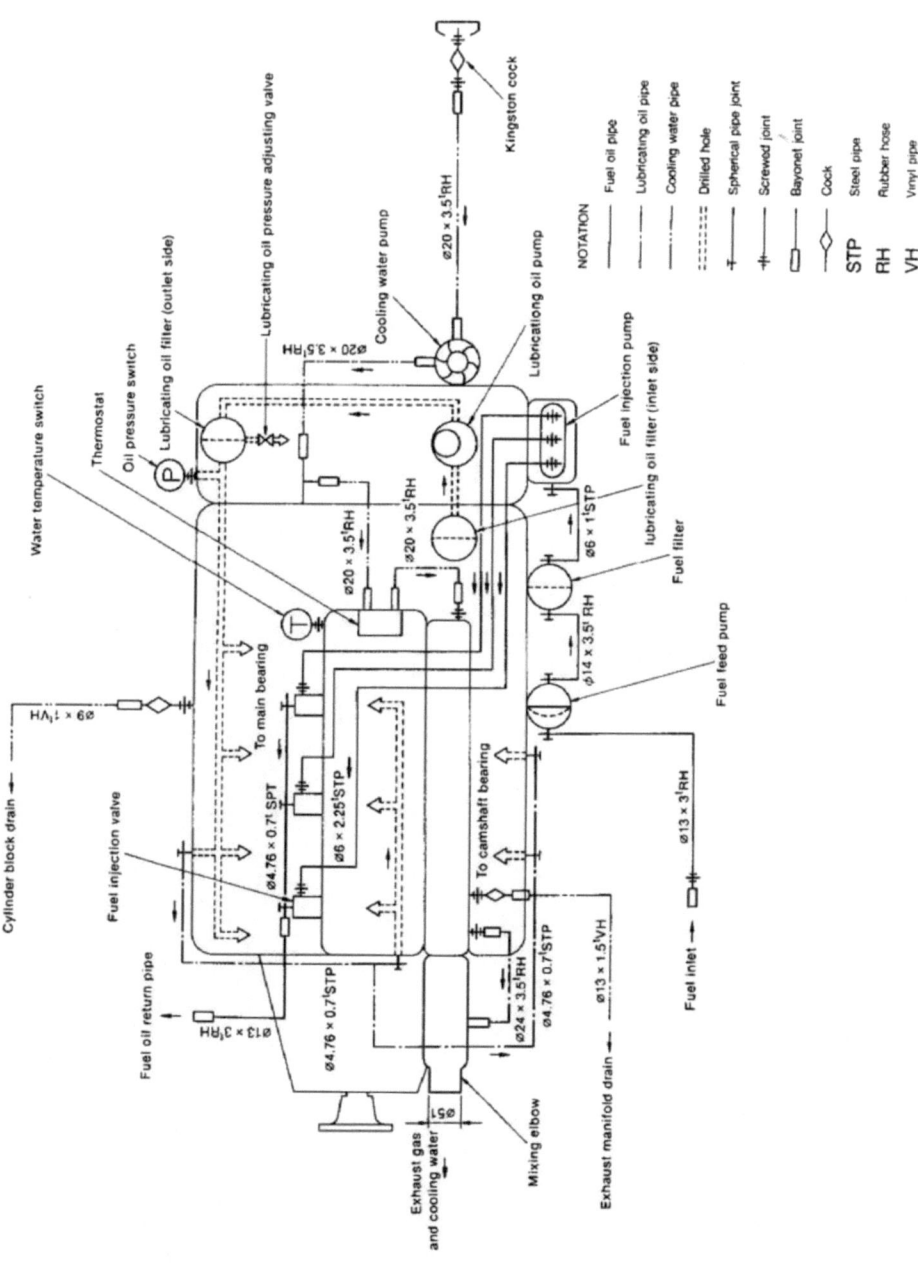

7-4 2GM20F

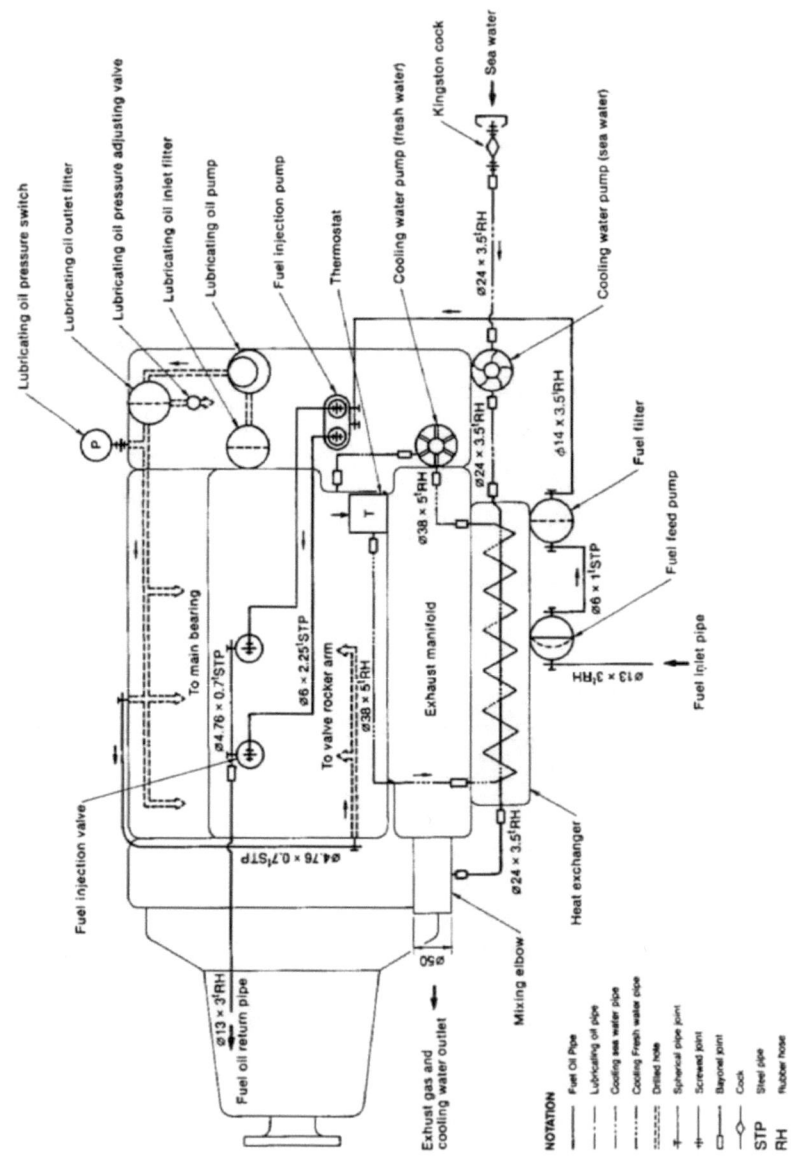

7-5 3GM30F and 3HM35F

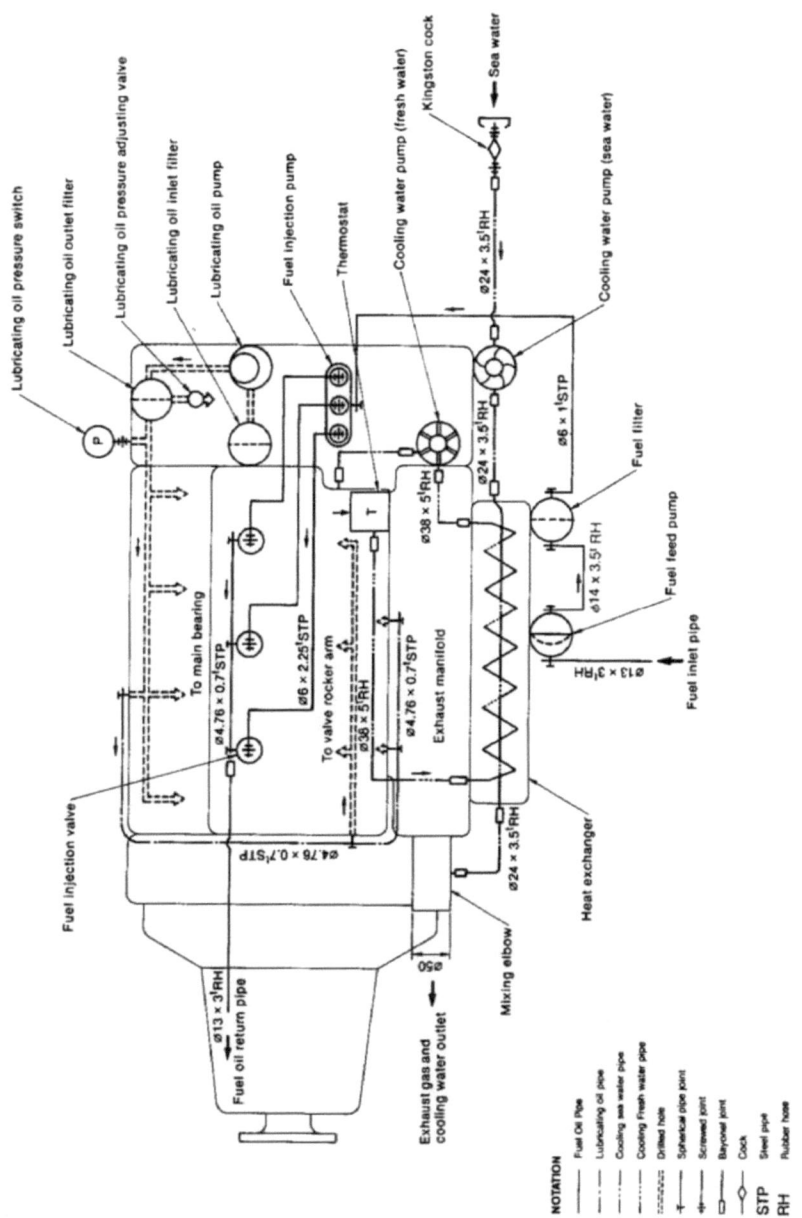

CHAPTER 2
BASIC ENGINE

1. Cylinder Block 2-1
2. Cylinder Head 2-9
3. Piston 2-28
4. Connecting Rod 2-34
5. Crankshaft 2-38
6. Flywheel and Housing 2-49
7. Camshaft 2-53
8. Timing Gear 2-59

1. Cylinder Block

1-1 Construction

The cylinder block comprises a single unit casting for the cylinder body without the use of cylinder liners.

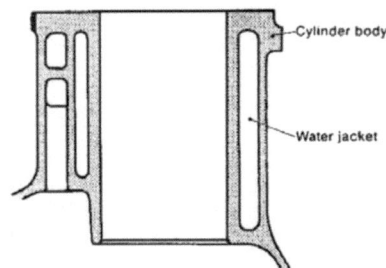

The cylinder block is a high-quality cast iron casting, with integral cylinders and deep skirt crankcase construction. As a result of stress analyses, the shape and thickness of each part has been optimized, and special ribs employed which not only increase the strength and rigidity of the block, but also reduce noise.

1-1.1 Cylinder of model 1GM10(C) engine

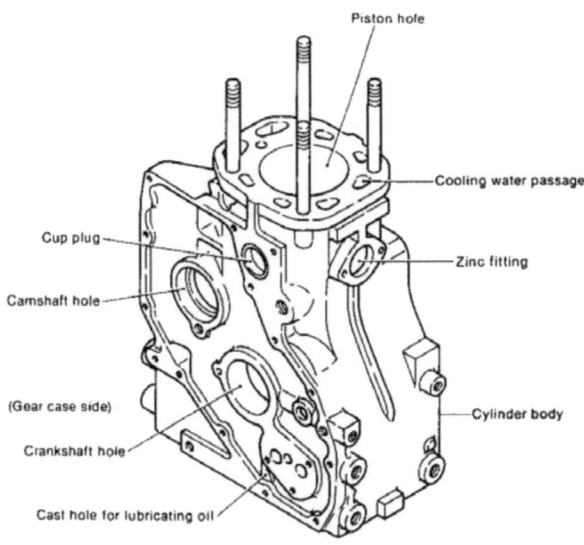

Chapter 2 Basic Engine
1. Cylinder Block

1-1.2 Cylinder of model 2GM20(F)(C) engine

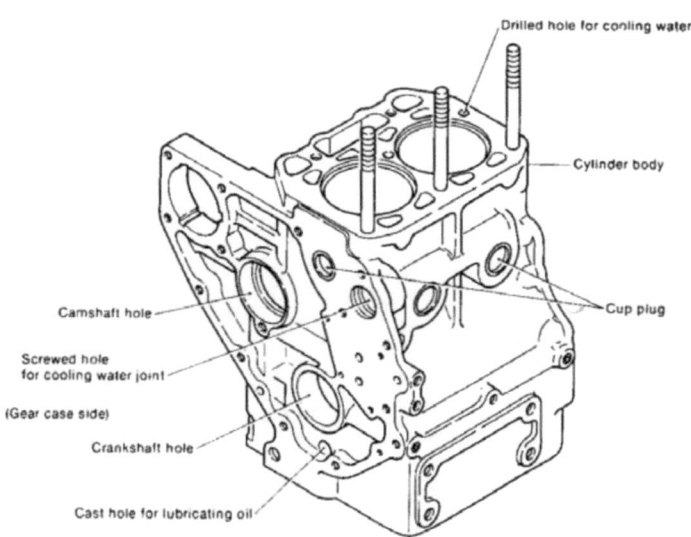

1-1.3 Cylinder of model 3GM30(F)(C) engine

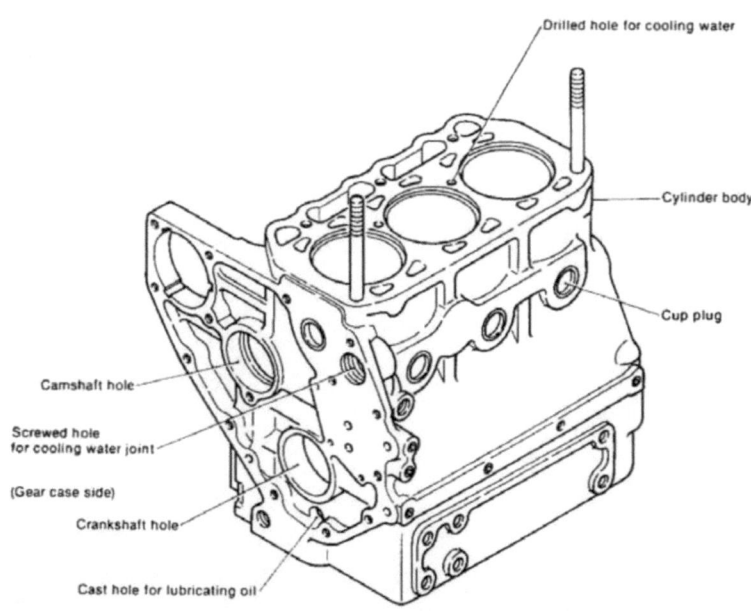

Chapter 2 Basic Engine
1. Cylinder Block

SM/GM(F)(C)·HM(F)(C)

1-1.4 Cylinder of model 3HM35(F)(C) engine

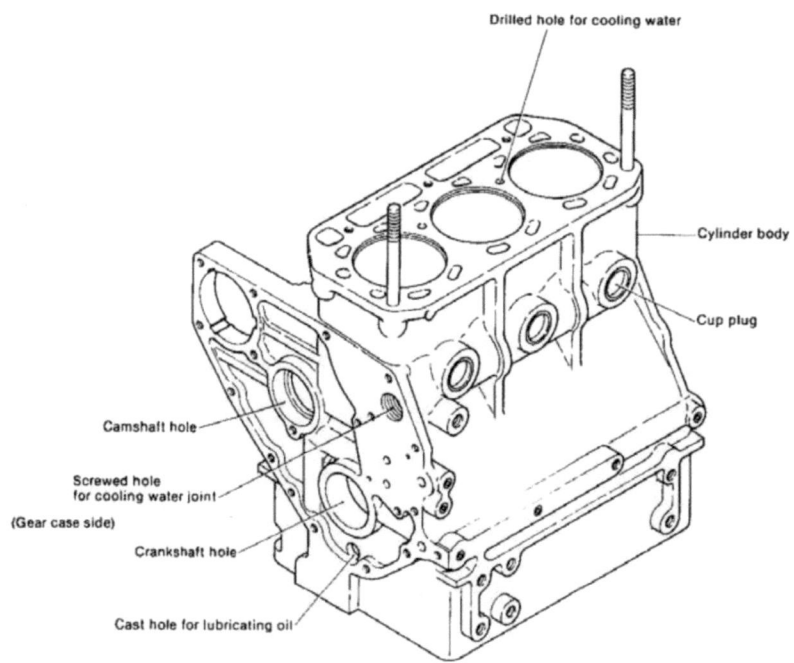

1-2 Cylinder block inspection

1-2.1 Inspecting each part for cracks

If the engine has been frozen or dropped, visually inspect it for cracks and other abnormalities before disassembling. If there are any abnormalities or the danger of any abnormalities occurring, make a color check.

1-2.2 Inspecting the water jacket of the cylinder for corrosion

Inspect the cooling water passages for sea water corrosion, scale, and rust. Replace the cylinder body if corrosion, scale or rust is severe.

1-2.3 Cylinder head stud bolts

Check for loose cylinder head bolts and for cracking caused by abnormal tightening, either by visual inspection or by a color check.
Replace the cylinder block if cracked.

1GM10(C)

Bolt diameter	M10
Pitch	1.5
Tightening torque	6.0kgf-m (43.4 ft-lb)

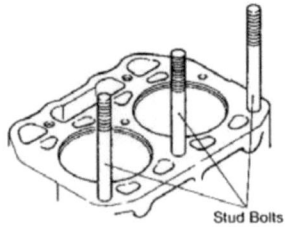

Stud Bolts

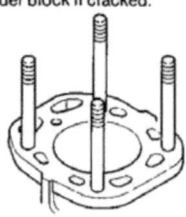

2GM20(F)(C)

Bolt diameter	M12
Pitch	1.25
Tightening torque	8.0kgf-m (57.9 ft-lb)

Chapter 2 Basic Engine
1. Cylinder Block

1-2.5 Color check flaw detection procedure

(1) Clean the inspection point thoroughly.
(2) Procure the dye penetration flaw detection agent. This agent comes in spray cans, and consists of a cleaner, penetrant, and developer in one set.

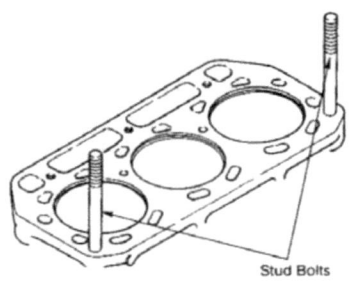
Stud Bolts

3GM30(F)(C), 3HM35(F)(C)   kgf-m(ft-lb)

	3GM30(F)(C)	3HM35(F)(C)
Bolt diameter	M12	
Pitch	1.25	
Tightening torque	8.0(57.9)	10.0(72.3)

(3) Pretreat the inspection surface with the cleaner. Spray the cleaner directly onto the inspection surface, or wipe the inspection surface with a cloth moistened with the cleaner.
(4) Spray the red penetration liquid onto the inspection surface. After cleaning the inspection surface, spray the red penetrant (dye penetration flaw detection agent) onto it and allow the liquid to penetrate for 5-10 minutes.
If the penetrant fails to penetrate the inspection surface on account of the ambient temperature or for other reasons, allow it to dry and respray the inspection surface.
(5) Spray the developer onto the inspection surface. After penetration processing, remove the residual penetrant from the inspection surface with the cleaner, and then spray the developer onto the inspection surface. If the inspection surface is flawed, red dots or lines will appear on the surface within several minutes. When spraying the developer onto the inspection surface, hold the can about 30—40cm from the surface and sweep the can slowly back and forth to obtain a uniform film.
(6) Reclean the inspection surface with the cleaner.
NOTE: *Before using the dye penetration flaw detection agent, read its usage instructions thoroughly.*

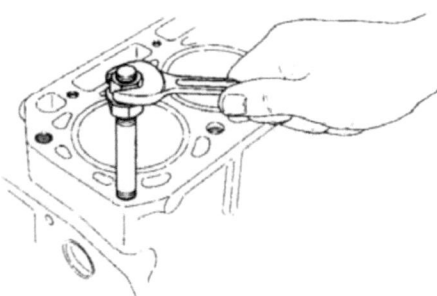

1-3 Cylinder bore measurement

Cylinder wear is measured with a cylinder gauge. The amount of cylinder wear becomes greater as the piston nears the top, and it becomes greatest at the position of the top ring when the piston is in top dead center. The reason for this is that when the piston is at the top position, lateral pressure is high due to the high explosive pressure, and lubrication is very difficult due to the high temperature. Therefore, the amount of wear must be measured in at least 3 positions, namely the top, middle and bottom positions of the cylinder.

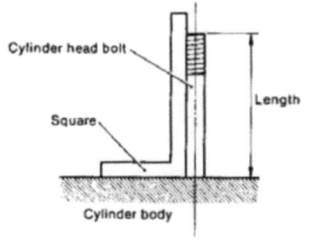

1-2.4 Oil and water passages

Check the oil and water passages for clogging and build-up of foreign matter.

Chapter 2 Basic Engine
1. Cylinder Block

SM/GM(F)(C)·HM(F)(C)

Although the greatest wear is at the top of the cylinder, the piston ring does not slide with the cylinder at the topmost position. Therefore, a step-like pattern is formed between the worn part and the non-worn part.
Furthermore, wear is liable to occur along the rotating direction of the crankshaft due to the lateral pressure of the piston. On the other hand, wear occurs in the direction of the crankshaft center line due to the thrust of the crankshaft and the angle of the connecting rod.
Therefore, the amount of wear must be measured in the directions of crankshaft rotation and the crankshaft center line. When the difference of these two values (i.e. circularity wear) is large, the cylinder must be repaired.

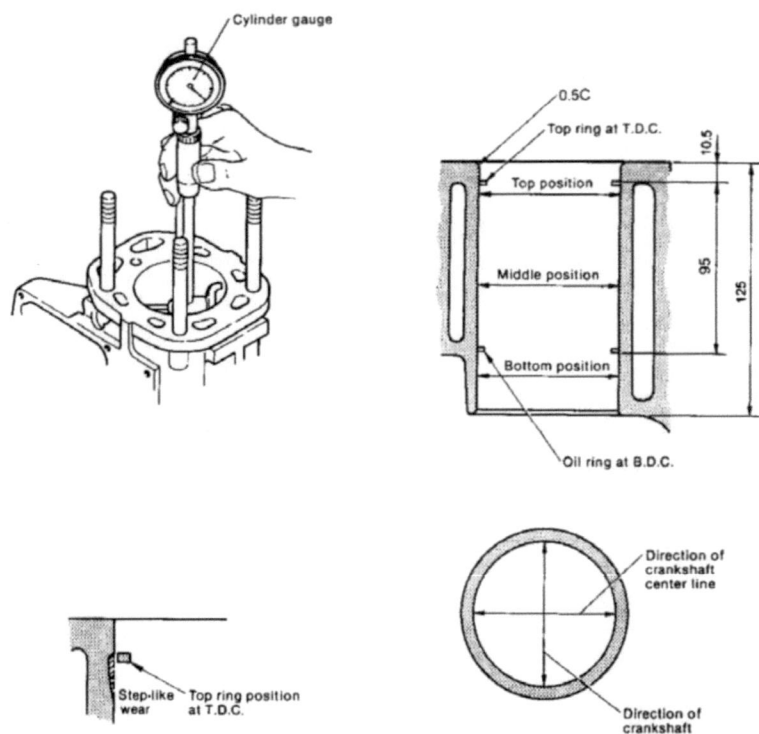

	1GM10(C),2GM20(F)(C),3GM30(F)(C)		3HM35(F)(C)	
	Maintenance standard	Wear limit	Maintenance standard	Wear limit
Cylinder diameter	⌀75.0~75.03 (2.9528~2.9540)	⌀75.10 (2.9567)	⌀80.0~80.03 (3.1496~3.1508)	⌀80.10 (3.1535)
Cylinder roundness	0~0.01 (0~0.0004)	0.02 (0.0008)	0~0.01 (0~0.0004)	0.02 (0.0008)

When the result indicates that eccentric and circularity wear exceed the specified limit, the cylinder must be rebored.

Chapter 2 Basic Engine
1. Cylinder Block

SM/GM(F)(C)-HM(F)(C)

1-3.2 Boring the cylinder

When wear on the inside of the cylinder is excessive, rectify by machining. This is what is known as boring.
When boring is carried out, note the following points.

(1) Dimension to be bored
The cylinder must be bored to the same dimension as an over-size piston.

Over-size piston   mm(in.)

ENG. MODEL	O.D. of standard piston	O.D. of over-size piston
1GM10(C) 2GM20(F)(C) 3GM30(F)(C)	φ75 (2.9528)	φ75.25 (2.9626)
3HM35(F)(C)	φ80 (3.1496)	φ80.25 (3.1594)

(2) Limit of cylinder's expanded I.D.
Never bore the cylinder beyond the limit of the expanded inner diameter, because no over-size piston is available for that dimension, besides which there is danger in having too thin a wall thickness.

Limit of cylinder's expanded I.D.   mm(in.)

ENG. MODEL	I.D. of standard cylinder	Limit of I.D. expansion
1GM10(C) 2GM20(F)(C) 3GM30(F)(C)	φ75.0~75.03 (2.9528~2.9540)	φ75.25~75.28 (2.9626~2.9638)
3HM35(F)(C)	φ80.0~80.03 (3.1496~3.1508)	φ80.25~80.28 (3.1595~3.1606)

Locater points of cylinder block
For the re-boring of the piston bore in the cylinder block, use the following locater positions. Before re-boring, be sure to remove packings and dust from the locater points.
1) 1GM10(C)
   Main locater: Oil pan side
   Sub locater: Timing gear case and F.O. feed pump side
2) 2GM20(F)(C), 3GM30(F)(C), 3HM35(F)(C)
   Oil pan side and φ2-pin holes

(3) Boring procedures
(1) 1GM10(C)
For processing the bore, face the oil pan side to the bottom and place the fixing faces of the gear case and the feed pump.

(2) 2GM20(F)(C), 3GM30(F)(C)
For processing bring the oil pan side to the bottom, and insert a pin to the 2-φ12⁺⁰·⁰¹⁸ (15mm depth) locater hole.

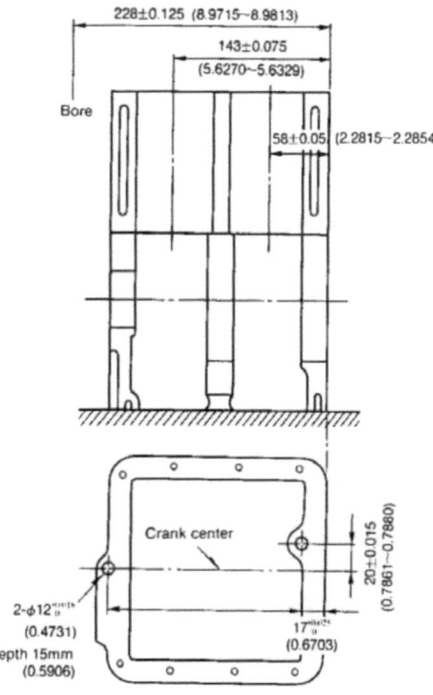

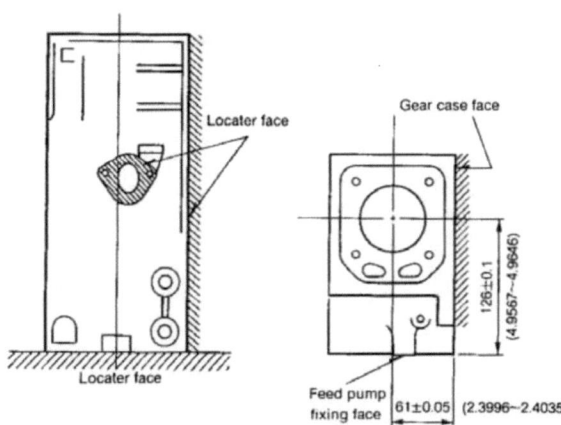

"l" dimension
2GM20: 172±0.05 (6.7697~6.7736)
3GM30: 257±0.05 (10.1161~10.1201)

Chapter 2 Basic Engine
1. Cylinder Block

(3) 3HM35
For processing bring the oil pan side to the bottom, and insert a pin to the 2-φ12$^{+0.018}_{0}$ (15mm depth) locator hole.

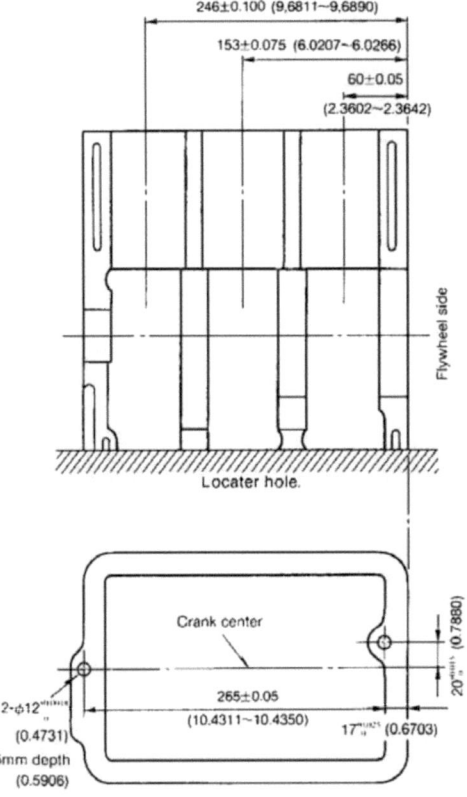

(4) Honing
The inside surface of the cylinder must be honed after being bored in order to remove machine tool marks.

1-4 Measurement of distortion on the upper surface of the cylinder

As the cylinder is repeatedly subjected to thermal expansion and high pressure it will not recover its original shape after the engine has stopped and cooled down and will be distorted. The distortion is mainly caused by construction and material differences of the cylinder, but may arise from the cylinder head bolts being tightened in the wrong order or an uneven tightening torque of the bolts when assembling. If there is any distortion at the upper surface of the cylinder, it will cause a compression pressure leakage, gas leakage or water leakage as a clearance is formed around the cylinder head even though the cylinder head is thoroughly secured.

(1) How to measure distortion on the upper surface of the cylinder
The amount of distortion is measured by placing a straight scale on the upper surface of the cylinder and inserting a thickness gauge between the upper surface of the cylinder and the straight scale.

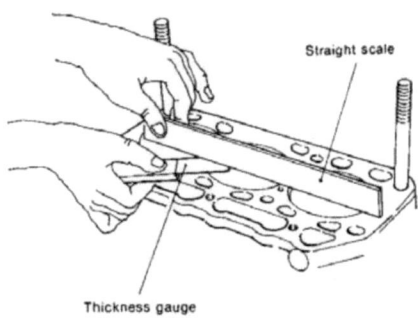

Measurement is to be carried out on the 4 sides and 2 diagonal lines as shown in the figure, and the largest value of clearance for each measurement is to be taken as the amount of distortion.

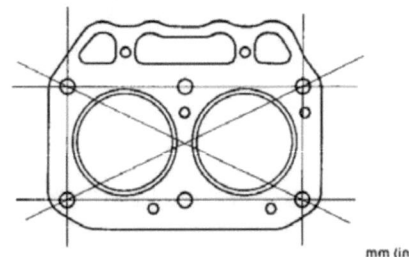

mm (in.)

	Allowable limit of distortion
1GM10(C)	0.05(0.002)
2GM20(F)(C)	0.05(0.002)
3GM30(F)(C) 3HM35(F)(C)	0.05(0.002)

2-7

Chapter 2 Basic Engine
1. Cylinder Block

SM/GM(F)(C)·HM(F)(C)

1-5 Cup plug
1-5.1 Purpose of cup plug

In order to minimize the danger of cylinder block breakage caused by the cooling water freezing, a cup plug is provided at the side of the cylinder block to prevent damage by frost.

In the event that cooing water freezing has caused the cup plug to come out repair in the following way.

In cold weather it is necessary to drain the cooling water completely from the inside of the cylinder block through the cooling water drain pipe.

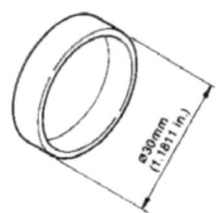

	1GM10(C)	2GM20(F)(C)	3GM30(F)(C)	3HM35(F)(C)
No. of plugs used	2	4	5	5
Part No.		105311-01090		

1-5.2 How to drive in the cup plug

Step No.	Description	Procedure	Tool or material used
1.	Clean and remove grease from the hole into which the cup plug is to be driven. (Remove scale and sealing material previously applied.)	Remove foreign materials with screw driver or saw blade.	•Screw driver or saw blade •Thinner
2.	Remove grease from the cup plug.	Visually check the nick around the plug.	•Thinner
3.	Apply Threebond No. 4 to the seat surface where the plug is to be driven in.	Apply over the whole outside of the plug.	•Threebond No. 4
4.	Insert the plug into the hole.	Insert the plug so that it sits correctly.	
5.	Place a driving tool on the cup plug and drive it in using a hammer. 2～3mm (0.0787～0.1181in.) *Using the special tool drive the cup plug to a depth where the edge of the plug is 2mm (0.0787in.) below the cylinder surface.	Drive in the plug parallel to the seating surface. ø30 $_{-0.1}^{0}$ mm (1.1772～1.1811in.) 3mm (0.1181in.) 100mm (3.9370in.) ø40mm (1.5748in.)	•Driving tool •Hammer

2-8

Printed in Japan
0000A0A1361

2. Cylinder Head

2-1. Construction

The cylinder head is an integral two/three cylinder type which is bolted to the block.

The unique Yanmar swirl type precombustion chambers are at an angle in the cylinder head, and form the combustion chambers, together with the intake and exhaust valves. Large diameter intake valves and smoothly shaped intake and exhaust ports provide high intake efficiency and superior combustion performance.

Special consideration has also been given to the shape of the cooling water passages so that the combustion surface and precombustion chamber are uniformly cooled by an ample water flow.

The thermostat is installed on the side surface of the cylinder at the timing gear case side. (On models 2GM20(C)(F), 3GM30(C)(F) and 3HM35(C)(F), it is integrated with the alternator bracket).

In addition, on models 2GM20(C), 3GM30(C) and 3HM35(C), the anticorrosion zinc is set on the side surface at the flywheel end, and prevents electrolytic corrosion.

2-1.1 Cylinder head of model 1GM10(C) engine

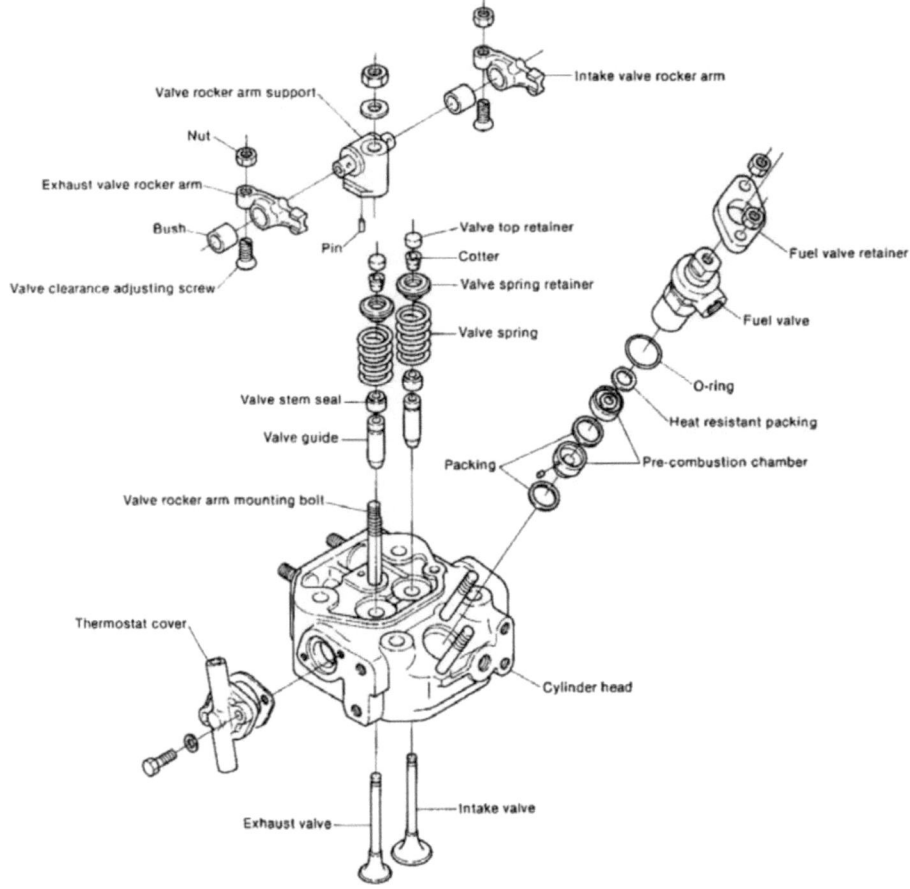

2-1.2 Cylinder head of model 2GM20(F)(C) engine

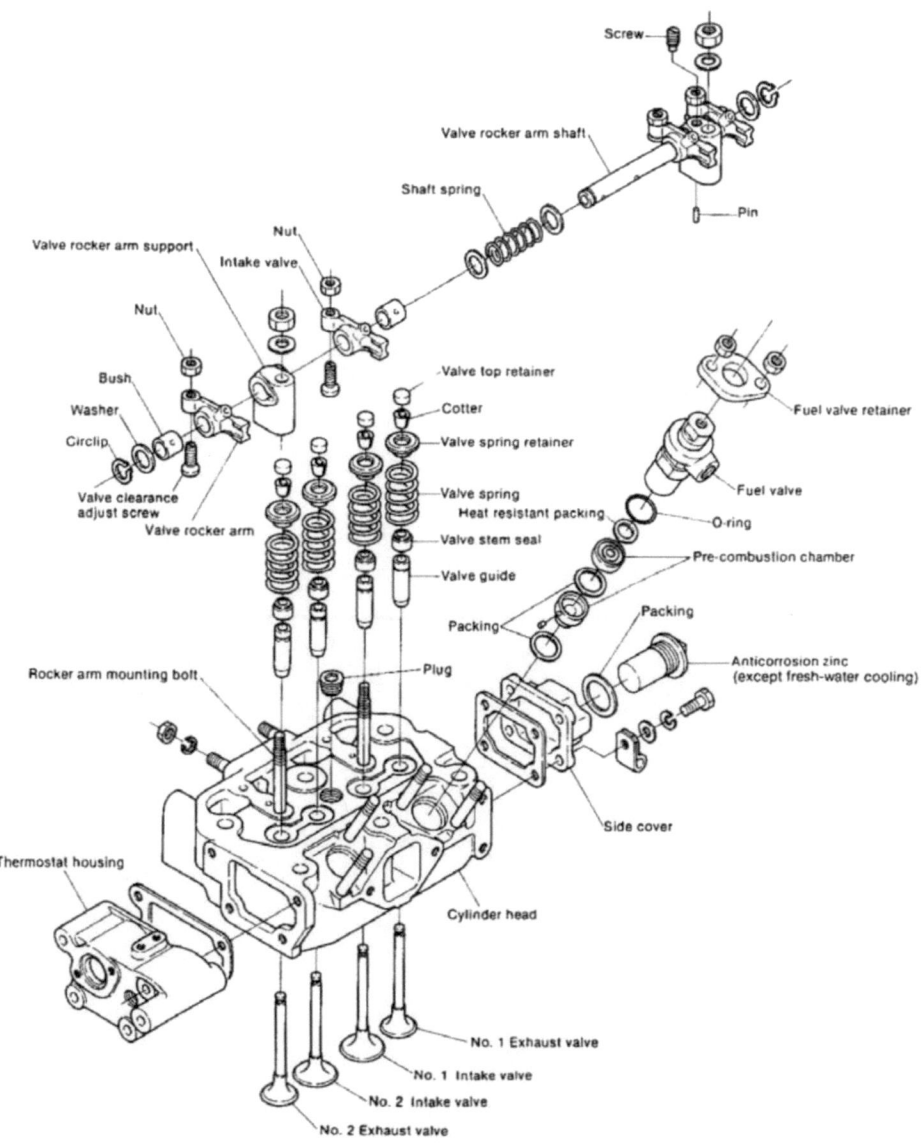

Chapter 2 Basic Engine
2. Cylinder Head

2-1.3 Cylinder head of models 3GM30(F)(C) and 3HM35 (F)(C)

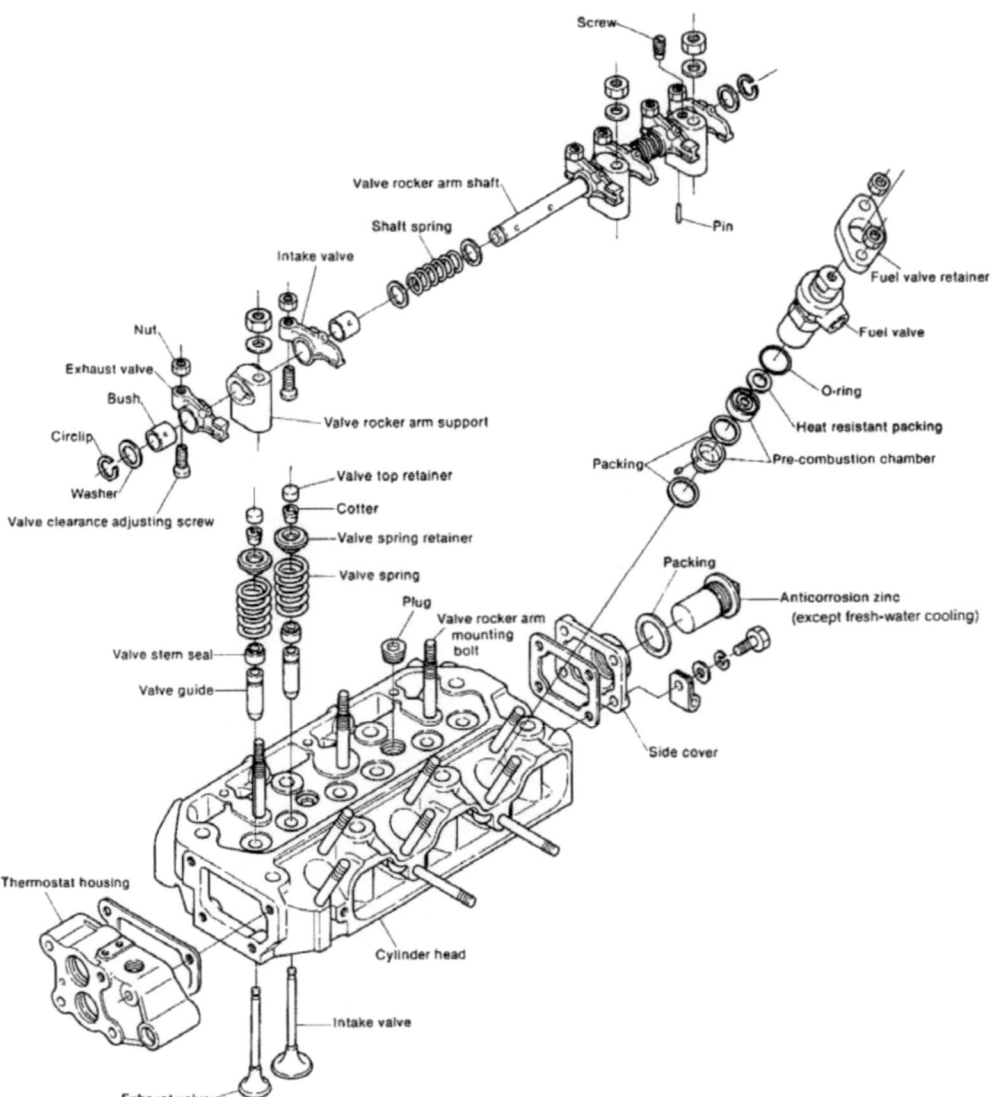

2-2 Cylinder head inspection and measurement

2-2.1 Measurement of carbon build-up at combustion surface and intake and exhaust ports

Visually check for carbon build-up around the combustion surface and the port near the intake and exhaust valve seats, and remove any build-up.
When a large amount of carbon has built up, check the top of the chamber combustion for oil flow at the intake and exhaust valve guides, and take suitable corrective action.

2-2.2 Deposit build-up in water passages

Check for build-up deposit in the water passages, and remove any deposit with a deposit remover. When a large amount of deposit has built up, check each part of the cooling system.

2-2.3 Inspection of corrosion in water passages and anti-corrosion zinc

Inspect the state of corrosion of the water passages, and replace the cylinder head when corrosion is severe.
   Corrosion pitting limit: 2mm (0.0787in.)
Inspect the anticorrosion zinc on the cylinder head cover, and replace the zinc when it is worn beyond the wear limit.
   Anticorrosion zinc wear limit: Volumetric ratio with new zinc = 1/2

2-2.4 Cracking of combustion surface

The combustion surface is exposed to high temperature, high pressure gas and low temperature air, and is repeatedly flexed during operation. Moreover, it is used under extremely severe conditions, such as the high temperature difference between the combustion surface and cooling water passages.
Inspect the combustion surface for cracking by the color check, and replace the cylinder head if any cracking is detected. At the same time, check for signs of overloading and check the cooling water flow.

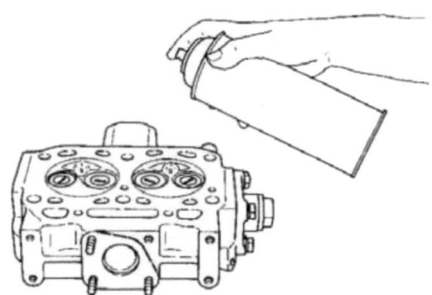

2-2.5 Cylinder head distortion

Distortion of the cylinder head causes gasket packing damage, compression leakage, change in compression, etc.
Measure the distortion as described below, and replace the cylinder head when the wear limit is exceeded. Since distortion of the cylinder head is caused by irregular tightening forces, faulty repair of the mounting face, and gasket packing damage, these must also be checked.

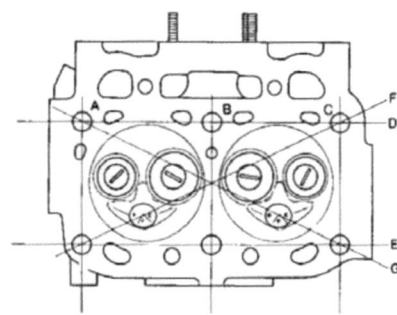

Cylinder head distortion   mm (in.)

	Wear limit
1GM10(C)	0.07 (0.0028)
2GM20(F)(C)	0.07 (0.0028)
3GM30(F)(C), 3HM35(F)(C)	0.07 (0.0028)

(1) Clean the cylinder head tightening surface.
(2) Place a straightedge across two symmetrical points at the four sides of the cylinder head, as shown in the figure.
(3) Insert feeler gauges between the straightedge and the cylinder head combustion face.

Measurement procedure

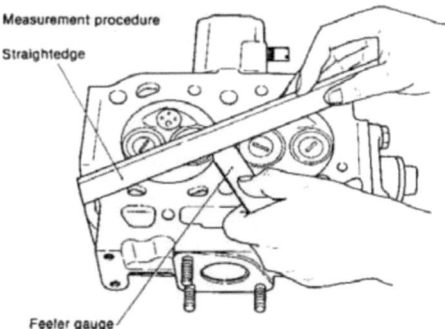

(4) The thickness of the largest feeler gauge that can be inserted is the amount of distortion.

2-2.6 Cylinder head valve seat

The valve seats become wider with use. If the seats become wider than the maintenance standard, carbon built-up at the seats will cause compression leakage. On the other hand, if the seats are too narrow, they will wear quickly and heat transmission efficiency will deteriorate. Clean the carbon and other foreign matter from the valve seats, and check that the seats are not scored or dented.

Chapter 2 Basic Engine
2. Cylinder Head

Measure the seat width with vernier calipers, and repair or replace the seat when the wear limit is exceeded.
When the valves have been lapped and/or ground, measure the amount of valve recess, and replace the valve when the wear limit is exceeded.

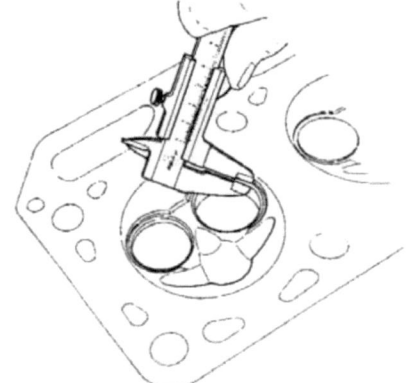

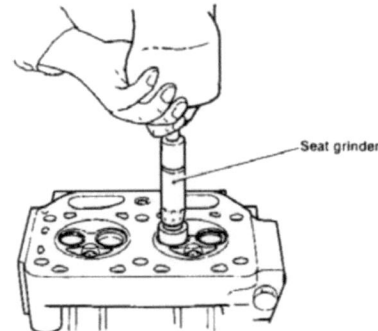

4) Mix the compound with oil, and lap the valve.
5) Finally, lap with oil.

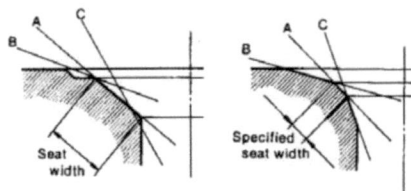

Before correction   After correction

(A) Grind with a 45° grinder
(B) Grind with a 15° grinder
(C) Grind with a 65° ~ 75° grinder

NOTE: *When the valve seat has been corrected with a seat grinder, insert an adjusting shim between the valve spring and cylinder head.*

2-2.7 Measuring valve sinkage

When the valve has been lapped many times, the valve will be recessed and will lower combustion performance. Therefore, measure the valve sinkage, and replace the valve and cylinder head when the wear limit is exceeded.

(Common to all models)   mm (in.)

	Maintenance standard	Wear limit
Seat width	1.77 (0.06969)	—
Seat angle	90°	—

(1) Lapping the valve seat.
  When scoring and pitting of the valve seat is slight, coat the seat with valve compound mixed oil, and lap the seat with a lapping tool.
  At this time, be sure that the compound does not flow into the valve stem and valve guide.

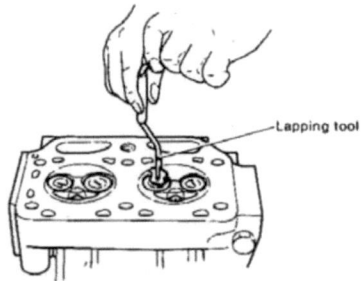

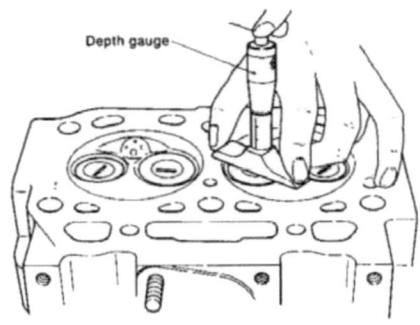

(2) Correcting valve seat width.
  When the valve seat is heavily pitted and when the seat width must be corrected, repair with a seat grinder.
  1) Repair pitting of the seat face with a 45° grinder.
  2) Since the valve seat is larger than the initial value, correct the seat width to the maintenance standard by grinding the inside face of the seat with a 70° grinder.
  3) Grind the outside face of the valve seat with a 15° grinder, and finish the seat width to the standard value.

Chapter 2 Basic Engine
2. Cylinder Head

SM/GM(F)(C)·HM(F)(C)

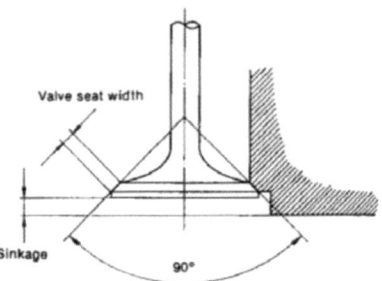

mm (in.)

	1GM10(C), 2GM20(F)(C), 3GM30(F)(C)		3HM35(F)(C)	
	Maintenance standard	Wear limit	Maintenance standard	Wear limit
Valve sinkage	0.95 (0.0374)	1.25 (0.0492)	1.25 (0.0492)	1.55 (0.0610)

2-2.8 Rocker arm support positioning pin
[for model 1GM10(C)]

Check if the guide pin is damaged or if the hole is clogged, and replace the pin if faulty.

2-3 Dismounting and remounting the cylinder head

When dismounting and remounting the cylinder head, the mounting bolts must be removed and installed gradually and in the prescribed sequence to prevent damaging the gasket packing and to prevent distortion of the cylinder head. Since the tightening torque and tightening sequence of the mounting bolts when remounting the cylinder head are especially important from the standpoint of engine performance, the following items must be strictly observed.

2-3.1 Cylinder head assembly sequence

(1) Check for loose cylinder head stud bolts, and lock any loose bolts with two nuts and then tighten to the prescribed torque.
The cylinder head is fitted to the engine with 4 stud bolts in model 1GM10(C), but in other engine models both stud bolts and collar head bolts are used.

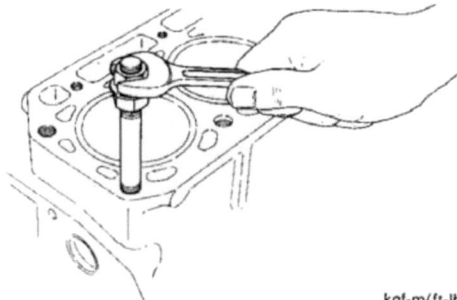

kgf-m(ft-lb)

	1GM10(C)	2GM20(F)(C), 3GM30(F)(C)	3HM35(F)(C)
Stud bolt diameter of cylinder head	M10	M12	M12
Cylinder head stud bolt tightening torque	6.0 (43.4)	8.0 (57.9)	10.0 (72.3)

(2) Checking the gasket packing mounting face.
Confirm correct alignment of the front and rear of the gasket packing, and install the packing by coating both sides with Three Bond 50.
Assemble the gasket packing keeping the flat surface upward (cylinder head side). Make sure that the gasket hole aligns with the drilled hole in the cooling water passage in the cylinder block.

1) For Model 1GM10(C)

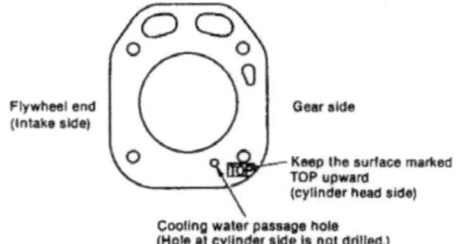

Chapter 2 Basic Engine
2. Cylinder Head

SM/GM(F)(C)-HM(F)(C)

2) For model 2GM20(F)(C)

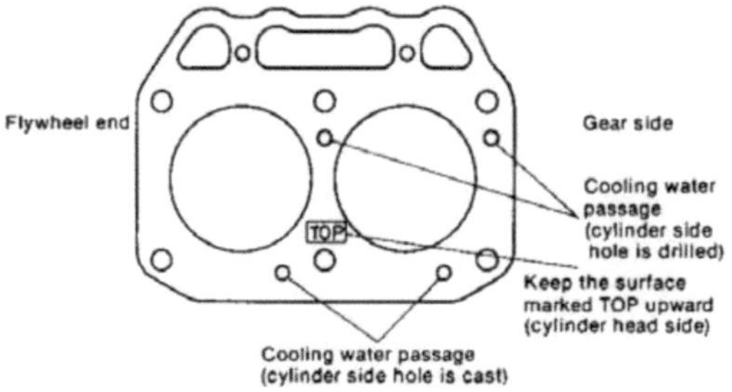

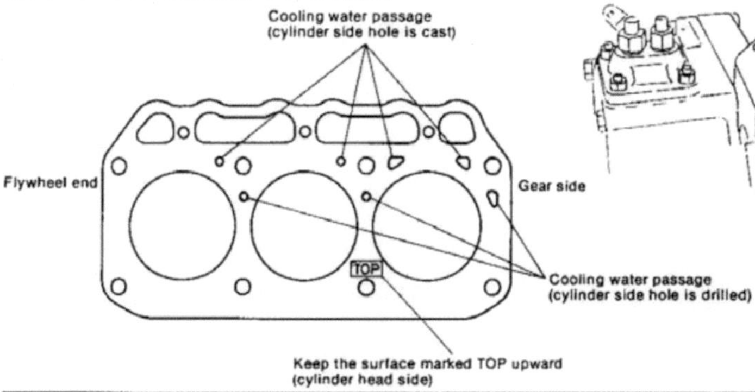

(3) Installing the cylinder head ass'y.
Position the cylinder head ass'y parallel to the top of the cylinder block, and install the ass'y on the block, being careful that the cylinder head ass'y does not touch the threads of the cylinder head bolts.

Flywheel end — Gear side
Cooling water passage (cylinder side hole is drilled)
Keep the surface marked TOP upward (cylinder head side)
Cooling water passage (cylinder side hole is cast)

3) For models 3GM30(F)(C) and 3HM35(F)(C)

Cooling water passage (cylinder side hole is cast)

Flywheel end — Gear side
Cooling water passage (cylinder side hole is drilled)
Keep the surface marked TOP upward (cylinder head side)

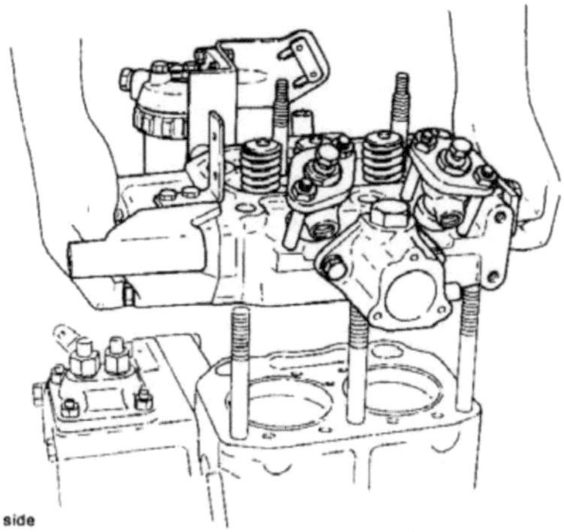

2-3.2 Tightening the cylinder head bolts and nuts

(1) Kinds of cylinder head fixing nuts and bolts, tightening torque, tightening sequence

1) Model 1GM10(C)

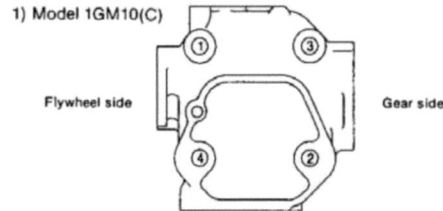

Flywheel side   Gear side

kgf-m(ft-lb)

Tightening sequence	Kinds of fixing	Dia.	Torque
1			
2	Stud bolt fixing nut	M10	7.5kgf-m (54.2 ft-lb)
3			
4			

2-15

Chapter 2 Basic Engine
2. Cylinder Head

2) Model 2GM20(F)(C)

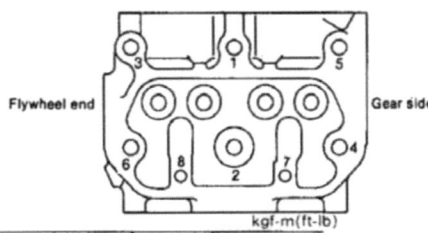

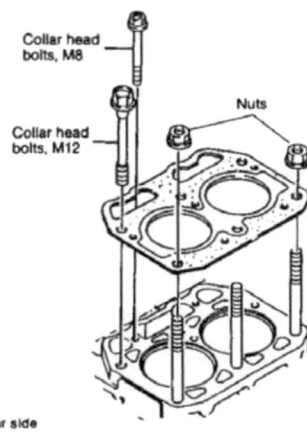

kgf-m(ft-lb)

Tightening sequence	Nut & bolt to be tightend	Dia.	Tightening Torque
1, 3, 5	Stud bolt fixing nut	M12	12.0(86.8)
7, 8	Collar head bolts	M8	3.0(21.7)
2, 4, 6		M12	12.0(86.8)

3) Models 3GM30(F)(C) and 3HM35(F)(C)

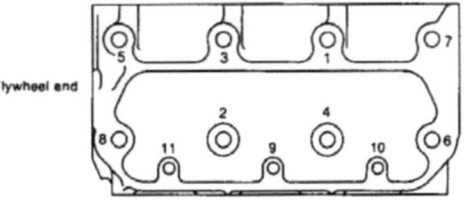

kgf-m(ft-lb)

Tightening sequence	Nut & bolt to be tightend	Dia.	Tightening torque	
			3GM30(F)(C)	3HM35(F)(C)
5, 7	Stud bolt fixing nut	M12	12.0(86.8)	13 (94.0)
9, 10, 11	Collar head bolts	M8	3.0(21.7)	3 (21.7)
1, 2, 3, 4, 6, 8		M12	12.0(86.8)	13 (94.0)

(2) Cylinder head nut tightening sequence
 1) Coat the threads of the cylinder head bolts with lubricating oil, and screw the cylinder head nuts onto the bolts.

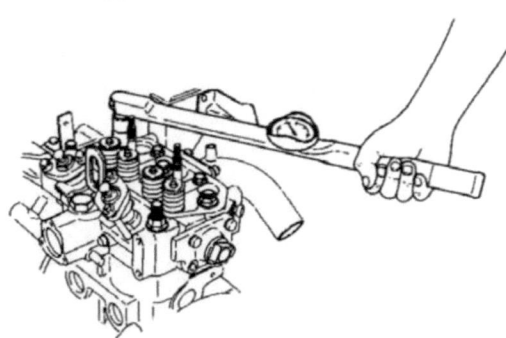

2) First, tighten the nuts sequentially to 1/3 of the prescribed torque.
3) Second, tighten the nuts sequentially to 2/3 of the prescribed torque.
4) Third, tighten the nuts to the prescribed torque.
5) Recheck that all the nuts have been properly tightened.
NOTE: After tightening, valve clearance must be adjusted.

2-3.3 Cylinder head nut loosening sequence

When loosening the cylinder head nuts, reverse the tightening sequence. The cylinder head nut loosening sequence is shown in the figure.

2-4 Intake and exhaust valves, valve guide and valve spring

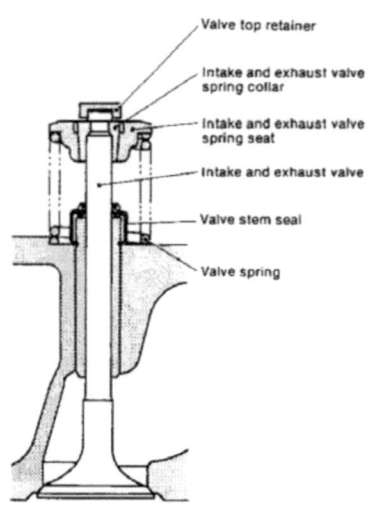

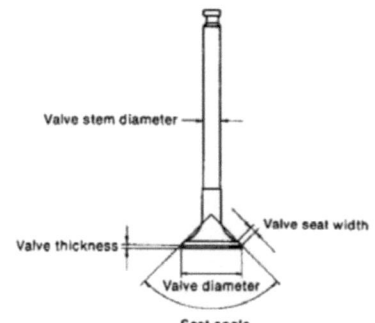

	1GM10(C) 2GM20(F)(C) 3GM30(F)(C)	3HM35(F)(C)
Intake valve diameter	⌀32 (1.2598)	⌀32 (1.2598)
Exhaust valve diameter	⌀26 (1.0236)	⌀27 (1.0630)
Valve seat width	3.15 (0.1240)	3.04 (0.1197)
Valve seat angle	90°	90°

NOTE: Note that the intake valve and exhaust valve have a different diameter.

mm (in.)

	1GM10(C), 2GM20(F)(C) 3GM30(F)(C)		3HM35(F)(C)	
	Maintenance standard	Wear limit	Maintenance standard	Wear limit
Valve thickness	0.75~1.15 (0.0295 ~0.0453)	—	0.85~1.15 (0.0335 ~0.0453)	—

2-4.1 Inspecting and measuring the intake and exhaust valves

(1) Valve seat wear and contact width.
Inspect valve seats for carbon build-up and heavy wear. Also check if each valve seat contact width is suitable. If the valve seat contact width is narrower than the valve seat width, the seat angle must be checked and corrected.

(2) Valve stem bending and wear.
Check for valve stem wear and strain, and repair when such damage is light. Measure the outside diameter and bend, and replace the valve when the wear limit is exceeded.

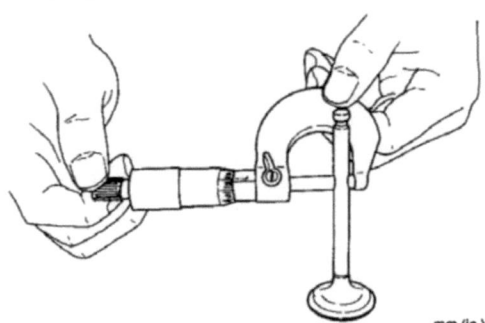

mm (in.)

	1GM10(C), 2GM20(F)(C), 3GM30(F)(C)		3HM35(F)(C)	
	Maintenance standard	Wear limit	Maintenance standard	Wear limit
Valve stem outside diameter	⌀7 (0.2756)	⌀6.9 (0.2717)	⌀7 (0.2756)	⌀6.9 (0.2717)
Valve stem bend	—	0.03 (0.0012)	—	0.03 (0.0012)

Chapter 2 Basic Engine
2. Cylinder Head

(3) Valve seat hairline cracks.
Inspect the valve seat by the color check, and replace the seat if cracked.

2-4.2 Inspecting and measuring valve guides

The same valve guide is used both for intake and exhaust valves in the model 1GM10(C) engine. It has a gas blow opening cut in the inner face at the bottom.
As for models 2GM20(F)(C), 3GM30(F)(C) and 3HM35 (F)(C), the valve guide is different for the intake valve and exhaust valve in that the inner face of the exhaust valve guide has a gas blow opening cut.
Be sure that the correct one is used when replacing the guides.

For model 1GM10(C)

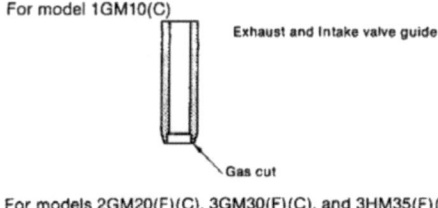

For models 2GM20(F)(C), 3GM30(F)(C), and 3HM35(F)(C)

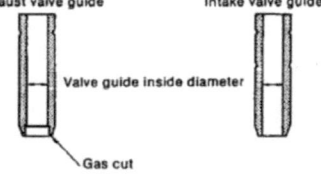

(1) Floating of the intake and exhaust valve guides.
Check for intake and exhaust valve guide looseness and floating with a test hammer, and replace loose or floating guides with guides having an oversize outside diameter.

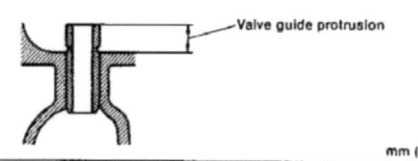

	mm (in.)
	All models
Valve guide protrusion	7 (0.2756)

(2) Measuring the valve guide inside diameter.
Measure the valve guide inside diameter and clearance, and replace the guide when wear exceeds the wear limit.

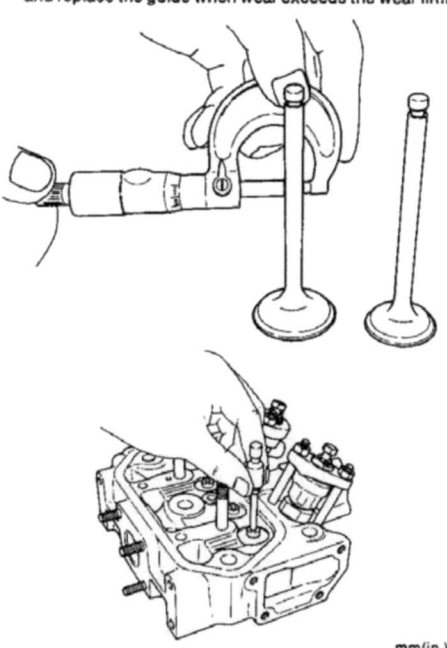

mm(in.)

			Maintenance standard	Clearance at assembly	Maximum allowable clearance	Wear limit
1GM10(C)	Intake	Valve guide inside diameter (after assembly)	ϕ7 (0.2756)	0.045–0.070 (0.0018–0.00028)	0.15 (0.0059)	ϕ7.08 (0.2787)
		Stem outside diameter	ϕ7 (0.2756)			ϕ6.9 (0.2717)
	Exhaust	Valve guide inside diameter (after assembly)	ϕ7 (0.2756)	0.045–0.070 (0.0018–0.0028)	0.15 (0.0059)	ϕ7.08 (0.2787)
		Stem outside diameter	ϕ7 (0.2756)			ϕ6.9 (0.2717)
2GM20(F)(C) 3GM30(F)(C) 3HM35(F)(C)	Intake	Valv guide inside diameter (after assembly)	ϕ7 (0.2756)	0.040–0.065 (0.0016–0.0026)	0.15 (0.0059)	ϕ7.08 (0.2787)
		m outside diameter	ϕ7 (0.2756)			ϕ6.9 (0.2717)
	Exhaust	Valve guide inside diameter (after assembly)	ϕ7 (0.2756)	0.045–0.0070 (0.0018–0.0028)	0.15 (0.0059)	ϕ7.08 (0.2787)
		Stem outside diameter	ϕ7 (0.2756)			ϕ6.9 (0.2717)

Chapter 2 Basic Engine
2. Cylinder Head

(3) Replacing the intake/exhaust valve guide

1) Using a special tool for extracting and inserting the valve guide, extract the valve guide.

2) Using the above tool, drive the valve guide into position by starting from the valve spring side and finish the inside diameter with a reamer.

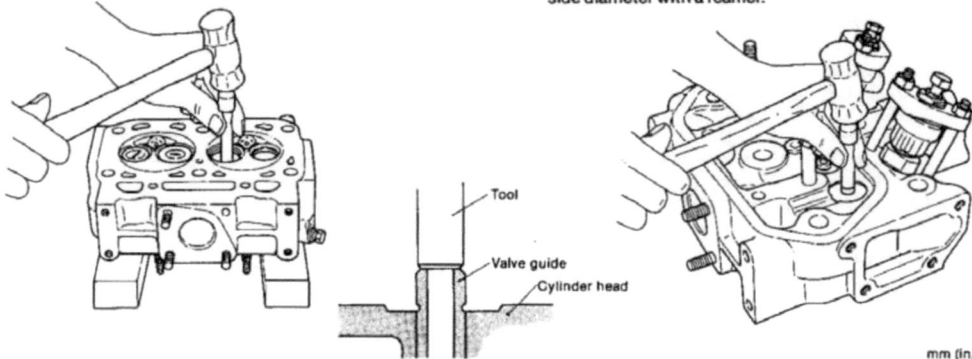

mm (in.)

Amount of interference of valve guide	1GM10(C)	2GM20(F)(C), 3GM30(F)(C)	3HM35(F)(C)
	0.005 ~ 0.034 (0.0002 ~ 0.0013)	0.018 ~ 0.047 (0.0007 ~ 0.0019)	0.018 ~ 0.047 (0.0007 ~ 0.0019)

Fit the intake and exhaust valve guides until the bottom of the groove around the outside of the valve guide is flush with the end of the cylinder head.
As the valve guide for model 1GM10(C) does not have a groove, fit it after checking its dimension and marking it.

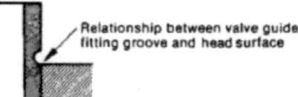

Relationship between valve guide fitting groove and head surface

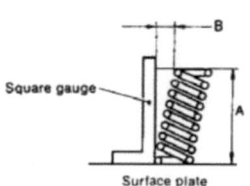

Square gauge

Surface plate

mm (in.)

	Maintenance standard
Valve spring free length (A)	38.5 (1.5157)

Allowable tilt value (B/A) is less than 0.035

2-4.3 Valve spring

(1) Valve spring inclination.
Since inclination of the valve spring is a direct cause of eccentric contact of the valve stem, always check it at disassembly.
Stand the valve upright on a stool, and check if the entire spring contacts the gauge when a square gauge is placed against the outside diameter of the valve spring.
If there is a gap between the gauge and spring, measure the gap with a feeler gauge.
When the valve spring inclination exceeds the wear limit, replace the spring.

(2) Valve spring free length.
Measure the free length of the valve spring, and replace the spring when the wear limit is exceeded.

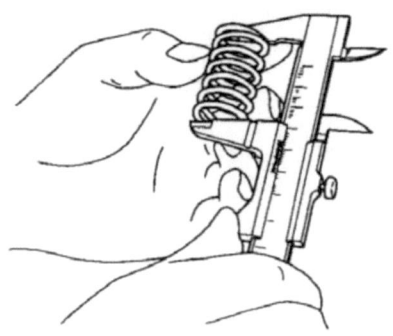

Also, measure the tension of the spring with a spring tester. If the tension is below the prescribed limit, replace the spring.

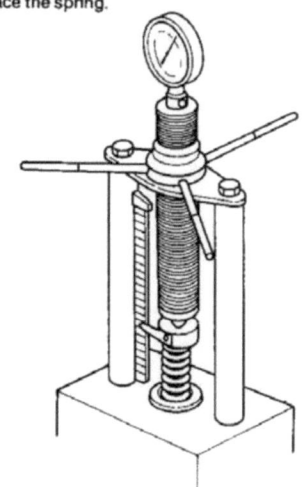

	1GM10(C), 2GM20(F)(C), 3GM30(F)(C)		3HM35(F)(C)	
	Maintenance standard	Wear limit	Maintenance standard	Wear limit
Valve spring free length	38.5mm (1.5157in.)	37mm (1.4567in.)	38.5mm (1.5157in.)	37mm (1.4567in.)
length when attached	29.2mm (1.1496in.)	—	30.2mm (1.1890in.)	—
Load applied attached	16.16kg (35.63lb)	13.7kg (30.20lb)	14.43kg (31.81lb)	12.2kg (26.90lb)

2-4.4 Valve stem seal

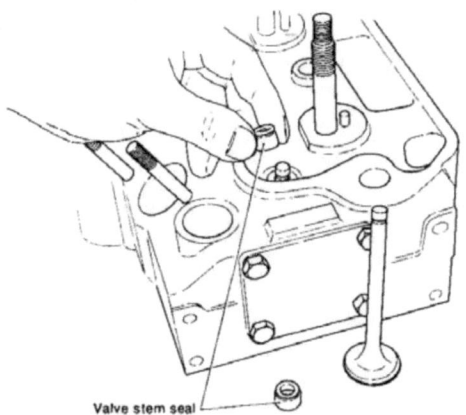

A valve stem seal is assembled at the top of the valve guide and the valve stem chamber oil is sucked into the combustion chamber through the valve guide (oil down) to prevent an increase in oil consumption.
The valve stem seal must always be replaced whenever it has been removed.
When assembling, coat the valve stem with engine oil before inserting.

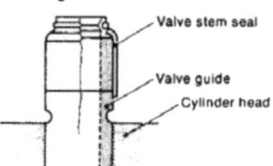

2-4.5 Spring retainer and spring cotter pin

Inspect the inside face of the spring retainer and the outside surface of the spring cotter pin, and the contact area of the spring cotter pin inside surface and the notch in the head of the valve stem. Replace the spring retainer and spring cotter pin when the contact area is less than 70% or when the spring cotter pin has been recessed because of wear.

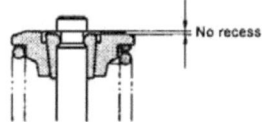

2-5 Precombustion chamber and top clearance

2-5.1 Precombustion chamber

Remove the packing and insulation packing at the precombustion chamber's front and rear chambers, and inspect.
Check for burning at the front end of the precombustion chamber front chamber, acid corrosion at the precombustion chamber rear chamber, and for burned packing.
Replace if faulty.

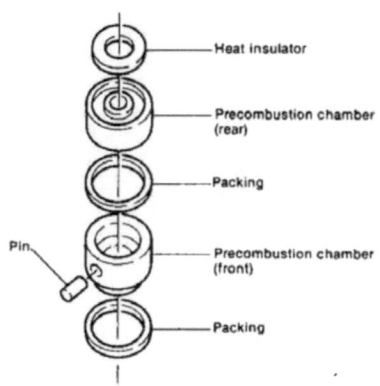

Chapter 2 Basic Engine
2. Cylinder Head

SM/GM(F)(C)·HM(F)(C)

2-5.2 Insulation packing

The insulation packing prevents transmission of heat from the precombustion chamber to the nozzle valve and serves to improve the nozzle's durability.
Always put in new insulation packing when it has been disassembled.

2-5.3 Top clearance

Top clearance is the size of the gap between the cylinder head combustion surface and the top of the piston at top dead center.
Since top clearance has considerable effect on the combustion performance and the starting characteristic of the engine, it must be checked periodically.

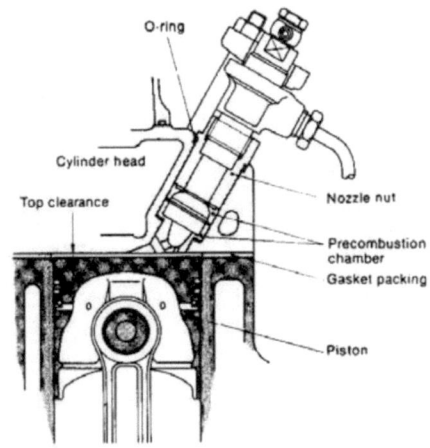

(1) Top clearance measurement
1) Check the cylinder head mounting bolts and tightening torque.
2) Remove the fuel injection valve and precombustion chamber.

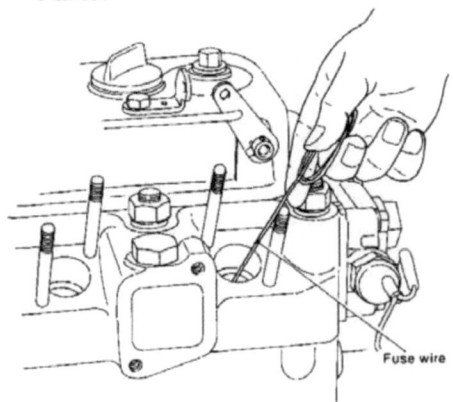

3) Lower the piston at the side to be measured.
4) Insert quality fuse wire (Ø1.2mm, 0.472in.) through the nozzle holder hole. (Be careful that the wire does not enter the intake and exhaust valve and the groove in the combustion surface.)
5) Crush the fuse wire by moving the piston to top dead center by slowly cranking the engine by hand.
6) Lower the piston by hand cranking the engine and remove the crushed fuse wire, being careful not to drop it.
7) Measure the thickness of the crushed part of the fuse wire with vernier calipers or a micrometer.

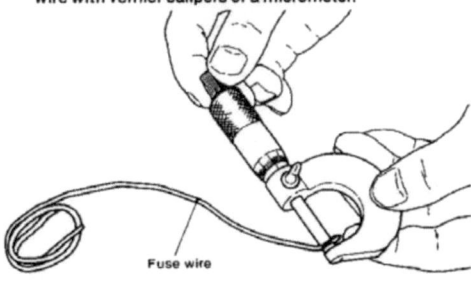

(2) Top clearance value.

		mm (in.)
	1GM10(C), 2GM20(F)(C), 3GM30(F)(C)	3HM35(F)(C)
Top clearance	0.68~0.88 (0.0268~0.0346)	0.66~0.86 (0.0260~0.0339)

When the top clearance value is not within the above range, check for damaged gasket packing, distortion of the cylinder head combustion surface, or other abnormal conditions.

2-6 Intake and exhaust valve rocker arm

Since the intake and exhaust valve rocker arm shaft and bushing clearance and valve head and push rod contact wear are directly related to the valve timing, and have an effect on engine performance, they must be carefully serviced.

2-6.1 Components of valve rocker arm

(1) Model 1GM10(C)

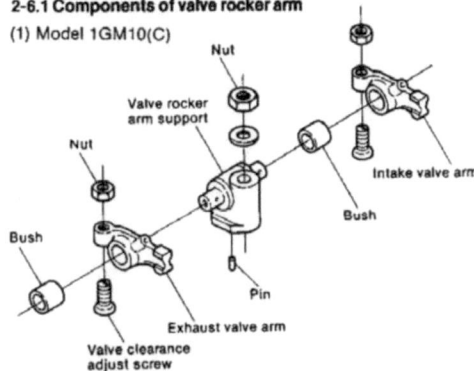

2-21

Chapter 2 Basic Engine
2. Cylinder Head

SM/GM(F)(C)-HM(F)(C)

The same part is used for both intake valve rocker arm and exhaust valve rocker arm. The bush is not fitted to the valve rocker arm.
In has a simple construction as the valve rocker arms are fitted to the valve rocker arm support from both sides without using the retainer. In the place of a retainer, the rib of the bonnet cover prevents the rocker arms from coming out.

NOTE: *Take care that the valve rocker arms do not get detached from the valve rocker arm shaft when dismantling or assembling. Replace the bonnet carefully when assembling.*

(2) Model 2GM20(F)(C)
The intake and exhaust valve rocker arms for two cylinders are fitted to a valve rocker arm shaft at both sides of the spring. The same part is used for both intake and exhaust valve rocker arms.

(3) Models 3GM30(F)(C) and 3HM35(F)(C)
The intake and exhaust valve rocker arms for three cylinders are fitted to a valve rocker arm shaft at both sides of the spring. The same intake and exhaust valve rocker arms, valve rocker arm support, spring and valve clearance adjusting screw are used for models 3GM30(F)(C) and 3HM35(F)(C).

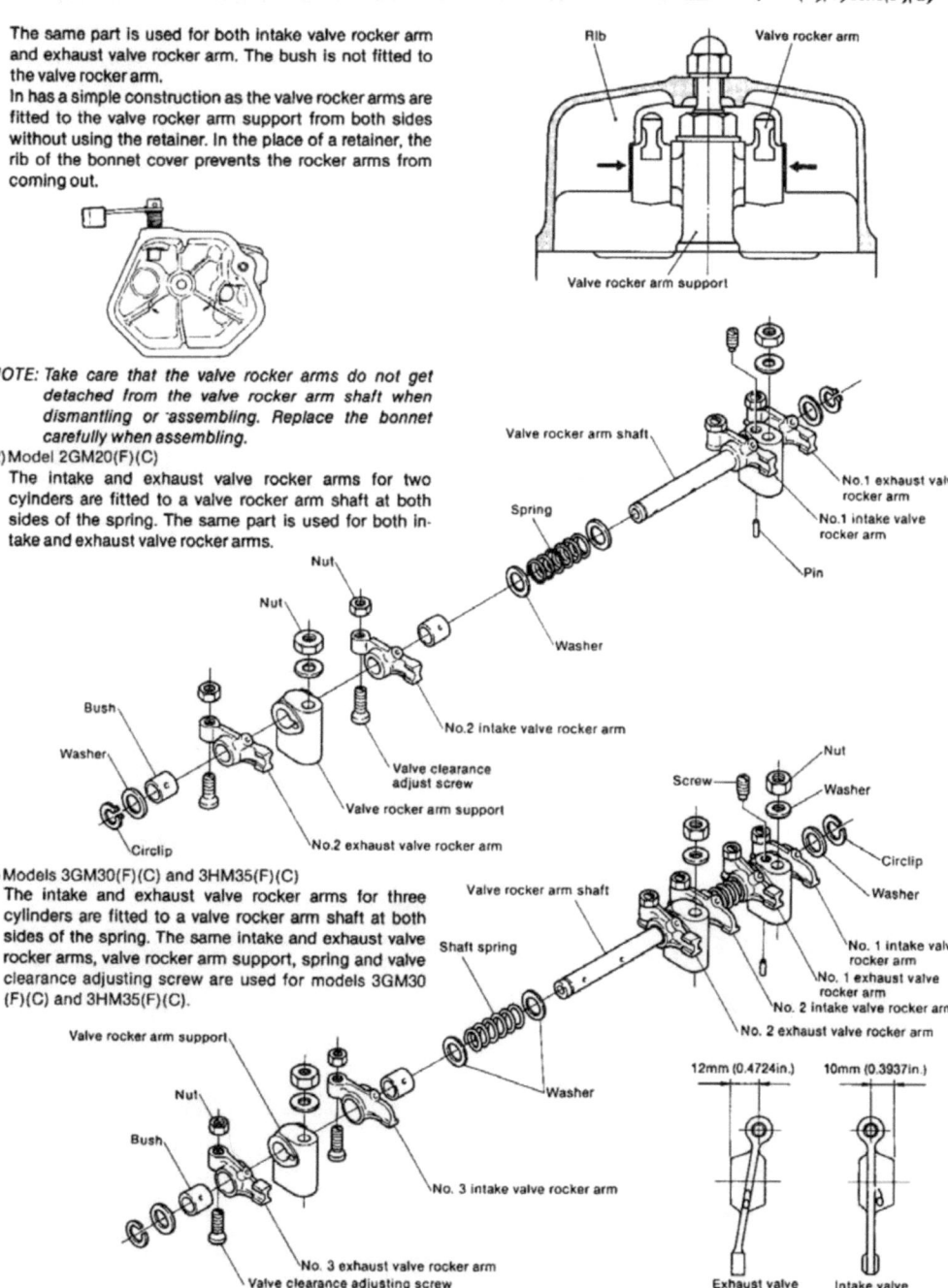

2-22

Printed in Japan
0000A0A1361

Chapter 2 Basic Engine
2. Cylinder Head

SM/GM(F)(C)-HM(F)(C)

2-6.2 Measuring the valve rocker arm shaft and bushing clearance

Measure the outside diameter of the valve rocker arm shaft and the inside diameter of the bushing, and replace the rocker arm or bushing if the measured value exceeds the wear limit.
Replace a loose valve rocker arm shaft bushing with a new bushing. However, when there is no tightening allowance, replace the valve rocker arm.

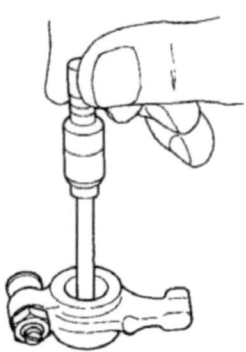

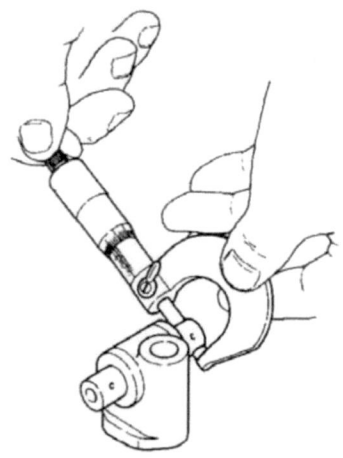

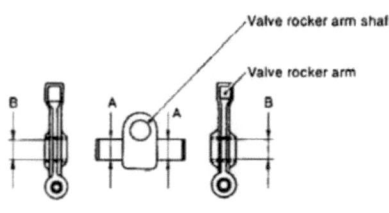

mm (in.)

			Maintenance standard	Clearance at assembly	Maximum allowable clearance	Wear limit
1GM10(C)	Intake and exhaust valve rocker arm shaft outside diameter	A	ø12 (0.4724)	0.016~0.052 (0.0006~0.0020)	0.15 (0.0059)	ø11.9 (0.4685)
	Intake and exhaust valve rocker arm bushing inside diameter (assembled)	B	ø12 (0.4724)			ø12.1 (0.4764)
2GM20(F)(C)	Intake and exhaust valve rocker arm shaft outside diameter	A	ø14 (0.5512)	0.016~0.052 (0.0006~0.0020)	0.15 (0.0059)	ø13.9 (0.5472)
	Intake and exhaust valve rocker arm bushing inside diameter (assembled)	B	ø14 (0.5512)			ø14.1 (0.5551)
3GM30(F)(C) 3HM35(F)(C)	Intake and exhaust valve rocker arm shaft outside diameter	A	ø14 (0.5512)	0.016~0.052 (0.0006~0.0020)	0.15 (0.0059)	ø13.9 (0.5472)
	Intake and exhaust valve rocker arm bushing inside diameter (assembled)	B	ø14 (0.5512)			ø14.1 (0.5551)

2-6.3 Valve rocker arm and valve top retainer contact and wear

Check the valve rocker arm and valve top retainer contact, and replace when there is any abnormal wear or peeling.

2-6.4 Valve clearance adjusting screw

Inspect the valve clearance adjusting screw and push rod contact, and replace when there is any abnormal wear or peeling.

2-6.5 Classification of the intake and exhaust valve rocker arms

Since the intake and exhaust valve rocker arms have different shapes, care must be exercised in service and assembly.

2-7 Adjusting intake and exhaust valve head clearance

Adjustment of the intake and exhaust valve head clearance governs the performance of the engine, and must be performed accurately. The intake and exhaust valve head clearance must always be checked and readjusted, as required, when the engine is disassembled and reassembled, and after every 300 hours of operation. Adjust the valve head clearance as described below.

2-7.1 Adjustment

Make this adjustment when the engine is cold.
(1) Remove the valve rocker arm cover.
(2) Crank the engine and set the piston to top dead center (TDC) on the compression stroke.

The matching mark is made at the setting hole of the starter motor on all models.

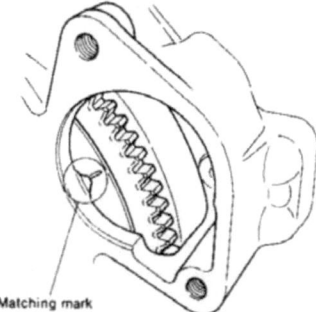
Matching mark

With respect to models 1GM10(C), 2GM20(F)(C) and 3GM30 (F)(C) only, a projection which serves as the matching mark is provided in the cast hole of the clutch housing.

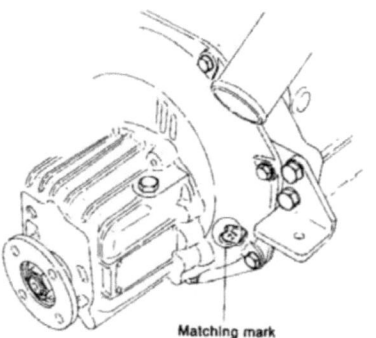
Matching mark

NOTE: Set to the position at which the valve rocker arm shaft does not move even when the crankshaft is turned to the left and right, centered around the matching mark.

(3) Check and adjust the intake and exhaust valve head clearances of the No. 1 piston.
Loosen the valve clearance adjusting screw lock nut, adjust the clearance to the maintenance standard with a feeler gauge, and retighten the lock nut.

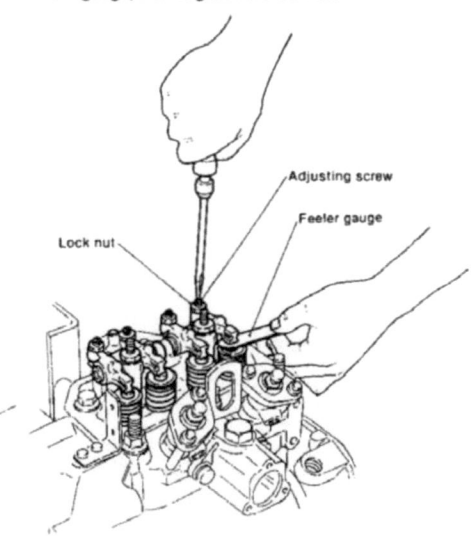

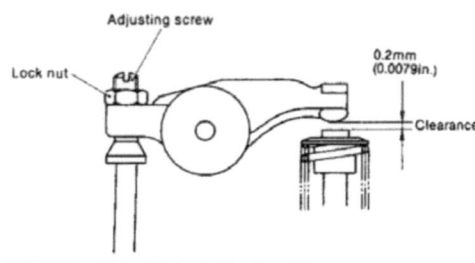

	All models
Intake and exhaust valve head clearance:	0.2mm (0.0079in.)

In the case of 2GM20(F)(C), adjust the valve head clearance of the No. 2 cylinder in the same manner after turning the crankshaft 180°.
In the case of 3GM30(F)(C), 3HM35(F)(C), adjust the valve head clearance on the No. 3 cylinder in the same manner after turning the crankshaft 240° and then adjust the No. 2 cylinder after turning the crankshaft another 240°.

NOTE: If you adjust the valve head clearance of the No. 2 cylinder first, turn the crankshaft 540°. Adjust the clearance of the No. 1 cylinder in the same manner on a 2 cylinder engine.

Chapter 2 Basic Engine
2. Cylinder Head

SM/GM(F)(C)-HM(F)(C)

3-7.2 Adjusting without a feeler gauge

Set the head clearance to zero by tightening the adjusting screw, being careful not to tighten the screw too tight.
Then adjust the valve clearance to the maintenance standard by backing off the adjusting screw by the angle given below.

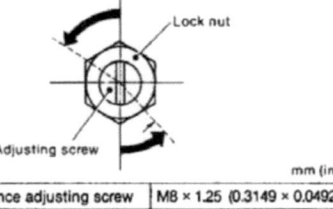

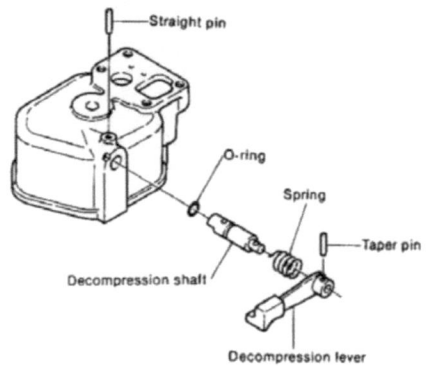

	mm (in.)
Valve clearance adjusting screw	M8 × 1.25 (0.3149 × 0.0492)
Adjusting screw backoff angle	Approx 58°

NOTE: Calculating the backoff angle.
　calculate the 0.2mm advance angle from 1.25mm advance at one turn = 360°
　$0.2/1.25 \times 360° = 58°$
　One side (60°) of the hexagonal nut should be used to measure.

2-8 Decompression mechanism

The decompression mechanism is used when the starter motor fails to rotate sufficiently because the battery is weak, and to facilitate starting in cold weather.
When the decompression lever is operated, the valve is pushed down, the engine is decompressed, the engine turns over easily and the flywheel inertia increases, thus making starting easy.

2-8.1 Model 1GM10(C)

2-8.2 Models 2GM20(F)(C), 3GM30(F)(C) and 3HM35(F)(C)

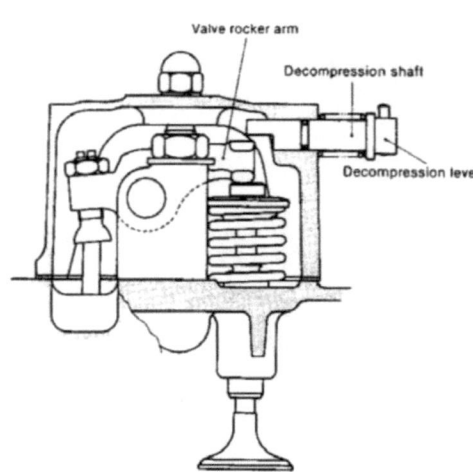

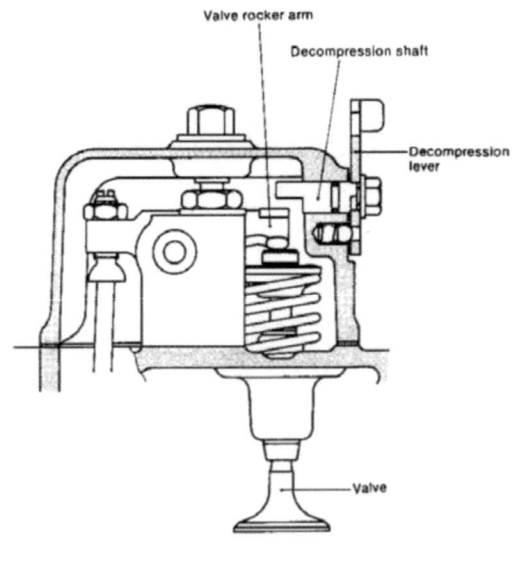

Chapter 2 Basic Engine
2. Cylinder Head

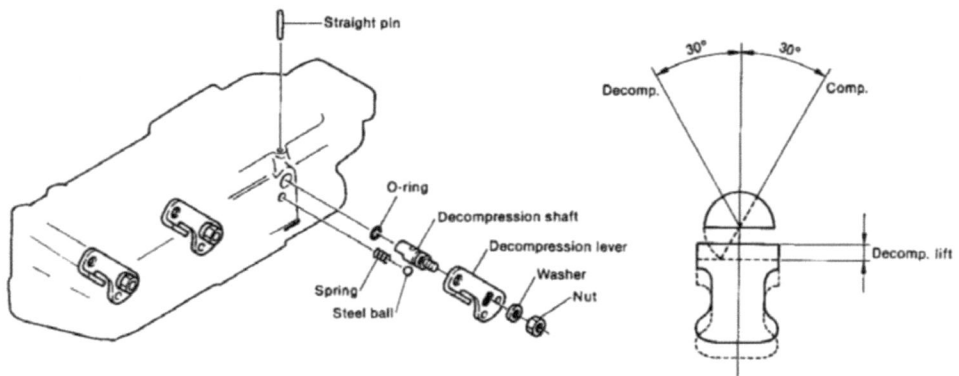

With this engine, there is no need to adjust the decompression lift.

2-9 Disassembling and reassembling the cylinder head

2-9.1 Disassembling the cylinder head

When disassembling the cylinder head, group the parts separately according to cylinder, intake or exhaust to avoid confusion.

(1) Disassembling the rocker arm ass'y
 1) Remove the rocker arm ass'y mounting nuts.
 2) Remove the rocker arm ass'y.

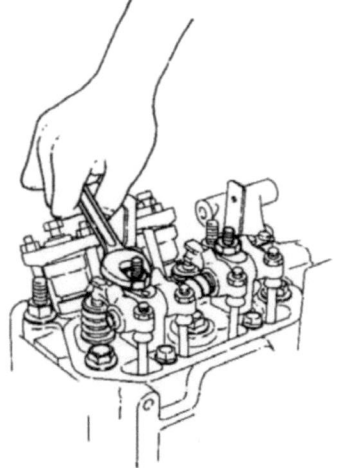

3) Remove the rocker arm retainer, and pull the rocker arm from the rocker arm support.

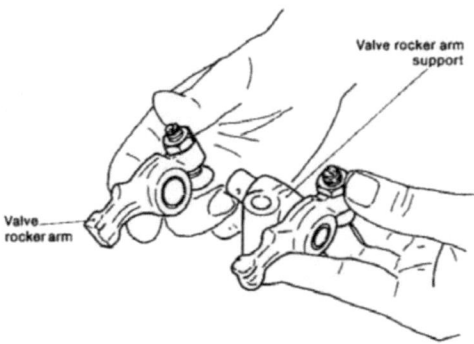

NOTE: *A retainer is not used for the valve rocker arm on model 1GM10(C) and is kept free, therefore the rocker arm can be removed directly.*

Chapter 2 Basic Engine
2. Cylinder Head

(2) Removing the precombustion chamber
1) Remove the rear precombustion chamber and packing.
2) Remove the front precombustion chamber and packing.

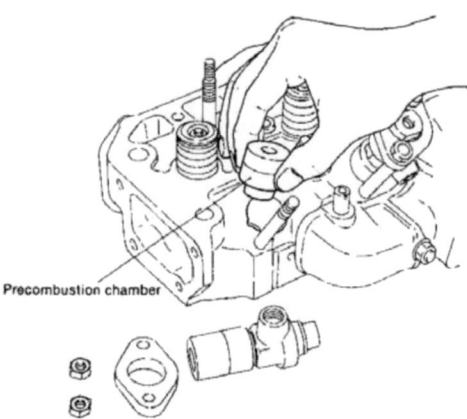
Precombustion chamber

2-9.2 Reassembling the cylinder head

Before reassembling the cylinder head, wash all the parts, inspect and measure the dimensions of each part, and repair or replace any parts that are abnormal. Be careful not to confuse the parts grouped by cylinder number and intake or exhaust.

(1) Assembling the intake and exhaust valves
1) Press the valve guide into the cylinder head.
2) Install the valve stem seal. (Always replace the valve stem seal with a new seal.)
3) Install the valve in the cylinder head.
4) Install the valve spring and valve spring seat.
5) Install the split collar.
   • Using the special tool
   • Using a wrench

(2) Installing the valve arm ass'y
1) Install the intake and exhaust rocker arms on the rocker arm support.
2) Install both the rocker arm supports and rocker arm retainers on the cylinder head, then tighten them with nuts.

(3) Installing the precombustion chamber
1) Install the front precombustion chamber and packing.
2) Install the rear precombustion chamber and packing. (Always replace the insulation packing.)

(3) Removing the intake and exhaust valve ass'y

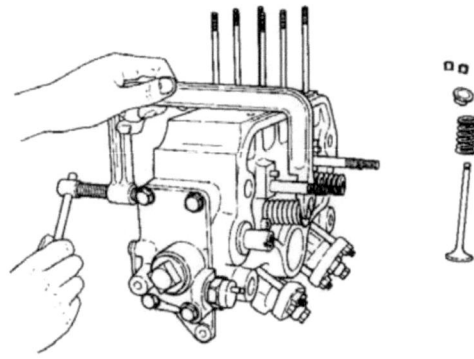

1) Set the special tool at the intake and exhaust valve ass'y and depress the valve spring by turning the lever.
2) When the special tool is not available, depress the valve spring with a wrench.
3) Remove the spring cotter pin.
4) Turn the lever of the special tool in the loosening direction, release the valve spring retainer, and remove the valve spring retainer and valve spring.
5) Pull the valve from the cylinder head.
6) Remove the valve stem seal.
7) Remove the valve guide.

3. Piston

3-1 Piston assembly construction

The pistons are made of LO-EX (AC8A-T6) for lightness and are designed for reduced vibration. The outside of the piston is machined to a special oval shape. During operation, thermal expansion is small, the optimum clearance between the piston and cylinder liner is maintained, and a stable supply of lubricating oil is assured.

Detail of A (heat dam)

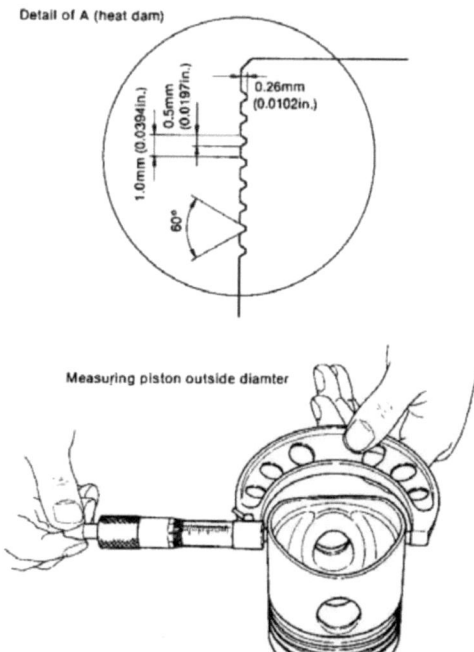

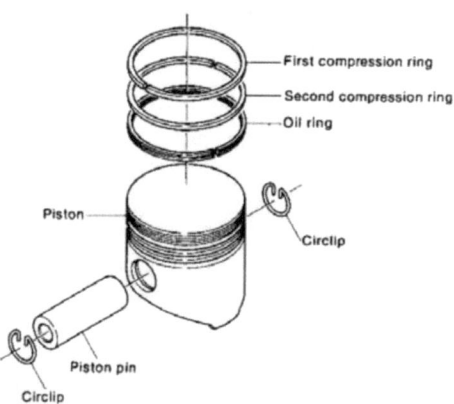

A complete set of piston rings consists of two compression rings and one oil ring.
To improve the rigidity of the piston skirt no ring is installed on the skirt itself so that the piston seldom becomes deformed and retains stable contact.
The piston pin is of the floating type. Both its ends are fastened with circlips.
Grooves called a heat dam are cut round the top section of the piston. These grooves help to dissipate heat and prevent scuffing.

3-2 Piston

3-2.1 Inspection

(1) Measuring important dimensions
   Measure each important dimension, and replace the piston when the wear limit is exceeded.

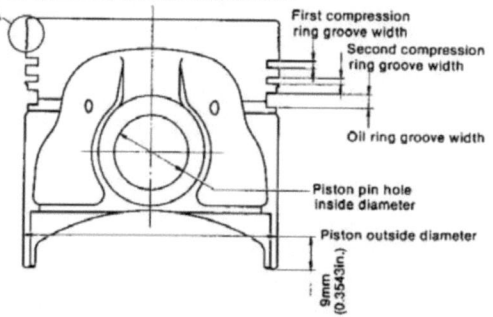

Measuring piston outside diamter

(2) Measure the clearance between the piston ring or oil ring and the ring groove with a thickness gauge.

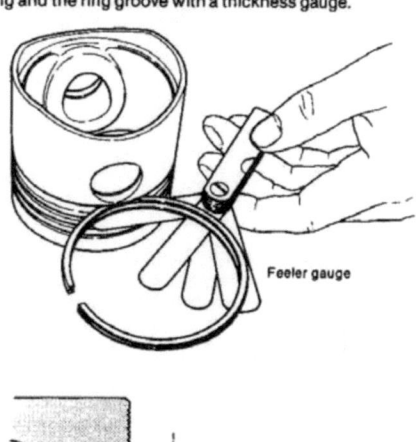

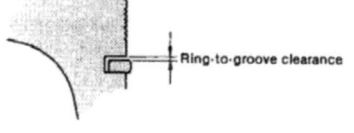

Chapter 2 Basic Engine
3. Piston

SM/GM(F)(C)-HM(F)(C)

mm (in.)

	1GM10(C), 2GM20(F)(C), 3GM30(F)(C)		3HM35(F)(C)	
	Maintenance standard	Wear limit	Maintenance standard	Wear limit
Piston outside diameter (At right angles to the piston pin, at a point 9.0mm (0.3543in.) from the bottom	φ74.91~74.94 (φ2.9492~2.9504)	74.85 (2.9468)	φ79.902~79.932 (φ3.1457~3.1470)	79.84 (3.1433)
Piston pin hole inside diameter	φ19.995~20.008 (0.7872~0.7877)	—	φ22.995~23.008 (0.9053~0.9058)	—
First compression piston ring-to-groove clearance	0.065~0.10 (0.0026~0.0039)	0.20 (0.0079)	0.065~0.10 (0.0026~0.0039)	0.20 (0.0079)
Second compression piston ring-to-groove clearance	0.035~0.07 (0.0014~0.0028)	0.20 (0.0079)	0.035~0.07 (0.0014~0.0028)	0.20 (0.0079)
Oil ring-to-groove clearance	0.02~0.055 (0.0008~0.0022)	0.15 (0.0059)	0.020~0.055 (0.0008~0.0022)	0.15 (0.0059)

(3) Piston pin outside contact and ring groove carbon build-up.
check if the piston ring grooves are clogged with carbon, if the rings move freely, and for abnormal contact around the outside of the piston. Repair or replace the piston if faulty.

3-2.2 Replacing a piston

If the dimension of any part is worn past the wear limit or the outside of the piston is scored, replace the piston.
(1) Replacement
 1) Install the piston pin circlip at one side only.
 2) Immerse the piston in 80°C oil for 10~15 minutes.

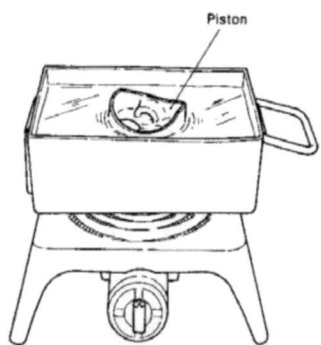
Piston

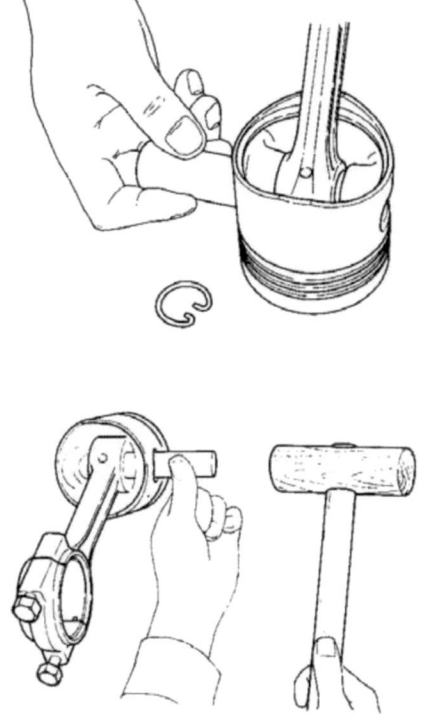

 3) Remove the piston from the hot oil and place it on a bench with the piston head at the bottom.
 4) Insert the small end of the connecting rod into the piston, insert the piston pin with a rotating motion, and install the other piston pin circlip.
 Use wooden hammer if necessary.

(2) Precautions
 1) Before inserting, check whether the piston pin is in the connecting rod.
 2) Coat the piston pin with oil to facilitate insertion.
 3) Check that the connecting rod and piston move freely.
 4) Insert the pin quickly, before the piston cools.

3-3 Piston pin and piston pin bushing

3-3.1 Piston pin

Measure the dimensions of the piston pin, and replace the pin if it is worn past the wear limit or severely scored.

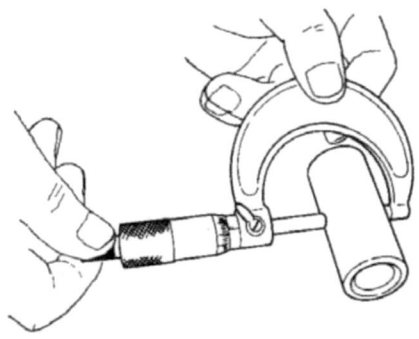

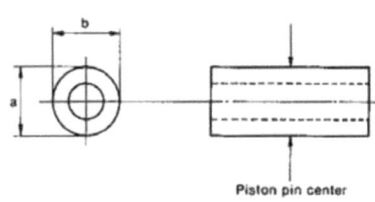

Piston pin center

Maximum wear measured in (a) (b) directions at central position marked*

mm (in.)

	1GM10(C), 2GM20(F)(C), 3GM30(F)(C)		3HM35(F)(C)	
	Maintenance standard	Wear limit	Maintenance standard	Wear limit
Piston pin outside diameter	ø20 $_{-0.009}^{0}$ (0.7870 ~ 0.7874)	ø19.98 (0.7866)	ø23 $_{-0.009}^{0}$ (0.9052 ~ 0.9055)	ø22.98 (0.9047)
Piston pin hole and piston pin tightening allowance	−0.005 ~ +0.017 (−0.0002 ~ +0.0007)	—	−0.005 ~ +0.017 (−0.0002 ~ +0.0007)	—

3-3.2 Piston pin bushing

A copper alloy wound bushing is pressed onto the piston pin.
Since a metallic sound will be produced if the piston pin and piston pin bushing wear is excessive, replace the bushing when the wear limit is exceeded.
The piston pin bushing can be easily removed and installed with a press. However, when installing the bushing, be careful that it is not tilted.

mm (in.)

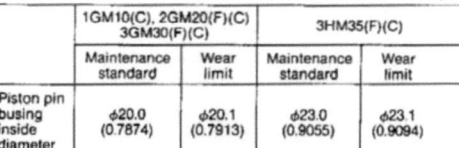

	1GM10(C), 2GM20(F)(C) 3GM30(F)(C)		3HM35(F)(C)	
	Maintenance standard	Wear limit	Maintenance standard	Wear limit
Piston pin busing inside diameter	ø20.0 (0.7874)	ø20.1 (0.7913)	ø23.0 (0.9055)	ø23.1 (0.9094)

NOTE: "Piston pin bushing inside diameter" is the dimension after pressing onto the connecting rod.

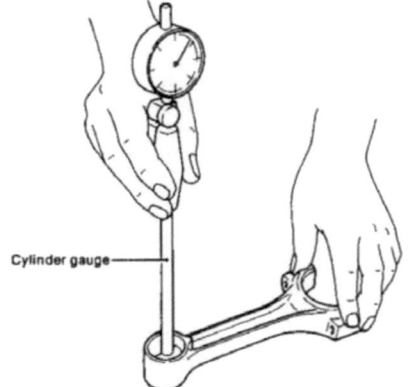

Cylinder gauge

3-4 Piston rings

3-3.1 Piston ring configuration

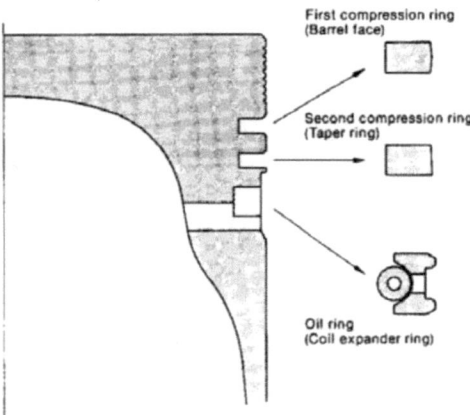

First compression ring (Barrel face)
Second compression ring (Taper ring)
Oil ring (Coil expander ring)

(1) The first compression ring is a barrel face ring that effectively prevents abnormal wear caused by engine loading and combustion gas blowby at initial run-in.
The sliding surface is hard chromium plated.

Model 3HM35(F)(C)

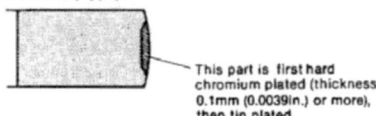

This part is first hard chromium plated (thickness 0.1mm (0.0039in.) or more), then tin plated.

Models 1GM10(C), 2GM20(F)(C), 3GM30(F)(C)

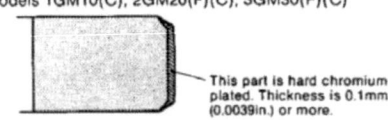

This part is hard chromium plated. Thickness is 0.1mm (0.0039in.) or more.

(2) The second compression ring is a taper ring having a sliding face taper of 30' ~ 1°30'. Since the cylinder liner is straight, and the contact area at initial operation is small, it is easily seated to the cylinder liner.
Moreover, the bottom of the sliding face is sharp, and oil splash is excellent and air-tightness is superb.

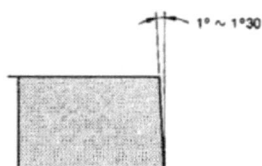

1° ~ 1°30

(3) The oil ring is a chrome-plated coil expander having a small contacting face, and exerts high pressure against the cylinder liner wall. Oil splash at the bottom of the sliding face is excellent, and its oil control effect is high.

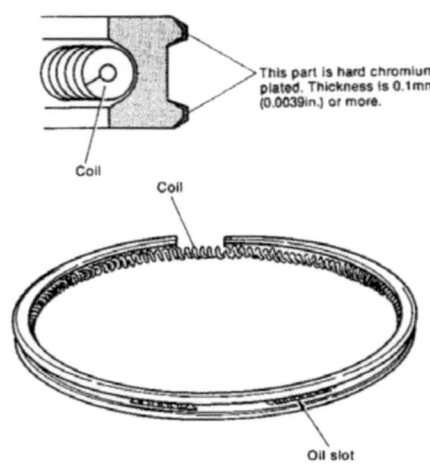

This part is hard chromium plated. Thickness is 0.1mm (0.0039in.) or more.
Coil
Coil
Oil slot

3-4.2 Inspection

(1) Piston ring contact
Inspect the piston ring contact, and replace the ring when contact is faulty. Since the oil ring side contact is closely related to oil consumption, it must be checked with particular care.

(2) Measuring the piston ring gap
Insert the piston into the cylinder or cylinder liner by pushing the piston ring at the head of the piston as shown in the figure, and measure the piston ring gap with a feeler gauge. Measure the gap at a point about 100mm (3.9370in.) from the top of the cylinder.
Measure by inserting a thickness gauge

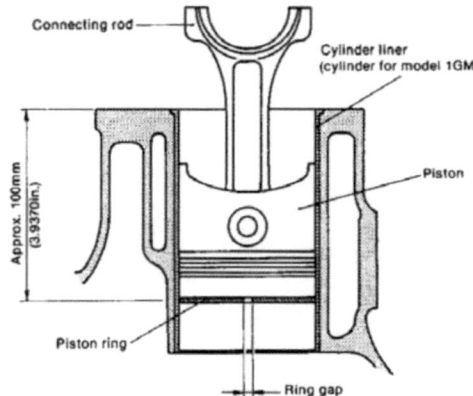

Connecting rod
Cylinder liner (cylinder for model 1GM
Approx. 100mm (3.9370in.)
Piston
Piston ring
Ring gap

Chapter 2 Basic Engine
3. Piston

SM/GM(F)(C)·HM(F)(C)

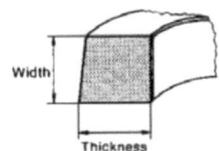
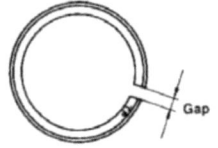

mm (in.)

		1GM10(C), 2GM20(F)(C), 3GM30(F)(C)		3HM35(F)(C)	
		Maintenance standard	Wear limit	Maintenance standard	Wear limit
1st. Piston ring	Width	1.97~1.99 (0.0776~0.0783)	1.90 (0.0748)	1.97~1.99 (0.0776~0.0783)	1.90 (0.0748)
	Thickness	3.10~3.30 (0.1220~0.1299)	—	3.20~3.40 (0.1260~0.1339)	—
2nd. Piston ring	Width	1.97~1.99 (0.0776~0.0783)	1.90 (0.0748)	1.97~1.99 (0.0776~0.0783)	1.90 (0.0748)
	Thickness	3.10~3.30 (0.1220~0.1299)	—	3.40~3.60 (0.1339~0.1417)	—
Oil ring	Width	3.97~3.99 (0.1563~0.1571)	3.90 (0.1535)	3.97~3.99 (0.1563~0.1571)	3.90 (0.1535)
	Thickness	2.40~2.80 (0.0945~0.1102)	—	2.70~3.10 (0.1063~0.1220)	—
1st. Piston ring gap		0.20~0.40 (0.0079~0.0157)	1.5 (0.0591)	0.25~0.45 (0.0098~0.0177)	1.75 (0.0689)
2nd. Piston ring gap		0.20~0.40 (0.0079~0.0157)	1.5 (0.0591)	0.20~0.40 (0.0079~0.0157)	1.5 (0.0591)
Oil ring gap		0.20~0.40 (0.0079~0.0157)	1.5 (0.0591)	0.25~0.45 (0.0098~0.0177)	1.75 (0.0689)

(3) Piston ring replacement precautions
1) Clean the ring grooves carefully when replacing the rings.
2) When installing the rings, assemble the rings so that the manufacturer's mark near the gap is facing the top of the piston.

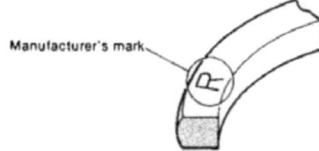

3) After assembly, check that the rings move freely in the grooves.
4) The rings must be installed so that the gaps are 120° apart. At this time, be careful that the ring gap is not lined up with the piston side pressure part.

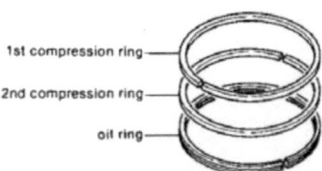

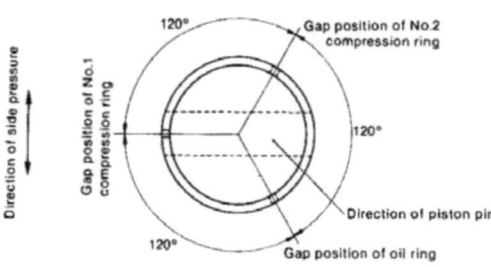

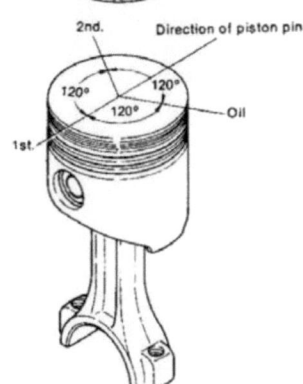

5) Since the oil ring is equipped with a coil expander, attach it to the piston so that the joint of the ring is opposite the gap of the coil expander.

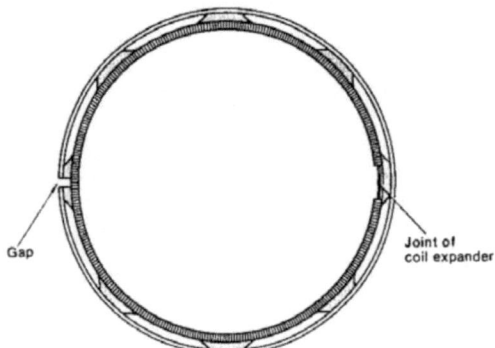

4. Connecting Rod

4-1 Connecting rod ass'y construction

The connecting rod connects the piston pin and crank pin and transmits the explosive force of the piston to the crankshaft. It is a stamp forging designed for extreme lightness and ample strength against bending. A kelmet bushing split at right angles is installed to the large end of the rod, and a round copper alloy is pressed onto the small end.

Pass a test bar through the large end and small end holes of the connecting rod, place the bars on a V-block on a stool and center the large end test bar. Then set the sensor of a dial indicator against the small end test bar and measure twist and parallelity. When the measured value exceeds the wear limit, replace the connecting rod. Twisting and poor parallelity will cause uneven contact of the piston and bushing and shifting of the piston rings, resulting in compression leakage.

Connecting rod twist and parallelity   mm (in.)

Maintenance standard	0.03/100 or less (0.00118/3.937)
Limit	0.08/100 (0.00315/3.937)

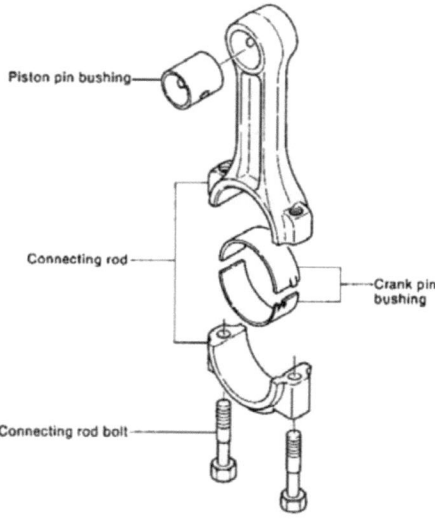

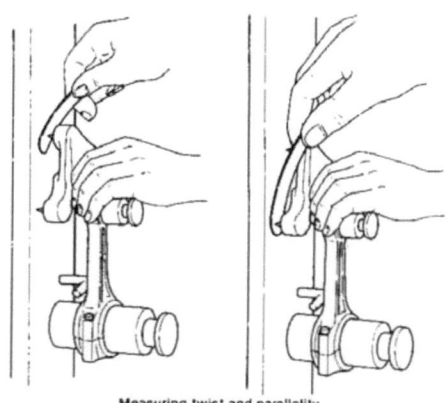

Measuring twist and parallelity

4-2 Inspection

4-2.1 Large and small end twist and parallelity

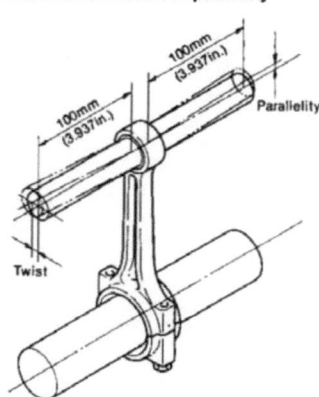

4-3 Crank pin bushing

Since the crank pin bushing slides while receiving the load from the piston, an easy-to-replace kelmet bushing with a wear-resistant overlay is used.

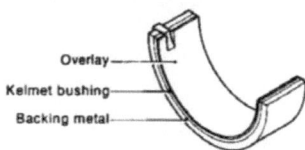

4-3.1 Crank pin bushing inside diameter

Tighten the large end of the connecting rod to the prescribed torque with the connecting rod bolts, and measure the inside diameter of the crank pin bushing. Replace the bushing if the inside diameter or the clearance at the crank pin part exceeds the wear limit.

Chapter 2 Basic Engine
4. Connecting Rod

SM/GM(F)(C)-HM(F)(C)

3) Tighten the connecting rod bolt to the prescribed tightening torque.

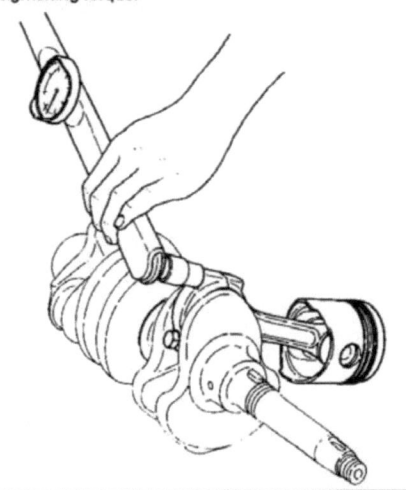

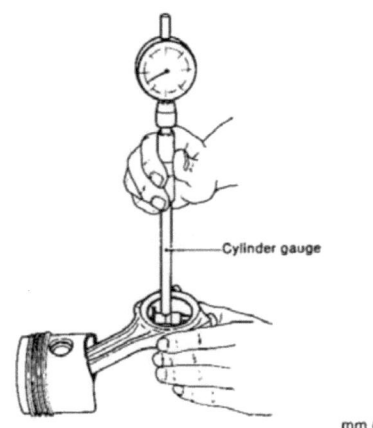

mm (in.)

	1GM10(C),2GM20(F)(C), 3GM30(F)(C)		3HM35(F)(C)	
	Maintenance standard	Wear limit	Maintenance standard	Wear limit
Crank pin bushing inside diameter	⌀40.0 (1.5748)	⌀40.10 (1.5787)	⌀44.0 (1.7323)	⌀44.10 (1.7362)
Crank pin and bushing oil clearance	0.028~0.086 (0.0011 ~0.0034)	0.13 (0.0051)	0.036~0.092 (0.0014 ~0.0036)	0.13 (0.0051)
Connecting rod bolt Thread diameter	M7 x P1.0 (0.2755 x 0.0393)		M9 x P1.0 (0.3543 x 0.0393)	
Connecting rod bolt tightening torque	2.5kgf-m (18.1ft-lb)		4.5kgf-m (32.5ft-lb)	

	1GM10(C) 2GM20(F)(C) 3GM30(F)(C)	3HM35(F)(C)
Connecting rod tightening torque	2.5kgf-m (18.1ft-lb)	4.5kgf-m (32.5ft-lb)
Hexagon width	12mm (0.4724 in.)	13mm (0.5118 in.)

4) Loosen the connecting rod bolt and slowly remove the connecting rod big end cap, then measure the crushed Plasti gauge with a gauge.

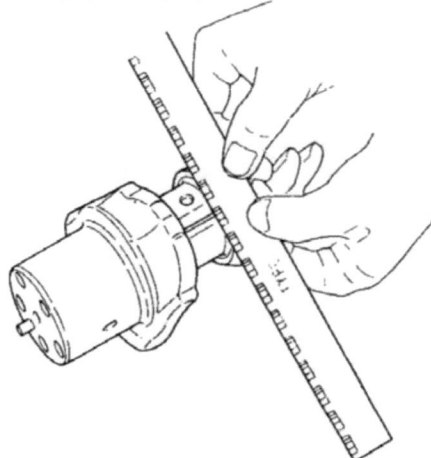

NOTE: The Crank pin bushing inside diameter must always be measured with the connecting rod bolts tightened to the prescribed torque.

4-3.2 Crank pin and bushing clearance (oil clearance)

Since the oil clearance affects both the durability of the bushing and lubricating oil pressure, it must always be the prescribed value. Replace the bushing when the oil clearance exceeds the wear limit.

(1) Measurement
 1) Thoroughly clean the inside surface and crank pin section of the crank pin bushing.
 2) Install the connecting rod on the crank pin section of the crankshaft and simultaneously fit a Plasti gauge on the inside surface of the crank pin bearing.

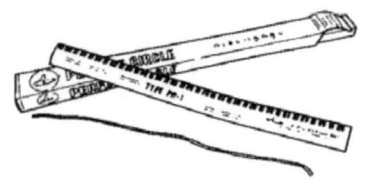

NOTE: Never adjust by shims or machine the crank pin bushing. Always replace the crank pin bushing with a new one.

Chapter 2 Basic Engine
4. Connecting Rod

SM/GM(F)(C)·HM(F)(C)

5) The crank pin and bushing clearance (oil clearance) may also be measured with a micrometer, in addition to measurement with a Plasti gauge. With this method, the outside diameter of the crankshaft crank pin section and the inside diameter of the connecting rod's big end bushing, when the connecting rod bolt has been tightened to the prescribed torque, are measured, and the difference between the large end bushing inside diameter and crank pin outside diameter is set as the oil clearance.

(2) Measurement precautions
  1) Be careful that the Plasti gauge does not enter the crank pin oil hole.
  2) Be sure that the crankshaft does not turn when tightening the connecting rod bolt.

4-3.3 Crank pin bushing replacement precautions

(1) Thoroughly clean the crank pin bushing and the rear of the crank pin bushing.
(2) Also clean the big end cap, install the crank pin bushing and check if the bushing contacts with the big end cap closely.
(3) When assembling the connecting rod, match the number of the big end section and the big end cap, coat the bolts with engine oil, and alternately tighten the bolts gradually to the prescribed tightening torque. If a torque wrench is not available, put matching marks (torque indication lines) on the bolt head and big end cap before disassembly and tighten the bolts until these two lines are aligned.

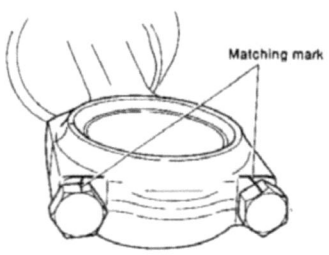
Matching mark

(4) Check that there is no sand or metal particles in the lubricating oil and that the crankshaft is not pitted. Clean the oil holes with particular care.

4-4 Tightening the connecting rod bolts

When tightening the connecting rod bolts, coat the threads of the bolts with engine oil.
Tighten the two bolts alternately and gradually to the prescribed tightening torque. If a torque wrench is not available, make matching marks (torque indication lines) on the head of the bolt and the big end cap and tighten the bolts until these two marks are aligned.

Models 1GM10(C), 2GM20(F)(C), 3GM30(F)(C)

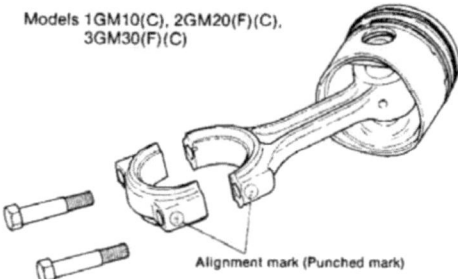

Alignment mark (Punched mark)

Model 3HM35(F)(C)

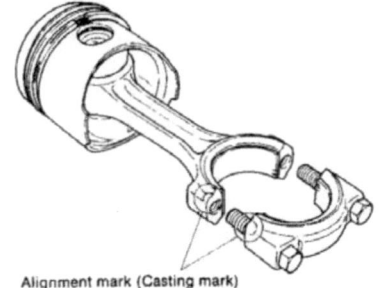
Alignment mark (Casting mark)

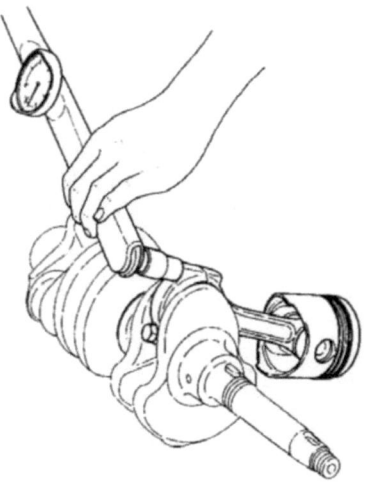

4-5 Connecting rod side clearance

After installing the connecting rod on the crankshaft, push the rod to one side and measure the side clearance by inserting a feeler gauge into the gap produced at the other side.

Chapter 2 Basic Engine
4. Connecting Rod

SM/GM(F)(C)-HM(F)(C)

The connecting rod bolts must also be tightened to the prescribed tightening torque in this case.

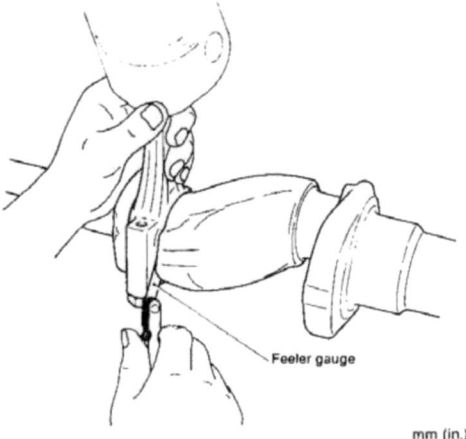

Feeler gauge

mm (in.)

	1GM10(C) 2GM20(F)(C) 3GM30(F)(C)	3HM35(F)(C)
Connecting rod side clearance	0.2~0.4 (0.0079~0.0157)	0.2~0.4 (0.0079~0.0157)

4-6 Piston bushing and piston pin

The piston bushing is a round copper alloy bushing driven onto the small end of the connecting rod. During use, the piston pin bushing and piston pin will wear. If this wear becomes excessive, a metallic sound will be produced and the engine will become noisy.

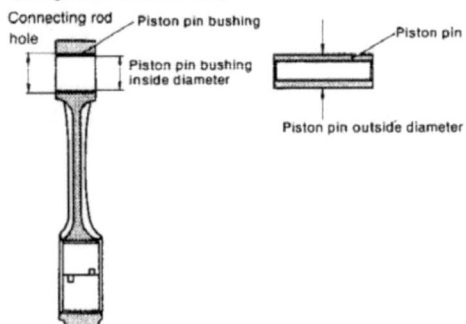

Connecting rod hole
Piston pin bushing
Piston pin bushing inside diameter
Piston pin
Piston pin outside diameter

mm (in.)

	1GM10(C),2GM20(F)(C), 3GM30(F)(C)		3HM35(F)(C)	
	Maintenance standard	Wear limit	Maintenance standard	Wear limit
Piston pin bushing inside diameter	ɸ20.0 (0.7874)	ɸ20.10 (0.7913)	ɸ23.0 (0.9055)	ɸ23.1 (0.9094)
Piston and bushing clearance	0.025~0.047 (0.0010 ~0.0019)	0.11 (0.0043)	0.025~0.047 (0.0010 ~0.0019)	0.11 (0.0043)
Connecting rod hole	ɸ22.0~22.021 (0.8661~0.8670)	—	ɸ23.0~23.021 (0.9055~0.9063)	—

Replacing the piston pin bushing

(1) When the bushing for the connecting rod piston pin is either worn out or damaged, replace it by using the "piston pin extracting tool" installed on a press.

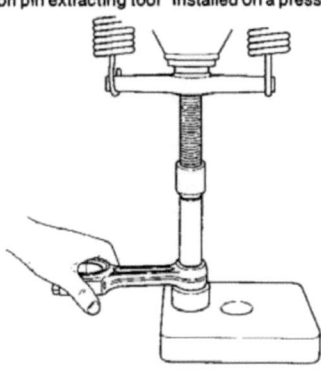

NOTE: Force the piston pin bushing into position so that its oil hole coincides with the hole on the small end of the connecting rod.

(2) After forcing the piston pin bushing into position, finish the inner surface of the bushing by using a pin honing machine or reamer so that it fits the piston pin to be used.

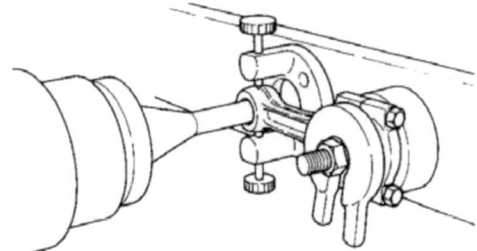

NOTE: Attach the bushing to the piston pin so that the pin, coated with engine oil can be pushed into position with your thumb.

5. Crankshaft

5-1 Crankshaft ass'y and bearing construction

The crankshaft is stamp-forged, and the crank pin and journal sections are high-frequency induction hardened, and ground and polished to a high precision finish. Therefore, the contact surface with the bushing is excellent and durability is superb.

The crankshaft is a balance weight integral type. Engine unbalance, which causes vibration, has been minimized by balancing the V-pulley, flywheel, and crankshaft.

The flywheel is fixed at the end of the crankshaft with hexagonal bolts and a locating pin. The crankshaft gear is fixed and keyed to the crankshaft inside the timing gear case, and the governor weight support is fixed with a hexagonal nut together with the crankshaft gear. It is so designed that the governor sleeve and the thrust bearing can be slid onto the crankshaft to get the gear side end of the crankshaft to perform as the governor shaft. The V-pulley is fitted outside the timing gear case and it drives the alternator and cooling water pump [on models 2GM20(F)(C), 3GM30(F)(C) and 3HM35(F)(C)].

5-1.1 Construction of model 1GM10(C)

Crankshaft assembly
The crankshaft is supported by the metal housing at the flywheel end, and by the bearing metal which is inserted into the cylinder body hole at the gear case end. Thrust metals are set at both sides of the bearing at the gear case end.

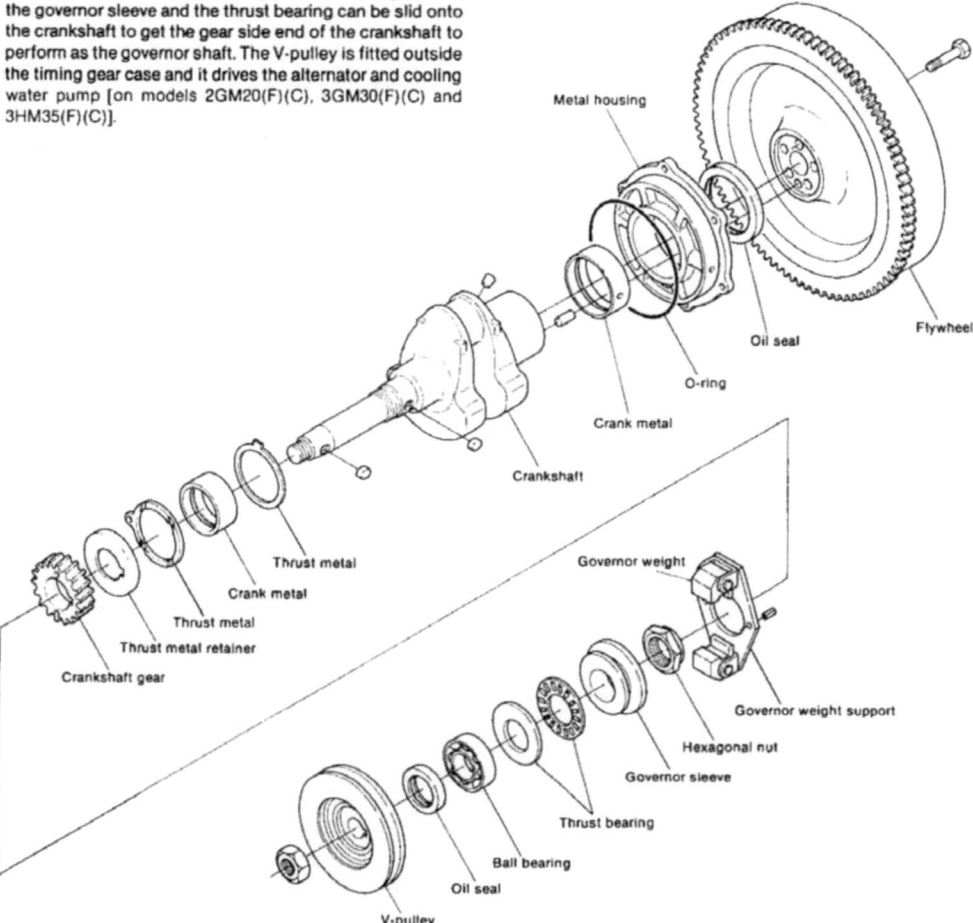

2-38

Chapter 2 Basic Engine
5. Crankshaft

5-1.2 Construction of models 2GM20(F)(C), 3GM30(F)(C) and 3HM35(F)(C) crankshaft assembly

The following figure shows the crankshaft assembly of model 3GM30(F)(C). On model 2GM20(F)(C) the intermediate bearing at the gear case end is not fitted. The construction of model 3HM35(F)(C) crankshaft assembly is the same as that of model 3GM30(F)(C).

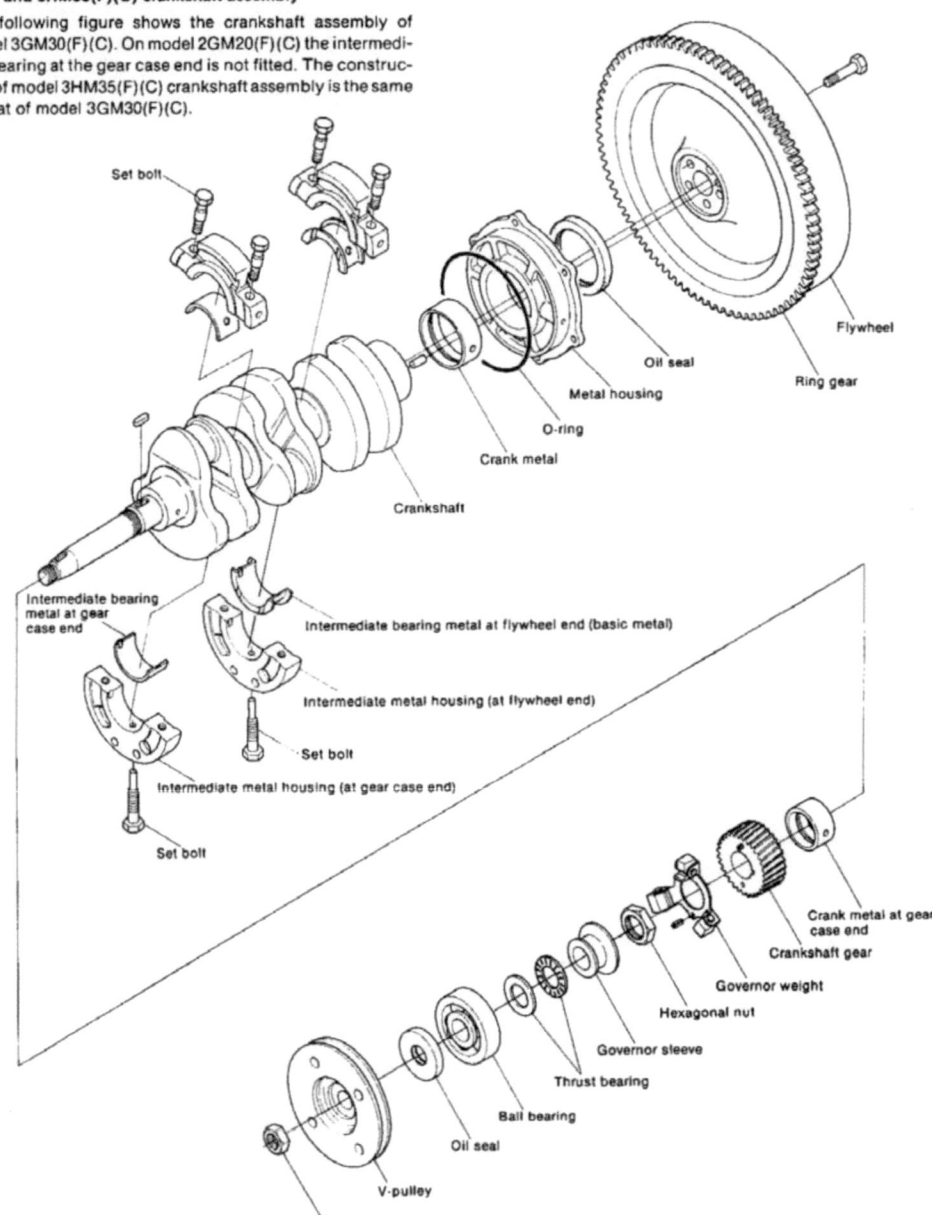

Hexagonal nut [left-handed thread for model 3HM35(F)(C)]

5-2 Inspection

5-2.1 Crank journal and crank pin

(1) Cracking

If cracking of the crank journal or crank pin is suspected, thoroughly clean the crankshaft and perform a color check on the shaft, or run a candle flame over the crankshaft and look for oil seepage from cracks. If any cracks are detected, replace the crankshaft.

(2) Crank pin and crank journal outside diameter measurement.

When the difference between the maximum wear and minimum wear of each bearing section exceeds the wear limit, replace the crankshaft. Also check each bearing section for scoring. If the scoring is light, repair it with emery cloth.

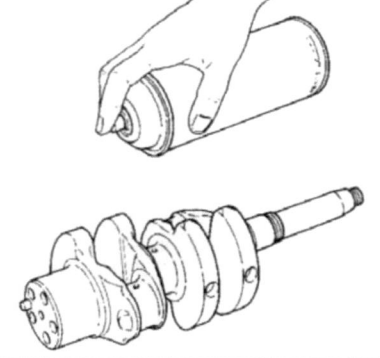
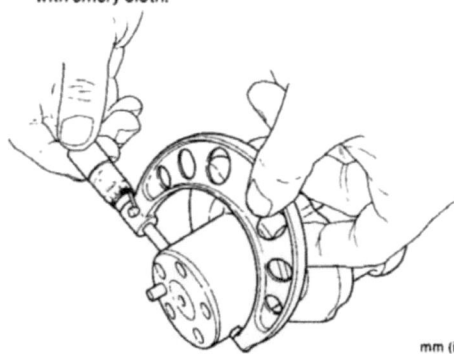

mm (in.)

			1GM10(C), 2GM20(F)(C), 3GM30(F)(C)		3HM35(F)(C)	
			Maintenance standard	Wear limit	Maintenance standard	Wear limit
Crank journal outside diameter	Gear case side	A	$\varnothing 44_{-0.050}^{-0.036}$ (1.7303 ~ 1.7309)	$\varnothing 43.90$ (1.7283)	$\varnothing 47_{-0.050}^{-0.036}$ (1.8484 ~ 1.8490)	$\varnothing 46.90$ (1.8465)
	Intermediate bearing	B	$\varnothing 44_{-0.050}^{-0.036}$ (1.7303 ~ 1.7309)	$\varnothing 43.90$ (1.7283)	$\varnothing 47_{-0.050}^{-0.036}$ (1.8484 ~ 1.8490)	$\varnothing 46.90$ (1.8465)
	Flywheel side	C	$\varnothing 60_{-0.050}^{-0.036}$ (2.3602 ~ 2.3608)	$\varnothing 59.90$ (2.3583)	$\varnothing 65_{-0.050}^{-0.036}$ (2.5571 ~ 2.5576)	$\varnothing 64.90$ (2.5551)
Crank pin outside diameter		D	$\varnothing 40_{-0.050}^{-0.036}$ (1.5728 ~ 1.5734)	$\varnothing 39.90$ (1.5709)	$\varnothing 44_{-0.050}^{-0.036}$ (1.7303 ~ 1.7309)	$\varnothing 43.90$ (1.7283)
Crank journal/pin eccentric wear			—	0.01 (0.0004)	—	0.01 (0.0004)
Crank journal and bushing oil clearance	Gear case side		0.036 ~ 0.092 (0.0014 ~ 0.0036)	0.15 (0.0059)	0.036 ~ 0.095 (0.0014 ~ 0.0037)	0.15 (0.0059)
	Intermediate bearing		0.036 ~ 0.092 (0.0014 ~ 0.0036)	0.15 (0.0059)	0.036 ~ 0.095 (0.0014 ~ 0.0037)	0.15 (0.0059)
	Flywheel side		0.036 ~ 0.095 (0.0014 ~ 0.0037)	0.15 (0.0059)	0.036 ~ 0.099 (0.0014 ~ 0.0039)	0.15 (0.0059)
Crank pin and crank pin bearing oil clearance			0.028 ~ 0.086 (0.0011 ~ 0.0034)	0.13 (0.0051)	0.036 ~ 0.092 (0.0014 ~ 0.0036)	0.13 (0.0051)

NOTE: The crankshaft of model 1GM10(C) does not have an intermediate bearing.

Measurement must be taken in at least 2 positions in the direction of crankshaft center line for each journal, and in each measurement, maximum and minimum wear directions must be measured. From these results, eccentric wear and maximum wear can be determined.

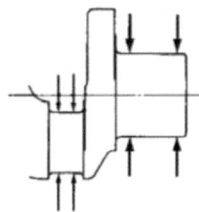
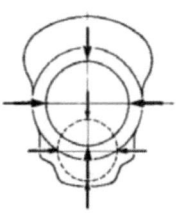

Chapter 2 Basic Engine
5. Crankshaft

SM/GM(F)(C)·HM(F)(C)

1GM10(C)

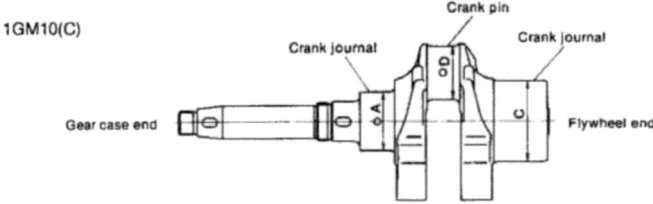

2GM20(F)(C)

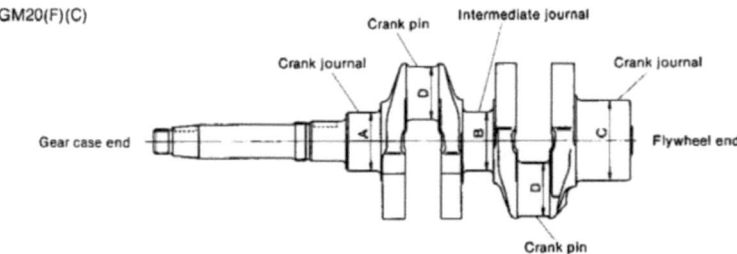

3GM30(F)(C), 3HM35(F)(C)

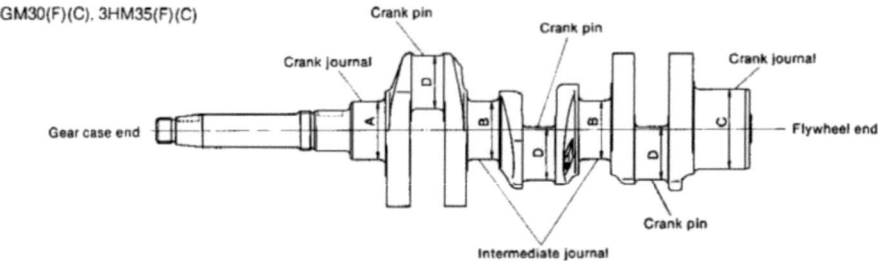

(3) Measuring the crankshaft bend [2GM20(F)(C), 3GM30(F)(C), 3HM35(F)(C)]

Measure on a surface plate. Place the journal parts of both ends of the crankshaft on a V-block and measure with a dial gauge while moving the crankshaft in an axial direction. If the deflection of the middle of the crankshaft exceeds the limit, replace the crankshaft.

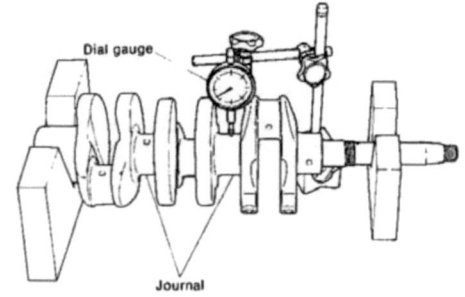

2-41

Chapter 2 Basic Engine
5. Crankshaft

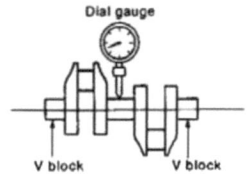

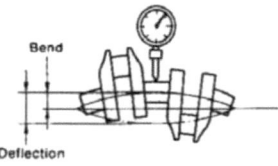

	Maintenance standard	Wear limit
Crankshaft bend	Less than 0.015 (0.0006)	0.15 (0.006)

mm (in.)

5-3 Crankshaft side gap

5-3.1 Side gap

The clearance in the axial direction after the crankshaft has been assembled is called the side gap.
If the side gap is too large, contact with pistons will be uneven, the clutch disengagement position will change, and other troubles will occur. If it is too small, the crankshaft sliding resistance will increase and cranking will become stiff.

For model 1GM10(C)
Adjust the side gap to the maintenance standard according to the thickness of the crankshaft thrust metal.
Thrust metals are installed on both sides of the crankcase and gear case.

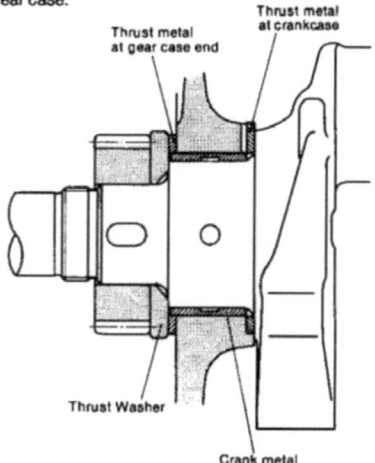

On models 2GM20(F)(C) and 3GM30(F)(C), the value of the side gap is the difference between the width of the basic bearing metal and the width of the journal. The basic bearing for model 2GM20(F)(C) is the intermediate bearing, and for models 3GM30(F)(C) and 3HM35(F)(C) it is the intermediate bearing at the flywheel end.

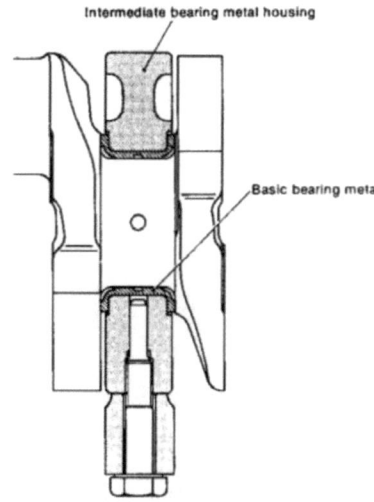

5-3.2 Measuring side gap

Set a dial indicator against the end of the crankshaft (or end of the flywheel) and measure the amount of movement of the crankshaft in the axial direction. If the measured value exceeds the wear limit, replace the crankshaft thrust washer. Main bearing housing packing of the prescribed thickness must be used.

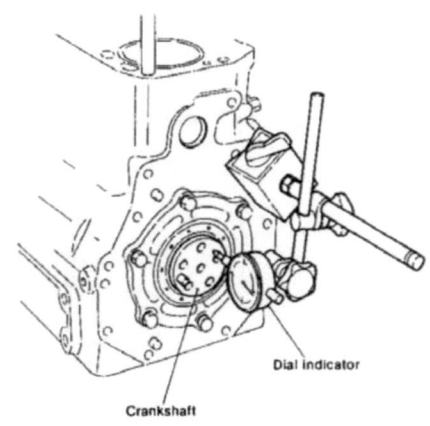

Chapter 2 Basic Engine
5. Crankshaft

SM/GM(F)(C)·HM(F)(C)

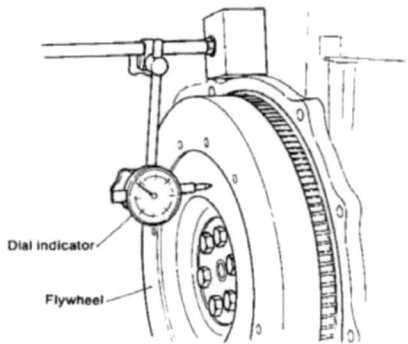

5-3.3 Side gap maintenance standard and wear limit

mm (in.)

Crank shaft side gap	1GM10(C)		2GM20(F)(C), 3GM30(F)(C)		3HM35(F)(C)	
	Maintenance standard	Wear limit	Maintenance standard	Wear limit	Maintenance standard	Wear limit
	0.06 ~ 0.19 (0.0024 ~ 0.0075)	0.30 (0.0012)	0.09 ~ 0.19 (0.0035 ~ 0.0075)	0.30 (0.0012)	0.09 ~ 0.18 (0.0035 ~ 0.0071)	0.30 (0.0012)

5-4 Disassembly of the crankshaft [2GM20(F)(C), 3GM30(F)(C), 3HM35(F)(C)]

For model 1GM10(C) see the chapter on disassembly and reassembly. Because there are points over which care must be taken in models 2GM20(F)(C), 3GM30(F)(C) and 3HM35(F)(C), disassembly and reassembly procedures are explained below.

5-4.1 Disassembly

(1) When disassembling, lay the cylinder down with the main bearing housing side on top so that the crankshaft will be vertical for easy operation.
   (*Remove the crank gear and flywheel beforehand.)
(2) Remove the main bearing housing.
(3) Attach a rope to the crankshaft, gradually lifting it with chain block etc. and remove the two set bolts of the intermediate main bearing housing. (If the crankshaft is lifted too much or not enough, the set bolts will be difficult to release.)
(4) Lift and remove the crankshaft (with the intermediate main bearing housing).
(5) Remove each intermediate main bearing housing from the crankshaft.

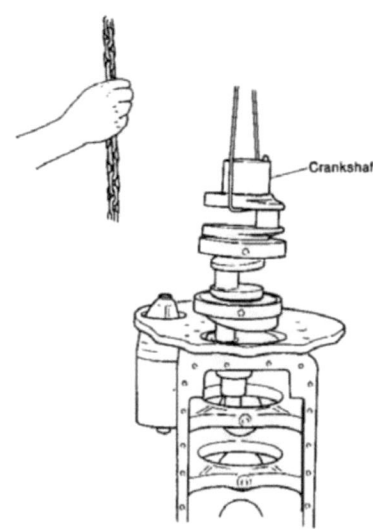

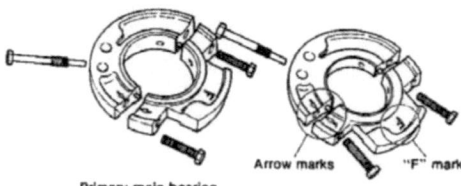

Primary main bearing

5-4.2 Reassembly

(1) Clean each part before reassembly.
(2) Attach the intermediate main bearing housing to the crankshaft and confirm that the crankshaft rotates smoothly.
  1) Assembling position and direction of the intermediate main bearing housing.
     • The "F" mark on the intermediate main bearing housing indicates the direction of assembly on the crankshaft flywheel.

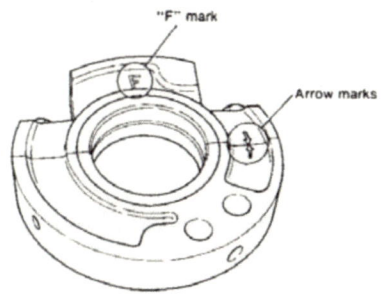
"F" mark
Arrow marks

  • Align the arrow marks pointing up and down on the side of the intermediate main bearing housing and assemble it so that the "F" mark is in the direction of the flywheel.
  • Assemble, integrated with thrust bearing, the intermediate main bearing on the flywheel side (between cylinder No. 1 and 2).
  2) Tightening torque of hexagonal bolts for affixing the top and bottom of the intermediate main bearing housing:

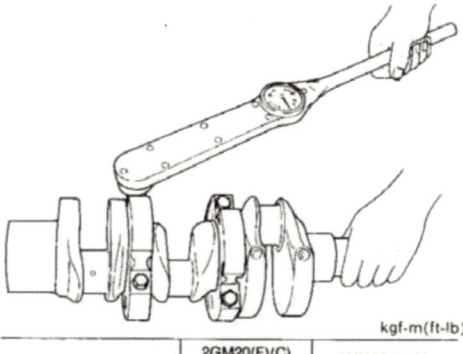

kgf-m(ft-lb)

	2GM20(F)(C) 3GM30(F)(C)	3HM35(F)(C)
Tightening torque	3.0~3.5 (21.7~25.3)	4.5~5.0 (32.5~36.2)

(3) Set the cylinder block up vertically, suspend the crankshaft and match the positions of the cylinder block oil hole and the intermediate main bearing housing set bolts to the intermediate main bearing housing.

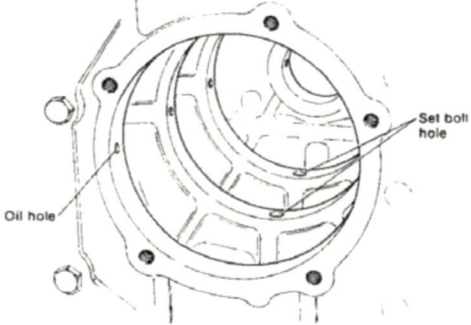
Set bolt hole
Oil hole

(4) Attaching the intermediate main bearing housing set bolts.
  1) First temporarily screw the set bolt in the intermediate main bearing housing on the timing gear housing side and with the prescribed tightening torque, start tightening from the intermediate main bearing housing on the flywheel side. After tightening the bolts confirm that the crankshaft rotates smoothly. (Each set bolt hole can be adjusted vertically.)

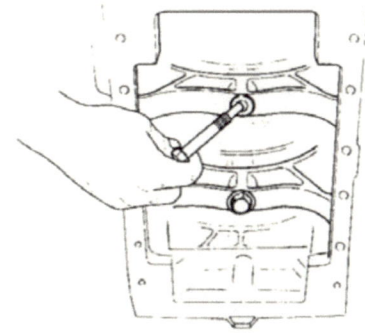

kgf-m(ft-lb)

	2GM20(F)(C) 3GM30(F)(C)	3HM35(F)(C)
Tightening torque of the set bolt	4.5~5.0 (32.5~36.2)	7.0~7.5 (50.6~54.2)

(5) Reassembly of the main bearing housing:
  1) Enclose a small amount of oil inside the oil seal and assemble after coating the bearing with oil.

Chapter 2 Basic Engine
5. Crankshaft

2) Be sure to place the "down" mark on the main bearing housing side in the downward direction.

kgf-m(ft-lb)

	2GM20(F)(C) 3GM30(F)(C)	3HM35(F)(C)
Main bearing housing tightening torque	2.5 (18.1)	2.5 (18.1)

5-5 Main bearing
5-5.1 Construction

(1) Model 1GM10(C)

The main bearing consists of a crank bearing and thrust metal. The crank bearing is a round copper-leak sintered alloy bearing featuring superior durability.
The crankshaft bearing at the gear case end is inserted into the cylinder block, and at the flywheel end it is fitted into the metal housing.
Two thrust metals are set on the bearing part at the gear case end; one is at the crankcase end and the other is at the gear case end.

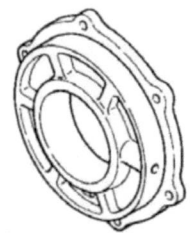

Metal housing for model 1GM10(C)

Thrust metal at gear case end for model 1GM10(C)

Thrust metal at crankcase end for model 1GM10(C)

(2) Models 2GM20(F)(C), 3GM30(F)(C) and 3HM35(F)(C)
For the intermediate main bearing on the flywheel side, a flange type bearing integrated with the thrust bearing is used. Because this is the primary main bearing, those without the thrust bearing on the sides of the flywheel and timing gear housing are whole circle bearings, while the intermediate main bearing on the timing gear housing side is the divided circle type.

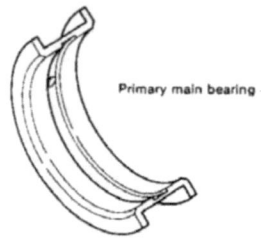
Primary main bearing

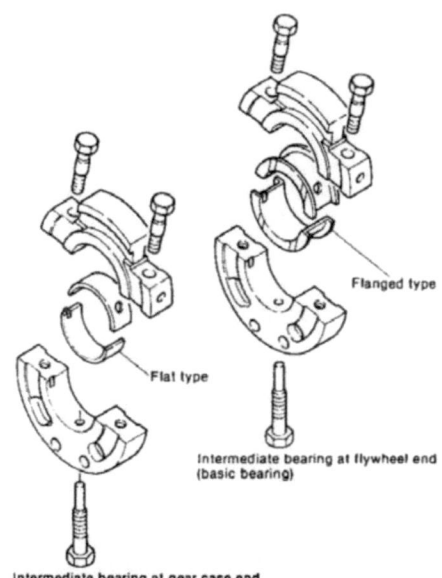

Flanged type

Flat type

Intermediate bearing at flywheel end (basic bearing)

Intermediate bearing at gear case end [models 3GM30(F)(C) and 3HM35(F)(C)]

5-5.2 Inspecting the crank bearing

(1) Check the crank bearing metal for scaling, deposited metal and seizure. Also check the condition of the contact surface. If defects are found, replace.
If the bearing metal contact is too unsymmetrical, carefully check all related component parts which might be responsible, and take proper measures.

(2) Determine the oil clearance by measuring the inside diameter of the crankshaft bearing and the outside diameter of the crankshaft.

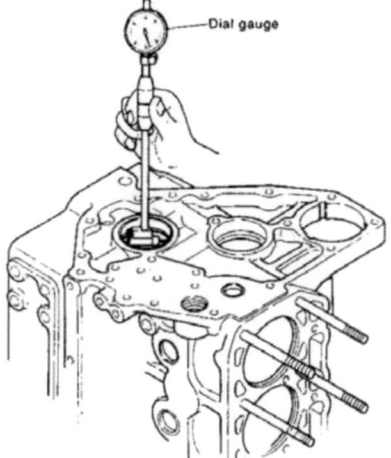

Dial gauge

Chapter 2 Basic Engine
5. Crankshaft

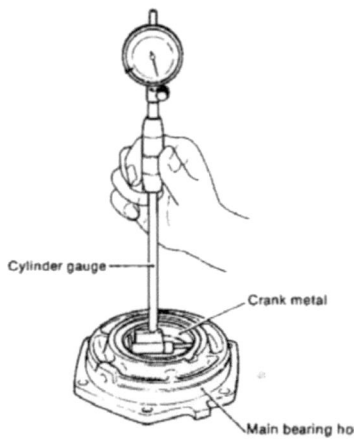

NOTES: 1) Measure the crank bearing at the four points shown in the figure and replace the bearing if the wear limit is exceeded at any of these points.
2) When measuring the inner diameter of the crank bearing, the crank bearing should be installed on the bearing housing and/or cylinder block.

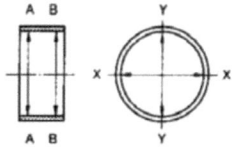

mm (in.)

		1GM10(C), 2GM20(F)(C), 3GM30(F)(C)		3HM35(F)(C)	
		Maintenance standard	Wear limit	Maintenance standard	Wear limit
Flywheel side	Main bearing inside diameter	⌀60.0 (2.3622)	⌀60.12 (2.3669)	⌀65.0 (2.5590)	⌀65.12 (2.5638)
	Crankshaft journal outside diameter	⌀60.0 (2.3622)	⌀59.90 (2.3583)	⌀65.0 (2.5590)	⌀64.90 (2.5551)
	Oil clearance	0.036 ~ 0.095 (0.0014 ~ 0.0037)	0.15 (0.0059)	0.036 ~ 0.099 (0.0014 ~ 0.0039)	0.15 (0.0059)
Opposite side of flywheel	Main bearing inside diameter	⌀44.0 (1.7323)	⌀44.12 (1.7370)	⌀47.0 (1.8504)	⌀47.12 (1.8551)
	Crankshaft journal outside diameter	⌀44.0 (1.7323)	⌀43.90 (1.7283)	⌀47.0 (1.8504)	⌀46.90 (1.8465)
	Oil clearance	0.036 ~ 0.092 (0.0014 ~ 0.0036)	0.15 (0.0059)	0.036 ~ 0.095 (0.0014 ~ 0.0037)	0.15 (0.0059)

5-5.3 Inspecting the thrust metal [for model 1GM10(C)]
Measure the thickness of the thrust metal and replace the metal when wear exceeds the wear limit.

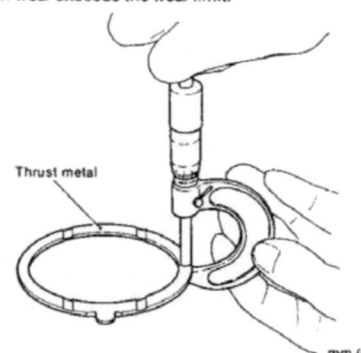

5-5.4 Inspecting the intermediate main bearing
[for models 2GM20(F)(C), 3GM30(F)(C) and 3HM35(F)(C)]

(1) Caution when inspecting
The intermediate main bearing is divided into two semicircles. Therefore, always measure after tightening the intermediate main bearing with the standard tightening torque. Measure at four places as in the main bearing, and replace it if it exceeds the wear limit.

kgf-m(ft-lb)

	2GM20(F)(C) 3GM30(F)(C)	3HM35(F)(C)
Tightening torque of the intermediate main bearing housing tightening bolt	3.0~3.5 (21.7~25.3)	4.5~5.0 (32.5~36.2)

mm (in.)

	Maintenance standard	Wear limit
Thrust metal at crankcase end	2.45 (0.0965)	2.25 (0.0886)
Thrust metal at gear case end	2.95 (0.1161)	2.75 (0.1083)

(2) Intermediate main bearing

The intermediate main bearing on the flywheel side is the primary main bearing. Because this is a flange type bearing, measure the flange width as well as the inside diameter. As the flange wears away the side gap of the crankshaft increases.

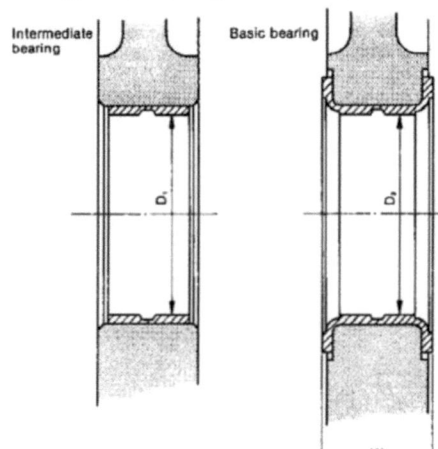

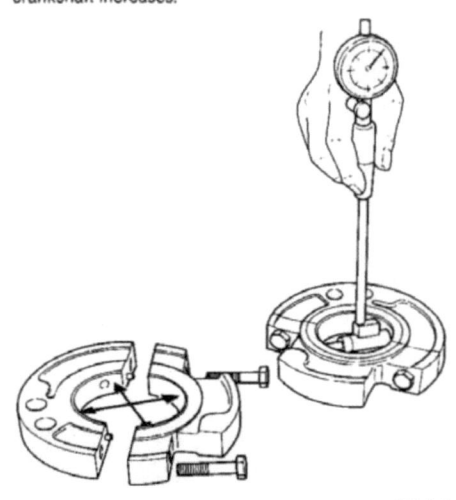

mm (in.)

	2GM20(F)(C), 3GM30(F)(C)		3HM35(F)(C)	
	Maintenance standard	Wear limit	Maintenance standard	Wear limit
Gear case side intermediate bearing inside diameter D	⌀44.0 (1.7323)	⌀44.12 (1.7370)	⌀47.0 (1.8504)	⌀47.12 (1.8551)
Flywheel side intermediate bearing inside diameter D_1	⌀44.0 (1.7323)	⌀44.12 (1.7370)	⌀47.0 (1.8504)	⌀47.12 (1.8551)
Width of intermediate bearing (Flywheel side) W	$25^{-0.09}_{-0.17}$ (0.9776 ~ 0.9807)	24.63 (0.9697)	$30^{-0.09}_{-0.17}$ (1.1744 ~ 1.1776)	29.63 (1.665)

NOTE: Only at the flywheel end for model 1GM10(C)

5-5.5 Replacing the crank bearing

Since the crank bearings at both ends of the crankshaft are attached to the cylinder block and bearing housing with a press, a force of approximately 1.0 ~ 1.5 tons (2200 ~ 3300 lbs.) is required to remove them.
Moreover, since the crankshaft will not rotate smoothly and other trouble may occur if the bearing is distorted, it must always be installed with the special tool.
(1) Removal

Assemble the spacer and plate A as shown in the figure, place the puller/extractor against the bearing from the opposite end and pull the bearing by tightening the nut of the special tool. Remove the oil seal before pulling the bearing pressed against the bearing housing.
(2) Installation
Coat the outside of the bearing with oil and align the positions of the bearing oil holes. Then press in plate B

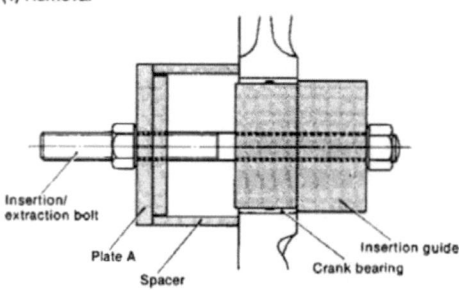

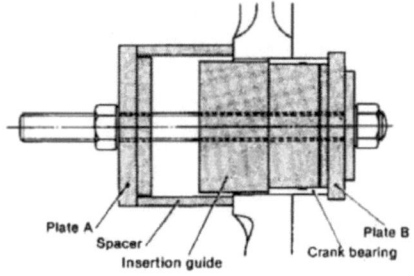

Chapter 2 Basic Engine
5. Crankshaft

SM/GM(F)(C)·HM(F)(C)

until it touches the cylinder block or bearing housing, using the puller/extractor as a guide, as shown in the figure.
After inserting the bearing, measure its outside diameter. If the bearing is distorted, remove it again and replace it with a new bearing.

(3) Crank bearing installation precautions

1) Pay careful attention to the crank bearing insertion direction. Insert the bearing so that the side with the outside fillet is on the outside.
2) Align the oil hole of the crank bearing with the oil holes of the cylinder block and bearing housing.
3) After inserting the crank bearing, check that the crankshaft rotates easily with the thrust metal and bearing housing installed.
4) Be careful that the bearing is not tilted during insertion.

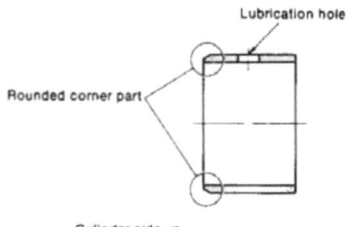

5-6 Crankshaft oil seal

5-6.1 Oil seal type and size

Spiral oil seals are employed at both ends of the crankshaft. This type of oil seal is pulled toward the oil pan by pump action while the engine is running so that there is no oil leakage.
Since the viscous pump action will be lost if the lip of the seal is coated with grease, coat the lip with oil when assembling.

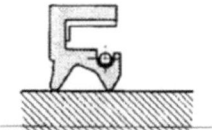 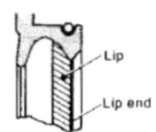

Oil seal	1GM10(C), 2GM20(F)(C), 3GM30(F)(C)			3HM35(F)(C)		
	Size	Spiral	Part No. (Yanmar)	Size	Spiral	Part No. (Yanmar)
For Main bearing metal housing	60829	Yes	124085-02220	65889	Yes	121551-02220
For gear case	25408	Yes	121450-01800	25408	Yes	121450-01800

5-6.2 Oil seal insertion precautions

(1) Clean the inside of the housing hole, ascertaining that the hole is not dented when the seal is removed.
(2) Be sure that the insertion direction of the oil seal is correct. Insert so that the main lip mounting on the spring is on the inside (oil side).

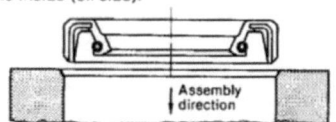

(3) Since the direction of rotation of the shaft is specified on a spiral oil seal, be sure that the rotating direction is correct.

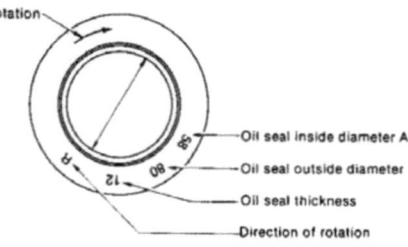

(4) Insert the oil seal with a press. However, when unavoidable, the seal may be installed by tapping the entire periphery of the seal with a hammer, using a block. In this case, be careful that the oil seal is not tilted. Never tap the oil seal directly.

GOOD

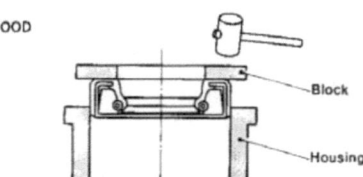

BAD

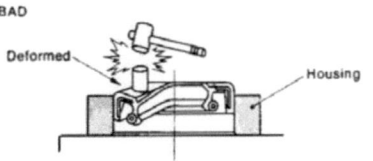

2-48

6. Flywheel and Housing

The function of the flywheel is, through inertia, to rotate the crankshaft in a uniform and smooth manner by absorbing the turning force created during the combustion stroke of the engine, and by compensating for the decrease in turning force during the other strokes.

The flywheel is mounted and secured by 5 bolts on the crankshaft end at the opposite end to the gear case; it is covered by the mounting flange (flywheel housing) which is bolted to the cylinder block.

On the crankshaft side of the flywheel is the fitting surface for the damper disc, through which the rotation of the crankshaft is transmitted to the input shaft of the reduction and reversing gear. The reduction and reversing gear is fitted to the mounting flange.

The flywheels unbalanced force on the shaft center must be kept below the specified value for the crankshaft as the flywheel rotates with the crankshaft at high speed. To achieve this, the balance is adjusted by drilling holes in the side of the flywheel, and the unbalanced moments are adjusted by drilling holes in the circumference.

The ring gear is shrink fitted onto the circumference of the flywheel, and this ring gear serves to start the engine by meshing with the starter motor pinion.

The stamped letter and line which show top dead center of each cylinder are positioned either on the flywheel at the crankshaft side or at the side of the reduction and reversing gear, and by matching these marks with the arrow mark at the setting hole of the starter motor or at the hole of the flywheel housing, the rotary position of the crankshaft can be ascertained in order to adjust tappet clearance or fuel injection timing.

6-1 Specifications of flywheel

			1GM10(C)	2GM20(F)(C)	3GM30(F)(C)	3HM35(F)(C)
Outside diameter of flywheel		mm		⌀252 $_{-0.2}^{0}$		⌀300
Width of flywheel		mm	70	70	70	44
Weight of flywheel (including ring gear)		kg	17.5	17.5	17.5	12.0
GD^2 value		kgf-m^2	0.7	0.7	0.70	0.70
Circumferential speed		m/s		47.5 (3600 rmp)		53.4 (3400 rpm)
Speed fluctuation rate		δ	1/71.2 (3600 rpm)	1/86.4 (3600 rpm)	1/116 (3600 rpm)	1/73.4 (3400 rpm)
Allowable amount of imbalance		gf-cm	30	30	30	25
Fixing part of damper disc	Pitch circle diameter of bolts	mm		150		170
	No. of bolts × bolt diameter			6-M8 thread equally spaced		
Fixing part of crankshaft	Pitch circle diameter of bolts	mm		41		46
	No. of thread holes	mm		5-M10		5-M10
	Fit joint diameter			⌀60M7		⌀65M7
Model of reduction and reversing gear				KM2-C	KM3A	KBW10E
Mounting flange No.				SAE No. 6 (in metric unit)		SAE No. 5 (in metric unit)
Ring gear	Center diameter	mm		246.38		289.56
	No. of teeth			Z = 97		Z = 114

Chapter 2 Basic Engine
6. Flywheel and Housing

6-2 Dimensions of flywheel and flywheel housing
6-2.1 For model 1GM10(C), 2GM20(F)(C), 3GM30(F)(C)

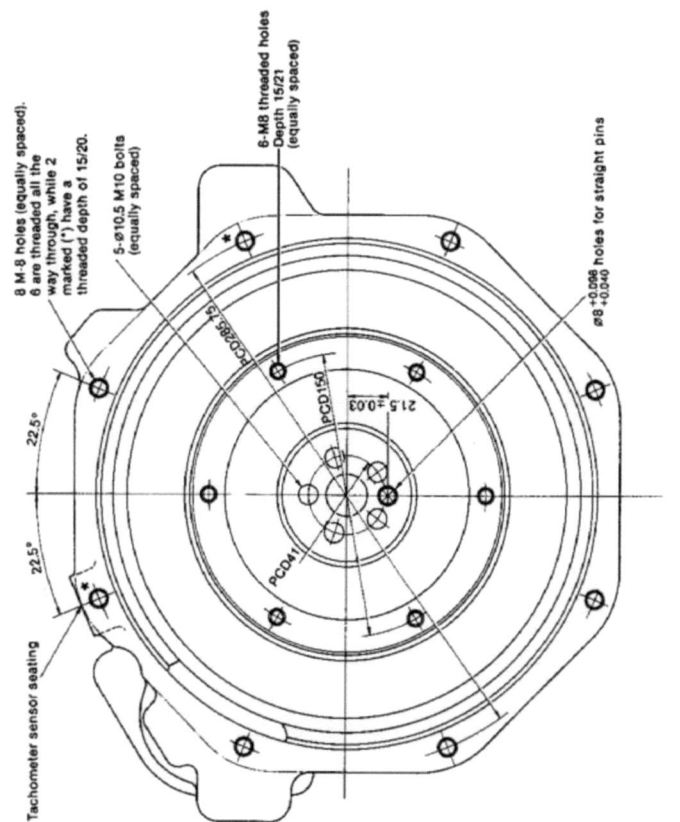

NOTE: Material of flywheel housing.
Sail-drive type: Cast iron
Marine gearbox type: Aluminum alloy

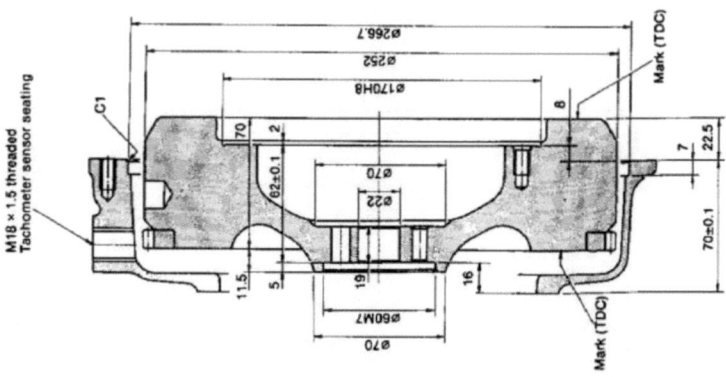

6-2.2 For model 3HM35(F)(C)

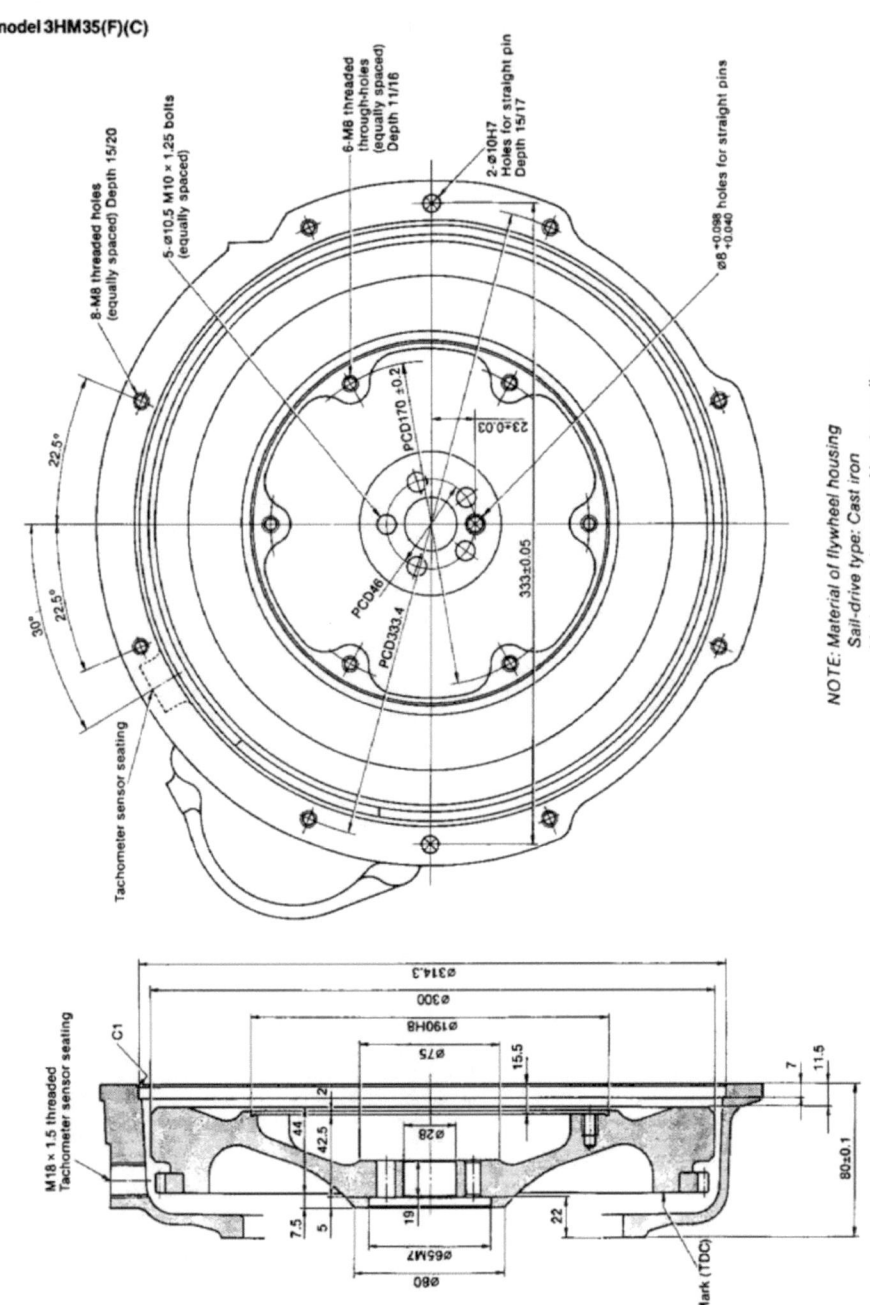

Chapter 2 Basic Engine
6. Flywheel and Housing

SM/GM(F)(C)·HM(F)(C)

6-3 Ring gear
When replacing the ring gear due to excessive wear or damaged teeth, heat the ring gear evenly at its circumference, and after it has expanded drive it gradually off the flywheel by tapping it with a hammer a copper bar or something similar around the whole circumference.

mm (in.)

	1GM10(C), 2GM20(F)(C) 3GM30(F)(C)	3HM35(F)(C)
Interference of ring gear	0.188~0.348 (0.0074~0.0137)	0.188~0.348 (0.0074~0.0137)

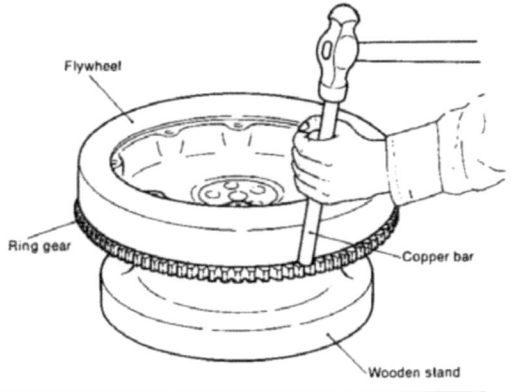

6-4 Position of top dead center
(1) Marking

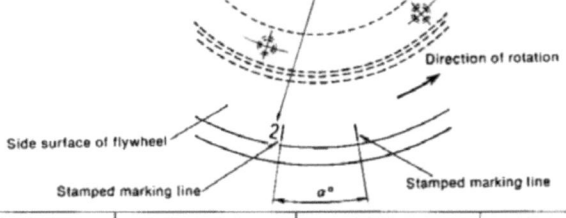

	1GM10(C)	2GM20(F)(C)	3GM30(F)(C)	3HM35(F)(C)
Stamped letter	1	1, 2	1, 3	1, 2, 3
Angle α of Stamped lines	15°	15°	18°	21°
Stamped surfaces	Both surfaces	Both surfaces	Both surfaces	Crankshaft side

(2) Matching mark
The matching mark is made at the setting hole of the starter motor on all models.

With respect to models 1GM10(C), 2GM20(F)(C) and 3GM30(F)(C) only, a projection which serves as the matching mark is provided in the cast hole of the clutch housing.

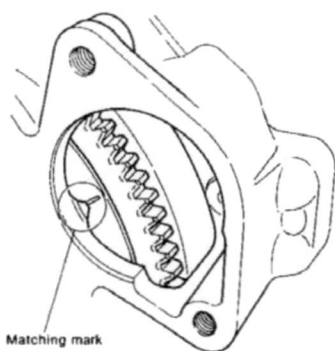
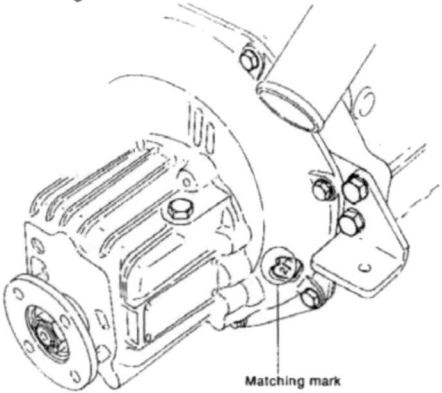

Chapter 2 Basic Engine
7. Camshaft

SM/GM(F)(C)-HM(F)(C)

7. Camshaft

7-1 Construction of the camshaft

The camshaft, an integral camshaft with intake and exhaust cams, is driven by the camshaft gear and may be timed individually.

On top of the intake and exhaust cams a tappet is mounted guided by the cylinder block. The tappet moves up and down with the rotation of the cam and opens and closes the intake and exhaust valves with the pushrod and rocker arm.

During high speed operation the cam surface is exposed to a strong force of inertia from moving valves and spring load, and comes in contact with the tappet at high surface pressure. Therefore, to reduce wear the surface is tempered by high frequency hardening, as well as a cam form selected to decrease the force of inertia. Since the intake and exhaust cam profile of this engine is a parabolic acceleration cam with a buffering curve, movement of the valve at high speed is smooth, improving the durability of the intake and exhaust valve seats.

The camshaft on models 1GM10(C) and 2GM20(F)(C) does not have an intermediate bearing. The camshaft on models 3GM30(F)(C) and 3HM35(F)(C) however is supported by two intermediate bearings in order to avoid deflection of the camshaft.

models and it is inserted into the camshaft together with the camshaft gear by matching the key and slot and is fixed by an end nut.

The cam for the fuel feed pump is integrated with the camshaft and it is machine finished. The cam is located between the intake and exhaust valve cams of No.1 cylinder at the flywheel end in all engine models.

7-1.1 Camshaft of engine model 1GM10(C)

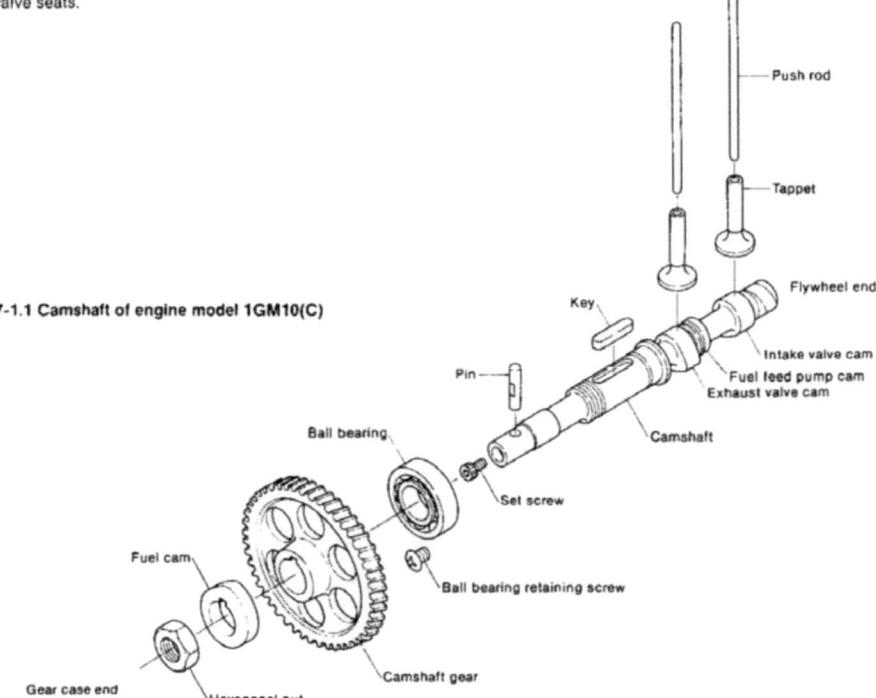

Chapter 2 Basic Engine
7. Camshaft

SM/GM(F)(C)·HM(F)(C)

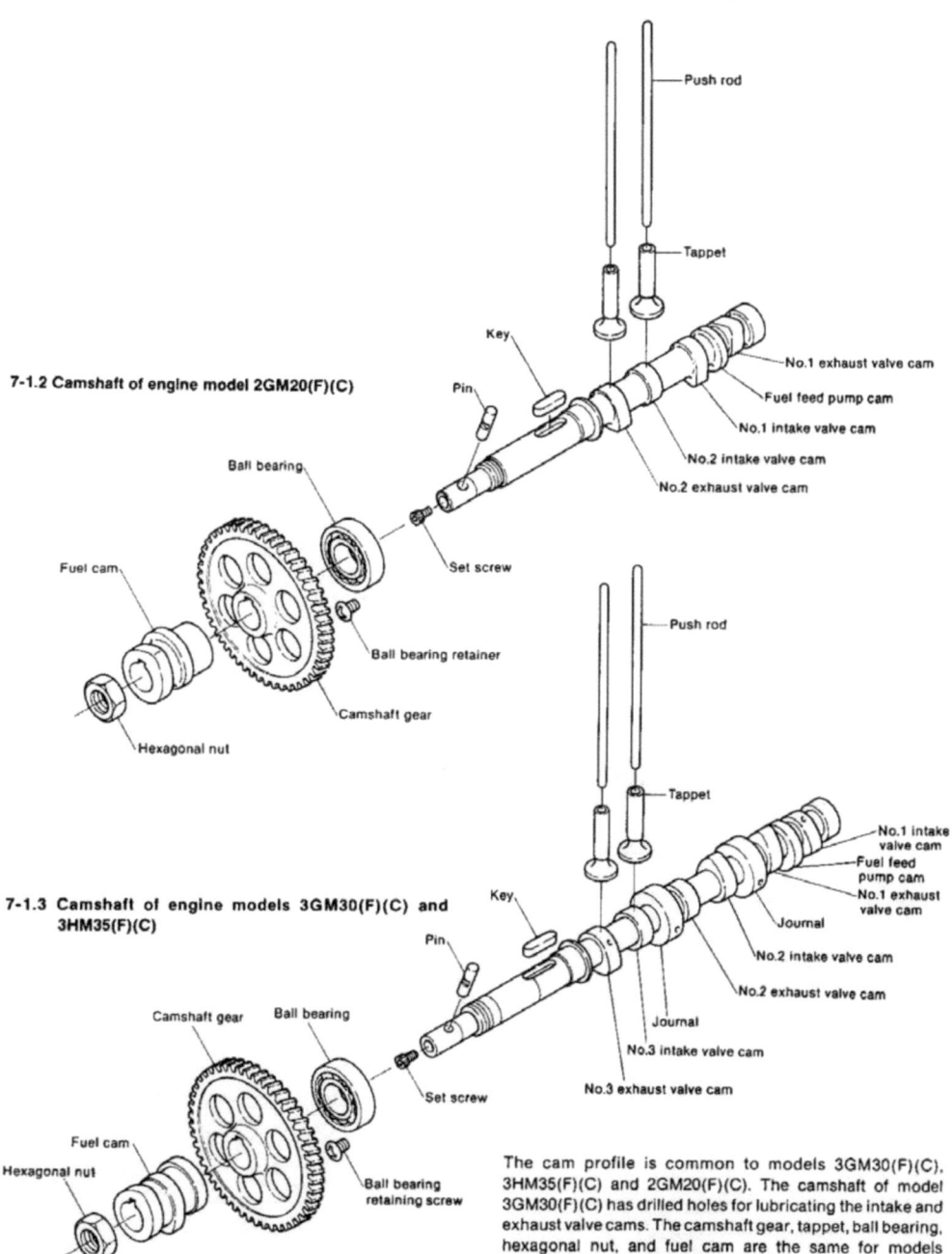

7-1.2 Camshaft of engine model 2GM20(F)(C)

7-1.3 Camshaft of engine models 3GM30(F)(C) and 3HM35(F)(C)

The cam profile is common to models 3GM30(F)(C), 3HM35(F)(C) and 2GM20(F)(C). The camshaft of model 3GM30(F)(C) has drilled holes for lubricating the intake and exhaust valve cams. The camshaft gear, tappet, ball bearing, hexagonal nut, and fuel cam are the same for models 3GM30(F)(C) and 3HM35(F)(C).

Chapter 2 Basic Engine
7. Camshaft

SM/GM(F)(C)·HM(F)(C)

7-2 Valve timing diagram

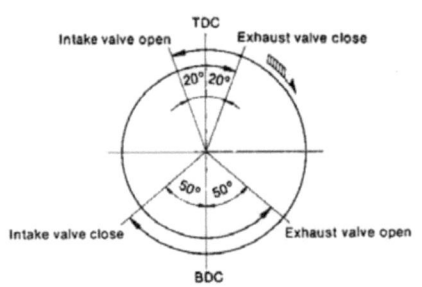

	All models
Intake and exhaust valve head clearance	0.2mm (0.0079in.)
Intake valve open b. TDC	20°
Intake valve close a. BDC	50°
Exhaust valve open b. BDC	50°
Exhaust valve close a. TDC	20°

7-3 Inspection

Visually check for steps or wear on the cam surface and replace if excessive.
Since the cam surface is tempered and ground, there is almost no wear. However, measure the height of the intake and exhaust cams, and replace the camshaft when the measured value exceeds the wear limit.

7-3.1 Camshaft height

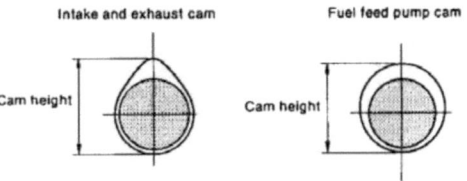

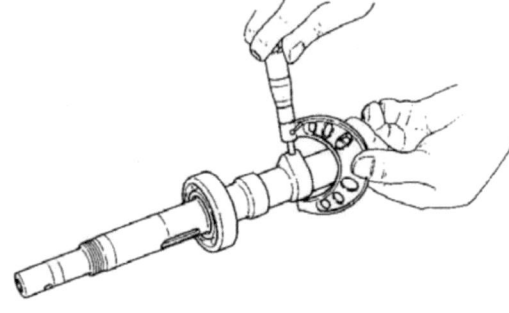

			mm (in.)
		Maintenance standard	Wear limit
Intake and exhaust cam	1GM10(C)	29 (1.1417)	28.70 (1.1292)
	2GM20(F)(C) 3GM30(F)(C) 3HM35(F)(C)	35 (1.3780)	34.70 (1.3661)
Fuel feed pump cam	1GM10(C)	22 (0.8661)	—
	2GM20(F)(C) 3GM30(F)(C)	33 (1.2992)	—
	3HM35(F)(C)	33.5 (1.3189)	—

7-3.2 Journals of camshaft

Measure the amount of wear and eccentricity of the camshaft journal. Measurements must be carried out in at least two directions for each position.
Replace the camshaft with a new one if the value exceeds the allowable limit.

mm (in.)

		Maintenance standard	Clearance at assembly	Maximum allowable clearance
Flywheel side	1GM10(C)	⌀20 (0.7874)	0.050~0.100 (0.0020~0.0039)	0.15 (0.0059)
	2GM20(F)(C), 3GM30(F)(C), 3HM35(F)(C)	⌀30 (1.1811)		
Center	3GM30(F)(C), 3HM35(F)(C)	⌀41.5 (1.6339)	0.050~0.100 (0.0020~0.0039)	0.15 (0.0059)

7-3.3 Camshaft deflection [models 3GM30(F)(C) and 3HM(F)(C)]

Support the camshaft at both ends on V-blocks, and measure the concentricity of the intermediate journal with a dial gauge. If the camshaft is excessively bent, replace it.
NOTE: Indicated value on the dial gauge is the amount of swing, and the amount of bend is half the reading given.

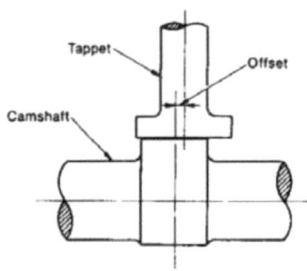

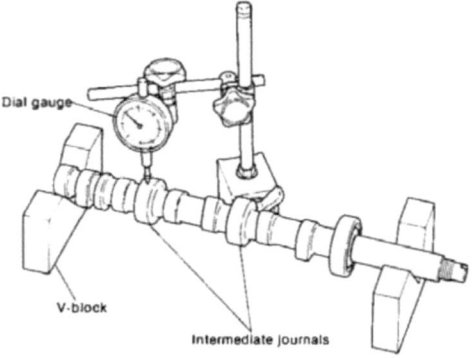

7-5.1 Tappet disassembly precautions

The cylinder number and intake and exhaust must be clearly indicated when disassembling the camshaft and tappets.

7-5.2 Tappet stem wear and contact

Measure the outside diameter of the tappet stem, and replace the tappet when the wear limit is exceeded or contact is uneven.

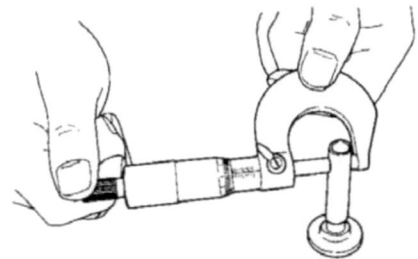

mm (in.)

		Maintenance standard	Wear limit
Camshaft deflection	3GM30(F)(C)	—	0.02 (0.0008)
	3HM35(F)(C)	—	0.02 (0.0008)

7-4 Camshaft ball bearing

The camshaft bearing is a single row deep groove ball bearing. The construction and material of this ball bearing is such that it can withstand the radial load, thrust loads in both directions, and a combinaiton of both of these loads. When the ball bearing does not rotate smoothly, or when the axial direction play is large, replace the bearing.

Ball bearing type

For model 1GM10(C)	6005
For models 2GM20(F)(C), 3GM30(F)(C), 3HM35(F)(C)	6205

mm (in.)

		Maintenance standard	Wear limit
Tappet stem outside diameter	1GM10(C)	φ10.0 (0.3937)	φ9.95 (0.3917)
	2GM20(F)(C) 3GM30(F)(C) 3HM35(F)(C)	φ10.0 (0.3937)	φ9.95 (0.3917
Tappet stem and guide hole clearance	1GM10(C)	0.025~0.060 (0.0010~0.0024)	0.10 (0.0039)
	2GM20(F)(C) 3GM30(F)(C) 3HM35(F)(C)	0.010~0.040 (0.0004~0.0016)	0.10 (0.0039)

7-5.3 Tappet and cam contact surface

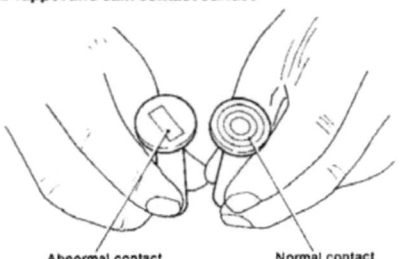

Abnormal contact   Normal contact

7-5 Tappets

These mushroom type tappets feature a special iron casting with chill-hardened contact surfaces for high wear resistance. The center of the cam surface width and the center of the tappet are offset to prevent eccentric wear of the contact surface.

Since the tappet and cam are offset, the tappet rotates in an up and down movement during operation, so there is no uneven contact.
Since eccentric wear will occur if cam tappet contact is poor, replace the tappet if there is any uneven contact or deformation.

Contact surface conditions are shown in the following:

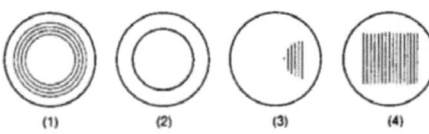

(1), (2) *Traces when the tappet is rotating normally.*
(3), (4) *Traces when the tappet does not rotate, the contact surface remains still and the point of contact wears away excessively. Discover the reason for the lack or rotation and replace the tappet.*

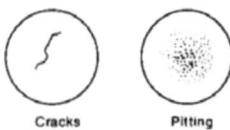

Also, there may be perforated pittings or cracks on the contact surface of the tappet. In such cases, discover the reason for abnormality and replace the tappet.

7-6 Push rods

The push rods are sufficiently rigid and strong to prevent bending.
Place the push rod on a stool or flat surface and measure the clearance between the center of the push rod and the flat surface, and replace the push rod if the wear limit is exceeded.

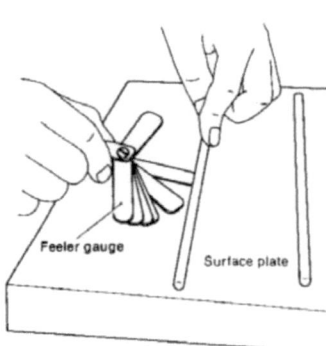

Check both ends for wear and peeling, and replace the push rod if faulty.

mm (in.)

		Maintenance standard	Wear limit
Push rod bend		0.03 or less (0.00118 or less)	0.3 (0.0118)
Push rod length	1GM10(C)	143 (5.6299)	—
	2GM20(F)(C) 3GM30(F)(C)	136 (5.3543)	—
	3HM35(F)(C)	171 (6.7323)	—

7-7 Fuel cam

7-7.1 Fuel cam check

The fuel cam is separate from the intake and exhaust valve cams and is secured to the camshaft together with the camshaft gear by a key. The cam drives the fuel pump.
The fuel cam like the intake and exhaust valve cams is ground-finished after being quenched. Therefore, it is almost free from wear. However, if step or eccentric wear is found to be excessive, replace the cam.

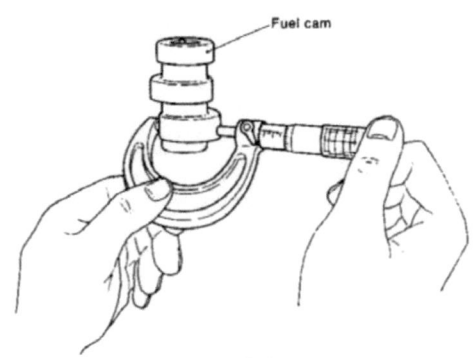

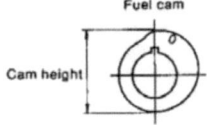

mm (in.)

	Maintenance standard	Wear limit
Fuel cam height - All models -	45 (1.7717)	44.90 (1.7677)

7-7.2 Fuel cam assembly precautions

Install the fuel cam by aligning it with the key of the camshaft. If the installation direction is not correct, the fuel injection timing will be considerably off and the engine will not start.
When assembling the fuel cam, be sure that the "0" mark side of the cam is opposite the camshaft gear.

Chapter 2 Basic Engine
7. Camshaft

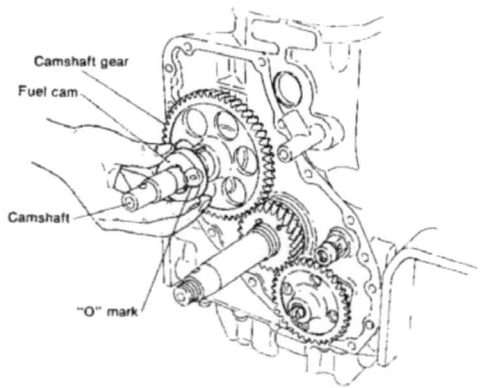

8. Timing Gear

8-1 Timing gear train construction

The camshaft, which is the basic component of the valve opening and closing mechanism, and the fuel cam, which determines the fuel injection timing, are driven by the timing gear.

The timing gear consists of the crankshaft gear and the camshaft gear.

The crankshaft gear also drives the governor weight and the lubricating oil pump by meshing with the lubricating oil pump gear.

For the timing gears, helical gears are used.

The timing gear case, which covers these gears, is fitted to the cylinder body with bolts.

8-1.1 Timing gear of model 1GM10(C)

The timing gear of model 1GM10(C) is as shown in the figure. The slit, which is at the end of the rotor shaft of the lubricating oil pump, is provided to connect with the shaft of the cooling water pump.

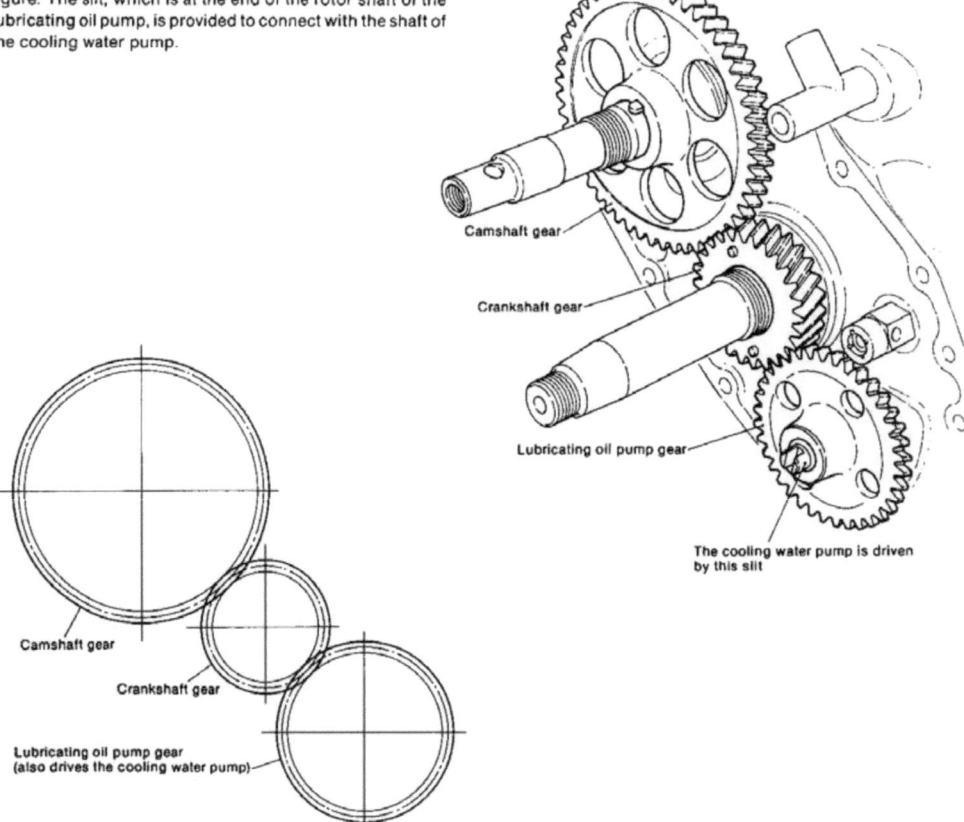

Camshaft gear

Crankshaft gear

Lubricating oil pump gear

The cooling water pump is driven by this slit

Camshaft gear

Crankshaft gear

Lubricating oil pump gear
(also drives the cooling water pump)

1GM10(C)

	Module (m)	Tooth profile	No. of teeth	Center distance
Camshaft gear	2.0	Full depth	52	$84^{+0.048}_{0}$ mm (3.3071 ~ 3.3090in.)
Crankshaft gear	2.0	Full depth	26	$66^{+0.046}_{0}$ mm (2.5984 ~ 2.6002in.)
Lubricating oil pump gear	2.0	Full depth	36	

Chapter 2 Basic Engine
8. Timing Gear

SM/GM(F)(C)-HM(F)(C)

8-1.2 Timing gear of models 2GM20(F)(C), 3GM30(F)(C) and 3HM 35(F)(C)

The same crankshaft gear and camshaft gears are used for these three models. Only on the lubricating oil pump gear for model 3HM35(F)(C) is a different gear used, but it has the same number of teeth and the gear train itself is of the same construction as that of these three models.
Helical gears are used as in model 1GM10(C).

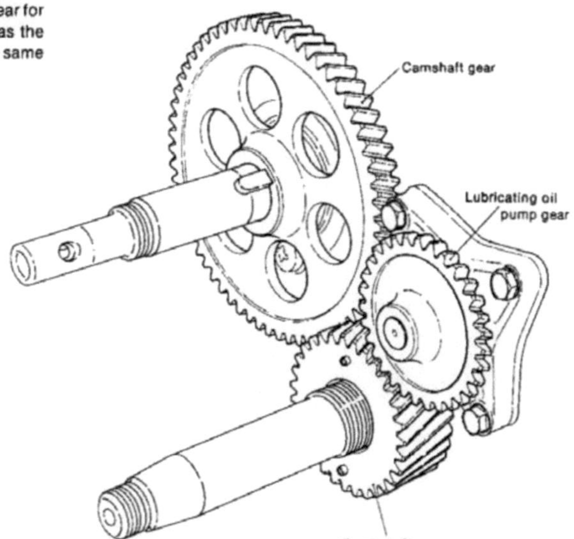

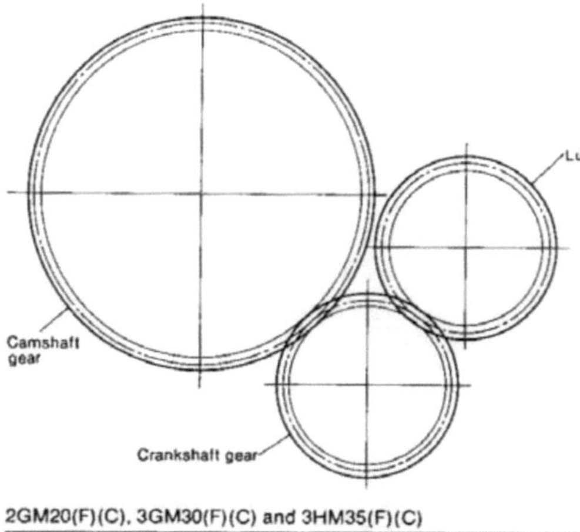

2GM20(F)(C), 3GM30(F)(C) and 3HM35(F)(C)

	Module (m)	Tooth profile	No. of teeth	Center distance
Camshaft gear	2.0	Full depth	62	$99^{+0.048}_{0}$mm (3.8976 ~ 3.8995in.)
Crankshaft gear	2.0	Full depth	31	$65.98^{+0.046}_{0}$mm (2.5976 ~ 2.5995in.)
Lubricating oil pump gear	2.0	Full depth	31	

2-60

Chapter 2 Basic Engine
8. Timing Gear

SM/GM(F)(C)·HM(F)(C)

8-2 Disassembly and reassembly of the timing gear

8-2.1 Disassembly

(1) Remove the alternator.
(2) Remove the rubber hose by loosening the hose clip on the cooling water pump.
NOTE: For models 2GM20(F)(C), 3GM30(F)(C) and 3HM35 (F)(C), the cooling water pump does not need to be removed. Model 1GM10(C) can be dismantled without removing the cooling water pump. However, when assembling, it is difficult to connect it with the rotor shaft of the lubricating oil pump if the gear case has not been previously assembled.
(3) Remove the crankshaft V-pulley.
(4) Remove the fuel injection pump
NOTE: Remove the cap of the oil supply port in model 1GM, or the cap at the timing gear case end in other models, and remove the fuel injection pump by moving the governor second lever while observing through the hole.
(5) Loosen the hexagonal bolt with the hole, and remove the straight pin from the manual starting handle.
(6) Remove the gear case.

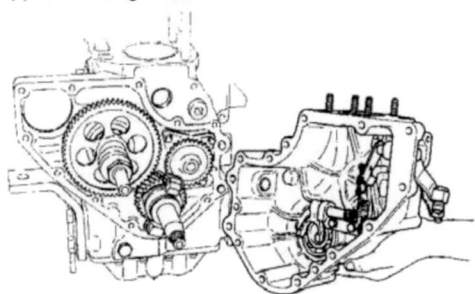

(7) Remove the governor sleeve and needle bearing collar.
(8) Loosen the hexagonal nut, and remove the governor weight support.
(9) Remove the camshaft nut, and take out the fuel cam.
(10) Remove the camshaft gear, crankshaft gear and lubricating oil pump.

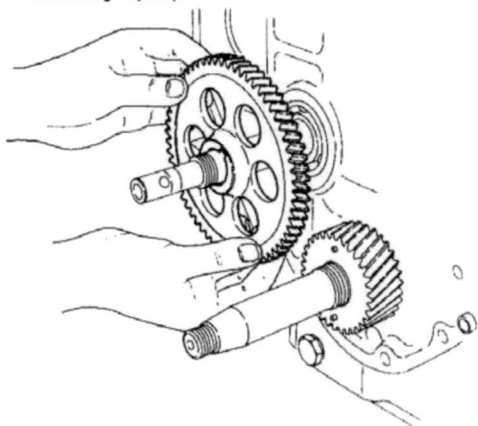

8-2.2 Disassembly and reassembly precautions

Reassemble in the reverse order of disassembly.
Pay attention to the following points when assembling.

(1) Timing mark
    A timing mark is provided on the crankshaft gear and camshaft gear to adjust the timing between opening and closing of the intake and exhaust valves and fuel injection when the piston is operated.
    Always check that these timing marks are aligned when disassembling and reassembling the timing gear.
    First, fit the crankshaft gear to the crankshaft by matching the key and slot. Next, by rotating the camshaft fit the camshaft gear in the position where the marks on the camshaft gear and the crankshaft gear align.

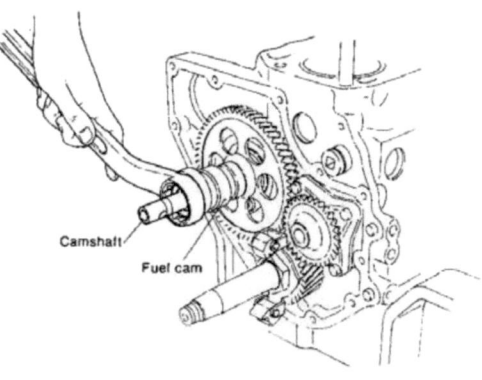

Camshaft
Fuel cam

(2) Fuel cam
    When the fuel cam is fitted to the camshaft, assemble it keeping the surface marked 'O' towards the front.
    (Refer to 2-57)
(3) Tightening torque of nut

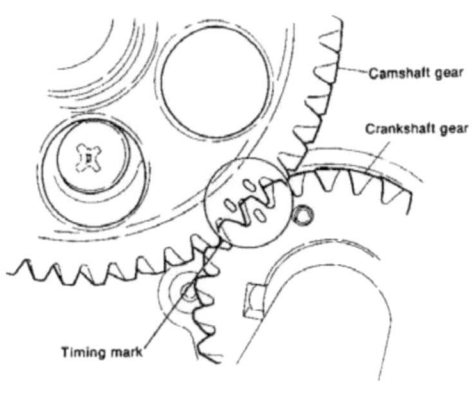

Camshaft gear
Crankshaft gear
Timing mark

2-61

Chapter 2 Basic Engine
8. Timing Gear

	kgf-m(ft-lb)
	All models
Camshaft end nut	7.0 ~ 8.0 (50.6 ~ 57.9)
Crankshaft nut	8.0 ~ 10.0 (57.9 ~ 72.3)

NOTE: When tightening or loosening the crankshaft nut, take care that the spanner does not touch the governor weight or weight support.

(4) assembling model 1GM10(C) cooling water pump
When model 1GM10(C) cooling water pump is assembled, ensure that the pump shaft engages with the slit of the rotor shaft end of the lubricating oil pump and with the bearing. Check by rotating the crankshaft.

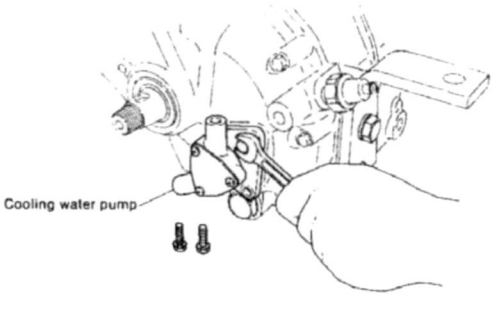
Cooling water pump

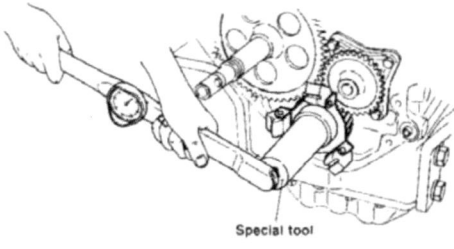
Special tool

8-3 Inspection
8-3.1 Backlash

Unsuitable backlash will cause excessive wear or damage at the tooth top and abnormal noise during operation. Moreover, in extreme cases, the valve and fuel injection timing will deviate and the engine will not run smoothly. When the backlash exceeds the wear limit, repair or relace the gears as a set.

Backlash

mm (in.)

	1GM10(C)		2GM20(F)(C), 3GM30(F)(C), 3HM35(F)(C)	
	Maintenance standard	Wear limit	Maintenance standard	Wear limit
Crankshaft gear and camshaft gear backlash	0.05 ~ 0.13 (0.0020 ~ 0.0051)	0.3 (0.0118)	0.05 ~ 0.13 (0.0020 ~ 0.0051)	0.3 (0.0118)
Crankshaft gear and lubricating oil pump driven gear backlash	0.05 ~ 0.13 (0.0020 ~ 0.0051)	0.3 (0.0118)	0.05 ~ 0.13 (0.0020 ~ 0.0051)	0.3 (0.0118)

Measuring backlash
(1) Lock one of the two gears to be measured and measure the amount of movement of the other gear by placing a dial gauge on the tooth surface.

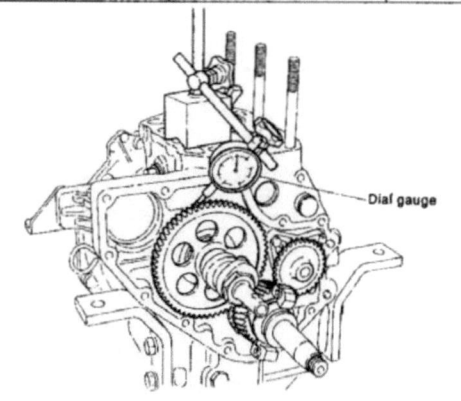
Dial gauge

(2) Insert a piece of quality solder between the gears to be measured and turn the gears. The backlash can be measured by measuring the thickness of the crushed part of the solder.

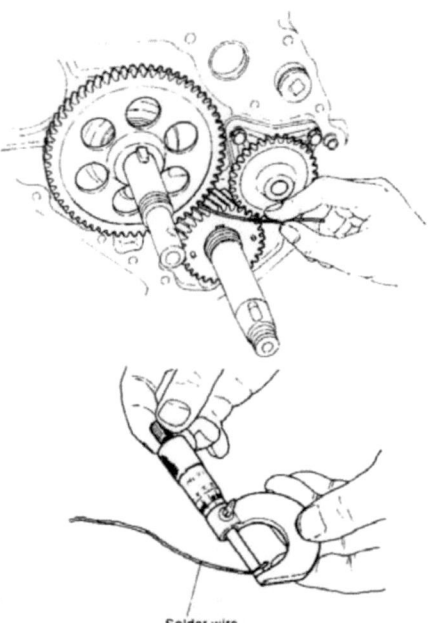

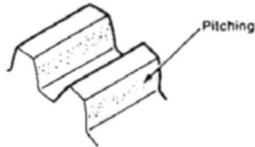
Solder wire

8-3.2 Inspecting the gear tooth surface

Check the tooth surface for damage caused by pitching and check tooth contact. Repair if the damage is light. Also inspect the gears for cracking and corrosion.
When gear noise becomes high because of wear or damage, replace the gears as a set.

8-3.3 Inspecting the gear boss

Check for play between each gear and the gear shaft, burning caused by play, key damage, and for cracking at the edge of the key groove. Replace the gears when faulty.

CHAPTER 3
FUEL SYSTEM

1. Fuel Injection System 3-1
2. Injection Pump 3-3
3. Injection Nozzle 3-25
4. Fuel Filter 3-29
5. Fuel Feed Pump 3-30
6. Fuel Tank (Option) 3-33

Chapter 3 Fuel System
1. Fuel Injection System

1. Fuel Injection System

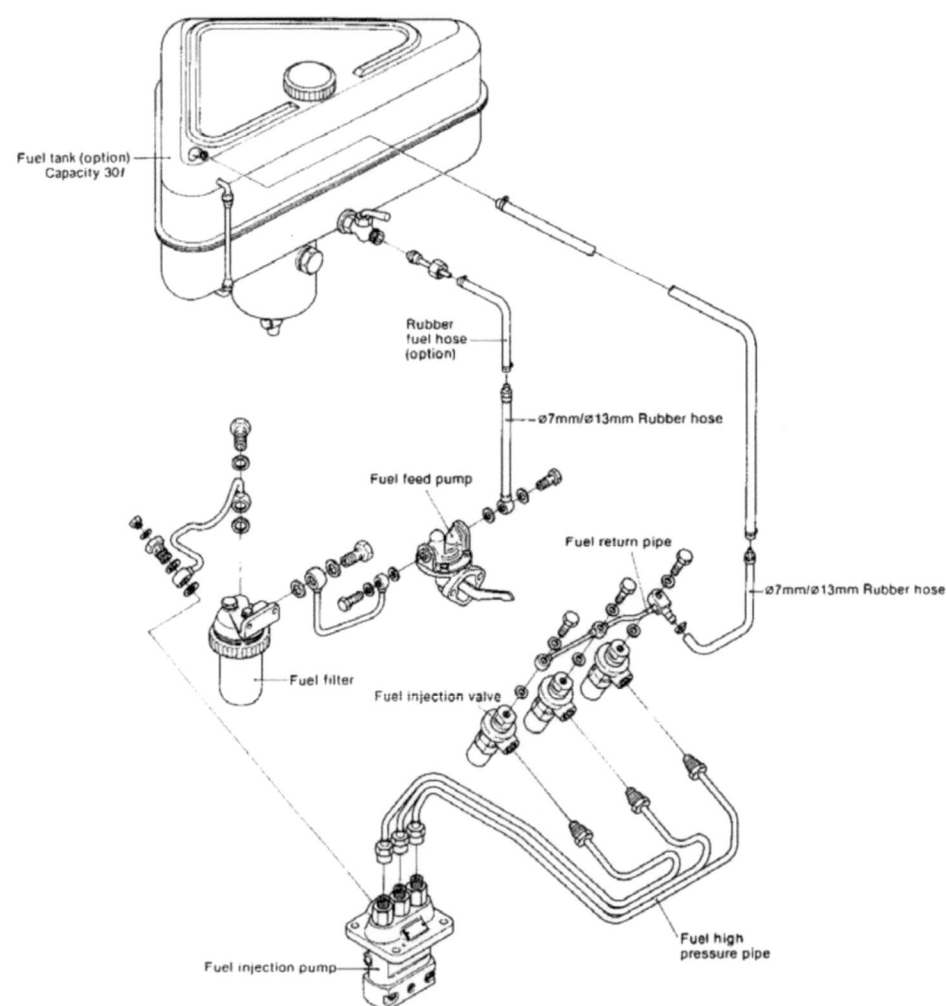

Chapter 3 Fuel System
1. Fuel Injection System

SM/GM(F)(C)-HM(F)(C)

1-1 Construction

The fuel system consists mainly of an injection pump, injection pipe, and an injection nozzle, plus a fuel tank, feed pump, fuel filter and other associated parts. The injection pump is driven by a fuel cam mounted on the camshaft and is controlled by a governor. Fuel stored in the fuel tank is fed to the fuel filter through the feed pump. (The feed pump is indispensable when the fuel tank is installed lower than the injection pump.)

Dirt and other impurities in the fuel are removed by the filter and the clean fuel is sent to the injection pump, which applies the necessary pressure for injection to the fuel and atomizes the fuel by passing it through the injection nozzle. The injection pump also controls the amount of fuel injected and the injection timing according to the engine load and speed by means of a governor.

The injection pump feeds the fuel to the injection nozzle through a high pressure pipe. The pressurized fuel is atomized and injected by the injection nozzle into the precombustion chamber.

Fuel that overflows the injection nozzle is returned to the fuel tank through the fuel return pipe. The quality of the equipment and parts comprising the fuel injection system directly affects combustion performance and has a considerable effect on engine performance. Therefore, this system must be inspected and serviced regularly to ensure top performance.

The pipework diagram of the fuel system is for the model 3GM30(F)(C) engine. Models 1GM10(C) and 2GM20(F)(C) are the same except for the shape of the fuel injection pump and fuel feed pump, and the number of fuel injection valves. It is also the same for models 3GM30(F)(C) and 3HM35(F)(C) except for the fuel injection pump and fuel injection valve.

1-2 Fuel injection system specifications

	1GM10(C)	2GM20(F)(C)	3GM30(F)(C)	3HM35(F)(C)
Type of injection pump	YPFR-07 07-1	YPFR-07 07-2	YPFR-07 07	YPFR-07 07
Type of injection nozzle	YDN-OSDYD1(Throttle)			YDN-OSDYD1(Throttle)
Injection pressure	170 kgf/cm^2(2418 lb/in.^{2})			160 kgf/cm^2(2276 lb/in.^{2})
Plunger diameter × stroke	ϕ6mm(0.2362in.) × 7mm(0.2756in.)			ϕ6.5mm(0.2559in.) × 7mm(0.2756in.)
Delivery valve suction capacity	23.5mm^3/st(0.0014in.^{3}/st)			23.5mm^3/st(0.0014in.^{3}/st)
fuel feed pressure	0.1kgf/cm^2(1.4224 lb/in.^{2})			0.1kgf/cm^2(1.4224 lb/in.^{2})

2. Injection Pump

The injection pump is the most important part of the fuel system. This pump feeds the proper amount of fuel to the engine at the proper time in accordance with the engine load.

This engine uses a Bosch integral type injection pump for two/three cylinders. It is designed and manufactured by Yanmar, and is ideal for the fuel system of this engine.

Since the injection pump is subjected to extremely high pressures and must be accurate as well as deformation and wear-free, stringently selected materials are used and precision finished after undergoing heat treatment.

The injection pump must be handled carefully. Since the delivery valve and delivery valve holder and the plunger and plunger barrel are lapped, they must be changed as pairs.

The fuel injection pump is constructed from the following main parts.
(1) Pump parts which compress and deliver the fuel: plunger, plunger barrel.
(2) Parts which move the plunger: camshaft, tappet, plunger spring, plunger spring retainer.
(3) Parts which control the injection amount: control rack, control pinion, control sleeve.
(4) Parts which prevent back flow and dripping during injection: delivery valve.

2-1 Construction
2-1.1 1GM10(C)

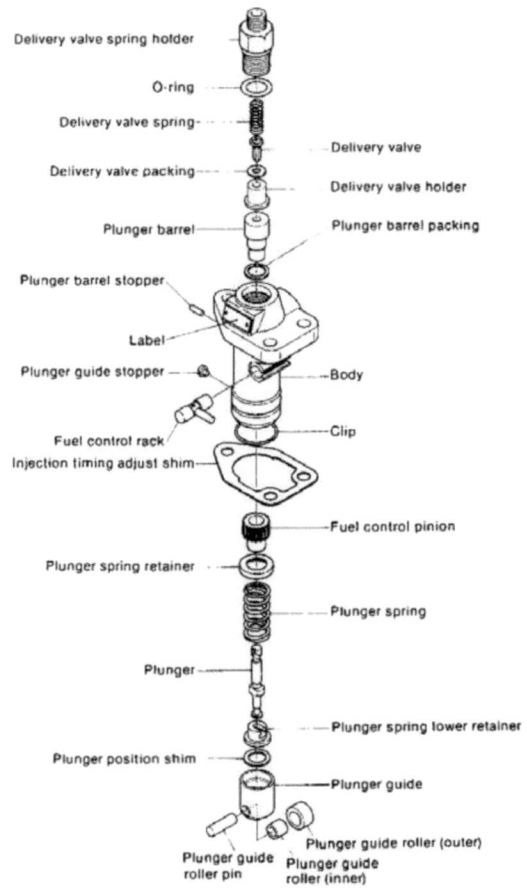

Chapter 3 Fuel System
2. Injection Pump

SM/GM(F)(C)·HM(F)(C)

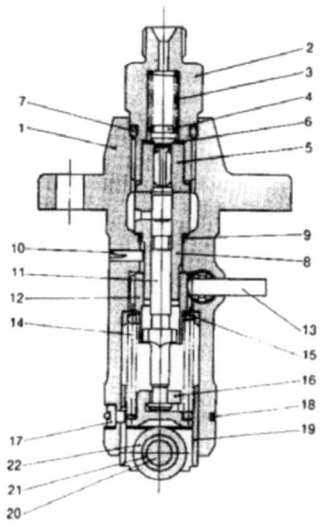
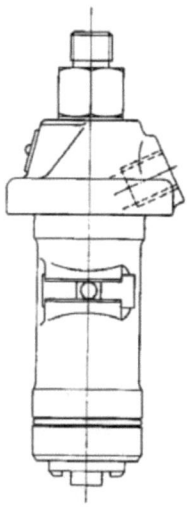

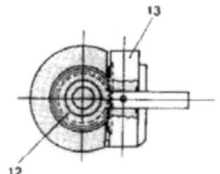

1 Body
2 Delivery valve spring holder
3 Delivery valve spring
4 Delivery valve
5 Delivery valve holder
6 Delivery valve packing
7 O-ring
8 Plunger barrel
9 Plunger barrel packing
10 Plunger barrel stopper
11 Plunger
12 Fuel control pinion
13 Fuel control rack
14 Plunger spring
15 Plunger spring retainer
16 Plunger spring lower retainer
17 Plunger guide stopper
18 Clip
19 Plunger guide
20 Plunger guide roller pin
21 Plunger guide roller (inner)
22 Plunger guide roller (outer)

Chapter 3 Fuel System
2. Injection Pump

2-1.2 2GM20(F)(C), 3GM30(F)(C), 3HM35(F)(C)

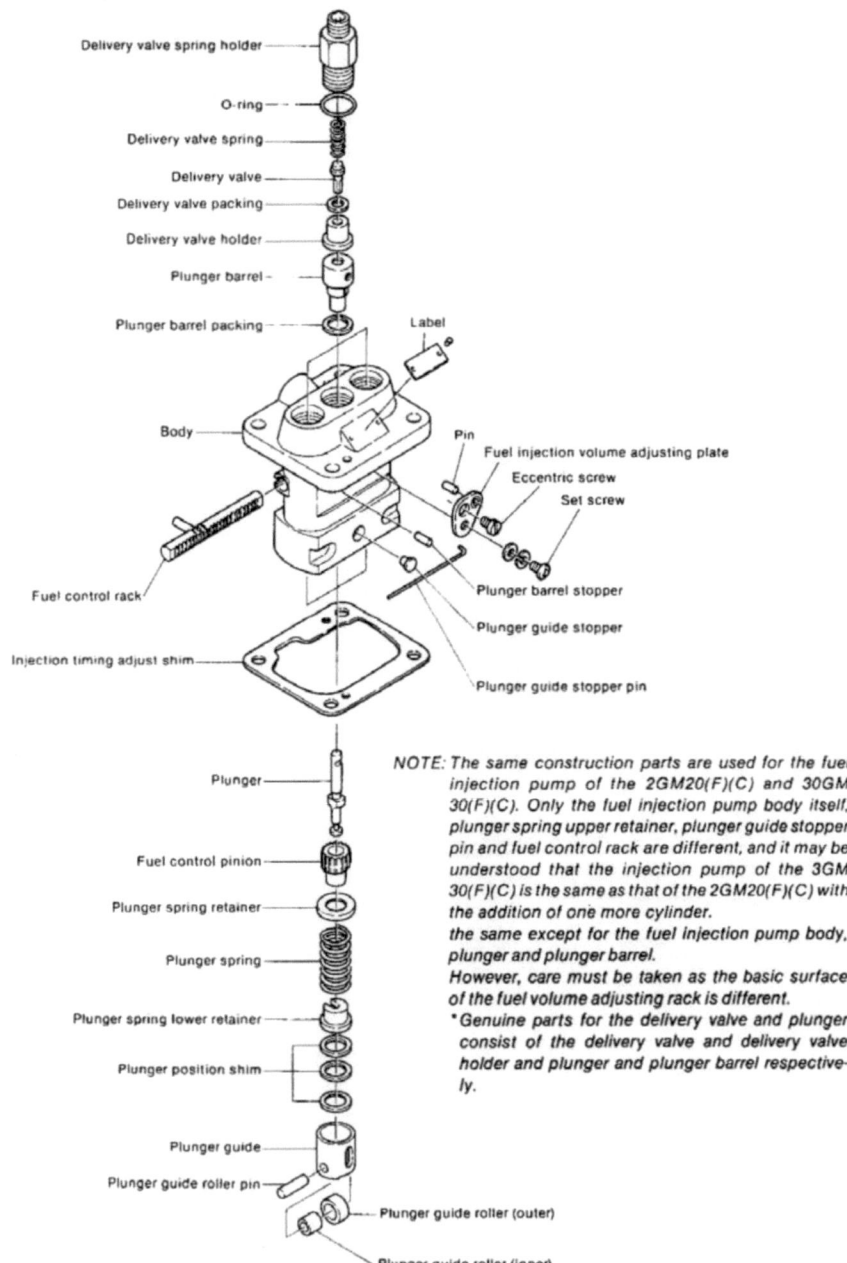

NOTE: The same construction parts are used for the fuel injection pump of the 2GM20(F)(C) and 30GM 30(F)(C). Only the fuel injection pump body itself, plunger spring upper retainer, plunger guide stopper pin and fuel control rack are different, and it may be understood that the injection pump of the 3GM 30(F)(C) is the same as that of the 2GM20(F)(C) with the addition of one more cylinder.
the same except for the fuel injection pump body, plunger and plunger barrel.
However, care must be taken as the basic surface of the fuel volume adjusting rack is different.
*Genuine parts for the delivery valve and plunger consist of the delivery valve and delivery valve holder and plunger and plunger barrel respectively.

Chapter 3 Fuel System
2. Injection Pump

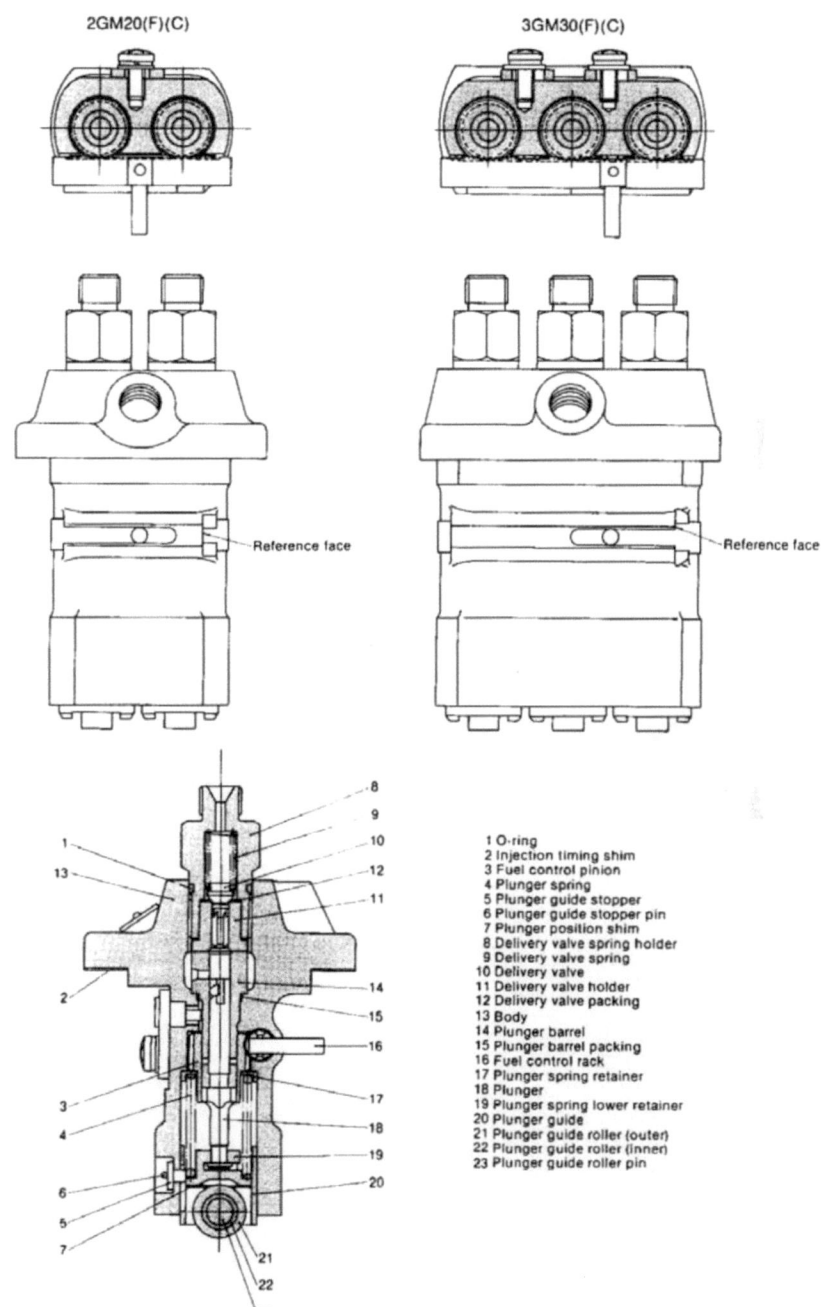

1. O-ring
2. Injection timing shim
3. Fuel control pinion
4. Plunger spring
5. Plunger guide stopper
6. Plunger guide stopper pin
7. Plunger position shim
8. Delivery valve spring holder
9. Delivery valve spring
10. Delivery valve
11. Delivery valve holder
12. Delivery valve packing
13. Body
14. Plunger barrel
15. Plunger barrel packing
16. Fuel control rack
17. Plunger spring retainer
18. Plunger
19. Plunger spring lower retainer
20. Plunger guide
21. Plunger guide roller (outer)
22. Plunger guide roller (inner)
23. Plunger guide roller pin

Chapter 3 Fuel System
2. Injection Pump

SM/GM(F)(C)·HM(F)(C)

3HM35(F)(C)

The construction is the same as the fuel injection pump on model 2GM20(F)(C) or 3GM30(F)(C) engines except for the differences of the plunger diameters, shape of plungers and plunger barrels. Take care as the position of the basic surface for adjusting the injection volume is different.

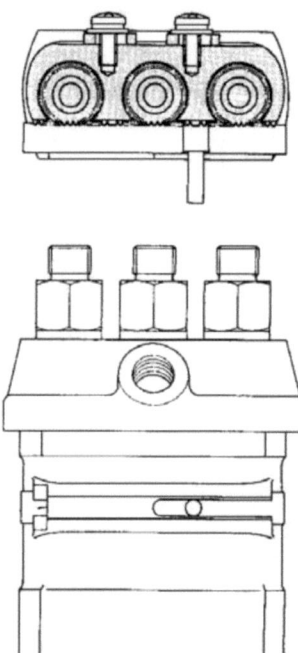

Reference face

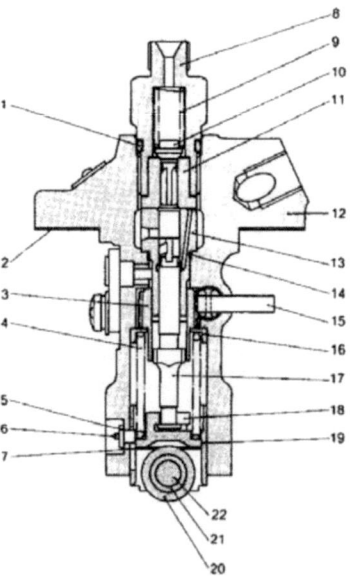

1 O-ring
2 Injection timing shim
3 Fuel control pinion
4 Plunger spring
5 Plunger guide stopper
6 Plunger guide stopper pin
7 Plunger position shim
8 Delivery valve spring holder
9 Delivery valve spring
10 Delivery valve
11 Delivery valve holder
12 Body
13 Plunger barrel
14 Plunger barrel packing
15 Fuel control rack
16 Plunger spring retainer
17 Plunger
18 Plunger spring lower retainer
19 Plunger guide
20 Plunger guide roller (outer)
21 Plunger guide roller (inner)
22 Plunger guide roller pin

Chapter 3 Fuel System
2. Injection Pump

SM/GM(F)(C)·HM(F)(C)

2-2 Specifications and performance of fuel injection pump
2-2.1 Specifications of fuel injection pump

			1GM10(C)	2GM20(F)(C)	3GM30(F)(C)	3HM35(F)(C)
Plunger diameter				6mm (0.2362in.)		6.5mm (0.2559in.)
Standard plunger stroke			7mm (0.2756in.)			
Static mechanical lift at injection			2.5mm (0.0984in.) [at starting 3.2mm (0.1260in.)]			
Sliding resistance of fuel volume adjusting rack (when pump stops)			60g (0.002 lb) or less			
Top clearance of plunger (at the set dimension of 76 ±0.05mm)			1.0mm (0.0394in.)			
Thickness of plunger position adjusting shim			0.1mm (0.0039in.), 0.2mm (0.0079in.), 0.3mm (0.0118in.)			
Plunger spring (124950-51190 commonly used)	Free length		35.5mm (1.3976in.)			
	Spring constant		1.93 kgf/cm (10.8 lb/in.)			
	Load	At upper limit	25.1 kg (55.3 lb)			
		At lower limit	11.6 kg (25.6 lb)			
		At static injection	16.4 kg (36.2 lb)			
Suction volume of delivery valve			23.5mm³ (0.0014in.³) (24.5 according to 1GM10(C) drawing)			
Opening pressure of delivery valve			Approx. 16.3 kgf/cm² (231.8 lb/in.²)			
Delivery valve spring (124550-51320 commonly used)	Free length		21.0mm (0.8268in.)			
	Spring constant		0.64 kgf/cm (9.1 lb/in.)			
Rack stroke			Approx. 15mm (0.5906in.)			

2-2.2 Injection volume characteristics of fuel injection pump

(1) Model 1GM10(C)

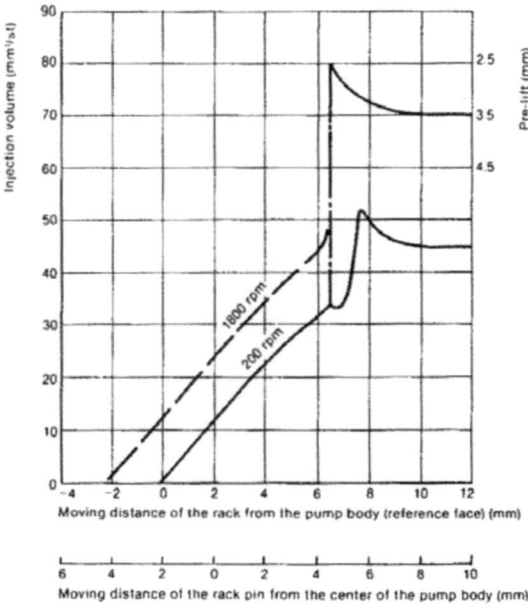

Moving distance of the rack from the pump body (reference face) (mm)

Moving distance of the rack pin from the center of the pump body (mm)

Chapter 3 Fuel System
2. Injection Pump

(2) Model 2GM20(F)(C)

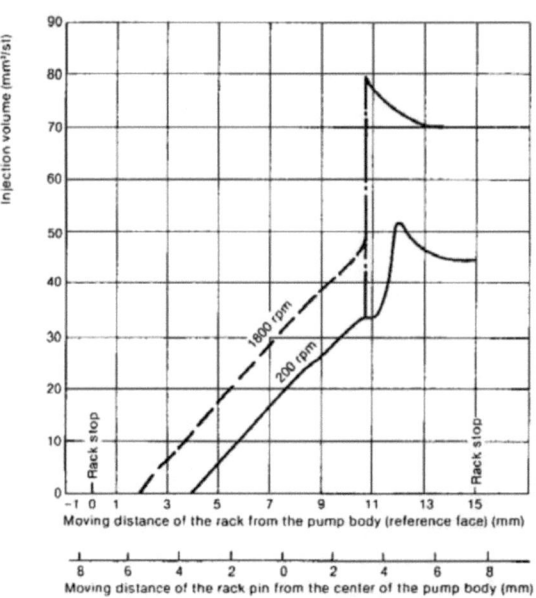

(3) Model 3GM30(F)(C)

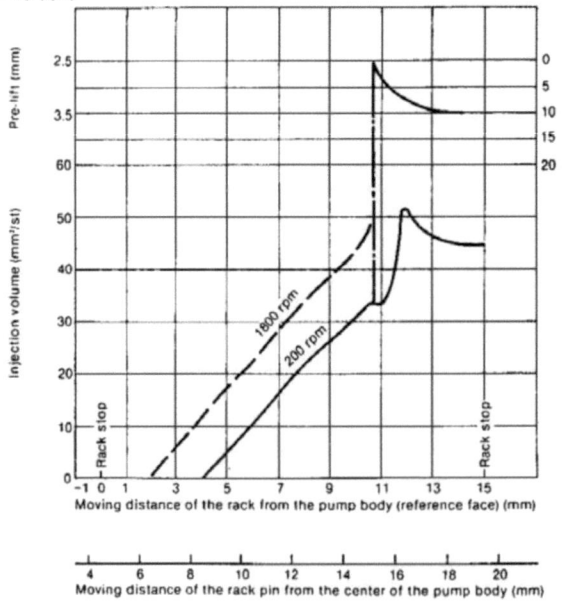

(4) Model 3HM35(F)(C)

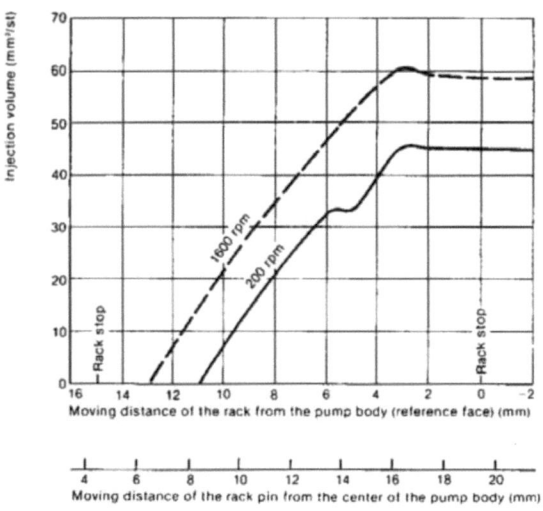

2-3 Operation of fuel injection pump

The fuel injection pump force-feeds the fuel by means of the plunger (1) which operates at a constant stroke. Since the plunger is lap fitted into the plunger barrel (2) for super precison, it can be replaced only as a set. The cylindrical surface of the plunger has an obliquely cut lead (3) and a groove which connects the lead to the plunger head. The plunger has an intake hole (4) through which the fuel passes and is force-fed by the plunger. Then the fuel opens the delivery valve (5), goes through the fuel injection tube, and is injected into the spiral-vortex type pre-combustion chamber from the injection valve. The plunger is fitted with the fuel control gear (6), and its flange (7) fits into the longitudinal groove which is cut in the inner surface of the lower end of the control gear. The fuel control gear is in mesh with the fuel control rack, the motion of which rotates the plunger to constantly vary the amount of fuel injected from zero to maximum.

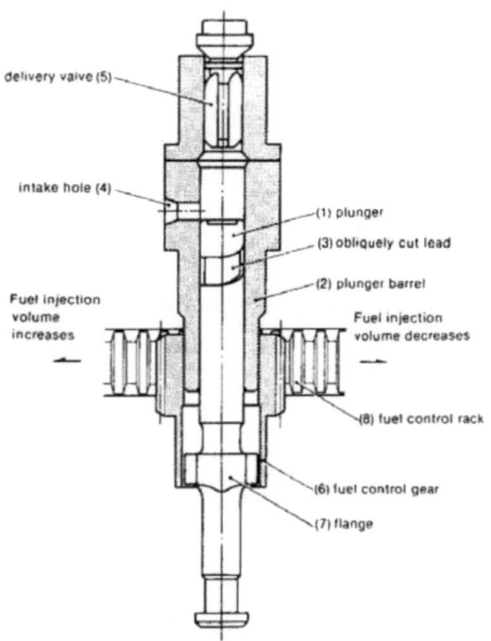

Chapter 3 Fuel System
2. Injection Pump

2-3.1 Fuel control

When the plunger (1) is at bottom dead center, the oil, which comes in through the oil hole, fills the delivery chamber (3) to above the plunger. The oil pressure then builds up as the plunger rises and closes the oil hole, and by opening the delivery valve, the oil is force-fed toward the fuel injection tube. As the plunger, pushed by the plunger guide, rises further, the pressure of the oil between the delivery chamber and the nozzle also increases. when this oil pressure builds up to 155 to 165 kgf/cm², the nozzle opens, and the fuel oil is injected into the spiral vortex type combustion chamber. However, if the plunger keeps rising and the lead groove(4)lines up with the oil hope(2) the oil under high pressure in the delivery chamber passes up the lead from the longitudinal groove and is driven back into the suction chamber from the oil hole. At the same time, force feeding of the fuel is suspended.

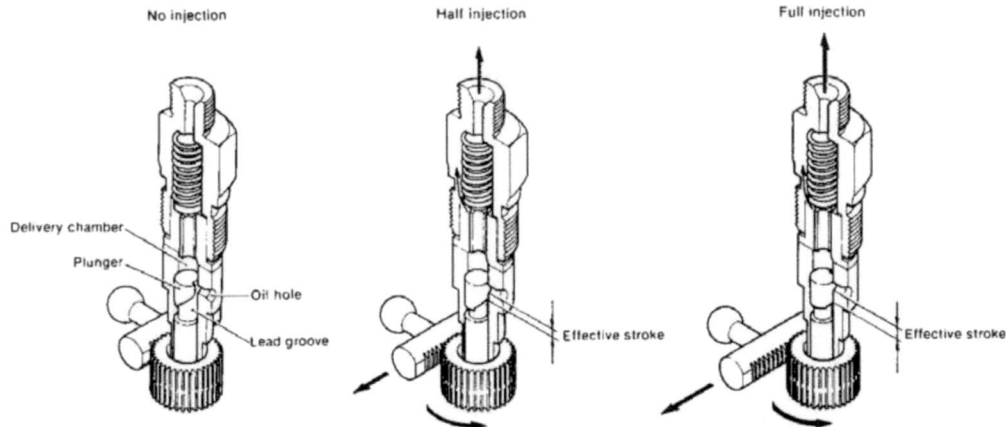

As a result of the above action, the plunger is rotated by the fuel control rack and the angle of this rotation changes the effective stroke of the plunger and controls the discharge of the pump. Also, when the fuel control rack lines up the longitudinal groove on the plunger with the oil hole, the oil hole does not close, despite the rise of the plunger, but rather the fuel is driven back to the suction chamber. As a result the fuel is not force-fed but the amount of injection is reduced to zero. At this time the fuel control rack is at the cylinder side end; when it reaches the opposite side end the maximum amount of fuel is injected. Before the maximum injection level is reached, the fuel injection control shaft regulates the amount of fuel injected to the normal operation level.

NOTE: *The plunger is an integral part of the plunger barrel and takes in and compresses fuel by reciprocating inside the plunger barrel. The plunger and plunger barrel are precisely machined, and because the plunger is driven in an extremely small space, the two should be used together and should not be changed with other cylinders.*

2-3.2 Action of the delivery valve and the sucking-back of fuel

The delivery valve on top of the plunger prevents the fuel inside the injection tube from flowing backward toward the plunger side and also serves to suck back the fuel to prevent the backward dripping of the nozzle valve. When the notch (lead) of the plunger comes up to the oil hole of the plunger barrel, the feeding pressure acting on the fuel oil drops, and the delivery valve falls due to the force of the spring. After the sucking-back collar has first shut off the fuel injection tube and the delivery chamber, the delivery valve drops further until comes in contact with the seat surface, in correspondence with the amount of fall (i.e., increase in volume), the fuel oil pressure within the injection tube drops, speeding up the closure of the nozzle valve, and sucking up the fuel before it drips back. This enhances the durability of the nozzle and improves fuel oil combustion.

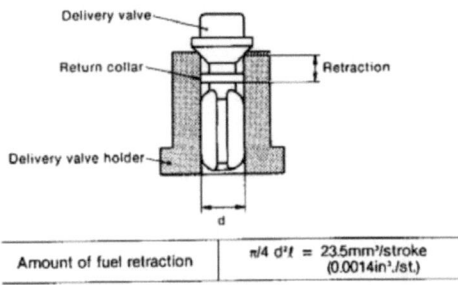

Amount of fuel retraction	$\pi/4\, d^2 \ell$ = 23.5mm³/stroke (0.0014in³./st.)

3-11

Chapter 3 Fuel System
2. Injection Pump

2-4 Disassembly of fuel injection pump

As a rule, the injection pump should not be disassembled, but when disassembly is unavoidable, proceed as described below.

2-4.1 Dismantling of fuel injection pump of model 1GM10(C) engine.

NOTES: 1) Before disassembly wash the pump in clean oil, and after assembly arrange all parts carefully.
2) Make sure the work area is exceptionally clean.

(1) Remove the plunger guide stopper pin with needle nose pliers.

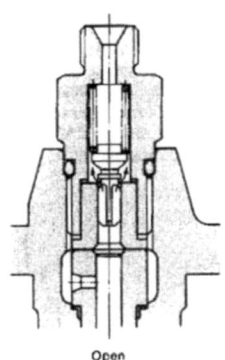
Open

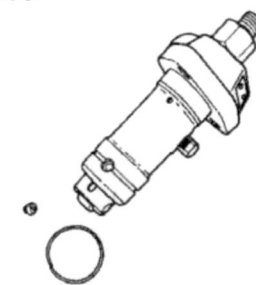

(2) Remove the plunger guide stopper.
The stopper can be removed by pushing the plunger guide down with the palm of your hand.
(3) Remove the plunger guide.
NOTE: Be careful not to lose the plunger stroke adjusting shim which is located inside the plunger guide.

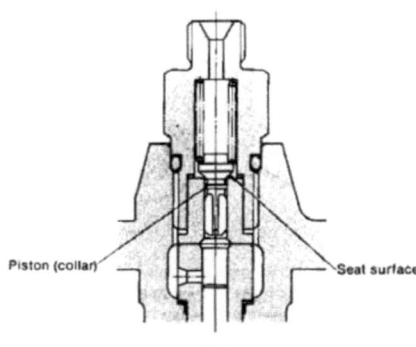
Piston (collar) — Seat surface
Close

Plunger stroke adjustment shim

(4) Remove the plunger and plunger spring lower retainer. Be careful not to damage the plunger.
(5) Remove the plunger spring, fuel control pinion and plunger spring upper retainer, using your fingers or tweezers.

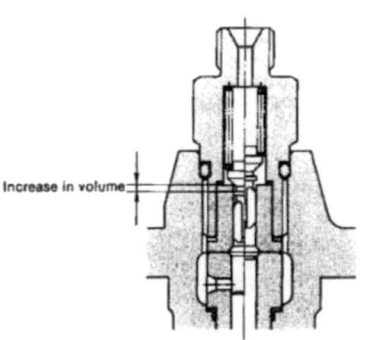
Increase in volume
Retraction of fuel

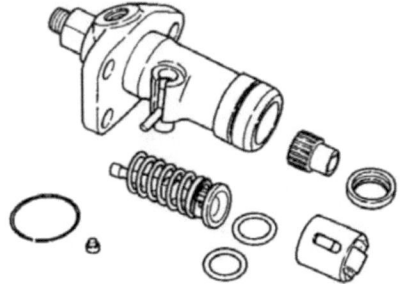

Chapter 3 Fuel System
2. Injection Pump

(6) Remove the fuel control rack.
(7) Remove the delivery valve holder; be careful not to damage the O-ring.
(8) Remove the delivery valve spring.
(9) Remove the delivery valve.

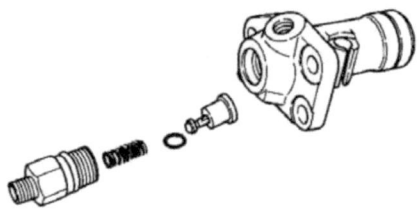

(10) Remove the plunger barrel by pushing it toward the delivery valve side.
(11) Remove the plunger barrel packing.

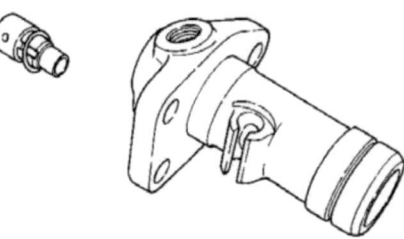

NOTES: 1) Line up the plunger barrel and the plunger, and put them in order.
2) Immerse the delivery valve, plunger, etc. in clean oil.
3) Do not loosen or remove the plunger barrel stopper, etc.

2-4.2 Dismantling of fuel pump of model 2GM20(F)(C), 3GM30(F)(C) and 3HM35(F)(C)

The cylinders are classified as No. 1, No. 2 and No. 3 from the left, when facing the name plate fitted on the upper part of the fuel injection pump. When dismantling, it is necessary to prepare pans or vessels in which to keep the dismantled parts from each cylinder; each part must be placed in the corresponding pan or vessel for each cylinder, namely, No. 1, No. 2 and No. 3 cylinder. If a part is placed in the wrong pan or vessel, reassembly becomes impossible without a pump tester. The following explanation is for to the pump of the 2 cylinder type engine [model 2GM20(F)(C)], but it applies equally to that of the 3 cylinder type engine [model 3GM30(F)(C)] which merely has an additional set. The construction of the fuel pump of model 3HM35(F)(C) engine is the same as that of model 3GM30(F)(C) engine except for the differences of plunger, plunger barrel, and the position of the injection volume adjusting rack.

(1) Remove the plunger guide stopper pin with needle nose pliers.

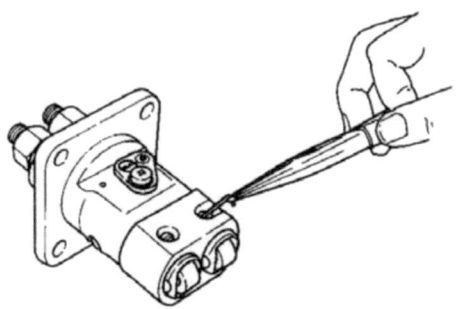

(2) Remove the No.1 plunger guide stopper.
The stopper can be removed by pushing the plunger guide down with the palm of your hand.
(3) Remove the No.1 plunger guide.

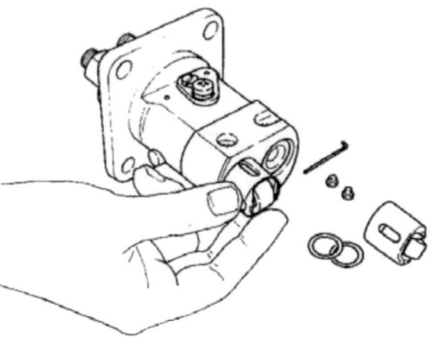

(4) Remove the No.1 plunger, plunger spring lower retainer and plunger shim; be careful not to damage the plunger.
(5) Remove the No.1 plunger spring.

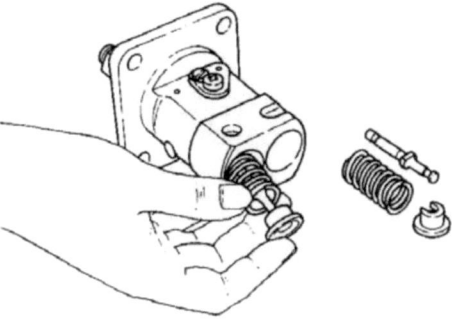

(6) Remove the No.1 plunger spring upper retainer, using your fingers or tweezers.
(7) Remove the No.1 control sleeve

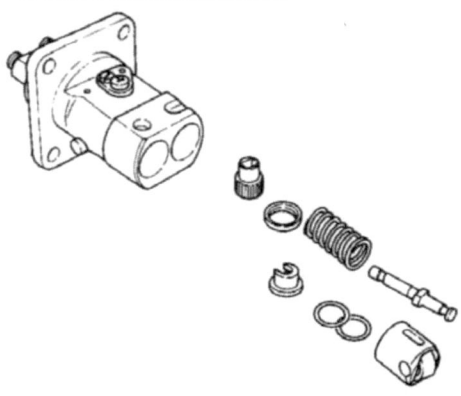

(8) Remove the No.1 delivery valve holder; be careful not to damage the O-ring.
(9) Remove the No.1 delivery valve spring.
(10) Remove the No.1 delivery valve, delivery valve seat and packing.

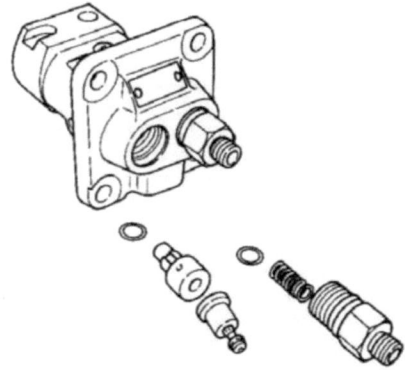

(11) Remove the No.1 plunger barrel; be careful not to damage the face that matches the delivery valve seat.
(12) Remove the No.1 plunger barrel packing.
(13) For No.2 cylinder, repeat the above steps (2) through (11).
(14) The above item also applies to No.3 cylinder for the 3 cylinder type engine.

(15) Remove the control rack.

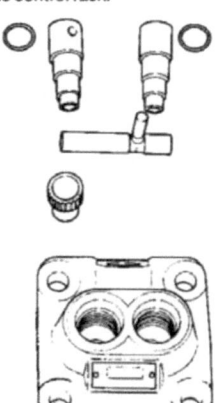

NOTES: 1) Line up the plunger valve and the plunger, and put them in order.
2) Immerse the delivery valve, plunger, etc. in clean oil.
3) Do not loosen or remove the injection control plate, etc.

2-5 Inspecting injection pump parts

2-5.1 Rinse each component part in clean light oil before inspecting it.

NOTE: Do not touch the sliding surface of the plunger and the delivery valve with your fingers during handling.

2-5.2 Tappet

Inspect the cam sliding surface of the tappet roller for wear, scoring and peeling; replace the tappet and roller assembly when the total tappet and roller play exceeds 0.3mm.

2-5.3 Control rack and pinion

(1) Check the control rack teeth and sliding surface for damage and abnormalities. If found, replace.

NOTE: When replacing the control rack, adjust fuel discharge with a fuel injection pump tester and stamp a rack mark.

Chapter 3 Fuel System
2. Injection Pump

(2) Replace pinion if teeth are damaged or worn unevenly.

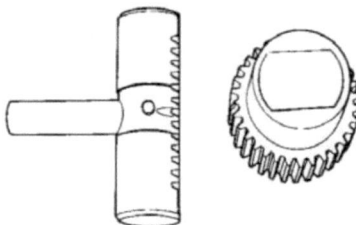

(3) If the control rack does not move smoothly when a force of within 60g is applied, replace the rack and pinion assembly.

2-5.4 Plunger

(1) Inspect the plunger for wear, scoring and discoloration around the lead. If any problems are found, conduct a pressure test and replace the plunger and plunger barrel assembly.

For models 1GM10(C), 2GM20(F)(C) and 3GM30(F)(C)

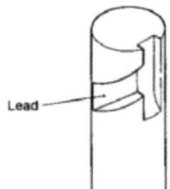

For model 3HM35(F)(C)

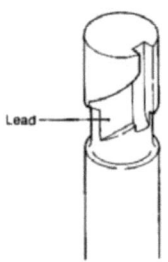

(2) Inspect the outside sliding surface of the plunger with a magnifying glass. Lap or replace the plunger and plunger barrel assembly when corrosion, hairline cracks, staining and/or scoring are detected.
(3) Check the clearance between the plunger collar and control sleeve groove. Replace these parts when wear exceeds the specified limit.

(4) After cleaning the plunger, tilt it approximately 60°, as shown in the figure, and slowly slide it down. Repeat this several times while rotating the plunger. The plunger should slide slowly and smoothly. If it slides too quickly, or binds along the way, repair or replace it.

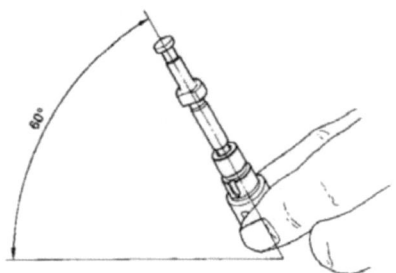

2-5.5 Delivery valve

(1) Replace the delivery valve if the return collar and seat are scored, dented or worn.

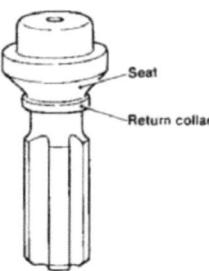

(2) Raise the delivery valve and put a finger over the hole on the valve seat bottom. Let go of the delivery valve. If it sinks quickly and stops at the position where the suck-back collar closes the valve seat hole, the delivery valve may be considered normal. If this is not the case, replace the delivery valve as a set.

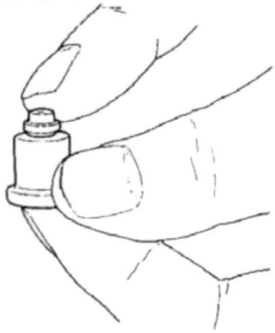

Chapter 3 Fuel System
2. Injection Pump

(3) Place your finger over the hole in the bottom of the valve seat and insert the valve into the valve body. If the valve returns to its original position when you remove your finger, the valve is okay. If some defect is found, replace with a new valve.

(4) If the valve closes completely by its own weight when you remove your finger from the hole on the bottom of the valve seat, the valve is okay. If it doesn't close perfectly replace with a new valve.

NOTE: When using a brand-new set, wash off the rustproof oil with clean oil or gasoline. Then, wash once more with clean oil, and follow the steps outlined above.

2-5.6 Plunger spring and delivery valve spring

Inspect the plunger spring and delivery valve spring for fractured coils, rust, inclination and permanent strain. Replace the spring when faulty.

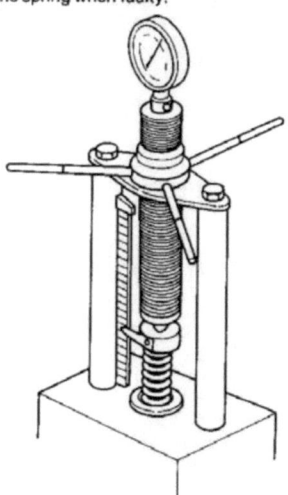

2-5.7 Plunger guide

Check the tappet roller (Inside and outside) and roller pin for damage and uneven wear, and replace if required.
Measure the clearance between the plunger and plunger guide. If the clearance exceeds the limit, replace.

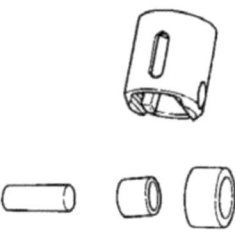

mm (in.)

Clearance limit	0.3 (0.0118)

2-6 Assembling the fuel injection pump

NOTES: 1) After inspection, divide the components into two groups, i.e. the components to be replaced, and those that are reusable. Rinse the components and store the two groups separately.
2) Replace the packing with a new one.

1GM10(C)

(1) While lining up the plunger barrel positioning groove with the dowel of the main unit, attach the plunger barrel to the main unit.

Attaching the plunger barrel to the main unit

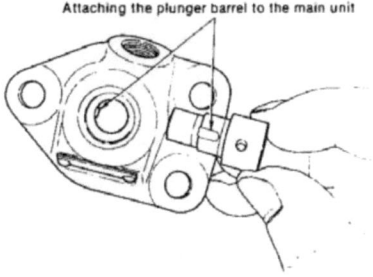

	Free length	Set length	Set load
Plunger spring	35.5mm (1.3976in.)	29.5mm (1.1614in.)	11.59 ±1.1 kg (23.13 ~ 27.98 lb)
Delivery valve spring	21mm (0.8268in.)	17.25mm (0.6791in.)	2.4 ±0.24 kg (4.76 ~ 5.82 lb)

(2) Attach the delivery valve seat and the delivery valve to the main unit.

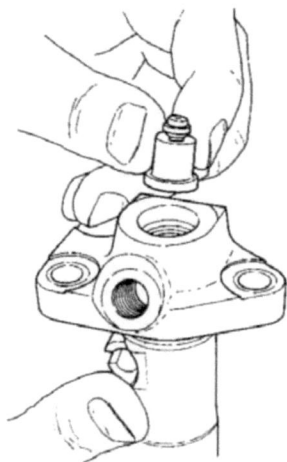

Attaching the delivery valve to the main unit

NOTE: *If the delivery valve tip projects noticeably above the top of the main unit of the pump, the plunger barrel has been installed incorrectly, and must be re-attached.*

(3) Attach the delivery valve packing and the delivery valve spring to the main unit and carefully tighten the delivery valve holder.

NOTE: *Tighten the delivery valve holder with a torque wrench after attaching the plunger and while checking the fuel control rack for sliding motion.*

1GM(10(C)	kgf-m(ft-lb)
Tightening torque	4.0 ~ 4.5 (28.92 ~ 32.54)

(4) With the matching mark of the fuel control rack directed towards the lower part of the main unit of the pump, attach the fuel control rack to the main unit.

NOTE: *Make sure the fuel control rack moves smoothly along its entire stroke.*

(5) By aligning the matching mark on the fuel control pinion with that on the fuel control rack, attach the fuel control pinion to the main unit.

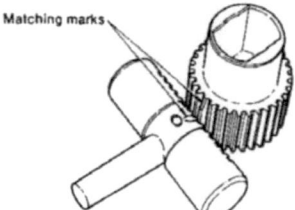

NOTE: *After attaching the fuel control pinion to the main unit, check its meshing by moving the fuel control rack.*

(6) Insert the plunger spring retainer and attach the plunger spring to the main unit.

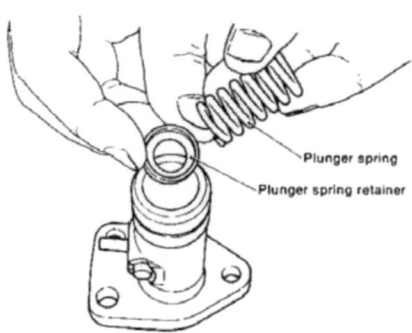

NOTE: *The plunger spring retainer should face the underside of the pump.*

(7) After aligning the matching mark on the plunger flange with that on the fuel control pinion, attach the plunger to the main unit.

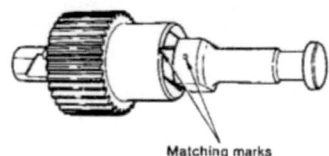

NOTE: *Invert and stand the main unit of the pump upright and attach the plunger to it carefully.*

(8) Mount the plunger lower retainer on the plunger.

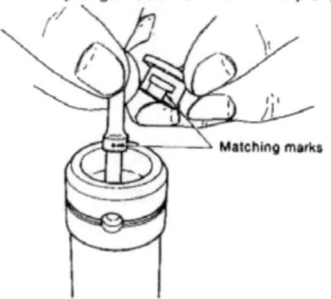

(9) Insert the plunger adjusting shims.

NOTE: *Insert the same number of shims with the same thickness as those inserted before disassembling the pump. After re-assembling the pump, measure and adjust the top clearance of the plunger.*

Chapter 3 Fuel System
2. Injection Pump

(10) While adjusting the direction of the plunger guide stopper hole for the plunger guide, insert the plunger guide carefully.
When the plunger guide stopper hole is lined up with the plunger guide, insert the plunger guide stopper. Then mount the retaining ring (clip).

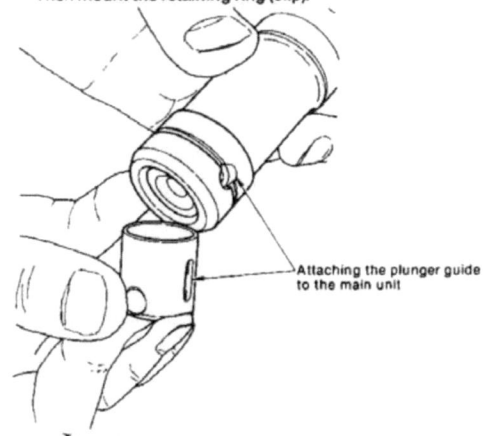

Attaching the plunger guide to the main unit

(11) After attaching tighten the delivery valve holder with a torque wrench.

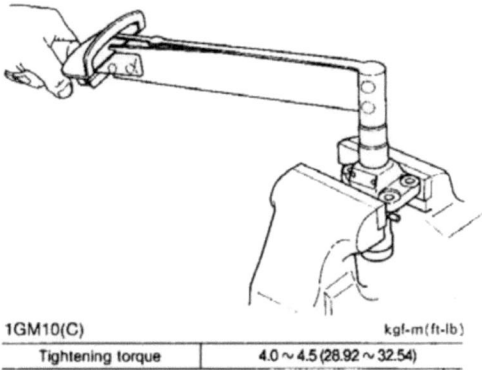

1GM10(C)

	kgf-m (ft-lb)
Tightening torque	4.0 ~ 4.5 (28.92 ~ 32.54)

2GM20(F)(C), 3GM30(F)(C) and 3HM35(F)(C)

To ensure that the injection pump is correctly reassembled, the following points must be kept in mind:
- The parts for each cylinder must not be mixed together.
- When parts are replaced, the parts for each cylinder must always be replaced at the same time.
- When assembling, parts must be washed in fuel oil and matching marks and scribe lines lined up.

(1) Install the No.1 plunger barrel packing.
(2) Insert the No.1 plunger barrel by aligning the groove of the barrel lock pin.

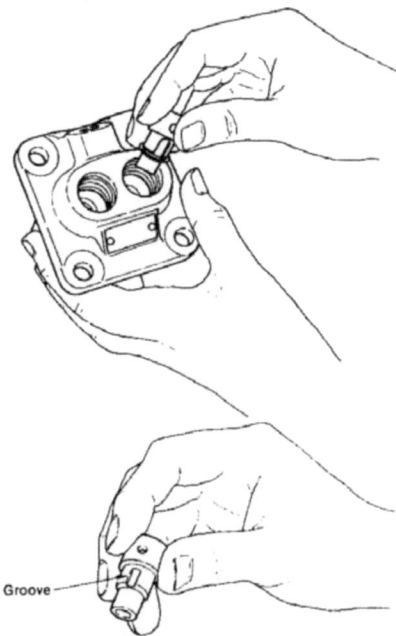

Groove

(3) Install the No.1 delivery valve, delivery valve seat and packing.

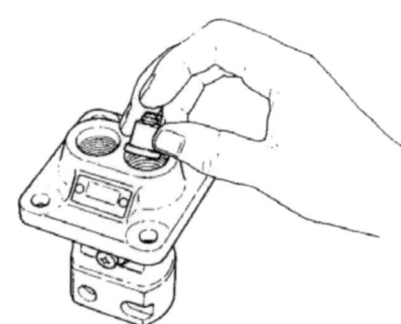

NOTE: If the delivery valve tip projects noticeably above the top of the main unit of the pump, the plunger barrel has been installed incorrectly, and must be re-attached.

(4) Insert the No.1 delivery valve spring.

Chapter 3 Fuel System
2. Injection Pump

SM/GM(F)(C)·HM(F)(C)

(5) Tighten the No.1 delivery valve holder.

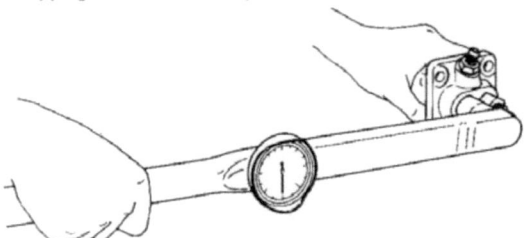

2GM20(F)(C), 3GM30(F)(C), 3HM35(F)(C)	kgf-m (ft-lb)
Tightening torque	4.0 ~ 4.5 (28.92 ~ 32.54)

NOTE: Tighten the delivery valve holder with a torque wrench after attaching the plunger and while checking the fuel control rack for sliding motion.

(6) With the matching mark of the fuel control rack directed towards the lower part of the main unit of the pump, attach the fuel control rack to the main unit.

NOTE: Make sure the fuel control rack moves smoothly along its entire stroke.

(7) By aligning the matching mark on the fuel control pinion with that on the fuel control rack, attach the fuel control pinion to the main unit.

2GM20(F)(C)

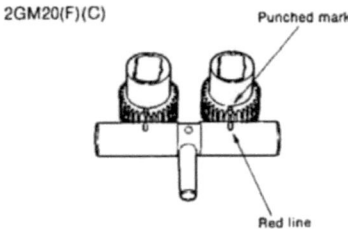

3GM30(F)(C), 3HM35(F)(C)

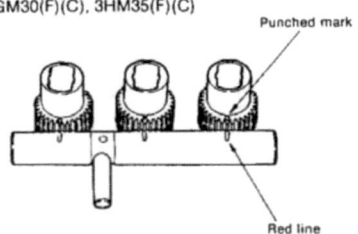

NOTE: After attaching the fuel control pinion to the main unit, check its meshing by moving the fuel control rack.

(8) Insert the No.1 plunger spring retainer and attach the plunger spring to the main unit.

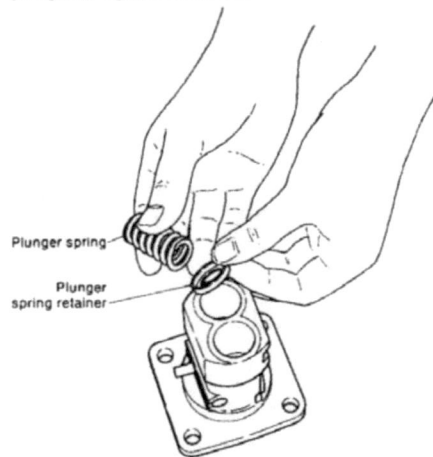

NOTE: The plunger spring retainer should face the underside the pump.

(9) After aligning the matching mark on the plunger flange with that on the fuel control pinion, attach the plunger to the main unit.

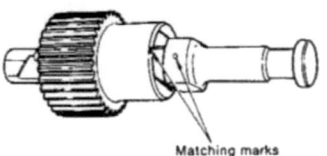

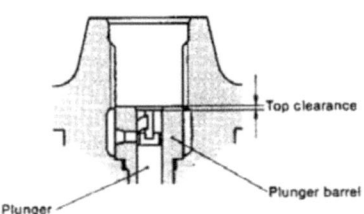

NOTE: By inverting and standing the main unit of the pump upright attach the plunger to it carefully.

(10) Install the No.1 plunger spring lower retainer. Make sure that it is not installed backwards.

Chapter 3 Fuel System
2. Injection Pump

(11) Insert the plunger shim.
NOTE: *Insert the same number of shims with the same thickness as those inserted before disassembling the pump. After re-assembling the pump, measure and adjust the top clearance of the plunger.*
(12) Insert the No.1 plunger guide.
(13) Insert the No.1 plunger guide stopper.
(14) For the pump of the 2 cylinder type engine, repeat the above steps for No.2 cylinder.
(15) For the pump of the 3 cylinder type engine, repeat the above steps for No.3 cylinder.
(16) Install the plunger guide stopper pin.
(17) After attachment tighten the delivery valve holder with a torque wrench.

2GM20(F)(C), 3GM30(F)(C), 3HM35(F)(C)   kgf-m(ft-lb)

Tightening torque	4.0 ~ 4.5 (28.92 ~ 32.54)

NOTE: *When the tightening torque of the delivery valve holder exceeds the prescribed torque, the plunger will be distorted, the sliding resistance of the control rack will increase, and proper performance will not be obtained. Moreover, excessive tightening will damage the pump body and delivery valve gasket, and cause a variety of other problems.*

2-7 Inspection after reassembly

When the engine doesn't run smoothly and the injection pump is suspected as being the cause, or when the pump has been disassembled and parts replaced, always conduct the following tests.

2-7.1 Control rack resistance test

After reassembling the pump, wash it in clean fuel, move the rack and check resistance as follows:
(1) This test is performed to determine the resistance of the control rack. When the resistance is large, the engine will run irregularly or race suddenly.
(2) Place the pump on its side, hold up the control rack and allow it to slide down by its own weight. The rack should slide smoothly over its entire stroke. Place the pump on end and perform the above test again; check for any abnormalities. [Resistance below 60g (0.132 lb)]
(3) Since a high sliding resistance is probably a result of the following, disassemble the pump and wash or repair it.

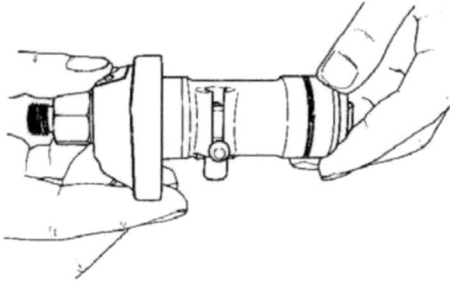

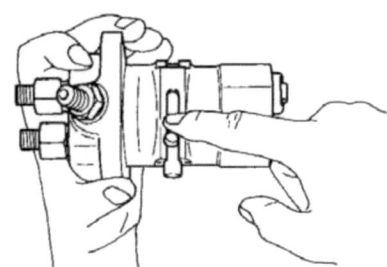

(a) Resistance of the rotating and sliding parts of the plunger assembly is too high.
(b) Delivery valve holder is too tight (plunger barrel distorted).
(c) Control rack or control pinion teeth and control rack outside circumference are dirty or damaged.
(d) Injection pump body control rack hole is damaged.
(e) Plunger barrel packing is not installed correctly and the barrel is distorted. (Since in this case fuel will leak into the crankcase and dilute the lubricating oil, special care must be taken).

2-7.2 Fuel injection timing

Fuel injection timing is adjusted by timing shims inserted between the pump body and gear case pump mounting seat.
The injection pump must be mounted on the engine, and each cylinder injection timing adjusted.
Adjusting the injection timing
(1) Remove the high pressure pipe from the pump.
(2) Install a measuring pipe if the injection pump does not have a nipple on the delivery side.
(3) Bleed the air from the injection pump.

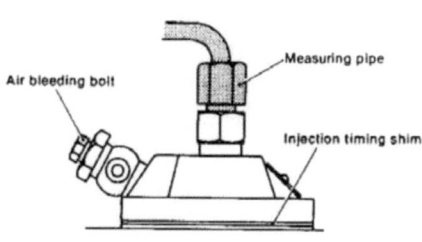

(4) Set the control rack to the middle fuel injection position (Pull the lever when setting the accelerator lever.)
(5) Turn the crankshaft slowly by hand, and read the timing mark (TD) on the flywheel the instant fuel appears at the measuring pipe or pipe joint nipple.
(FID+ Fuel injection from delivery valve.)

Chapter 3 Fuel System
2. Injection Pump

The thickness of the plunger location adjusting shim and the injection timing adjusting plate is 0.1 mm. With this the injection timing can be changed by approximately 1° on the crankshaft.

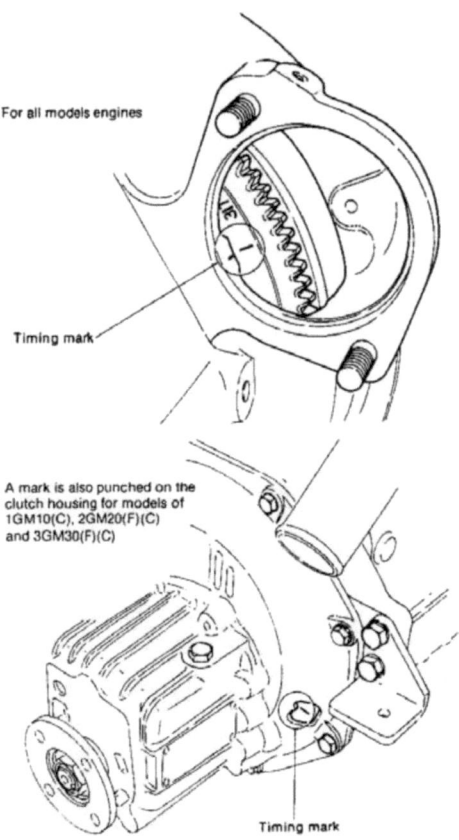

For all models engines

Timing mark

A mark is also punched on the clutch housing for models of 1GM10(C), 2GM20(F)(C) and 3GM30(F)(C)

Timing mark

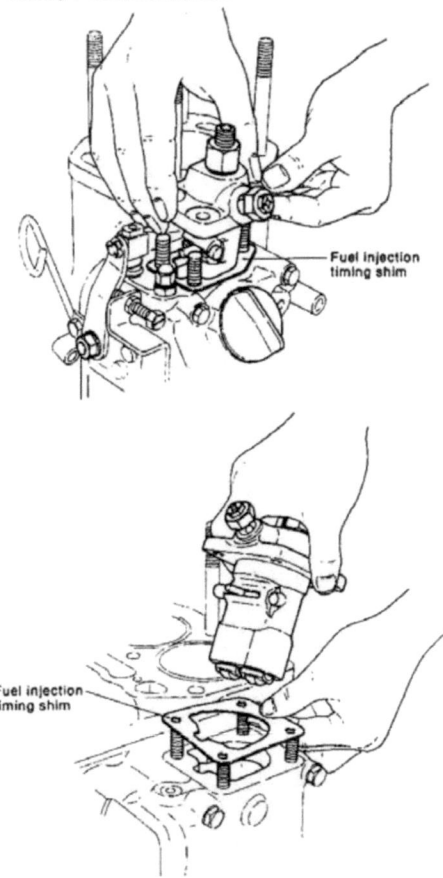

Fuel injection timing shim

Fuel injection timing shim

(6) If the injection timing is off, add plunger shims when the timing is slow, and remove shims when the timing is fast. Adjust the timing of every pump in the same manner. (Refer to item, "Plunger head gap adjustment".)

(7) After the injection timing of every pump has been matched, recheck the injection timing as described in item (5) above. If the injection timing is not properly set, adjust it with the timing shims.

(8) Finally, turn the crankshaft slowly and confirm that it turns easily. If it is stiff or does not rotate, the plunger head gap is too small.

		1GM10(C)	2GM20 (F)(C)	3GM30 (F)(C)	3HM35 (F)(C)
Fuel injection timing		bTDC15° (FID)	bTDC15° (FID)	bTDC18° (FID)	bTDC21° (F)(D)
Fuel inection timing shim	0.2mm (0.008in.)	1 shim 104271-01930	2 shims 124950-01931	2 shims 121450-01931	
	0.3mm (0.012in.)	1 shim 104271-01940	2 shims 124950-01941	2 shims 121450-01941	
	0.5mm (0.020)	—	1 shim 124950-01961	1 shim 124950-01961	
	Set No.	104271-01950	124950-01951	121450-01951	

3-21

2-8 Injection pump adjustment

The injection pump is adjusted with an injection pump tester after reassembly.

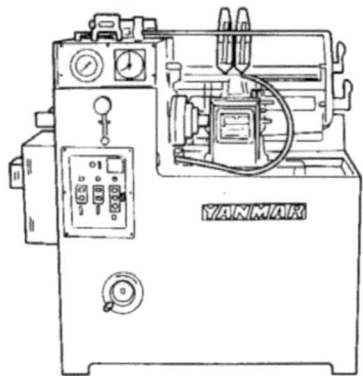

2-8.1 Setting pump on tester

(1) After the injection pump has been disassembled and reassembled, install it on a pump tester
...cam lift: 7mm (0.276in.).
(2) Confirm that the control rack slides smoothly. If it does not, inspect the injection pump and repair it so that the rack slides smoothly
...control rack full stroke: 15mm (0.5905in.).

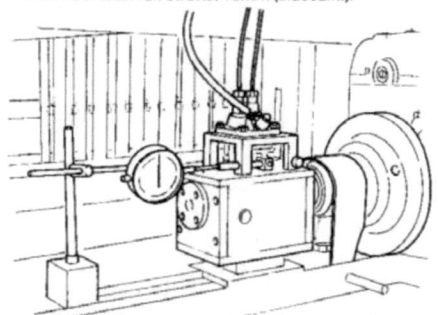

(3) Run the pump tester at low speed, loosen the air bleeder screw, and bleed the air from the injection pump.

2-8.2 Measuring the sliding resistance of the fuel control rack

Measure the sliding resistance of the fuel control rack with a spring scale (balance).
(1) Number of pump rotations/sliding resistance: 0rpm/less than 60 g. (0.132 lb)

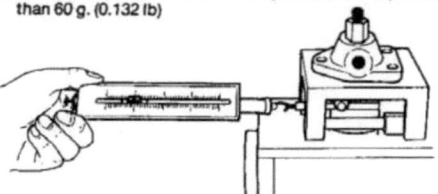

NOTE: If the sliding resistance is unsatisfactory, disassemble, inspect and repair the fuel control rack.

2-8.3 Adjusting the plunger top clearance

(1) Set the pump installation dimension (end of plunger barrel when the roller is on the cam base cycle) at 76 ±0.05mm (2.9902 ~ 2.9941in.), remove the delivery valve holder and delivery valve, and set the plunger to top dead center by turning the camshaft. Measure the difference in height (head gap) between the end of the plunger and the end of the plunger barrel using a dial gauge.

mm (in.)

Plunger top clearance	1.0 ±0.05 (0.0374 ~ 0.0398)

(2) Using the plunger top clearance measuring jig
1) Install a dial gauge on the measuring jig.
2) Stand the measuring jig on a stool and set the dial gauge pointer to O.
3) Remove the pump delivery valve and install the measuring jig.
4) Turn the camshaft to set the plunger to top dead center and read the dial gauge. The value given is the plunger top clearance.

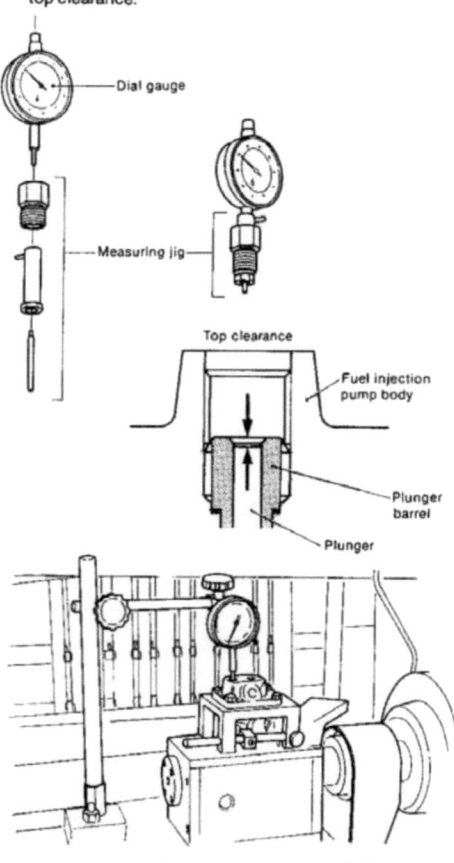

Chapter 3 Fuel System
2. Injection Pump

(3) When the plunger top clearance is larger than the prescribed value, remove the plunger guide and insert plunger shims between the plunger spring lower retainer and the plunger guide. Adjust each pump in the same manner.

Plunger shim thickness	0.1mm (0.004in.)	174307-51710
	0.2mm (0.008in.)	174307-51720
	0.3mm (0.012in.)	174307-51730

(4) After rechecking adjustment, install the delivery valve.

Delivery valve holder tightening torque	4.0 ~ 4.5 kgf-m (29 ~ 32.6 lb-ft)

2-8.4 Checking the cylinder injection interval

(1) Align the control rack punch mark with the pump reference face.

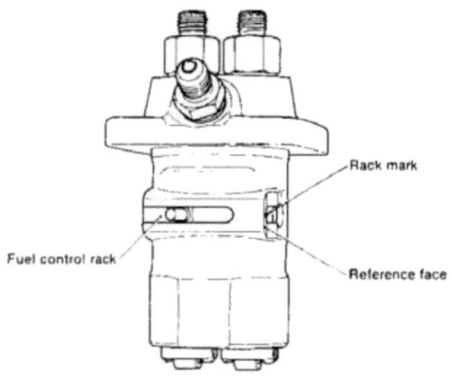

(2) Turn the pump by hand to check the No.1 cylinder injection timing.
(3) Turn the pump in the prescribed direction and check the No.2/3 cylinder injection timing.
(4) Using the plunger shims, adjust each cylinder injection timing interval.

	For crankshaft angle	For camshaft angle
2GM20(F)(C)	180° 540° 1 ~ 2 ~ 1	90° 270° 1 ~ 2 ~ 1
3GM30(F)(C), 3HM35(F)(C)	240° 240° 240° 1 ~ 3 ~ 2 ~ 1	120° 120° 120° 1 ~ 3 ~ 2 ~ 1

2-8.5 Delivery valve oil-tight test

(1) Install a 1,000 kgf/cm² (14,223 lb/in.²) pressure gauge on the delivery valve holder.
(2) Drive the fuel pump to apply a pressure of approximately 120 kgf/cm² (1,707 lb/in.²) and measure the time required for the pressure to drop from 100 kgf/cm² (1,422 lb/in.²) to 90 kgf/cm² (1280 lb/in.²)

Pump speed	200 rpm
Pressure drop standard	20 sec. or more
Pressure drop limit	5 sec. or less

(3) If both the plunger and the delivery valve fail the test, replace them.

2-8.6 Plunger pressure test

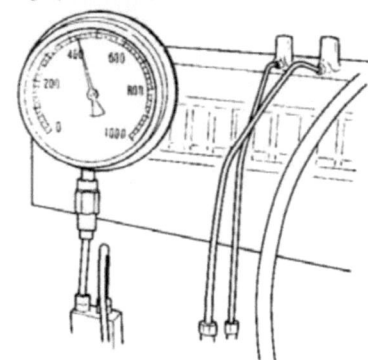

(1) Install a 1,000 kgf/cm² (14,223 lb/in.²) pressure gauge on the delivery valve holder.
(2) Check that there is no oil leaking from the delivery valve holder and high pressure pipe mountings, and that the pressure does not drop suddenly when raised to 500 Kgf/cm² (7,112 lb/in.²) or higher.
Pressure gauge AVT 1/2 × 150 × 1,000 kgf/cm²

2-8.7 Measuring the fuel injection volume

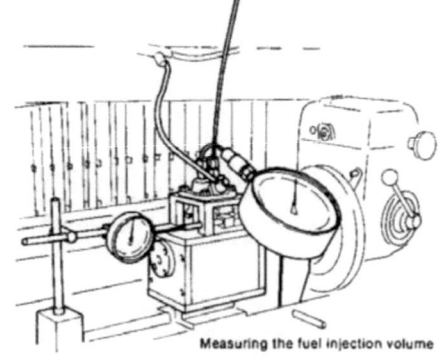

Measuring the fuel injection volume

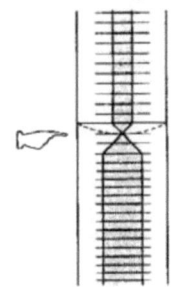

3-23

Chapter 3 Fuel System
2. Injection Pump

(1) Set the fuel pump camshaft speed.
(2) Check the injection nozzle.

	1GM10(C)	2GM20(F)(C), 3GM30(F)(C)	3HM35(F)(C)
Pump speed	1800 rpm		1700 rpm
Plunger diameter x stroke	⌀6 x 7mm (0.2362 x 0.2756in.)		⌀6.5 x 7mm (0.2559 x 0.2756in.)
Injection nozzle type	YDN-OSDYD1		YDN-OSDY1
Pressure of fuel injeciton	170kgf/cm² (2418 lb/in²)		160kgf/cm² (2276 lb/in²)
Amount of injection at rack mark position	22.5~23.5cc (1.37~1.43in.³)	21.5cc~22.5cc (1.31~1.37in.³)	27.5~28.5cc (1.68~1.74in.³)
Allowable error between cylinder	—	1cc (0.06in.³) or less	1cc (0.06in.³) or less
Stroke	1000		1000

NOTE: Mainting the pressure for feeding oil to the injection pump at 0.5 kgf/cm². (7.1 lb/in.³)

2-8.8 Adjustment of injection volume for each cylinder

(1) Fluctuation of injection volume

The injection volumes of each cylinder must be adjusted to within 3% of each other.

$$\text{Average injection volume} = \frac{\text{total volume of all cylinder injection}}{\text{number of cylinders}}$$

$$\text{Difference} = \frac{\text{Maximum injection volume} - \text{average injection volume}}{\text{Average injection volume}} \times 100$$

When the difference exceeds 3%, adjust the injection volume by sliding the control sleeve and pinion, when the difference exceeds 3%, the engine output will drop and/or one cylinder will overheat.

(2) Adjustment of injection volume

In order to adjust the fluctuation of injection volume for each cylinder, alter the position of the injection volume adjusting plate at the side of the fuel injection of pump body.

The injection volume adjusting plate is operated by the eccentric bolt which is integrated with the locking pin of the plunger barrel and changes the position of the plunger barrel. When the plunger barrel is turned, the relative position of the suction hole with respect to the lower lead of the plunger changes the injection volume.

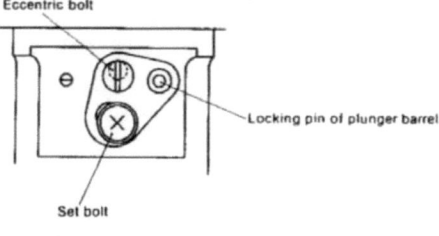

By loosening the set bolt and turning the eccentric bolt clockwise, the position of the pin moves to the left to increase the injection volume, and by turning the eccentric bolt counterclockwise, the pin moves to the right to decrease the injection volume.
After adjusting the injection volume, tighten the set bolt securely.

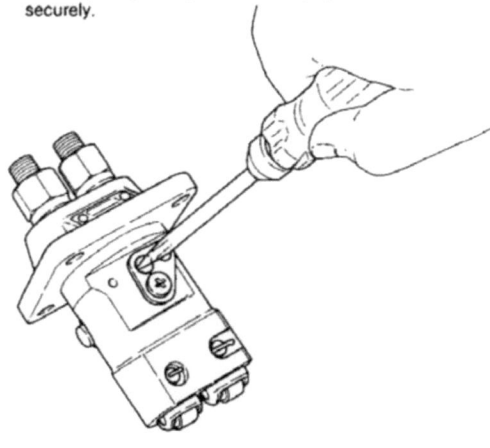

3. Injection Nozzle

3-1 Construction

The injection nozzle atomizes the fuel sent from the injection pump and injects it into the precombustion chamber in the prescribed injection pattern to obtain good combustion through optimum fuel/air mixing.

The main parts of the injection nozzle are the nozzle holder and nozzle body. Since both these parts are exposed to hot combustion gas, they must be extremely durable.

Moreover, since their operation is extremely sensitive to the pressure of the fuel, high precision is required. Both are made of quality alloy steel that has been specially heat treated and lapped, so they must always be handled as a pair.

Common parts are used for the fuel valve of models 1GM10(C), 2GM20(F)(C) and 3GM30(F)(C). The only difference between the GM model series and model 3HM35(F)(C) is the nozzle case nut.

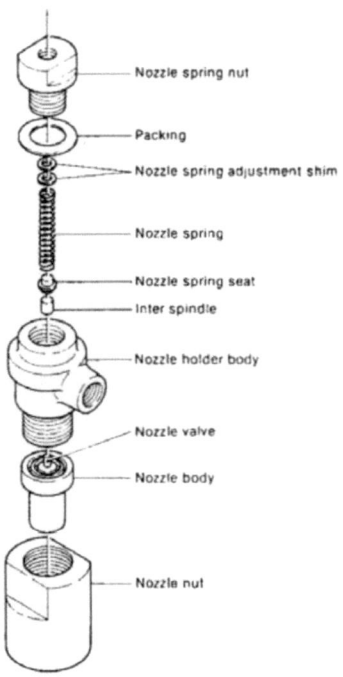
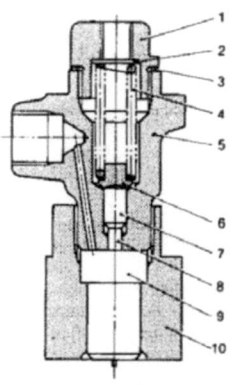

1. Nozzle spring nut
2. Nozzle spring adjustment shim
3. Packing
4. Nozzle spring
5. Nozzle holder body
6. Nozzle spring seat
7. Inter spindle
8. Nozzle valve
9. Nozzle body
10. Nozzle nut

3-2 Specifications for nozzle valve

Engine model			1GM10(C), 2GM20(F)(C), 3GM30(F)(C)	3HM35(F)(C)
Nozzle		Type of nozzle valve	YDN-OSDYD1 (Throttle)	
		Valve opening pressure	170±5 kgf/cm² (2347 ~ 2489 lb/in.²)	160±5 kgf/cm² (2205 ~ 2347 lb/in.²)
		Diameter of injection nozzle	ø1mm (0.0394in.)	
		Angle of injection	5° ~ 10°	
Nozzle spring		Free length	30.0mm (1.1811in.)	
		Mounted length	28.7mm (1.1299in.)	
		Mounted load	14.14 kg (31.17 lb)	
Nozzle spring adjusting plate (for adjusting nozzle opening pressure)			0.1mm (0.0039in.) 0.15mm (0.0059in.) 0.2mm (0.0079in.) 0.3mm (0.0118in.) 0.5mm (0.0197in.)	

3-3 Yanmar throttle nozzle

The semi-throttle nozzles used in this engine are designed and manufactured by Yanmar. A semi-throttle nozzle resembles a pintle nozzle, except that with the former the nozzle hole at the end of nozzle and nozzle body are longer and the end of the nozzle is tapered. This nozzle features a "throttling effect": relatively less fuel is injected into the precombustion chamber at the initial stage of injection, and the volume is increased as the nozzle rises. This type of throttle nozzle ideal for small, high-speed engines.

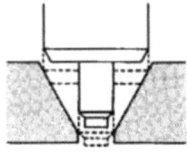
Pintle nozzle

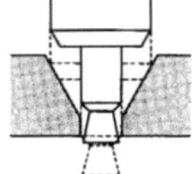
YANMAR semi-throttle nozzle

3-4 Nozzle operation

The nozzle is pushed down to its lowest position by the pressure-adjusting nozzle spring and contacts the valve seat of the nozzle body.

Under high pressure, fuel from the fuel pump passes through the hole drilled in the nozzle holder, enters the circular groove at the end of the nozzle body and then enters the pressure chamber at the bottom of the nozzle body.

When the force acting in the axial direction on the differential area of the nozzle on the pressure chamber overcomes the force of the spring, the nozzle is pushed up and the fuel is injected into the precombustion chamber through the throttle hole.

The nozzle is closed again when the pressure in the nozzle body's pressure chamber drops below the force of the spring.

This cycle is repeated at each opening and closing of the injection pump delivery valve.

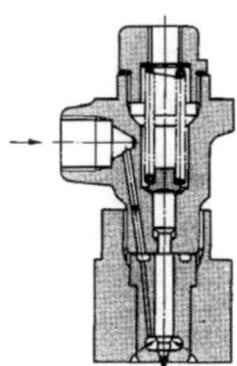

3-5 Disassembly and reassembly

3-5.1 Disassembly sequence

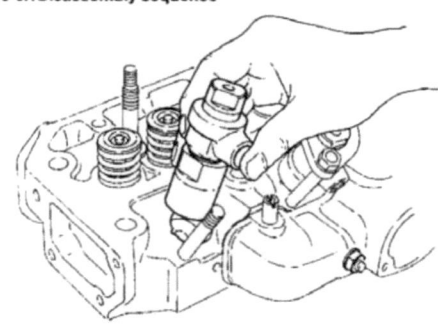

(1) Remove the carbon from the nozzle end.
(2) Loosen the nozzle spring holder.
(3) Remove the nozzle holder body from the nozzle mounting nut.

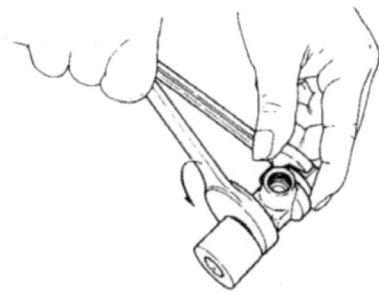

(4) Remove the nozzle body and nozzle ass'y from the nozzle mounting nut.
(5) Remove the nozzle spring retainer from the nozzle holder body, and remove the nozzle spring retainer, inter-spindle etc.

Reassemble in the reverse order of disassembly, paying special attention to the following items.

3-5.2 Disassembly and reassembly precautions

(1) The disassembled parts must be washed in fuel oil, and carbon must be completely removed from the end of the nozzle body, the nozzle body and the nozzle mounting nut fitting section.
If reassembled while any carbon remains, the nozzle will not tighten evenly, causing faulty injection.

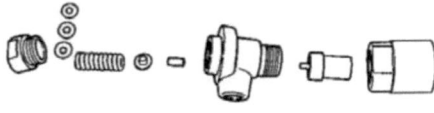

Chapter 3 Fuel System
3. Injection Nozzle

SM/GM(F)(C)-HM(F)(C)

(2) Parts for No.1 cylinder and No.2 cylinder must be kept separate. The nozzle body and nozzle must always be handled as a pair.
(3) Precautions when using a new nozzle.
First immerse the new nozzle in rust-preventive oil, and then seal it on the outside with seal peel. After removing the seal peel, immerse the nozzle in diesel oil and remove the rust-preventive oil from both inside and outside the nozzle.
Stand the nozzle holder upright, lift the nozzle about 1/3 of its length: it should drop smoothly by it own weight when released.

(4) The nozzle must be fitted on the nozzle holder with the nozzle spring retainer loosened.
If the nozzle is installed with the nozzle spring tightened, the nozzle mounting nut will be tightened unevenly and oil will leak from between the end of the nozzle holder body and the end of the nozzle mounting nut, causing faulty injection.

Nozzle tightening torque		kgf-m(ft-lb)
	Nozzle nut	10 (72.36)
	Nozzle spring nut	7.0~8.0 (50.65~57.89)

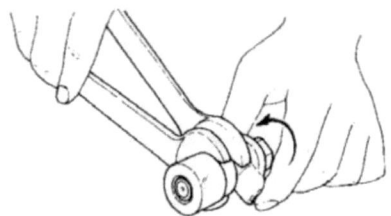

(5) When installing the injection nozzle on the cylinder head, tighten the nozzle holder nuts alternately, being careful to tighten them evenly.

	kgf-m(ft-lb)
Tightening torque	2 (14.5)

The nozzle holder must be installed with the notch side on the nozzle side.

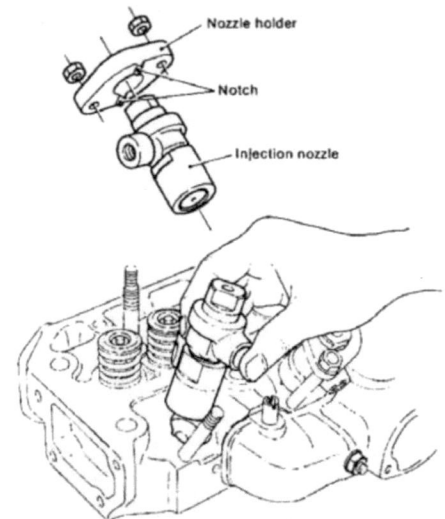

3-6 Injection nozzle inspection and adjustment
3-6.1 Carbon and corrosion on the nozzle body
Inspect the end and sides of the nozzle body for carbon build-up and corrosion. If there is considerable carbon build-up, check the properties of the fuel used, etc.
Replace the body if heavily corroded.

3-6.2 Checking nozzle action
Wash the nozzle in clean fuel oil and hold the nozzle body upright, then lift the nozzle about 1/3 of its length with one hand. The nozzle is in good condition if it drops smoothly by its own weight when released. If the nozzle slides stiffly, repair or replace it.

Chapter 3 Fuel System
3. Injection Nozzle

SM/GM(F)(C)·HM(F)(C)

3-6.3 Adjusting the nozzle injection pressure

Fit the injection nozzle to the high pressure pipe of a nozzle tester and slowly operate the lever of the tester. Read the pressure when instant injection from the nozzle begins.
If the injection pressure is lower than the prescribed pressure, remove the nozzle spring holder and adjust the pressure by adding nozzle spring shims.
The injection pressure increases about 10 kgf/cm² (142.2 lb/in.²) when a 0.1mm (0.004in.) shim is added.

Stream

- Injection pressure low
- Nozzle seized
- Nozzle spring broken
- Dirt on valve seat

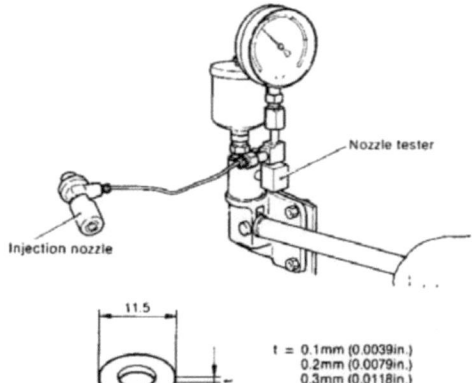

Spike

- Injection port damaged or dirty
- Carbon build-up
- Nozzle end abnormally worn

Spray

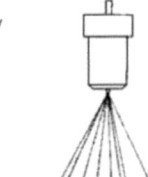
- Injection port worn
- Carbon build-up

	1GM10(C), 2GM20(F) 3GM30(F)(C)	3HM35(F)(C)
Injection pressure	170±5 kgf/cm² (2347 ~ 2489 lb/in.²)	160±5 kgf/cm² (2205 ~ 2347 lb/in.²)

3-6.4 Nozzle seat oil tightness check

After injecting the fuel several times by operating the lever of the nozzle tester, wipe the oil off the injection port. Then raise the pressure to 20 kgf/cm² (284.5 lb/in.²) 140kgf/cm² (1991 lb/in.²) lower than the prescribed injection pressure. The nozzle is faulty if oil drips from the nozzle. In this case, clean, repair or replace the nozzle.

Slanted

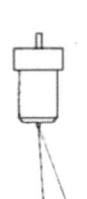
- Uneven seat contact
- Injection port damaged or worn
- Carbon build-up

3-6.5 Checking the spray condition

Adjust the nozzle injection pressure to the prescribed value and check the condition of the spray while operating the tester at 4—6 times/sec. Judge the condition of the spray by referring to the below figure.

3-6.6 Inspecting the nozzle spring

Inspect the nozzle spring for fractured coils, corrosion, and permanent strain, and replace the spring when faulty.

Normal

3-6.7 Inspecting the nozzle spring retainer and inter-spindle

Inspect the nozzle spring retainer and inter-spindle for wear and peeling of the contact face, and repair or replace the spring if faulty.

Chapter 3 Fuel System
4. Fuel Filter

4. Fuel Filter

4-1 Construction
The fuel filter is installed between the feed pump and injection pump, and serves to remove dirt and impurities from the oil fed from the fuel tank through the feed pump.

The fuel filter incorporates a replaceable filter paper element. Fuel from the fuel tank enters the outside of the element and passes through the element under its own pressure. As it passes through, the dirt and impurities in the fuel are filtered out, allowing only clean fuel to enter the interior of the element. The fuel exits from the outlet at the top center of the filter and is sent to the injection pump.

A cross-headed hexagonal bolt is fitted to the fuel filter body. Loosen the bolt with a cross-headed screw driver before starting or after dismantling and reassembly to bleed the air in the fuel system to the fuel oil filter.

4.2 Specification
(Common to all models)

Filtering Area	333cm² (20.3in.²)
Material of element	Cotton fiber
Filter mesh	10 ~ 15μ

4-3 Inspection
The fuel filter must be periodically inspected. If there is water and sediment in the filter, remove all dirt, rust, etc. by washing the filter with clean fuel.

The normal replacement interval for the element is 250 hours, but the element should be replaced whenever it is dirty or damaged, even if the 250 hour replacement period has not elapsed.

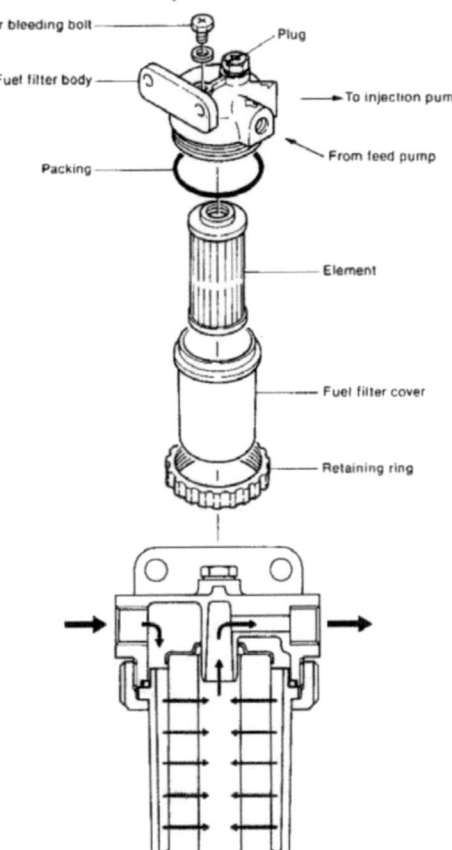
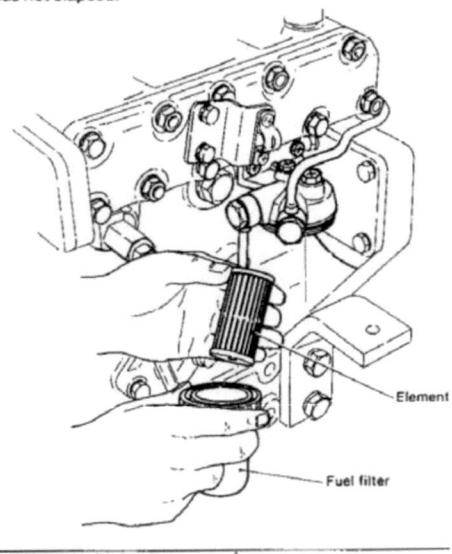

Filter cleaning	First time 50 hours
Filter element replacement	Every 250 hours

Chapter 3 Fuel System
5. Fuel Feed Pump

SM/GM(F)(C)-HM(F)(C)

5. Fuel Feed Pump

5-1 Construction

The fuel pump feeds the fuel from the fuel tank to the injection pump through the fuel filter. When the fuel tank is installed at a higher position than the fuel filter and injection pump, the fuel will be fed by its head pressure, but if the fuel tank is lower than the filter and injection pump, a fuel pump is required.

The fuel pump of this engine is a diaphragm type and is installed on the exhaust side of the cylinder body. The diaphragm is operated by the movement of a lever by the fuel feed pump cam at the cam shaft.

Specifications

	1GM10(C)	2GM20(F)(C), 3GM30(F)(C), 3HM35(F)(C)
Part No.	105582-52010	129301-52020
Suction head	Max.0.8m (3.15in.)	
Capacity	0.3 ℓ/min. at 1000rpm	
Feed Pressure	0.1kgf/cm² (1422 lb/in.²) at 600~1800rpm	
Suction pressure	−60 mmHG at 600rpm	

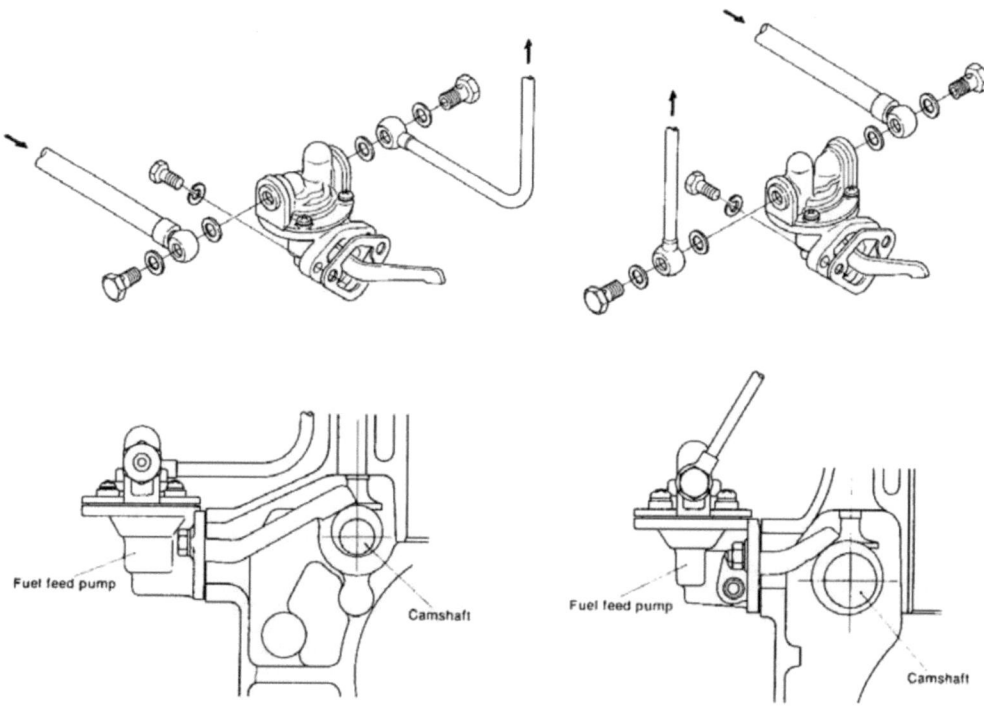

1GM10(C)

2GM20(F)(C), 3GM30(F)(C), 3HM35(F)(C)

Chapter 3 Fuel System
5. Fuel Feed Pump

SM/GM(F)(C)·HM(F)(C)

5-2 Disassembly and reassembly

5-2.1 Disassembly

Clean the outside of the pump, inscribe a matching mark on the upper body and lower body of the pump, disassemble and put the components in order.

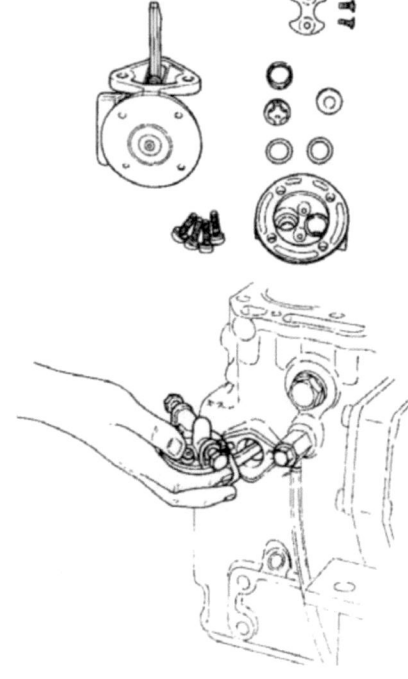

Component parts of fuel feed pump

5-2.2 Reassembly

Assemble the pump by reversing the disassembling procedures. Pay close attention to the following:
(1) Clean the components, blow compressed air against them, and inspect. Replace any defective components.
(2) Replace the packings, etc. with new ones.
(3) When mounting the valves, be careful not to mix up the inlet and outlet valves. Also, don't forget the valve packing.

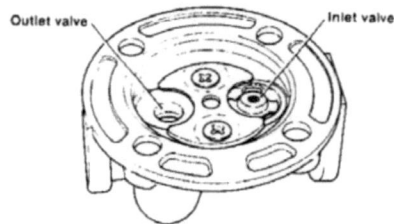

Outlet valve   Inlet valve

(4) Make sure the diaphragm mounting hole is in the correct position and gently attach the diaphragm to the pump body.
(5) Line up the matching marks on the pump body, and clamp on the pump body evenly.

Tightening torque of screw	30±10 kgf·cm (1.45 ~ 2.89 ft-lb)

5-3 Inspecting and adjusting the fuel feed pump

5-3.1 Checking the pump for fuel oil leaks

After removal, immerse the pump in kerosene, stop its outlet port with a finger and, by operating the rocker arm, check for bubbles.
If any bubbles are present, this indicates a defective point which should be replaced.

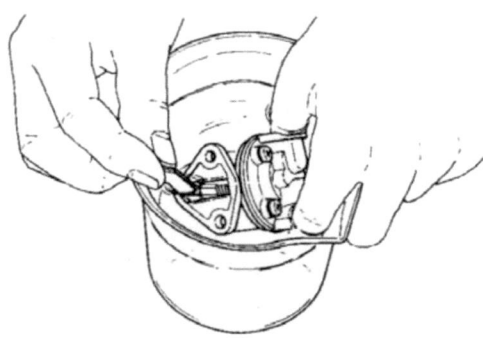

5-3.2 Checking the pump for engine oil leaks

Check pump mounting bolts for looseness and the pump packing for breaks. Retighten any loose bolts and replace defective packing.

5-3.3 Measuring the sucking power

Attach a piece of vinyl hose to the inlet port, keep the pump at a specified height (head) above the fuel oil level, and operate the rocker arm by hand. If the fuel oil spurts out from the outlet port, the pump is all right. A simpler method of testing pump power is as follows: cover the inlet port with a finger and, by operating the rocker arm by hand, estimate the pump's sucking power by judging the suction on the finger. Although this is not an exact method, it can at least confirm that the diaphragm, valves, etc. are operating.

Chapter 3 Fuel System
5. Fuel Feed Pump

SM/GM(F)(C)-HM(F)(C)

5-3.5 The contact area and mounting condition of valve

Test the valve seat as follows: Remove the valve and blow into the valve seat from the direction in which the valve spring is mounted. If air leaks, replace the seat with a new one. If fuel oil leaks as a result of dust, foreign objects, etc. caught in the valve seat, rinse it and clean it by blowing it with air.

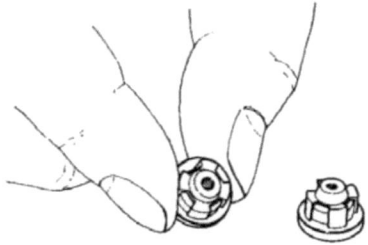

5-3.6 Diaphragm spring and rocker arm spring

Check the diaphragm spring and rocker arm spring for permanent deformation, and the rocker arm and rocker pin for wear. If any of these components are defective, replace them with new ones.

NOTE: When it becomes necessary to replace any of these parts, the entire fuel feed pump assembly should be replaced.

5-3.4 Aging, breakdown and cracking of the diaphragm

Since the diaphragm is constantly in motion, the cloth on its flexible parts becomes thin, cracked, and sometimes breaks down after long periods of use. A broken diaphragm causes fuel oil leakage and fragments of the diaphragm often contaminate the engine oil, seriously hampering fuel oil discharge or blocking it altogther.

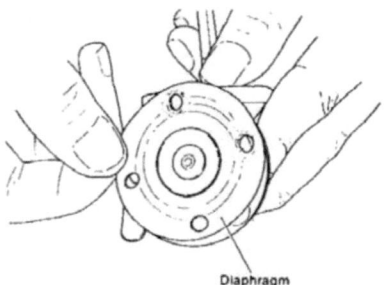

Diaphragm

6. Fuel Tank (Option)

The fuel tank is optionally available. Its capacity is 30 litres for all engine models and is triangular to fit compactly into the engine room. As an accessory, a rubber hose of 2m length is attached to feed fuel oil from the fuel tank to the fuel pump. A connection to return fuel oil is provided at the top of the fuel tank, and by connecting a rubber hose from the fuel valve, the overflow oil can be returned to the tank.

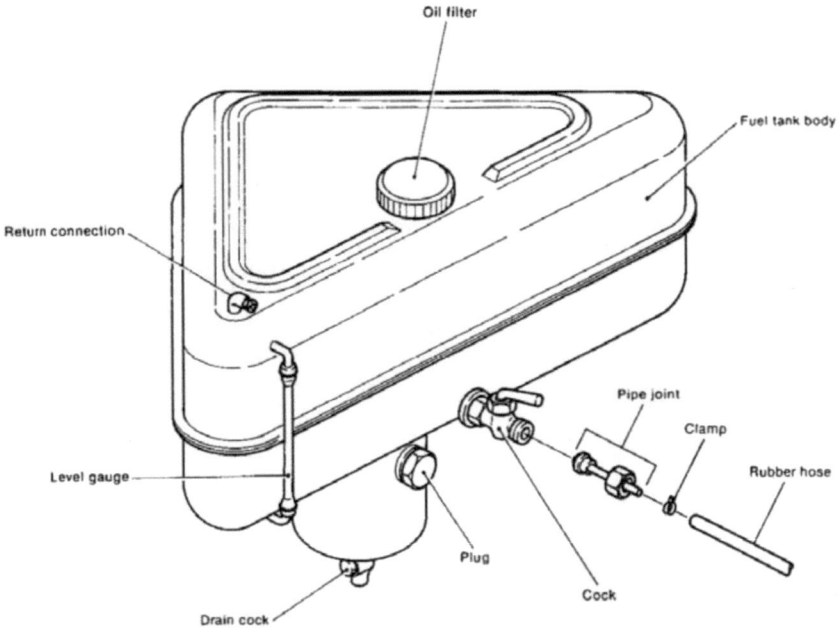

Material	Steel plate
Capacity	30 ℓ
Thread of outlet cock	PF 1/2
Size of rubber hose	ø7/ø13 × 2000mm (0.2756/0.5118 × 78.74in.)

Chapter 3 Fuel System
6. Fuel Tank (Option)

SM/GM(F)(C)·HM(F)(C)

Dimension

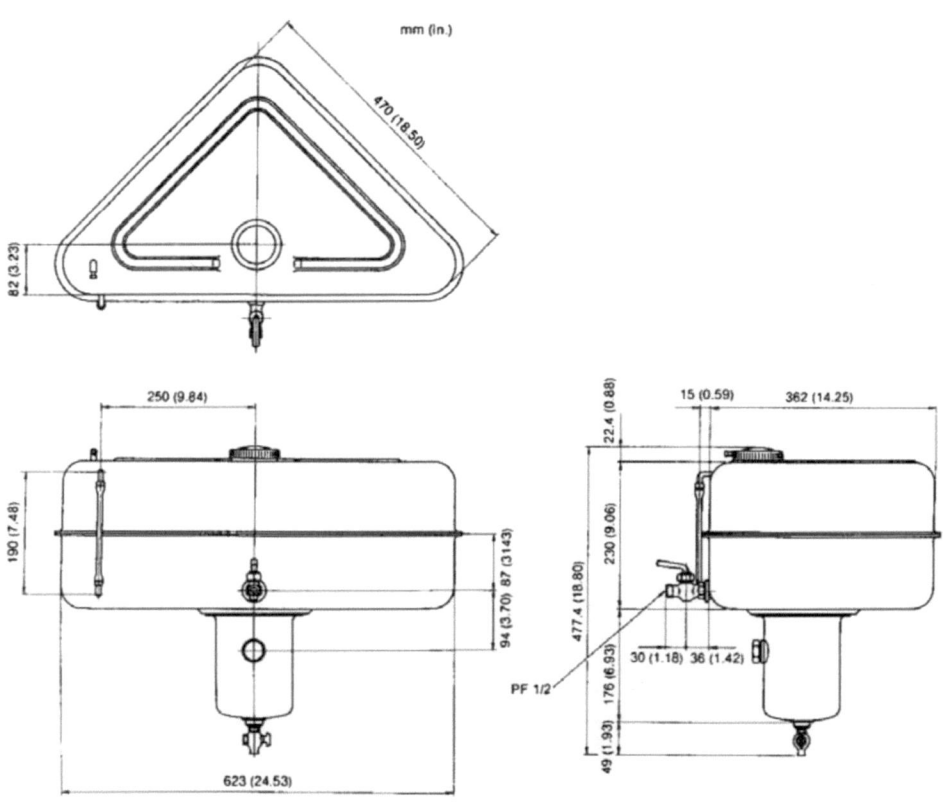

CHAPTER 4
GOVERNOR

1. Governor 4-1
2. Injection Limiter 4-9
3. No-Load Maximum Speed Limiter 4-11
4. Idling Adjuster 4-12
5. Engine Stop Lever 4-13

1. Governor

The governor serves to keep engine speed constant by automatically adjusting the amount of fuel supplied to the engine according to changes in the load. This protects the engine against sudden changes in the load, such as sudden disengagement of the clutch, the propeller leaving the water in rough weather, or other cases where the engine is suddenly accelerated.

This engine employs an all-speed governor in which the centrifugal force of the governor weight, produced by rotation of the crankshaft, and the load of the regulator spring are balanced.
The governor is remotely controlled by a wire. Refer to the "Control System" chapter for details.

1-1 Construction

(1) 1GM10(C)

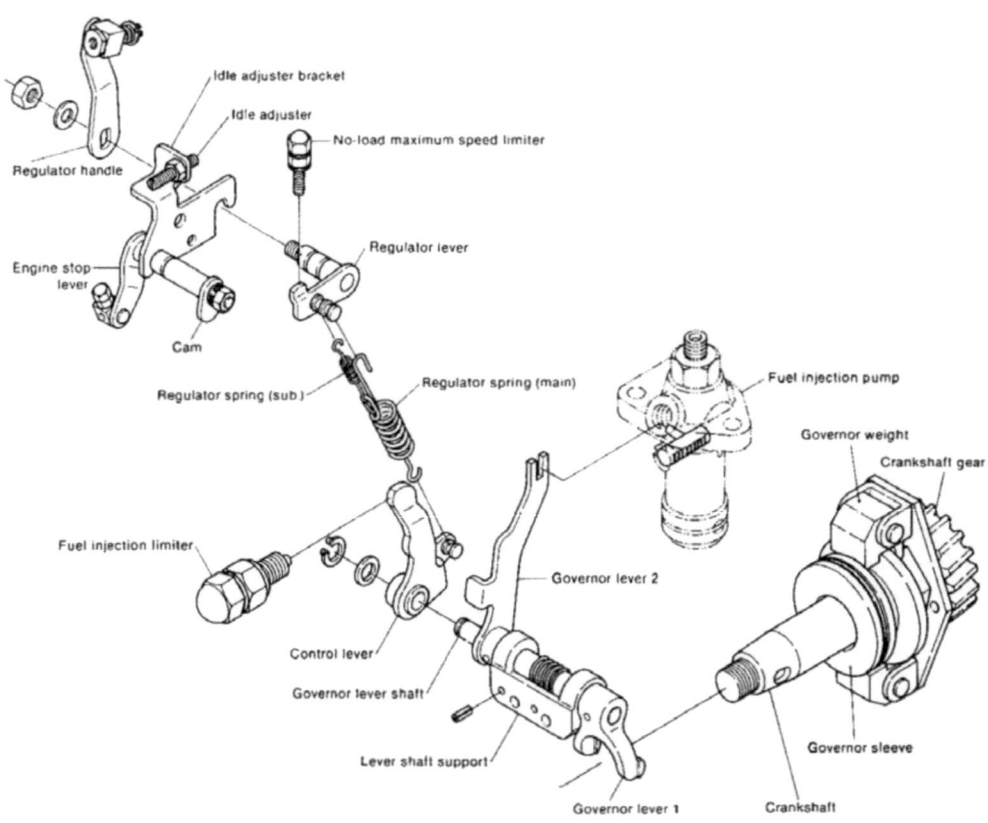

Chapter 4 Governor
1. Governor

(2) 2GM20(F)(C), 3GM30(F)(C), 3HM35(F)(C)

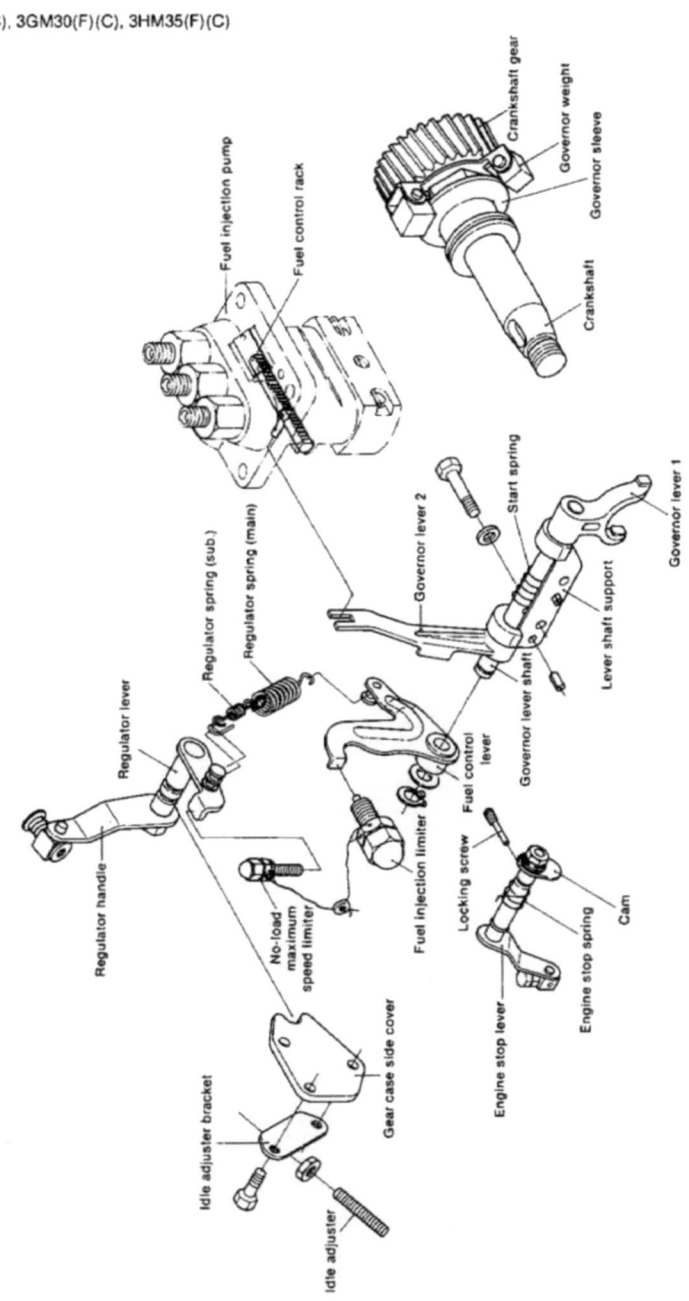

Chapter 4 Governor
1. Governor

SM/GM(F)(C)·HM(F)(C)

1-1.1 1GM10(C)

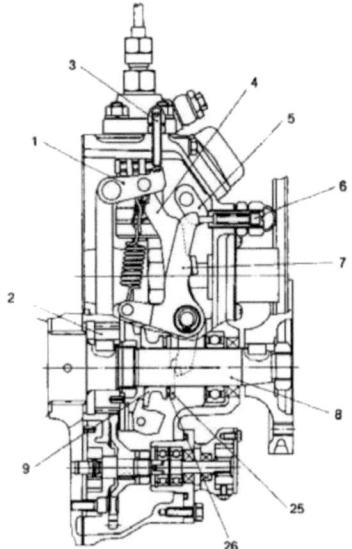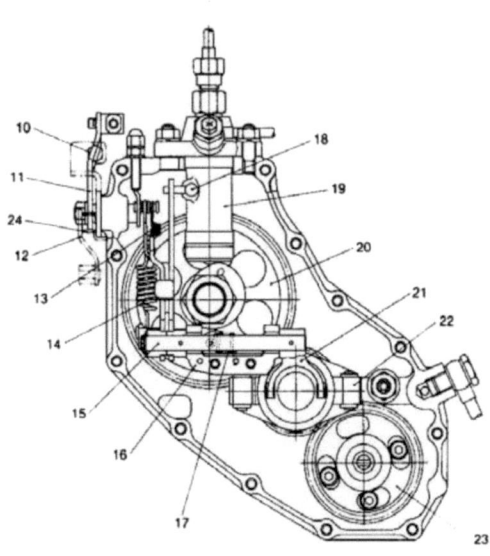

1 Regulator lever
2 Crankshaft gear
3 No-load maximum speed limiter
4 Governor lever 2
5 Engine stop cam
6 Fuel injection limiter
7 Fuel control lever
8 Crankshaft
9 Governor sleeve
10 Idle adjuster

11 Regulator handle
12 Engine stop lever
13 Regulator spring (sub.)
14 Regulator spring (main)
15 Governor lever shaft
16 Governor lever shaft support
17 Start spring
18 Fuel control rack
19 Fuel injection pump
20 Camshaft gear

21 Governor lever 1
22 Governor weight
23 Lubricating oil driving gear
24 Engine stop spring
25 Thrust collar
26 Thrust needle bearing

4-3

Chapter 4 Governor
1. Governor

SM/GM(F)(C)·HM(F)(C)

1-1.2 2GM20(F)(C), 3GM30(F)(C), 3HM35(F)(C)

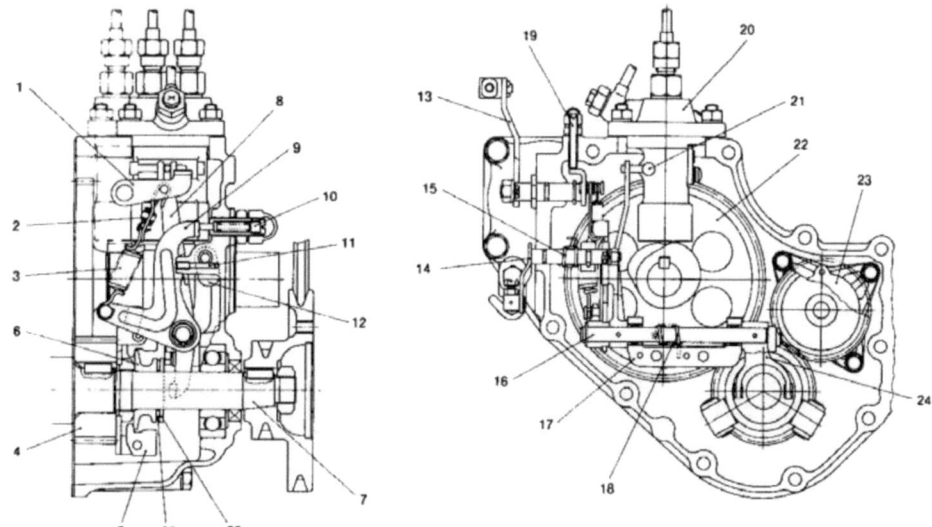

1 Regulator lever
2 Regulator spring (sub.)
3 Regulator spring (main)
4 Crankshaft gear
5 Governor weight
6 Governor sleeve
7 Crankshaft
8 Governor lever 2
9 Fuel control lever
10 Fuel injection limiter

11 Locking screw
12 Engine stop cam
13 Regulator handle
14 Engine stop lever
15 Engine stop spring
16 Governor lever shaft
17 Governor lever shaft support
18 Start spring
19 No-load maximum speed limiter
20 Fuel injection pump

21 Fuel control rack
22 Camshaft gear
23 Lubricating oil pump
24 Governor lever 1
25 Thrust needle bearing
26 Thrust collar

Chapter 4 Governor
1. Governor

1-2 Operation

The position of the two governor weights (open and closed) is regulated by the speed of the engine. The centrifugal force of the governor weights pivots around the governor weight pin and is converted into an axial force that acts on the sleeve. This force is transmitted to governor lever 2 through governor lever 1, and lever 1 shifts the fuel control rack to increase or decrease the fuel supply. The governor lever is stabilized at the point at which the force produced by the governor weight is balanced with the load of the regulator spring connecting the regulator lever and fuel control lever. When the speed is reduced by application of a load, the force of the regulator spring pushes the governor sleeve in the "fuel increase" direction, stabilizing the engine speed by changing the position of the regulator lever.

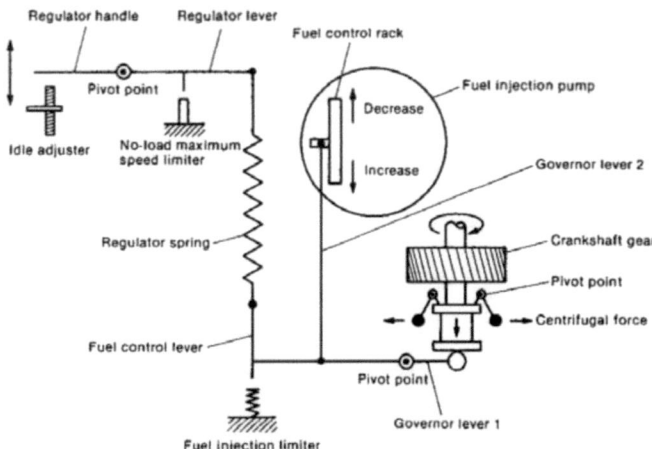

1-3 Performance

		1GM10(C), 2GM20(F)(C), 3GM30(F)(C)	3HM35(F)(C)
No-load maximum speed		$3825{}_{0}^{-50}$ rpm	3625±25 rpm
No-load minimum speed		850±25 rpm	
Instant speed regulation	δi	15% or less	
Stabilization time	ts	10 sec. or less	
Stabilized speed regulation	δs	6.5% or less	
Fluctuation of rotation		30 rpm or less	

Instant speed regulation $\quad \delta i = \left| \dfrac{ni - nr}{nr} \right| \times 100$

Stabilized speed regulation $\quad \delta s = \left| \dfrac{ns - nr}{nr} \right| \times 100$

ni: Instant maximum (minimum) speed:
  The maximum or minimum engine speed which is momentarily reached immediately after the load has been suddenly changed from the rated load to another load or from an arbitrary load to the rated load.
ns: Stabilized speed:
  The speed which is set according to the lapse of time after the load has been changed from a rated load to another load or from an arbitrary load to the rated load.

nr: Rated speed
ts: Stabilization time:
  The time it takes for engine to return to the set speed after a change.

(When load is suddenly changed from rated load to low load)

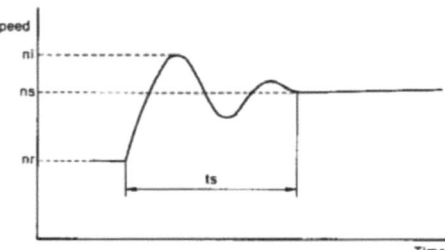

ni: Instant maximum speed (rpm)
ns: Stabilized speed (rpm)
nr: Rated speed (rpm)
ts: Stabilization time (sec.)

Chapter 4 Governor
1. Governor

1-4 Disassembly
1-4.1 Disassembly

(1) Remove the injection limiter and no-load maximum speed limiter from the gear case.
(2) Remove the idle adjuster and adjuster bracket.
(3) Remove the cover at the gear case end [oil supply port in the case of model 1GM10(C)] move the governor lever 2 to match the control rack to the pulled-out position of the fuel injection pump (indicated by a slot in the gear case to show the position); then take out the fuel injection pump.
(4) Remove the gear case from the cylinder block.
(5) Pull the thrust collar, the thrust needle bearing and the governor sleeve from the crankshaft.
(6) Loosen the end nut of crankshaft, and remove the governor weight assembly.
(7) Remove the regulator spring (main-sub.) from the regulator lever 2 and fuel control lever.
(8) Remove the circlip of the regulator lever, and remove the regulator lever and handle. [Without circlip in the case of model 1GM10(C)].
(9) Remove the governor lever shaft support bolt from the rear of the gear case, and take out the governor lever shaft assembly.
(10) Loosen the nut of engine stop lever, and pull the cam.
(11) Draw out the locking screw from the rear of the gear case, and remove the taper pin for setting the return spring.
(12) Remove the engine stop lever and the spring.

1-4.2 Reassembly and precautions

Reassemble in the reverse order of disassembly, paying special attention to the folowing items.
(1) Check the governor weight movement.
(2) Check for the movement of the governor sleeve sliding on the crankshaft.
(3) Since a common taper pin hole is drilled in the governor lever shaft and governor levers 1 and 2, they must be replaced as an ass'y.
(4) Since the movement and play of the governor lever have a direct effect on the governor's performance, they must be carefully checked.

1-5 Parts inspection and replacement
1-5.1 Regulator spring

(1) Inspect the spring for coil damage, corrosion and hook deformation, and replace if faulty.
(2) Measure the spring's dimensions and spring constant. Since the spring constant determines the governor's performance, it must be carefully checked.

Spring specifications
 1) Regulator spring (main)

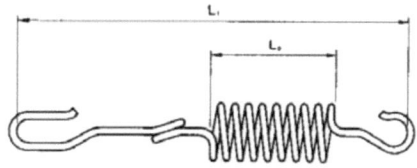

		1GM10(C)	2GM20(F)(C), 3GM30(F)(C), 3HM35(F)(C)
Wire diameter		φ1.8mm (0.0709in.)	φ2.3mm (0.0906in.)
Coil outside diameter		φ13.8mm (0.5433in.)	φ18.3mm (0.7205in.)
Nmber of coils		8.5	7.5
Spring constant		0.715kgf/mm (0.400 lb/in.)	0.922kgf/mm (0.516 lb/in.)
Free length	L₀	18mm (0.7087in.)	20mm (0.7874in.)
	M₁	76mm (2.992in.)	78mm (3.0709in.)

2) Regulator spring (sub)

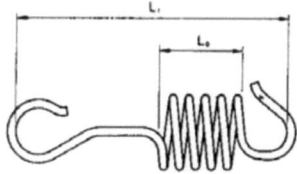

		1GM10(C)	2GM20(F)(C), 3GM30(F)(C), 3HM35(F)(C)
Wire diameter		φ1.8mm (0.0315in.)	φ1.2mm (0.0472in.)
Coil outside diameter		φ6.8mm (0.2677in.)	φ9.2mm (0.3622in.)
Nmber of coils		4	7
Spring constant		0.474kgf/mm (0.265 lb/in.)	0.578kgf/mm (0.3237 lb/in.)
Free length	L₀	5mm (0.1969in.)	10mm (0.3937in.)
	M₁	26mm (1.0236in.)	23mm (0.9055in.)

1-5.2 Sleeve

(1) Slide the sleeve on the crankshaft to check that it slides smoothly.
(2) Measure the clearance between the crankshaft and the inside of the sleeve, check the contact between the governor weight.

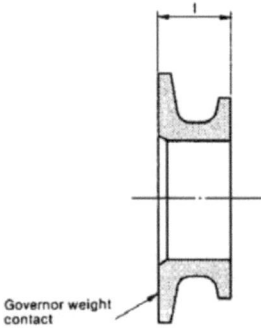

mm (in.)

	Maintenance standard	Clearance when assembled	Maximum allowable clearance	Wear limit
Crankshaft outside diameter	$\varnothing 25^{-0.007}_{-0.028}$ (0.9831 ~ 0.9840)	0.06 ~ 0.111 (0.0024 ~ 0.0044)	0.2 (0.0079)	—
Governor sleeve inside diameter	$\varnothing 25^{+0.063}_{+0.053}$ (0.9863 ~ 0.9875)			—
Governor sleeve overall length (ℓ)	15 ±0.1 (0.5866 ~ 0.5945)	—	—	14.8 (0.5827)

1-5.3 Thrust collar

Check the contact between the governor lever 1 and replace the collar when wear exceeds the wear limit.

mm(in.)

	Maintenance standard	Wear limit
Thrust collar thickness	3 (0.1181)	0.1 (0.0394)

1-5.4 Thrust needle bearing

Replace the bearing when wear exceeds the specified limit.

1-5.5 Governor weight

(1) Check contact with the sleeve and for wear.

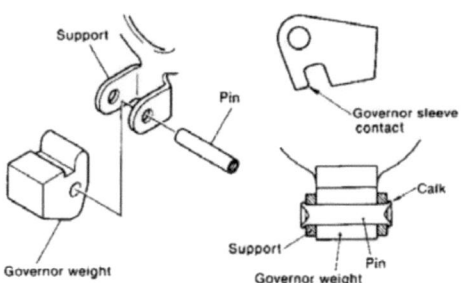

1-5.6 Governor lever shaft

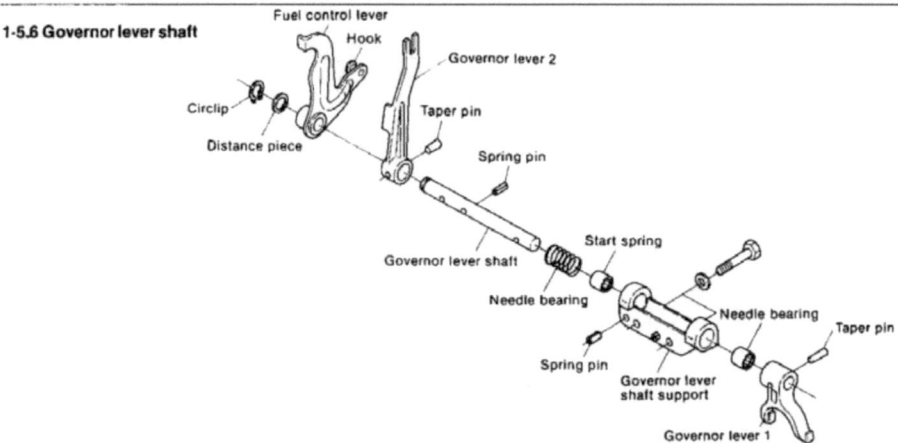

Chapter 4 Governor
1. Governor

(1) Replace the governor lever shaft if there is play between the shaft and needle bearing, play when the lever is moved, or if the shaft does not move smoothly.
(2) Repair or replace the shaft if there is play between lever 1, lever 2, fuel control lever or support and the shaft, or if the taper pin is loose.
(3) Inspect the contact between the governor lever 1 and the governor sleeve, replace it if it is too damaged.

1-5.7 Regulator lever and handle

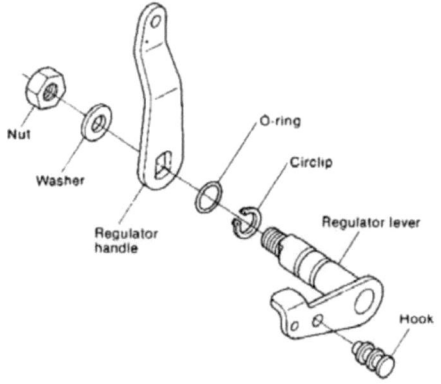

(1) Check for play in the regulator lever and regulator handle if faulty, replace them as a set.
(2) Check for O-ring damage. Replace if faulty.

2. Injection Limiter

2-1 Construction

Since surplus power is required from the standpoints of sudden overloads and durability, the engine is equipped with an injection control shaft that limits the amount of fuel injected into the precombustion chamber to a fixed amount. Since the injection control spring (torque spring) affects engine performance by adjusting engine torque, Yanmar has selected the best position for the operating conditions.

Pay close attention when handling the sealed-wire.
If the engine does not accelerate smoothly (i.e. the speed is not well controlled), turn the limiter slightly counterclockwise.

NOTE: If it is turned back too much, it will produce exhaust smoke.

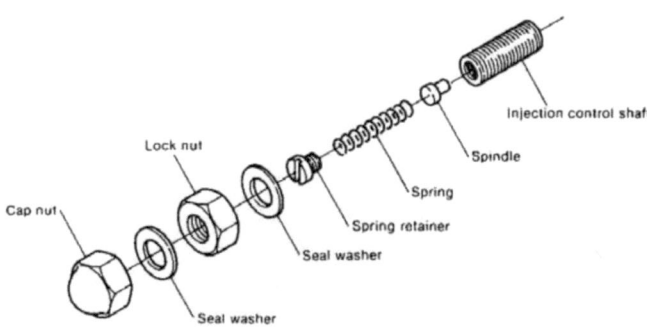

2-2 Inspection

(1) Hold the end of the spindle, and check it for smooth movement.
(2) Replace the spring if it is damaged, corroded or permanently strained.

2-3 Adjustment

In the case of models, 1GM10(C), 2GM20(F)(C), and 3GM30(F)(C)

(1) Set the governor lever to the free position and remove the injection pump adjustment cover [oil supply port in the case of model 1GM10(C)]
(2) Remove the injection control shaft cap nut, loosen the hexagonal lock nut, and loosen the injection control shaft (so that the spring inside the injection control shaft is disabled).
(3) Move governor lever 2 slowly to the left until the rack and injection control shaft touch lightly.
(4) Set the governor lever to the free position and push the rack by slowly turning the injection control shaft clockwise.
(5) Align the center mark of the rack with the reference face.
(6) Lock the injection control shaft with the hexagonal nut and cap nut.

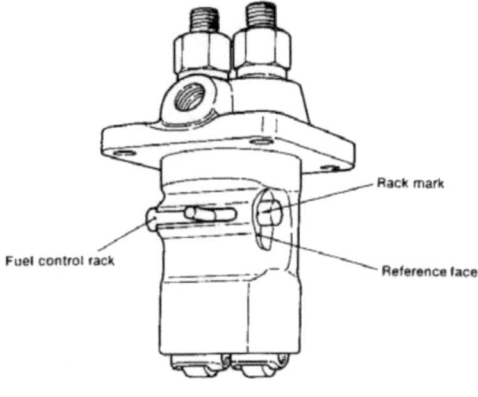

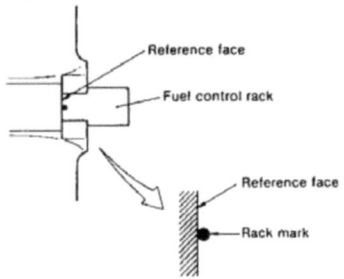

Chapter 4 Governor
2. Injection Limiter

SM/GM(F)(C)·HM(F)(C)

In the case of model 3HM35(F)(C)

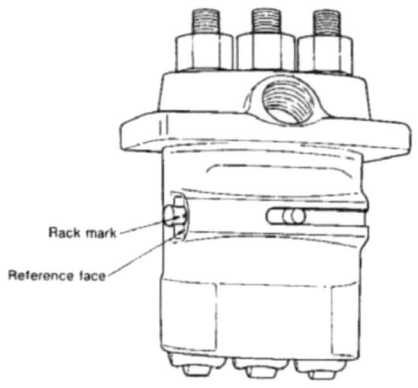
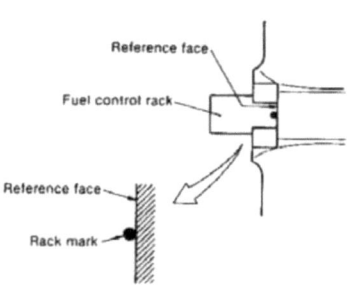

NOTE: When the engine is stopped, the control rack will automatically stay at the position which allows the maximum fuel injection volume.
Therefore, to match the rack mark, move the engine stop lever to the position where the mark is matched and fix the lever at that position, then adjust so that the fuel limiter comes into contact with the lever.

3. No-Load Maximum Speed Limiter

3-1 Construction
A stopper is installed on the regulator lever so that the engine speed at no-load does not exceed a fixed speed. The fuel control rack is stopped when the regulator lever contacts the stopper.

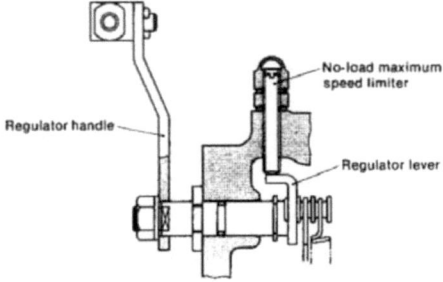

3-2 Handling precautions
The no-load maximum speed is adjusted during bench testing at the factory, and is locked with wire and sealed with lead. Care must be taken to keep the seal from being accidentally broken.

4. Idling Adjuster

When controlling the speed with the push-pull remote control, the idling adjustor operates so that the regulator handle does not move beyond the idling position in order to keep the engine running.

4-1 1GM10(C)

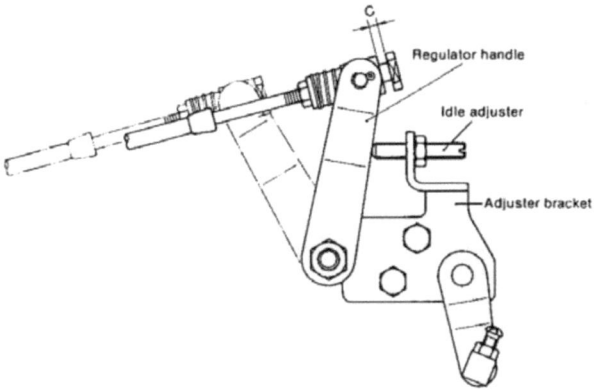

4-2 2GM20(F)(C), 3GM30(F)(C), and 3HM35(F)(C)

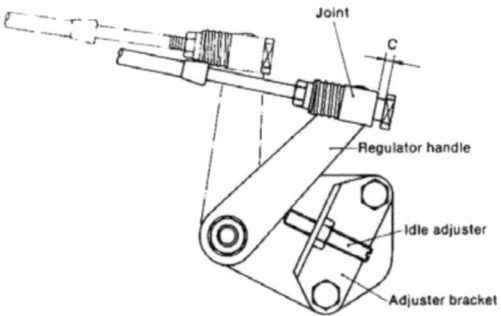

(1) When the control lever is in the neutral position, set the push-pull cable so that clearance C is 1 to 3mm (0.0397 ~ 0.1181in.).
(2) Take care not to fit the joint in the wrong direction.

5. Engine Stop Lever

5-1 Construction

With this device, governor lever 2 is moved by the cam of the engine stop lever shaft, regardless of the position of the regulator lever, so as to adjust the fuel control rack and reduce the supply of fuel.
This device can be remote-controlled.

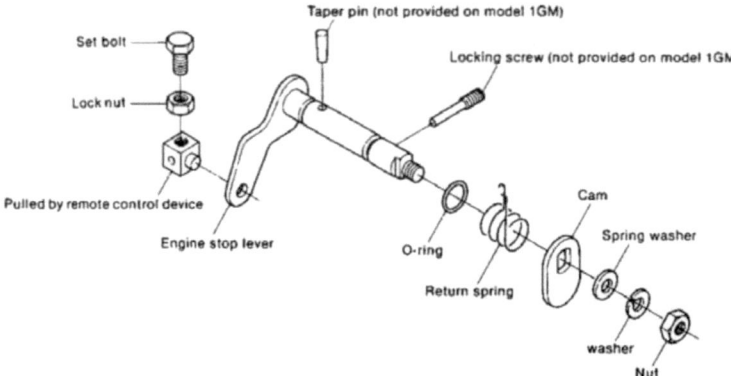

(1) 1GM10(C)

(2) 2GM20(F)(C), 3GM30(F)(C), 3HM35(F)(C)

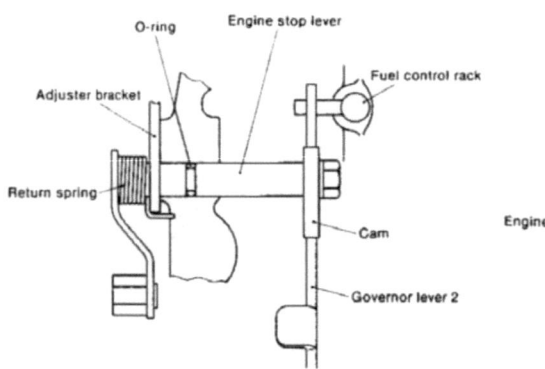

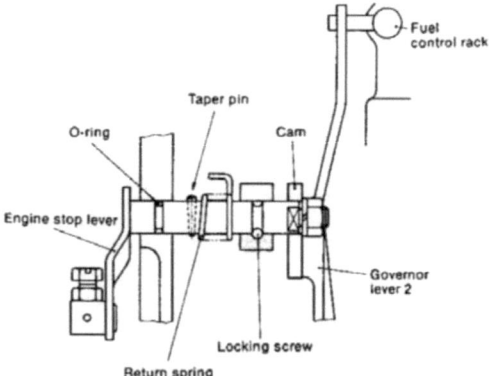

5-2 Inspection

(1) Check for play in the Cam or Taper pin and the engine stop lever. If faulty, replace them as a set.
(2) Check for O-ring damage. Replace if faulty.
(3) Inspect the spring for coil damage and corrosion and replace if faulty.
(4) Inspect the contact between the governor lever 2 and the dam. Replace the cam if it is too damaged.

CHAPTER 5
INTAKE AND EXHAUST SYSTEM

1. Intake and Exhaust System 5-1
2. Intake Silencer 5-3
3. Exhaust System 5-4
4. Breather 5-6

Chapter 5 Intake and Exhaust System
1. Intake and Exhaust System

SM/GM(F)(C)·HM(F)(C)

1. Intake and Exhaust System

The intake air silencer is installed at the intake side for the purpose of reducing noise and cleaning the air.
The exhaust system for model 1GM10(C) and 2GM20(F)(C) engines is so constructed that the mixing elbow is fitted directly to the cylinder head. The cooling water passs into this mixing elbow and is mixed with exhaust gas at the pipe outlet.

A water-cooled exhaust manifold is installed on engine models 3GM30(F)(C) and 3HM35(F)(C), and the mixing elbow is fitted to the outlet port of the exhaust manifold. The cooling water, after passing through the water jacket and cooling the exhaust gas, is mixed with the exhaust gas in the mixing elbow.

1-1 Intake and exhaust system of model 1GM10(C).

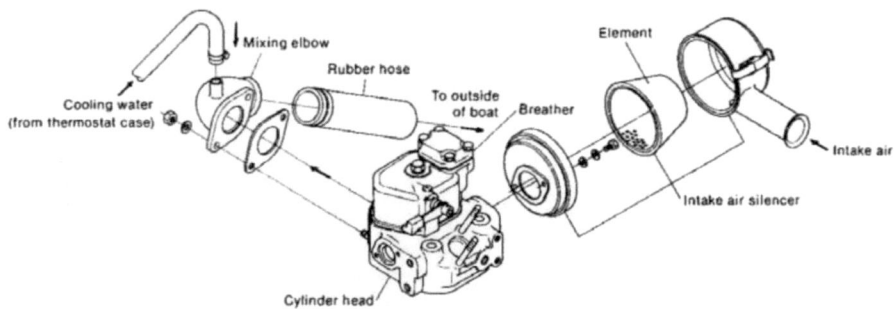

1-2 Intake and exhaust system of model 2GM20(F)(C)

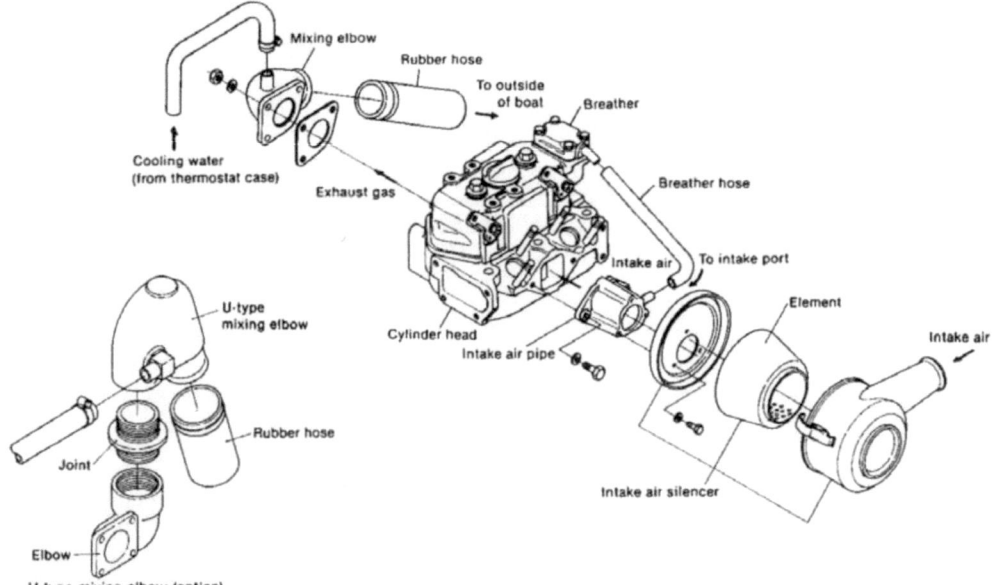

5-1

1-3 Intake and exhaust system of models 3GM30(F)(C) and 3HM35(F)(C)

The intake and exhaust system for models 3GM30(F)(C) and 3HM35(F)(C) is the same except for the construction of the breather.

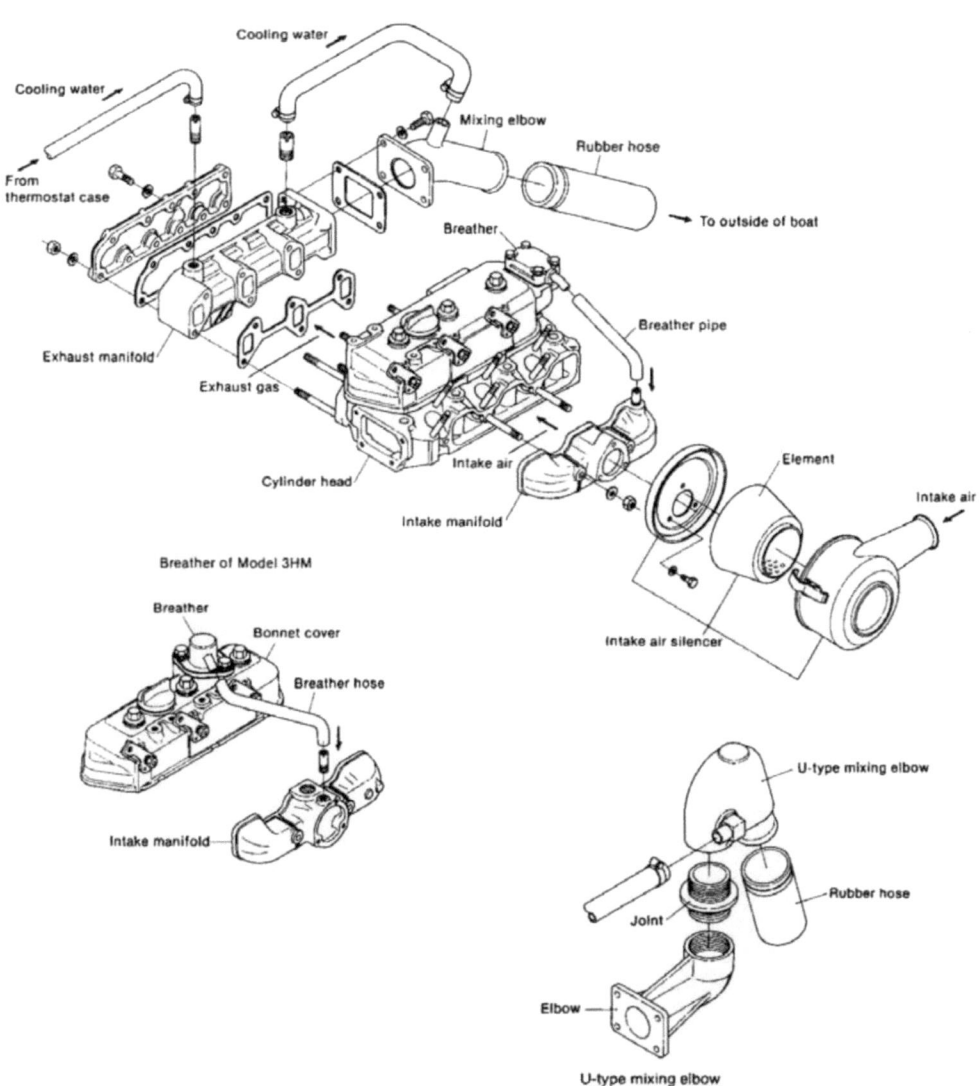

Chapter 5 Intake and Exhaust System
2. Intake Silencer

SM/GM(F)(C)-HM(F)(C)

2. Intake Silencer

2-1 Construction

A round polyurethane sound absorbing type intake silencer is employed to silence the intake air sucked into the cylinder head from the intake port.
Besides providing a silencing effect, the silencer also acts as an air cleaner.

	1GM10(C)	2GM20(F)(C) 3GM30(F)(C)	3HM35(F)(C)
Rated air volume (average)	150 ℓ/min	1560 ℓ/min	2800 ℓ/min
Draft resistance	150 mmAq	100 mmAq	150 mmAq

2-2 Inspection of the intake silencer

Occasionally, disassemble the intake silencer, remove the polyurethane element and inspect it. Because the element filters the air, if it is used over a long period of time it will become clogged and this decreases the amount of intake air, and may also be a cause of decreased output.

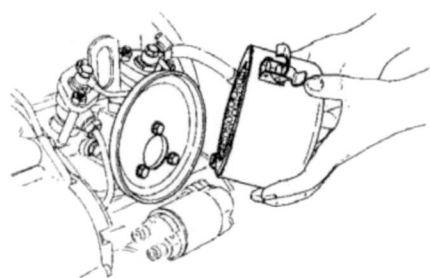

2-3 Washing the intake silencer element

Wash the air intake silencer element with a neutral detergent.

Washing period	Every 250 hours

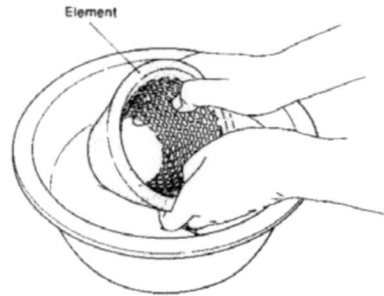

Element

3. Exhaust System

The mixing elbow of models 1GM10(C) and 2GM20(F)(C) is fitted directly to the outlet port of the cylinder head instead of being fitted to the exhaust manifold.
However, on models 3GM30(F)(C) and 3HM35(F)(C), an exhaust manifold is also installed and the mixing elbow is fitted to the manifold outlet port.

3-1 Exhaust manifold and mixing elbow

The high temperature, high pressure exhaust gas intermittently emitted from the cylinders at the speed of sound enters the exhaust manifold where it is muffled by expansion and water cooling. It is then mixed with the cooling water at the mixing elbow to lower its temperature and muffle it further, and is discharged.
A water-cooled exhaust manifold is employed for a high muffling effect.

3-1.1 For models 1GM10(C) and 2GM20(F)(C)

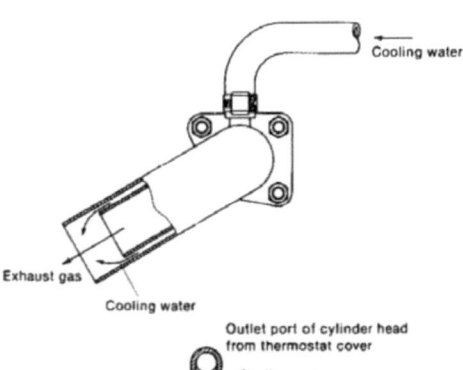

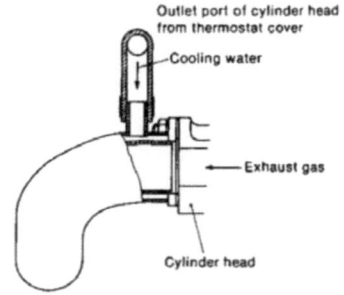

3-1.2 For models 3GM30(F)(C) and 3HM35(F)(C)

Both exhaust manifold and mixing elbow are installed.

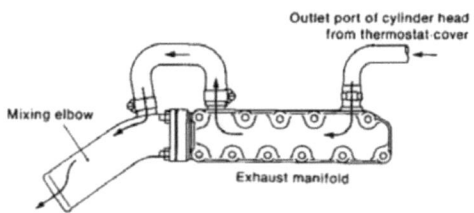

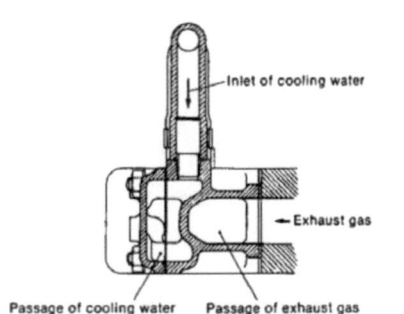

The construction of the exhaust manifold is shown in the figure, and a water chamber is formed between the exhaust manifold and the cover to cool the exhaust gas. The construction of the mixing elbow is the same for models 1GM10(C) and 2GM20(F)(C)

3-1.3 U type mixing elbow (optional)

For model 2GM20(F)(C)

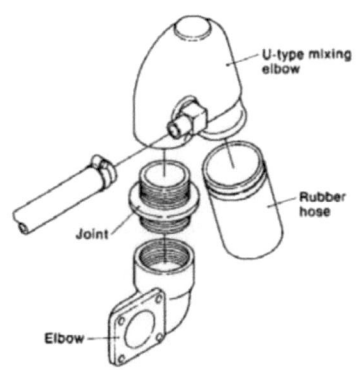

As shown in the figure, the construction for models 1GM10(C) and 2GM20(F)(C) is such that there is no exhaust manifold and the mixing elbow is fitted to the exhaust gas outlet port. A double construction technique has been adopted for the mixing elbow; as the exhaust gas passes through it the cooling water passes round the outside to cool the exhaust gas and then the gas and water mix close to the outlet port.

Chapter 5 Intake and Exhaust System
3. Exhaust System

For models 3GM30(F)(C) and 3HM35(F)(C)

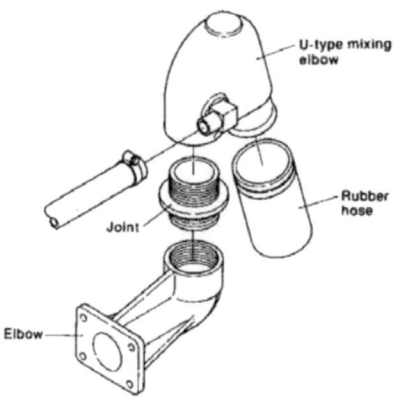

3-2 Exhaust manifold inspection

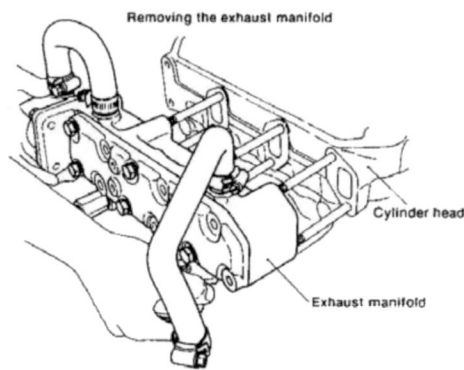

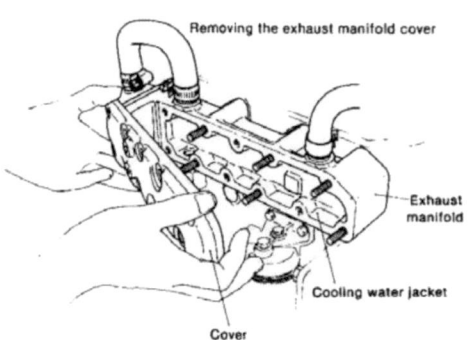

(1) Gasket packing
Inspect the gasket packing and replace if damaged.
(2) Carbon build-up in the exhaust passage
Remove the exhaust manifold elbow and cover and check carbon build-up in the exhaust passage. Remove any carbon in the passage. If carbon build-up becomes heavy, the exhaust pressure will rise, causing overheating of the cylinders and difficult starting.
(3) Corrosion and scale at the cooling water jacket
Inspect the water passage for the build-up of scale and foreign matter and remove if found. Also check for corrosion of the anticorrosion zinc installed on the cylinder head and the cylinder head water jacket and replace if corrosion is severe. Also, replace the cylinder head if it has been cracked by local overheating.
(4) Drain cock
Inspect the drain cock for clogging and check its action. Repair or replace if faulty.

3-3 Mixing elbow inspection

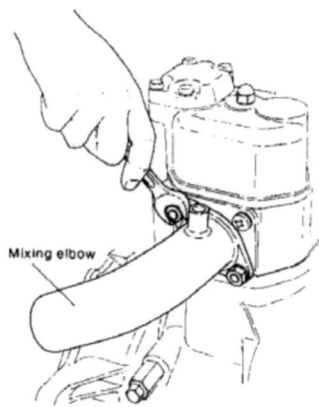

Check for carbon build-up and for corrosion inside the pipe, and repair or replace the pipe if faulty.
Also, inspect the mixing elbow mounting threads for cracking and corrosion.
This section is affected by exhaust gas and vibration.
NOTE: *The part where high temperature gas and cooling water are mixed is especially likely to corrode, so it must be inspected with special care.*

4. Breather

4-1 Construction of breather

The same construction is adopted for each model of engine in that the breather device is fitted to the bonnet cover, and the vapor in the crank case is sucked into the intake port or intake manifold through the tappet hole and breather. However, the construction of the breather itself differs from model to model.

NOTE: If trouble is experienced with the breather, take care that the engine does not jolt when running as the lubricating oil may enter from the inlet port and mix with the fuel oil.

4-1.1 Breather for model 1GM10(C)

The vapor which lifts up the leaf spring fitted at the top of the bonnet then enters the other air chamber, and is sucked through the intake port.

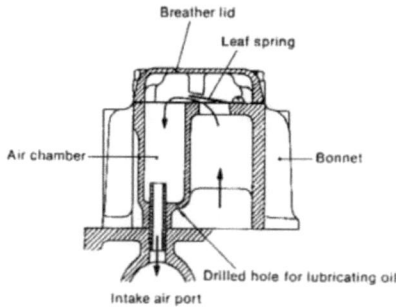

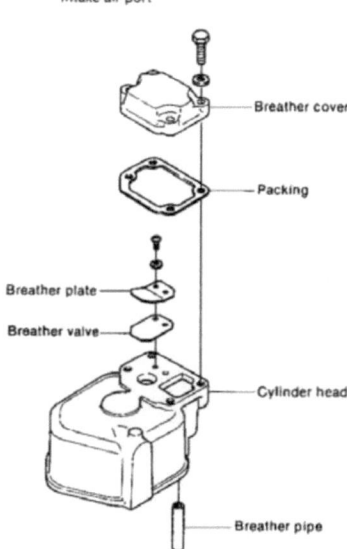

4-1.2 Breather for models 2GM20(F)(C) and 3GM30(F)(C)

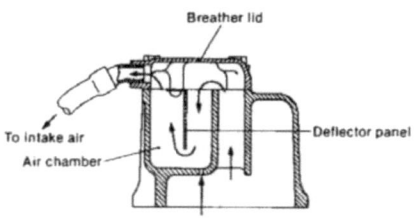

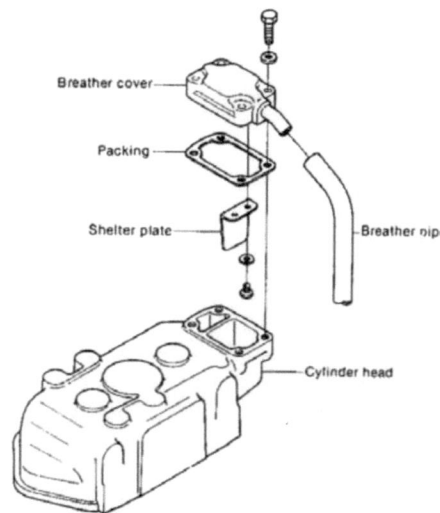

The deflector panel fitted to the breather lid enters the air chamber, and forces air circulation.

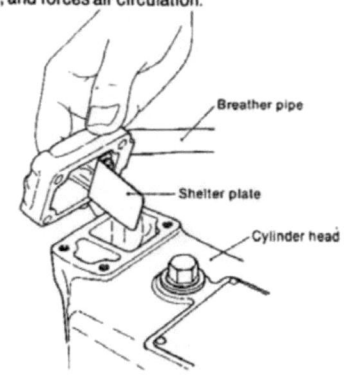

4-1.3 Breather for model 3HM35(F)(C)

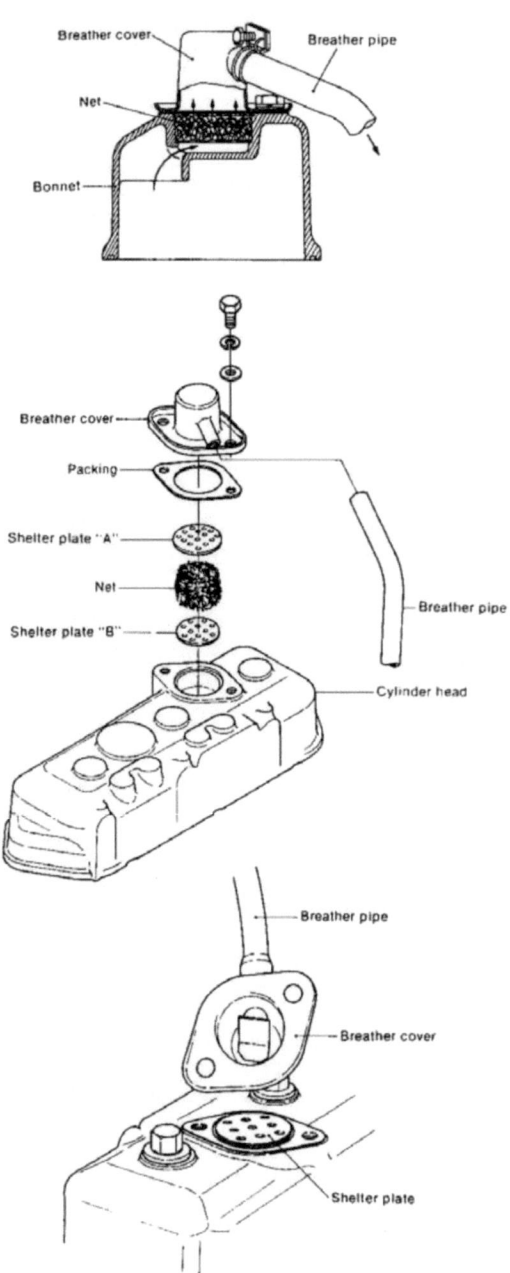

CHAPTER 6
LUBRICATION SYSTEM

1. Lubrication System 6-1
2. Oil Pump 6-5
3. Oil Filter 6-9
4. Oil Pressure Regulator Valve 6-12
5. Oil Pressure Measurement 6-14

1. Lubrication System

Engine parts are lubricated by a trochoid pump forced lubrication system. To keep the engine exterior uncluttered and to eliminate vibration damage to piping, exterior piping has been minimized by transporting the lubricating oil through passages drilled in the cylinders and timing gear case.

1-1 Lubricating oil passage of model 1GM10(C)

The lubrication oil filling port is located at the top of the timing gear case, and the lubrication oil poured into the filler is stored in the oil sump after passing through the casting hole in the cylinder wall. The lubricating oil in the oil sump is drawn up the suction pipe through the drilled hole in the cylinder by the action of the trochoid pump, and it is then fed to the lubricating oil filter after passing through the drilled hole in the filter mounting base. The lubricating oil which has passed through the filter is fed through a pipe to the main gallery of the cylinder, and then fed to the main bearing through the oil pressure regulator valve.

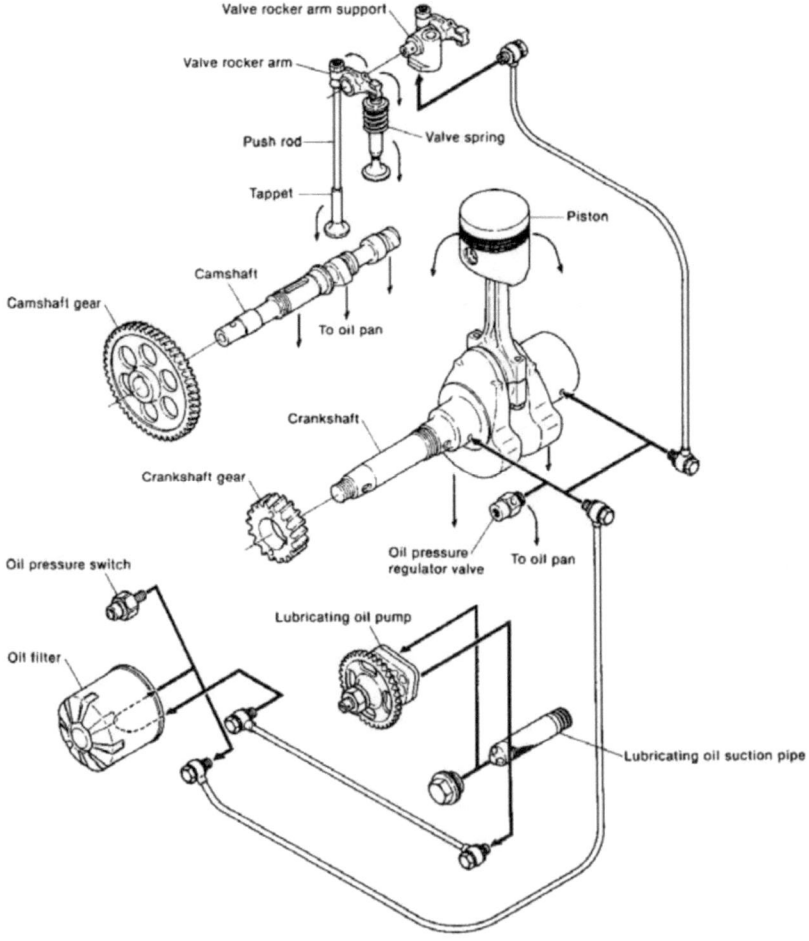

1-2 Lubrication oil passage of model 2GM20(F)(C)

The lubricating oil supplied from the oil filter in the rocker arm cover is collected through the tappet holes in the oil pan at the bottom of the cylinder block.

The lubricating oil is drawn back up through the lubricating oil suction pipe by the trochoid pump and fed to the oil filter, where impurities are filtered out. Then it is adjusted to the prescribed pressure by the oil pressure regulating valve and sent to the main bearing.

The lubricating oil sent to the gear side main bearing flows in two paths: one from the main bearing to lubricate the crank pin through the hole drilled through the crankshaft.

The lubricating oil sent to the flywheel side main bearing also flows in two paths: one from the main bearing to lubricate the crank pin through the hole drilled through the crankshaft, and the other to the rocker arm shaft through the hole drilled through the cylinders and cylinder head. From the rocker arm shaft, the lubricating oil flows through the small hole in the rocker arm to lubricate the push rods and part of the valve head.

The oil that has dropped to the push rod chamber from the rocker arm chamber lubricates the tappets, cam and cam bearing, and returns to the oil pan.

The pistons, piston pins and contact faces of the cylinder liners are splash lubricated by the oil that has lubricated the crank pin. Moreover, an oil pressure switch is provided in the lubricating system to monitor normal circulation and the pressure of the lubricating oil. When the lubricating oil pressure drops $0.5 kgf/cm^2$ (7.114 lb/in.^{2}), the oil pressure switch illuminates the oil pressure lamp on the instrument panel to notify the operator.

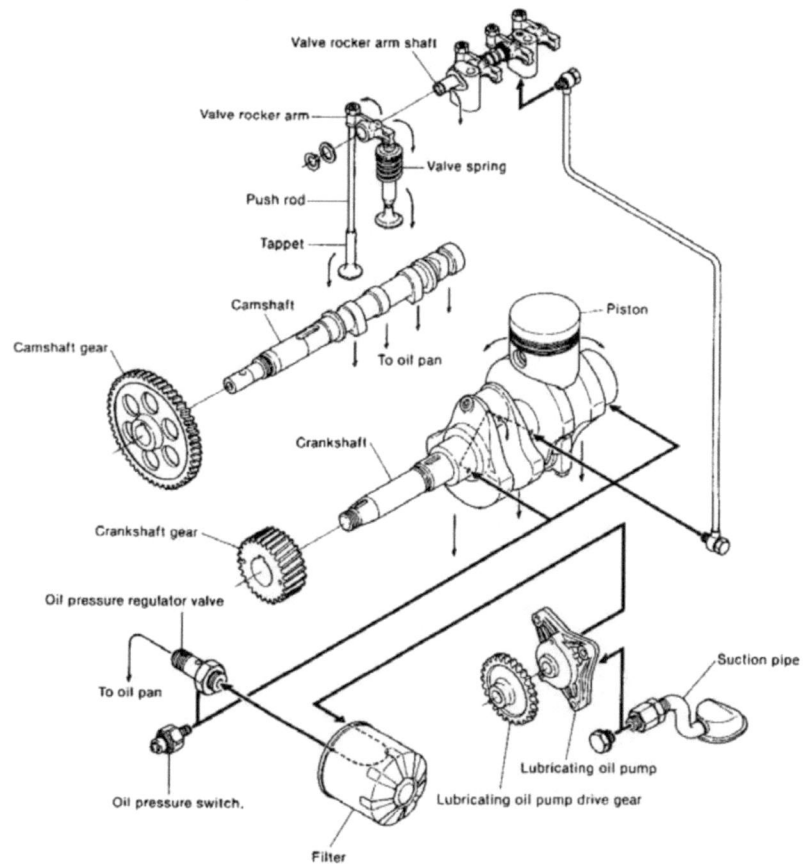

Chapter 6 Lubrication System
1. Lubrication System

1-3 Lubrication oil passage of model 3GM30(F)(C) and 3HM35(F)(C)

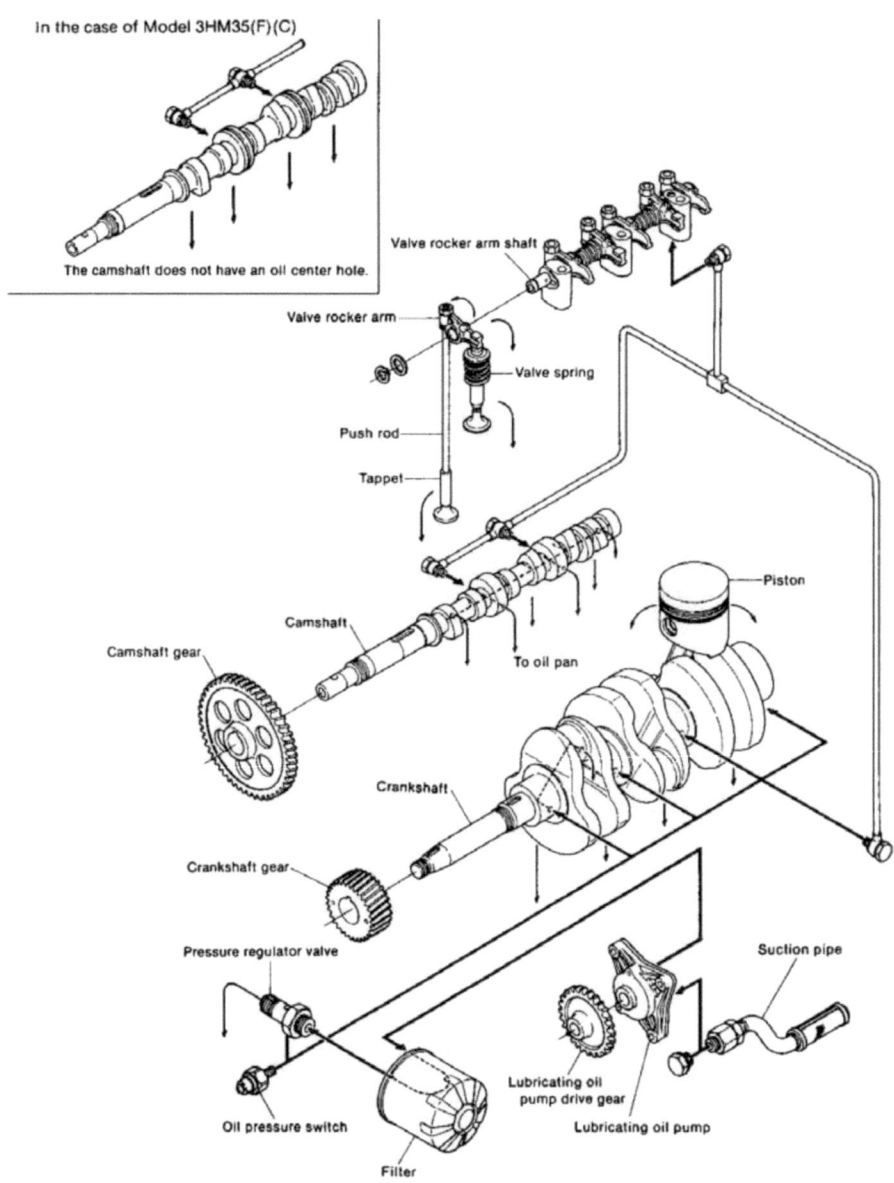

Chapter 6 Lubrication System
1. Lubrication System

SM/GM(F)(C)·HM(F)(C)

1-4 Table of capacity for lubricating oil system

			1GM10(C)	2GM20(F)(C)	3GM30(F)(C)	3HM35(F)(C)
Lubricating oil pump	Pump speed	rpm	2600	3600	3600	3400
	Discharge volume	ℓ/min ℓ/h	3.9 234	12.5 760	12.5 760	12 720
	Discharge pressure	kgf-cm^2 (lb/in.^{2})	\multicolumn{4}{c}{3.5 ±0.5 (42.67 ~ 56.89)}			
Lubricating oil filter	Filter capacity					
	Discharge pressure	kgf-cm^2 (lb/in.^{2})	\multicolumn{4}{c}{1 (14.22)}			
Oil pressure regulator valve	Standard pressure	kgf-cm^2 (lb/in.^{2})	\multicolumn{4}{c}{3.5 ±0.5 (42.67 ~ 56.89)}			
	Full open pressure (Max)	kgf-cm^2 (lb/in.^{2})	\multicolumn{4}{c}{4 (56.89)}			
Lubricating oil pressure alarm switch	ON	kgf-cm^2 (lb/in.^{2})		0.2 ±0.1 (1.422 ~ 4.266)	0.2 ±0.1 (1.422 ~ 4.266)	0.5 ±0.1 (5.689 ~ 8.534)
Lubricating oil tank	Crankcase oil capacity, Total (effective)	ℓ	1.3 (0.6)	2.0 (1.3)	2.6 (1.6)	5.4 (2.7)

2. Oil Pump

2-1 Construction

The oil pump is a compact, low pressure variation trochoid pump comprising trochoid curve inner and outer rotors. Pumping pressure is provided by the change in volume between the two rotors caused by rotation of the rotor shaft.

The lubricating oil pump is installed on the cylinder body at the timing gear case end, and its rotor shaft gear is driven by the crankshaft gear.
The lubricating oil is drawn in and discharged through drilled holes in the cylinder body.

2-1.1 Lubricating oil pump on model 1GM10(C)

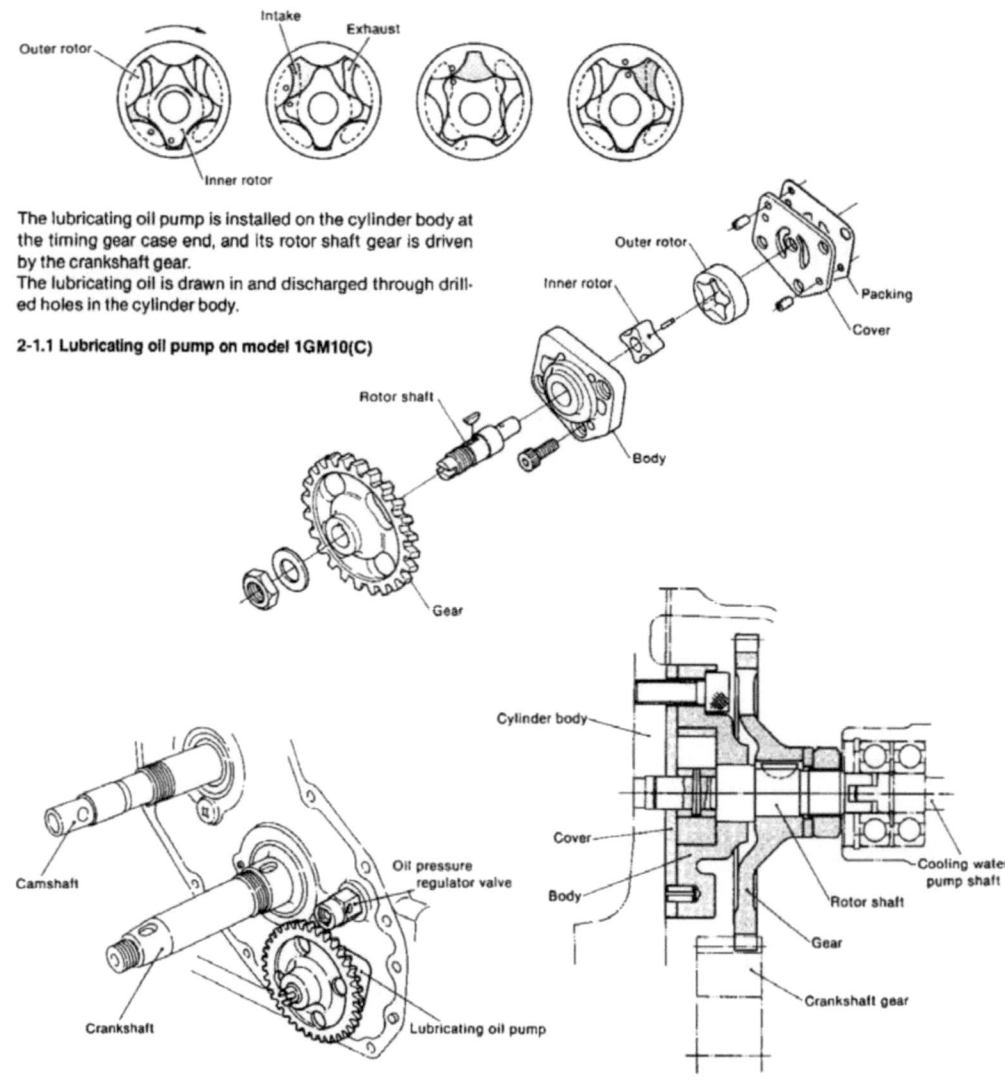

Chapter 6 Lubrication System
2. Oil Pump

2-1.2 Lubricating oil pump on models 2GM20(F)(C), 3GM30(F)(C) and 3HM35(F)(C)

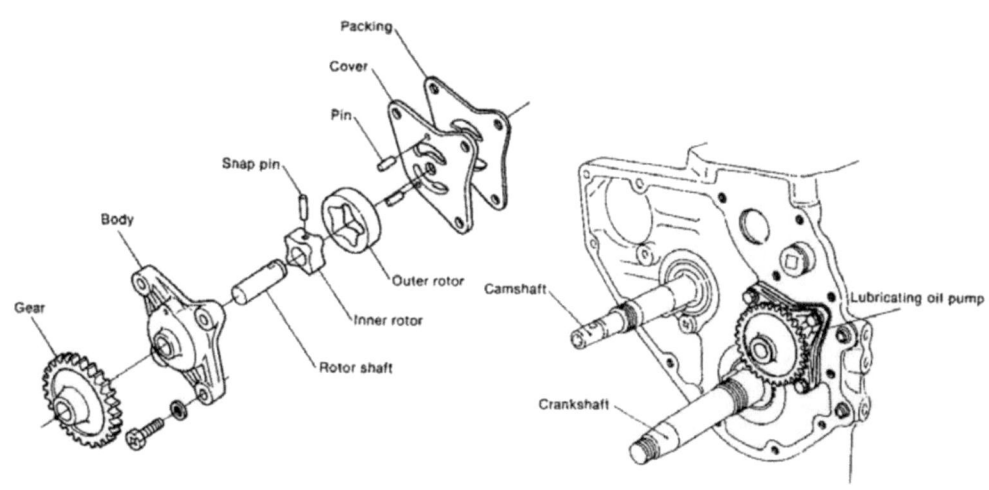

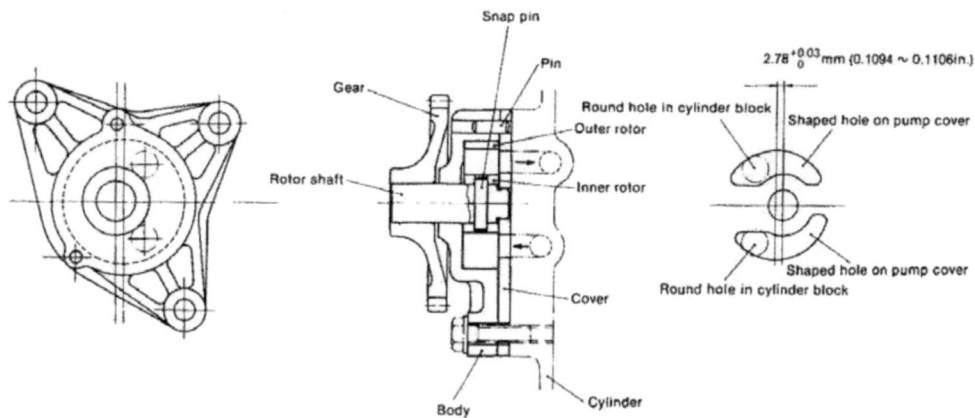

2-1.3 Specifications of lubrication oil pump

	1GM10(C)	2GM20(F)(C), 3GM30(F)(C)	3HM35(F)(C)
Engine speed	3600 rpm	3600 rpm	3400 rpm
Pump speed	2600 rpm	3600 rpm	3400 rpm
Discharge volume	3.9 ℓ/min 234 ℓ/h	12.5 ℓ/min 760 ℓ/h	12 ℓ/min 720 ℓ/h
Discharge pressure	3.5±0.5 kgf/cm² (42.67 ~ 56.89 lb/in.²)	3.5±0.5 kgf/cm² (42.67 ~ 56.89 lb/in.²)	3.5±0.5 kgf/cm² (42.67 ~ 56.89 lb/in.²)

Chapter 6 Lubrication System
2. Oil Pump

2-2 Disassembly
2-2.1 Model 1GM10(C)
(1) Remove the timing gear case

2-2.2 Models 2GM20(F)(C), 3GM30(F)(C) and 3HM35(F)(C)
(1) Remove the timing gear case.

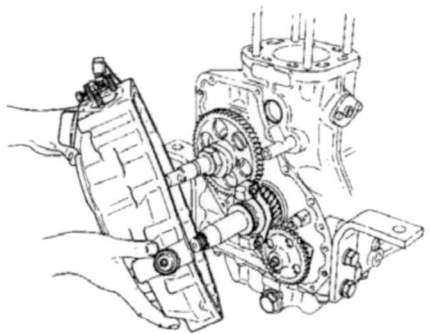

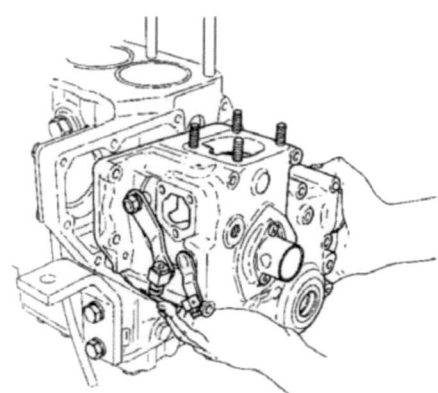

(1') Remove gear case
(2) Withdraw the governor sleeve and thrust bearing, and also take out the governor weight support after removing the hexagonal nut.
NOTE: The lubricating oil pump drive gear cannot be removed without removing the governor weight support.
(3) Remove the hexagonal nut of the lubricating oil pump rotor shaft, then remove the pump drive gear.
(4) Remove the pump body from the cylinder by removing the fixing bolt with a hexagonal bar spanner.
(4') Remove the loosening bolt with a hexagonal bar spanner.

(2) Remove the lubricating oil pump driving gear and pump assembly.

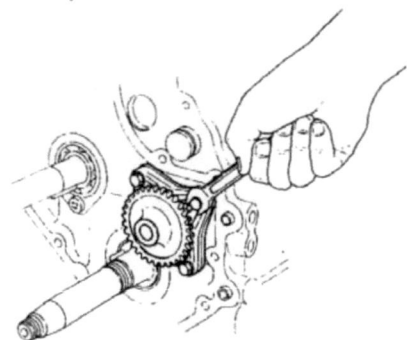

NOTE: Do not separate the lubricating pump gear from the rotor shaft. If removed, it cannot be used again. When any part is unusable, replace it as a complete assembly.

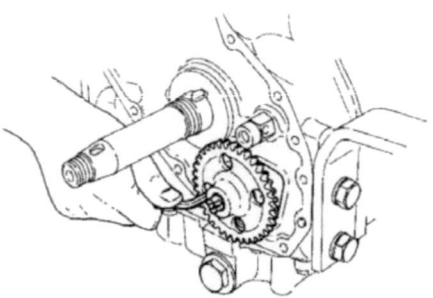

(5) Remove the pump cover.
(6) Take out the outer rotor and the assembly of the inner rotor and rotor shaft.

Chapter 6 Lubrication System
2. Oil Pump

2-3 Inspection
When the discharge pressure of the oil pump is extremely low, check the oil level. If it is within the prescribed range, the oil pump must be inspected.

(1) Outer rotor and pump body clearance
Measure the clearance by inserting a feeler gauge between the outside of the outer rotor and the pump body casing. If the clearance exceeds the wear limit, replace the outer rotor and pump body as a set.

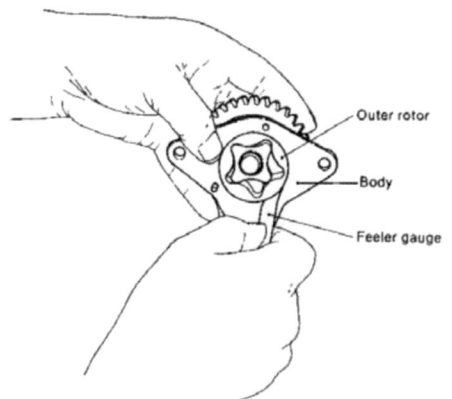

	mm (in.)
Maintenance standard	0.050 ~ 0.105 (0.00197 ~ 0.00413)
Wear limit	0.15 (0.00591)

(2) Outer rotor and inner rotor clearance
Fit one of the teeth of the inner rotor to one of the grooves of the outer rotor and measure the clearance at the point where the teeth of both rotors are aligned. Replace the inner rotor and outer rotor ass'y if the wear limit is exceeded.

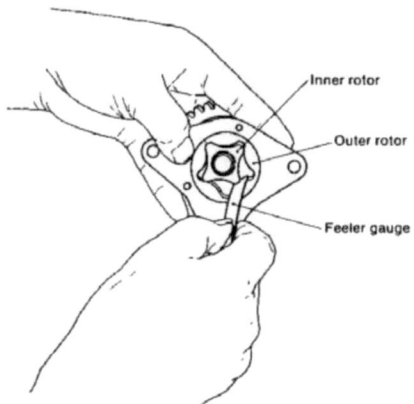

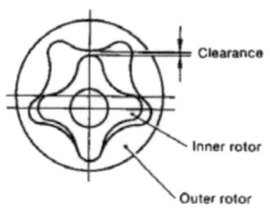

	mm (in.)
Maintenance standard	0.050 ~ 0.105 (0.00197 ~ 0.00413)
Wear limit	0.15 (0.00591)

(3) Pump body and inner rotor, outer rotor side clearance
Install the inner rotor and outer rotor into the pump body casing so that they fit snugly.
Check the clearance by placing a ruler against the end of the body and inserting a feeler gauge between the ruler and the end of the rotor. Replace as a set if the wear limit is exceeded.

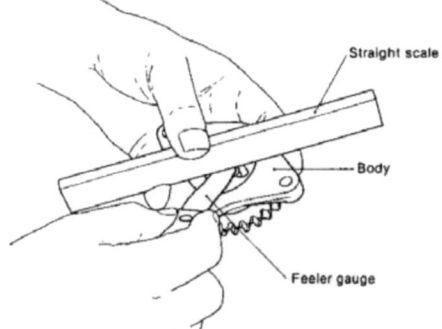

	1GM10(C)	2GM20(F)(C), 3GM30(F)(C), 3HM35(F)(C)
Maintenance standard	0.03~0.08 (0.0012~0.0031)	0.03~0.07 (0.0012~0.0031)
Wear limit	0.13 (0.0051)	0.13 (0.0051)

(4) Rotor shaft and body clearance
Measure the outside diameter of the rotor shaft and the inside diameter of the body shaft hole, and replace the rotor shaft and body as an ass'y if the clearance exceeds the wear limit.

mm (in.)

	1GM10(C)		
		Clearance when assembled	Maximum allowable clearance
	Maintenance standard		
Rotor shaft outside diameter	ø14 (0.5512)	0.015~0.050 (0.0006~0.0020)	0.2 (0.0079)
Rotor shaft hole inside diameter	ø14 (0.5512)		

6-8

3. Oil Filter

3-1 Construction

The oil filter removes the dirt and metal particles from the lubricating oil to minimize wear of moving parts. The construction of the oil filter is shown below.

The lubricating oil from the oil pump is passed through the filter paper and distributed to each part as shown by arrow A in the figure.

After extended use, the filter paper will become clogged and its filter performance will drop. When the pressure loss caused by the filter paper exceeds $1 kgf/cm^2$ ($14.22 lb/in.^2$), the bypass valve inside the filter opens and the lubricating oil is sent to each part automatically as an emergency measure, without passing through the filter, as shown by arrow B.

The oil filter is located at the fitted position of the oil pressure regulator valve on the side surface of the gear case together with the oil pressure valve for engine models 2GM20(F)(C), 3GM30(F)(C) and 3HM35(F)(C). However, in the case of engine model 1GM10(C), the filter alone is fitted on its mounting base at the gear case end, cylinder end surface. The oil pressure regulator valve is installed separately on the end surface of the cylinder, in the gear case.

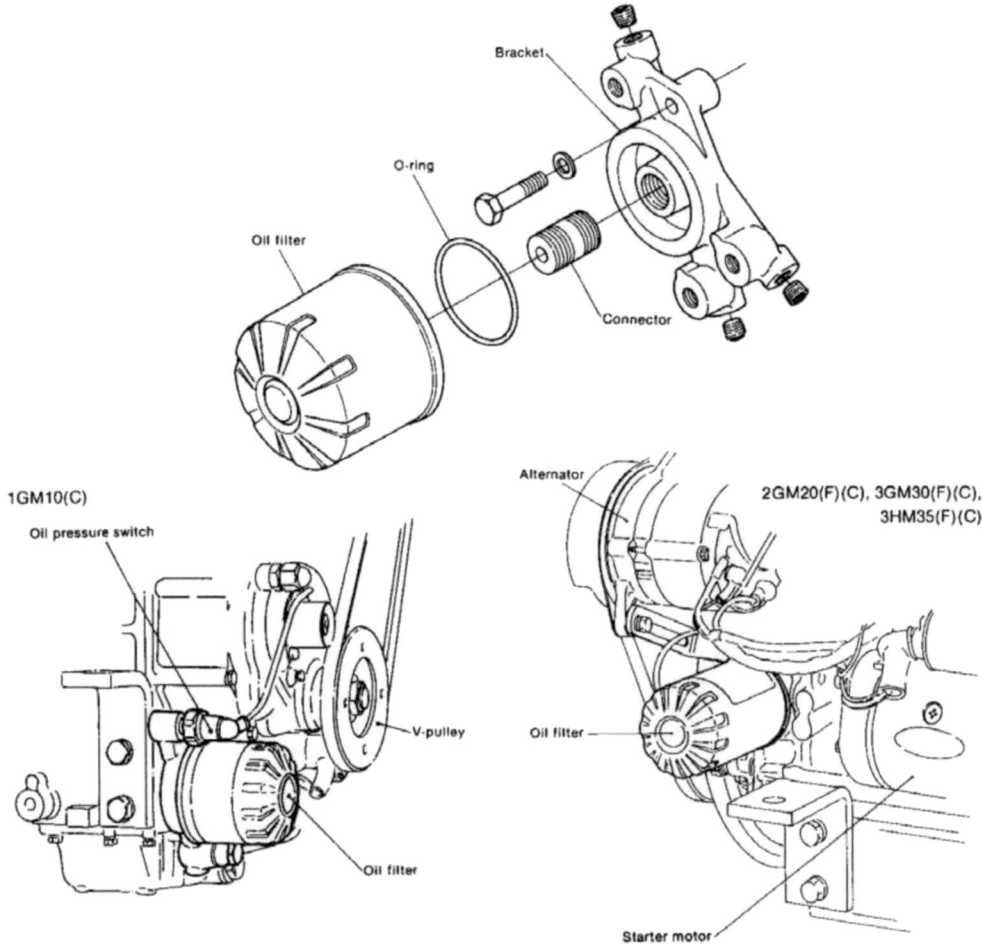

Chapter 6 Lubrication System
3. Oil Filter

1GM10(C)

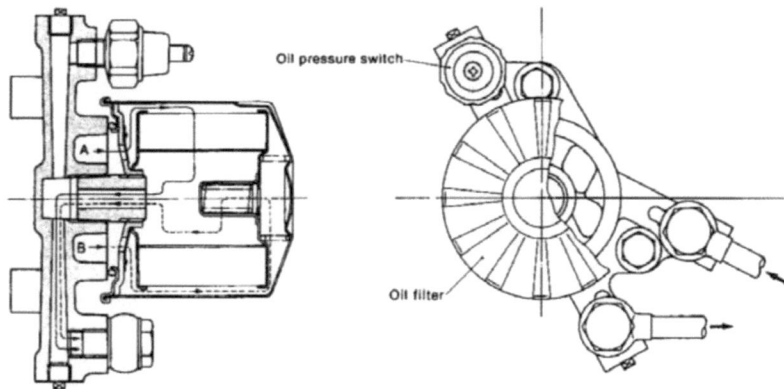

2GM20(F)(C), 3GM30(F)(C) and 3HM35(F)(C)

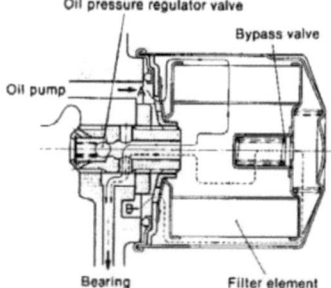

3-2 Replacement

When the oil filter has been used for an extended period, the filter paper will become clogged, unfiltered lubricating oil will be sent directly to each part from the bypass circuit, and wear of moving parts will be accelerated. Therefore, it is important that the filter be periodically replaced.
Because this is a cartridge type oil filter, it is replaced as a complete unit.

Oil filter replacement period	Every 300 hours of engine operation

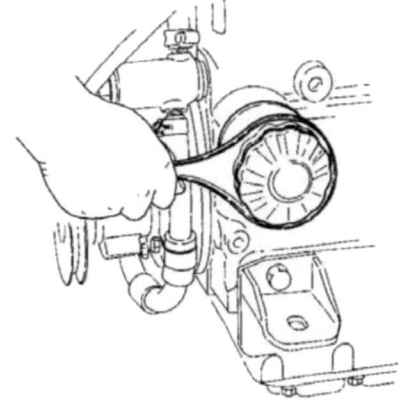

Chapter 6 Lubrication System
3. Oil Filter

1GM10(C)

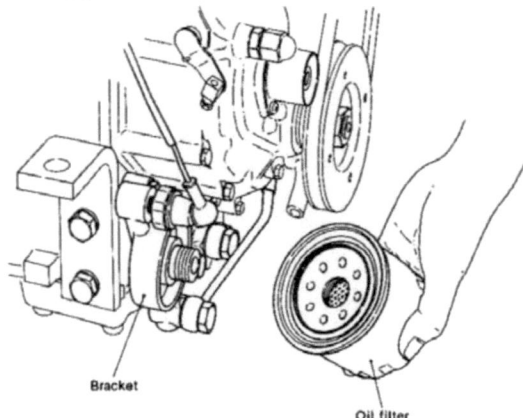

Bracket
Oil filter

2GM20(F)(C), 3GM30(F)(C) and 3HM35(F)(C)

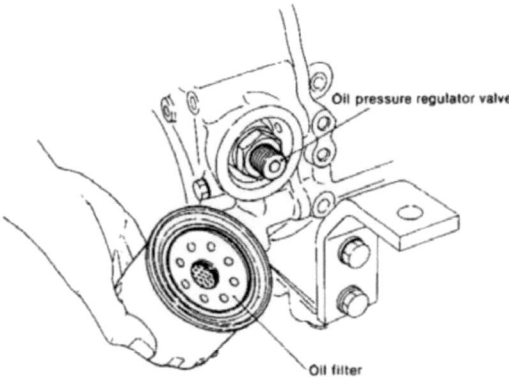

Oil pressure regulator valve
Oil filter

3-2.1 Replacing the oil filter
(1) Clean the oil filter mounting face on the cylinder block.
(2) Before installing the new filter, coat the rubber packing with a thin coat of lubricating oil.
(3) Turn the filter gently until it contacts the rubber packing of the seal surface, then tighten another 2/3 turn.
(4) After installation, run the engine and check the packing face for oil leakage.

3-2.2 In case of oil leakage
If there is oil leakage, remove the oil filter and replace the packing. At the same time, inspect the cylinder block mounting face and repair the face with an oil stone if it is scored.

4. Oil Pressure Regulator Valve

4-1 Construction

The oil pressure regulator valve serves to adjust the pressure of the lubricating oil to the prescribed pressure during operation. When the pressure of the lubricating oil from the oil filter exceeds the force of the spring, the metal ball is pushed away from the valve seat and the lubricating oil flows to the oil pan through the gap between the ball and seat. The spring's force is adjusted with a shim.

In engine model 1GM10(C), the oil pressure regulator valve is located at the end surface of the cylinder in the gear case and the pressure is regulated at the intermediate section of the oil passageway between the lubricating oil main gallery and the main bearing at the gear end.

The regulator valve is located in the mounting position of the lubricating oil filter of the timing gear case for engine models 2GM20(F)(C), 3GM30(F)(C) and 3HM35(F)(C)

4-1.1 Model 1GM10(C)

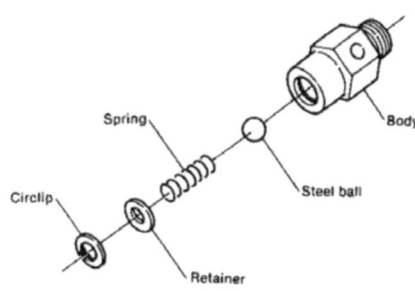

Mounting position for model 1GM10(C)

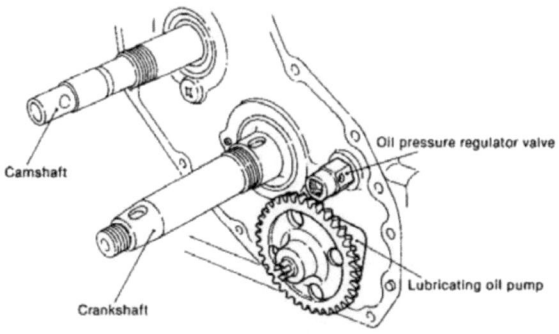

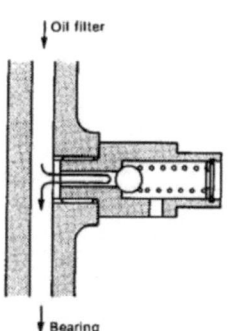

When the pressure is lower than the regulated pressure

Mounting position for model 2GM20(F)(C)

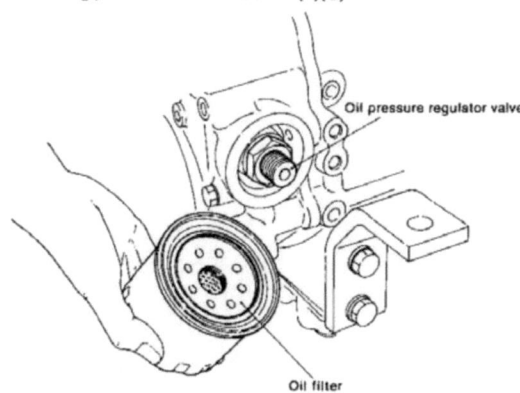

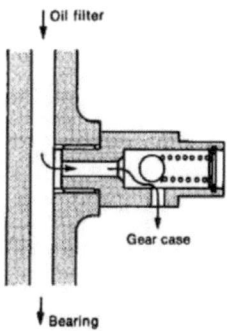

When the pressure is higher that the regulated pressure

Chapter 6 Lubrication System
4. Oil Pressure Regulator Valve

4-1.2 Models 2GM20(F)(C), 3GM30(F)(C) and 3HM35(F)(C)

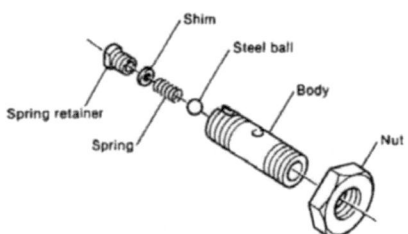

4-1.3 Specifications

	1GM10(C)	2GM20(F)(C), 3GM30(F)(C), 3HM35(F)(C)
Standard pressure	3.5±0.5 kgf/cm² (42.67 ~ 56.89 lb/in.²)	3.5±0.5 kgf/cm² (42.67 ~ 56.89 lb/in.²)

As the lubricating oil pressure regulator valve has been calked during manufacture so that it cannot be dismantled, replace it as a unit if any replacement becomes necessary.

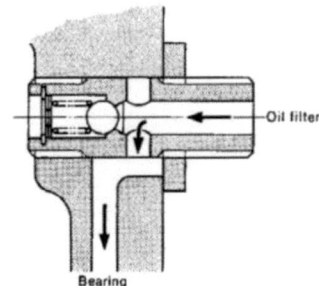

When the pressure is lower that the regulated pessure

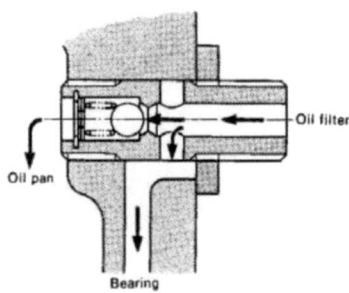

When the pressure is higher that the regulated pressure

6-13

5. Oil Pressure Measurement

The lubricating oil pressure is monitored by a pilot lamp, but it must also be measured using a pressure gauge. Connect the oil pressure gauge to the pilot lamp unit for primary pressure and to the lubricating oil pipe connector for secondary pressure, as shown in the figure.

Secondary oil pressure is especially important. Idle the engine at medium speed when measuring the oil pressure. Also check whether the oil pressure rises smoothly and to the standard value.

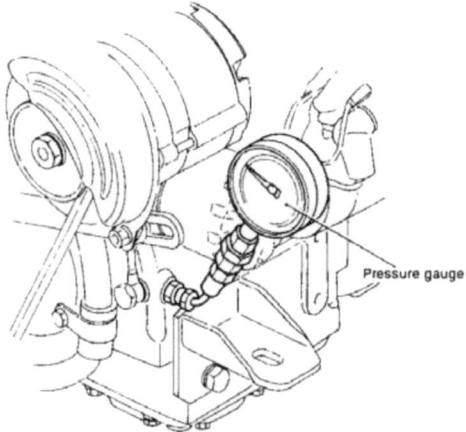
Pressure gauge

kgf/cm²(lb/in.²)

	1GM10(C), 2GM20(F)(C), 3GM30(F)(C)		3HM35(F)(C)	
	850 rpm	3600 rpm	850 rpm	3400 rpm
Secondary pressure standard value	0.5 (7.11)	3.5±0.5 (42.67~56.89)	0.5 (7.11)	3.5±0.5 (42.67~56.89)

If the oil pressure is lower than the standard value, probable causes are:
(1) Clearance of lubricated bearings in the lubricating oil circuit is too large (Shaft or bearing is worn).
(2) Excessive oil escaping from rocker arm support.
Therefore, inspection and repair of the bearings and rocker arm support are required.

CHAPTER 7

DIRECT SEA-WATER COOLING SYSTEM

1. Cooling System 7-1
2. Water Pump 7-5
3. Thermostat 7-11
4. Anticorrosion Zinc 7-14
5. Kingston Cock (Optional) 7-16
6. Bilge Pump and Bilge Strainer (Optional) 7-17

Chapter 7 Direct Sea-Water Cooling System
1. Cooling System

SM/GM(F)(C)·HM(F)(C)

1. Cooling System

1-1 Composition

(1) A sea water direct cooling system incorporating a rubber impeller pump is employed.
(2) A thermostat is installed and a bypass circuit is provided to keep the cooling water temperature constant at all times.
This not only prevents overcooling at initial operation, but also improves the combustion performance and increases the durability of moving parts by keeping the temperature constant.
(3) Anticorrosion zinc is provided at the cylinder and the cylinder head to prevent electrolytic corrosion of the cylinder jacket and cylinder head by the sea water.
(4) A cooling water temperature sender is installed so that any abnormal rise in cooling water temperature is indicated by the lamp on the instrument panel.
(5) A scoop strainer is provided at the water intake Kingston cock to remove dirt and vinyl from the water.
(6) Rubber hoses are used for all interior piping. This eliminates pipe brazing damage due to engine vibration and simplifies the engine's vibration mounting.

1-2 Cooling water route

The cooling water is sucked up by the water pump through a Kingston cock installed on the hull. The water delivered from the water pump branches in two directions at the cylinder intake coupling. Some of the water enters the cylinder jacket and the rest bypasses the cylinder jacket and enters the mixing elbow or the exhaust manifold.
The water that enters the cylinder jacket cools the cylinders and then rises to the cylinder head through the passage between the cylinder and cylinder head and to cool the cylinder head.
The cooling water from the cylinder head, after passing the thermostat, enters the mixing elbow in models 1GM10(C) and 2GM20(C). However, in models 3GM30(C) and 3HM35(C), it first passes to the exhaust manifold to cool the exhaust gas and then enters the mixing elbow.
After that, the water is discharged to the outside of the boat through the rubber hose from the mixing elbow.
The thermostat is closed until the cooling water temperature reaches a fixed temperature (42°C), making the flow go to the cylinder head and then through the bypass circuit.
When the cooling water temperature exceeds 42°C, the thermostat opens, and the cooling water begins to flow through the entire system. At 52°C, the thermostat valve is fully opened and the cooling water temperature is maintained at that level.

1-3 Piping

To simplify the cooling system piping and eliminate cracking of the brazed parts by vibration, rubber or vinyl hoses connected with hose clips are adapted for this engine.
Therefore, the following items must be distorted when inspecting the cooling system:
(1) There must be no extreme bends in the piping.
(2) The cross section of the piping must not be changed by heavy objects on the piping.
(3) There must be no fractures or cracks which allow water leakage.
(4) Piping must not touch high temperature parts, and piping must be securely clamped.
(5) Hose clips must be securely tightened and there must be no leakage from the insertion sections.

Chapter 7 Direct Sea-Water Cooling System
1. Cooling System

SM/GM(F)(C)·HM(F)(C)

1-2.1 Cooling water passage of engine model 1GM10(C)

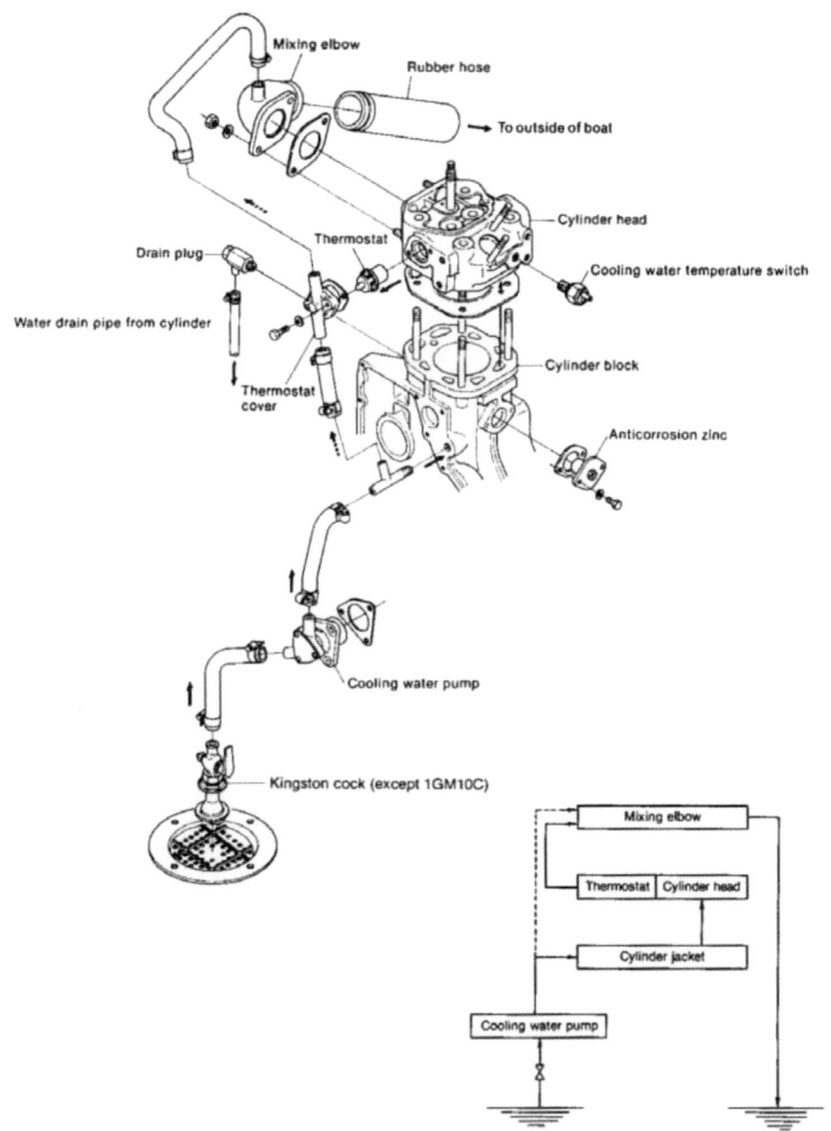

⟵— When the cooling water is at the correct temperature
⟵··· When the cooling water temperature is lower than the correct temperature

Chapter 7 Direct Sea-Water Cooling System
1. Cooling System

SM/GM(F)(C)·HM(F)(C)

1-2.2 Cooling water passage of engine model 2GM20(C)

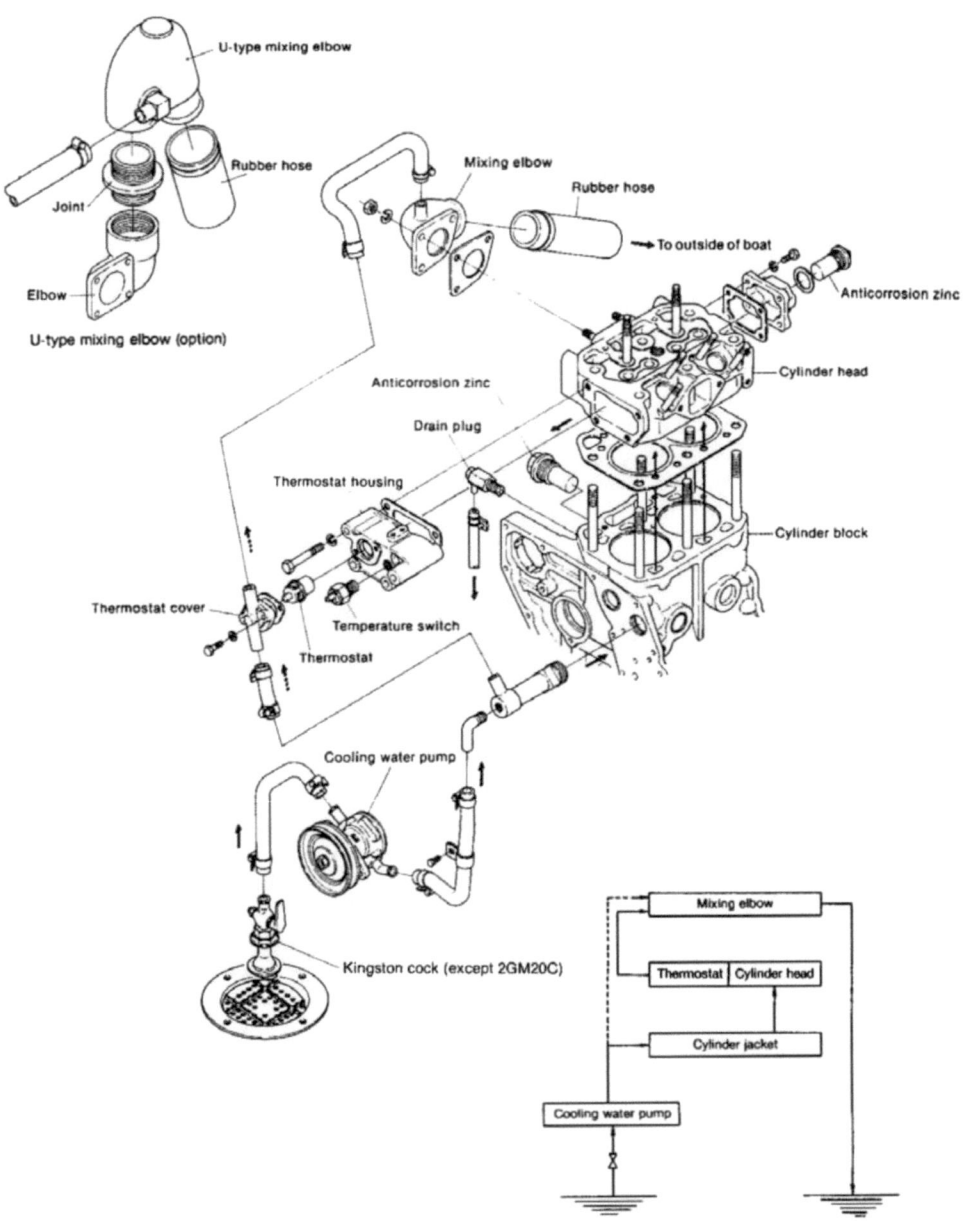

← When the cooling water is at the correct temperature
←---- When the cooling water temperature is lower than the correct temperature

Chapter 7 Direct Sea-Water Cooling System
1. Cooling System

SM/GM(F)(C)·HM(F)(C)

1-2.3 Cooling water passage of engine models 3GM30(C) and 3HM35(C)

The construction of model 3HM35(C) is almost the same as that of model 3GM30(C). As well as the different shape of the thermostat cover, model 3HM35(C) has 2 thermostats.

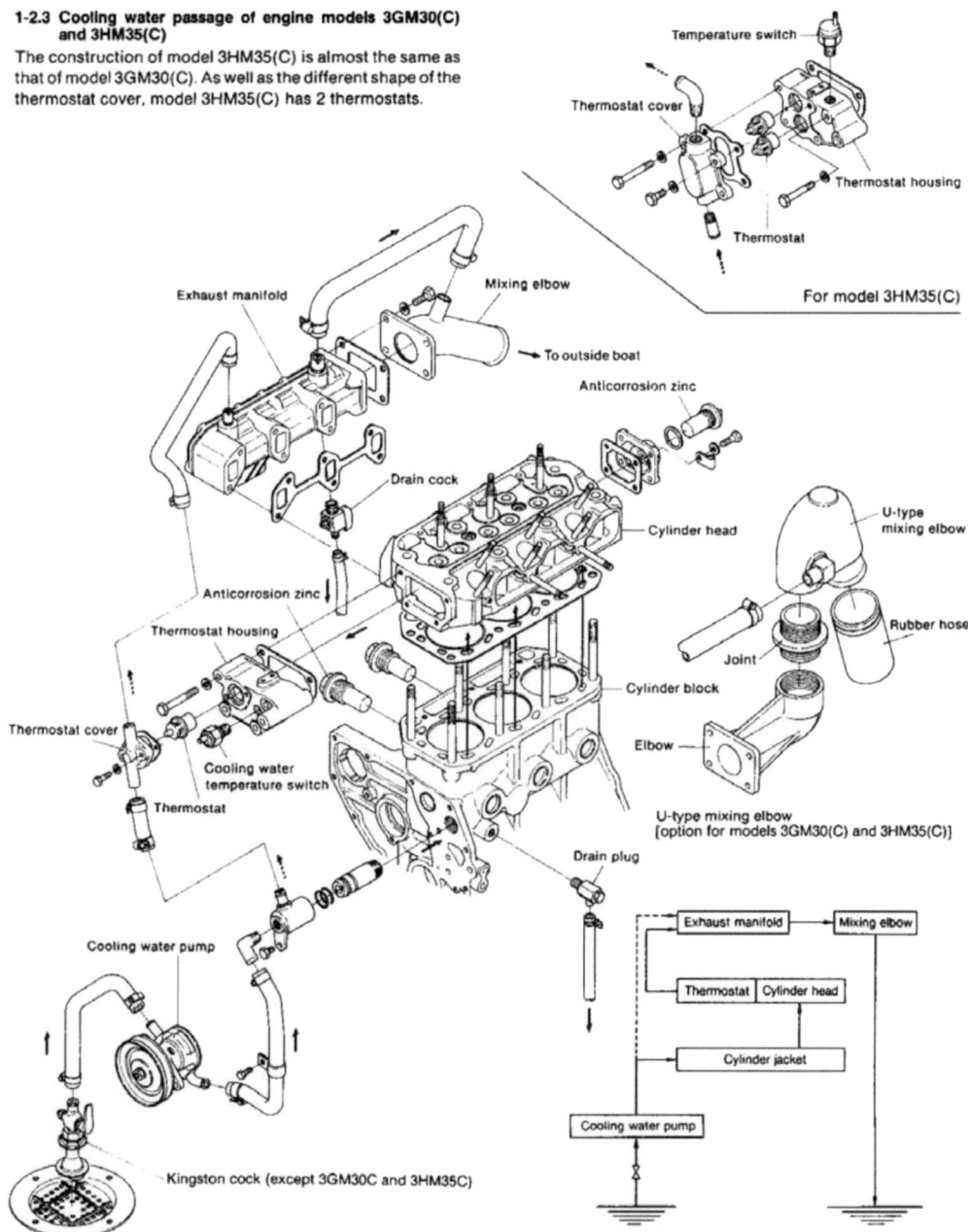

← When the cooling water is at the correct temperature
←··· When the cooling water temperature is lower than the correct temperature

7-4

2. Water Pump

2-1 Construction and operation

The water pump is a rubber impeller type pump. The rubber impeller, which has ample elasticity, is deformed by the off-set plate inside the casing, causing the water to be discharged. This pump is ideal for small, high-speed engines.

The cooling water pump of engine model 1GM10(C) is driven by connecting the cooling water pump shaft to the slit on the end of the lubricating oil pump drive shaft.

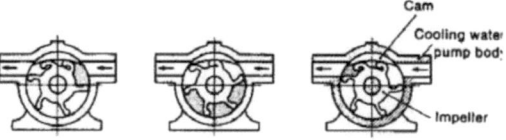

2-1.1 Cooling water pump of engine model 1GM10(C)

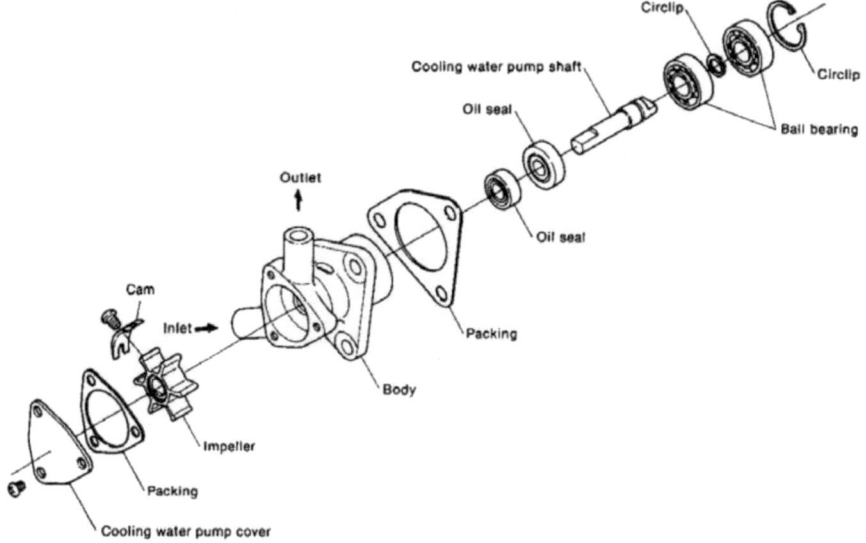

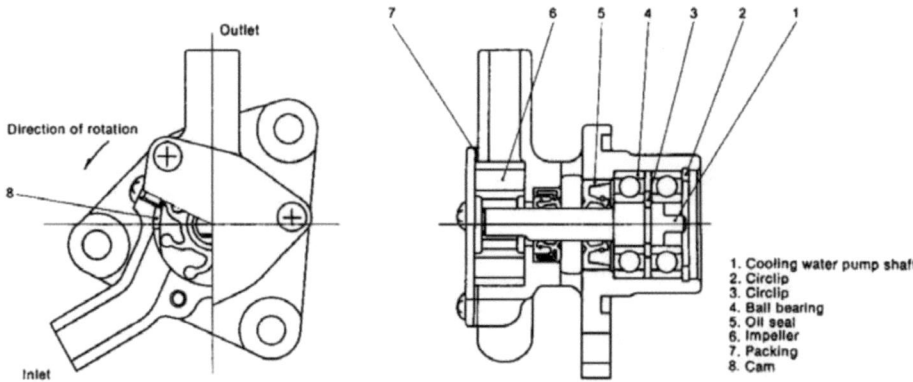

1. Cooling water pump shaft
2. Circlip
3. Circlip
4. Ball bearing
5. Oil seal
6. Impeller
7. Packing
8. Cam

Chapter 7 Direct Sea-Water Cooling System
2. Water Pump

2-1.2 Cooling water pump of engine models 2GM20(C) and 3GM30(C)

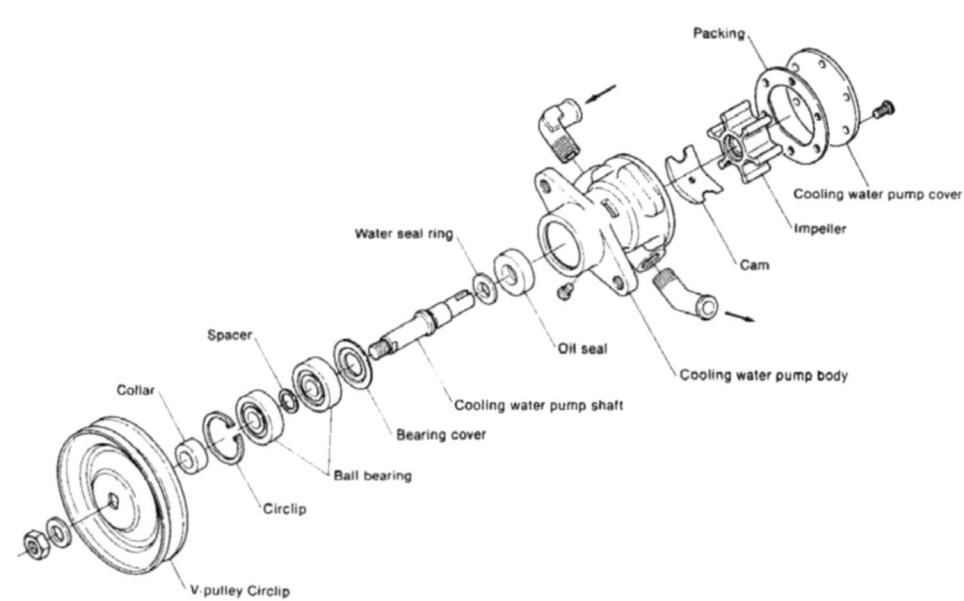

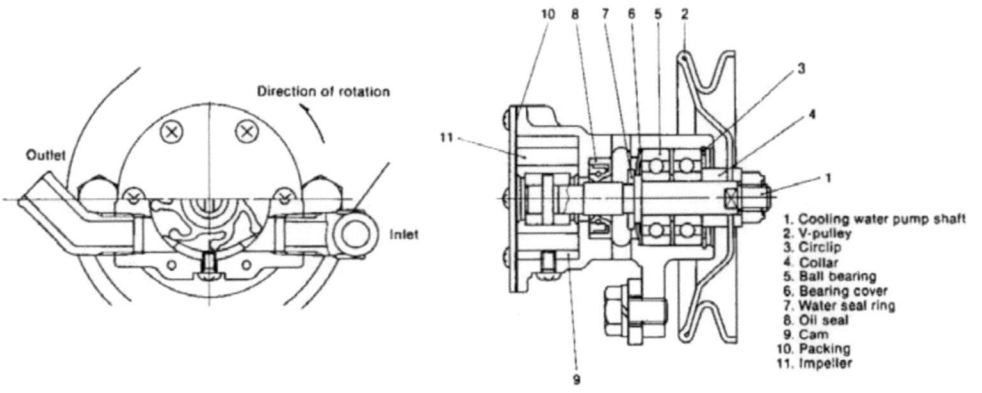

1. Cooling water pump shaft
2. V-pulley
3. Circlip
4. Collar
5. Ball bearing
6. Bearing cover
7. Water seal ring
8. Oil seal
9. Cam
10. Packing
11. Impeller

Chapter 7 Direct Sea-Water Cooling System
2. Water Pump

2-1.3 Cooling water pump of engine model 3HM35(C)

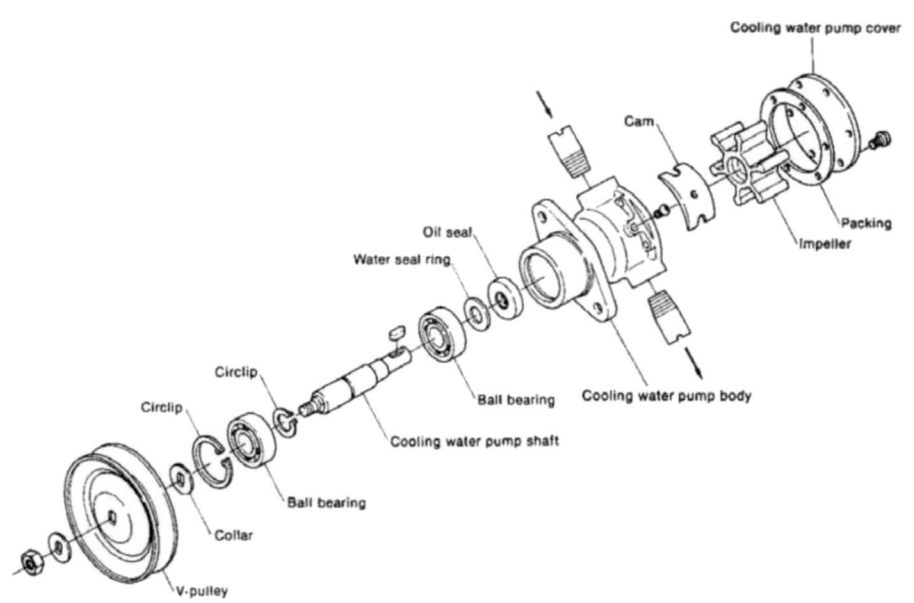

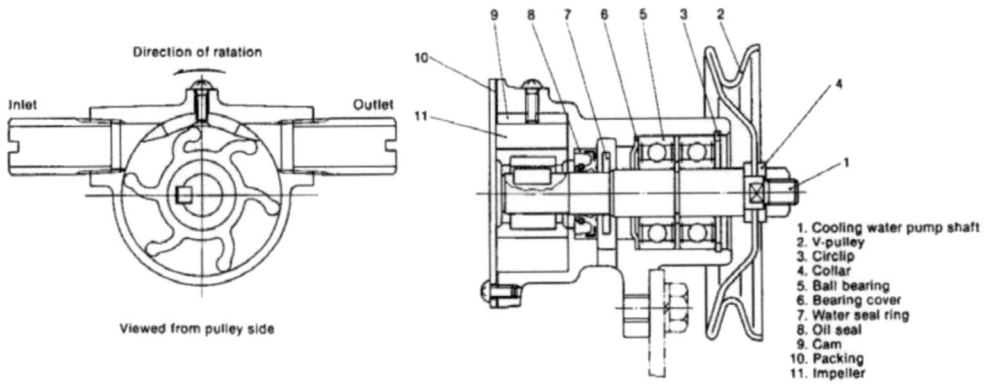

1. Cooling water pump shaft
2. V-pulley
3. Circlip
4. Collar
5. Ball bearing
6. Bearing cover
7. Water seal ring
8. Oil seal
9. Cam
10. Packing
11. Impeller

Chapter 7 Direct Sea-Water Cooling System
2. Water Pump

2-1.4 Specifications

	1GM10(C)	2GM20(C), 3GM30(C)	3HM35(C)
Rated speed	2600rpm	2720rpm	2660rpm
Suction head	0.5m (1.64 ft)	1.0m (3.28 ft)	1.0m (3.28 ft)
Total head	3.0m (9.84 ft)	3.0m (9.84 ft)	4.0m (13.12 ft)
Delivery capacity	300 ℓ/h	700 ℓ/h	1500 ℓ/h

2-2 Disassembly
2-2.1 For model 1GM10(C)
(1) Loosen the water pump mounting bolts, remove the water pump ass'y from the timing gear case.

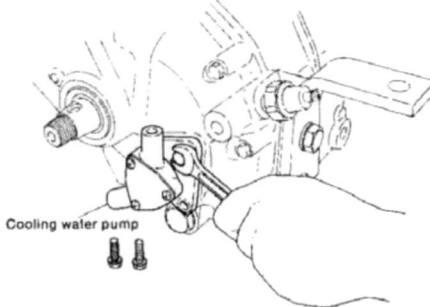
Cooling water pump

(2) Remove the cooling water pump cover and packing by removing the 3 screws which secure the cooling water pump cover.
(3) Pull the water pump impeller.
(4) Remove the set screw and remove the offset plate.
(5) Remove the bearing snap ring and remove the impeller shaft and bearing ass'y while tapping the impeller side of the impeller shaft lightly.
(6) Pull the oil seal from the pump body.
(7) Pull the ball bearing and spacer from the impeller shaft.

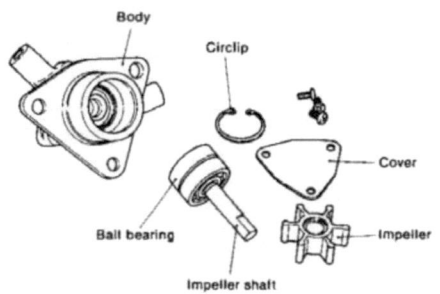

2-2.2 for models 2GM20(C), 3GM30(C) and 3HM35(C)
(1) After removing the V-belt by loosening the mounting bolts of the cooling water pump bracket, remove the cooling water pump assembly.

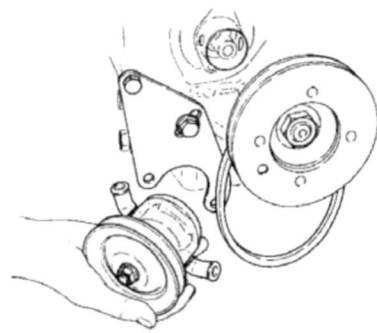

(2) Remove the cooling water pump bracket.
(3) Remove the V-pulley mounting bolt and V-pulley.
(4) Remove the cooling water pump cover fixing screws, and then remove the cooling water pump cover and packing.
(5) Pull the water pump impeller.
(6) Remove the set screw and remove the offset plate.
(7) In engine model 3HM35(C), remove the key from the impeller shaft.
(8) Remove the bearing snap ring and remove the impeller shaft and bearing ass'y while tapping the impeller side of the impeller shaft lightly.
At the same time, the bearing cover and seal ring can be removed together with the impeller shaft.
(9) Pull the oil seal from the pump body.
(10) Pull the ball bearing and spacer from the impeller shaft.

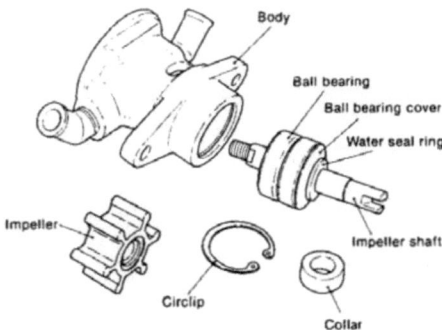

2-3 Reassembly precautions

(1) Before inserting the rubber impeller into the casing, coat the sliding face, pump shaft and impeller fitting section with grease or Monton X.

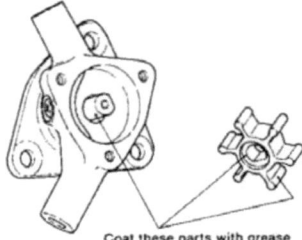

Coat these parts with grease

(2) Be sure that the direction of curving of the impeller is correct.
The impeller is curved in the direction opposite the direction of rotation.

Model 1GM10(C)

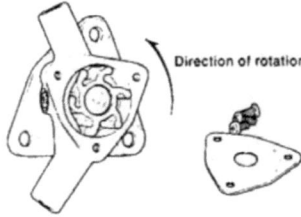

Direction of rotation

Models 2GM20(C), 3GM30(C) and 3HM35(C)

Direction of rotation

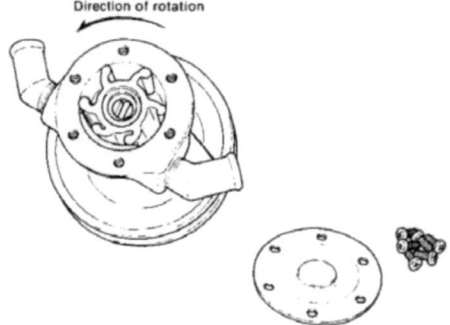

(3) Adjust the V-belt tension. [for models 2GM20(C), 3GM30(C) and 3HM35(C)]
If the V-belt tension is slack, the discharge of the cooling water will diminish; if it is too tight, the play of the pump bearings and the wear of the wear plate will be accelerated. Adjust the tension to the specified value. Check the deflection of the V-belt by pressing it in the center with your fingers.

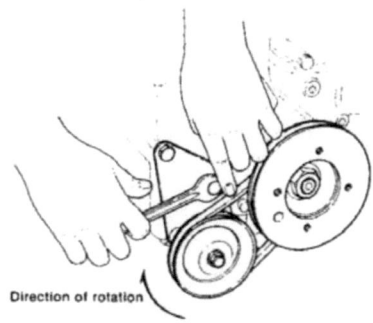

Direction of rotation

	2GM20(C)	3GM30(C)	3HM35(C)
V-belt tension	To be 5 ~ 7mm (0.1964 ~ 0.2756in.) deflection when pushed by the thumb with a force of 10kg (22.0 lb)		
Type of V-belt	M19in.		
V-belt part No.	104511-78780		

NOTE: Mount the belt in the direction of pump rotation.
(4) If the sliding surface of the V-belt is cracked, worn or is stained with oil, etc., replace it with a new one.
(5) Check after assembly
After assembly, attach the belt and run the engine to ascertain whether or not it provides the specified discharge.

2-4 Handling precautions

(1) Never operate the water pump dry as this will damage the rubber impeller.
(2) Always turn the engine in the correct direction of rotation. Turning the engine in the opposite direction will damage the rubber impeller.
(3) Inspect the pump after every 1,500 hours of operation and replace if faulty.

2-5 Inspection

(1) Inspect the rubber impeller for fractures, cracks and other damage, and replace if faulty.

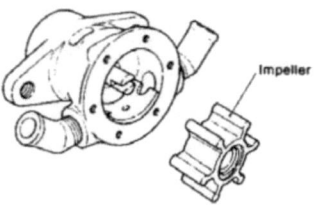

Impeller

Chapter 7 Direct Sea-Water Cooling System
2. Water Pump

SM/GM(F)(C)·HM(F)(C)

(2) Rubber impeller side wear

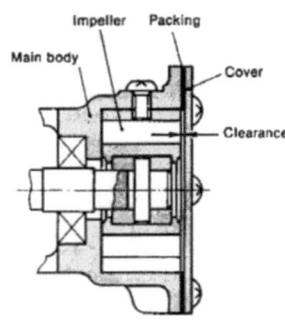

(3) Water pump impeller shaft oil seal section wear.

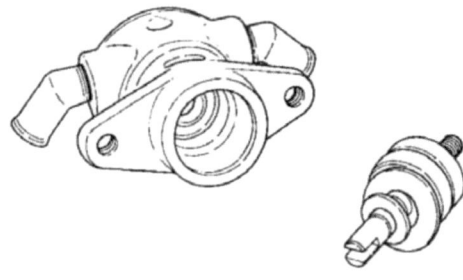

mm (in.)

	Maintenance standard	Wear limit
Oil seal section shaft diameter	10.0 (0.3937)	9.9 (0.3898)

If water leakage increases while the engine is running, or if the components are found to be defective when disassembled, replace them.

(4) Inspect the bearing for play and check for seizing at the impeller shaft fitting section. Replace the bearing if there is any play.

1) Model 1GM10(C) mm (in.)

	Maintenance standard	Clearance at assembly	Maximum allowable clearance	Wear limit
Impeller width	12 ±0.1 (0.4685 ~ 0.4764)	0.2 (0.0079)	0.4 (0.0157)	—
Housing width	(without packing) 11.9 (0.4685)			—
Wear plate wear	—			0.2 (0.0079)

2) Models 2GM20(C) and 3GM30(C) mm (in.)

	Maintenance standard	Clearance at assembly	Maximum allowable clearance	Wear limit
Impeller width	19 ±0.1 (0.744 ~ 0.752)	0.2 (0.0079)	0.4 (0.0157)	—
Housing width	18.9 (0.7441) (without packing) 19.2 (0.7559) (with packing)			—
Wear plate wear	—			0.2 (0.0079)

3) Model 3HM35(C) mm (in.)

	Maintenance standard	Clearance at assembly	Maximum allowable clearance	Wear limit
Impeller width	22.1 ±0.1 (0.8661 ~ 0.8740)	0.2 (0.0079)	0.4 (0.0157)	—
Housing width	(without packing) 22 (0.8661)			—
Wear plate wear	—			0.2 (0.0079)

3. Thermostat

3-1 Construction and operation

The thermostat remains closed until the cooling water temperature reaches a fixed temperature. Until the cooling water reaches this fixed temperature, it collects at the cylinder head and the water flowing from the water pump is discharged through the bypass circuit. When the cooling water temperature exceeds the fixed temperature, the thermostat opens and the cooling water flows through the main circuit of the cylinder and cylinder head. The thermostat serves to prevent overcooling and improve combustion performance by maintaining the cooling water temperature at the specified level.

In engine model 1GM10(C), the thermostat is mounted on the cylinder head at the gear case end. In engine models 2GM20(C) 3GM30(C) and 3HM35(C), it is mounted on the thermostat housing which is combined with the generator mounting base on the cylinder head at the gear case end.

Model 1GM10(C)

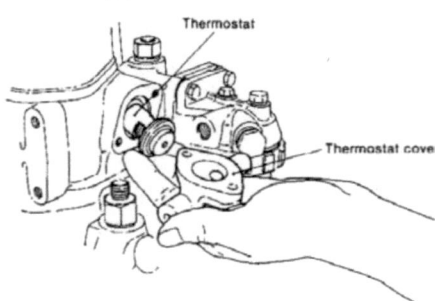

Models 2GM20(C) and 3GM30(C)

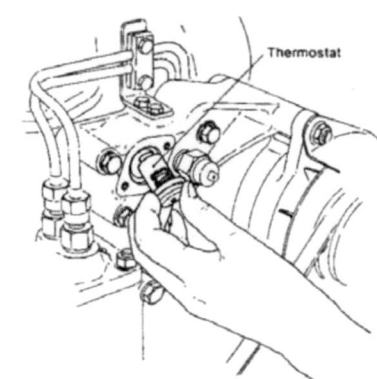

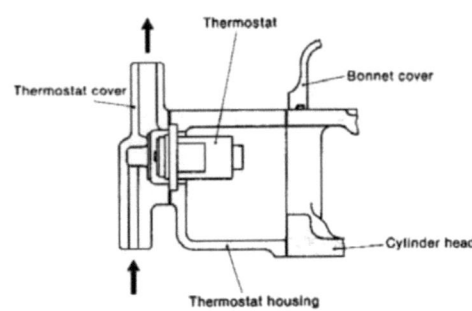

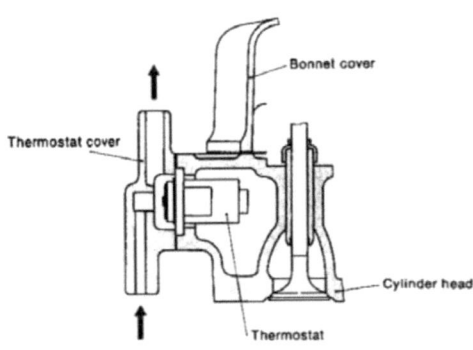

Model 3HM35(C)

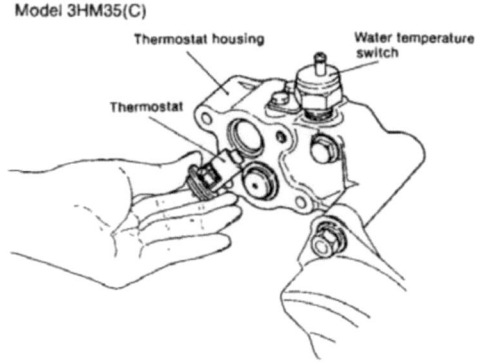

Chapter 7 Direct Sea-Water Cooling System
3. Thermostat

SM/GM(F)(C)·HM(F)(C)

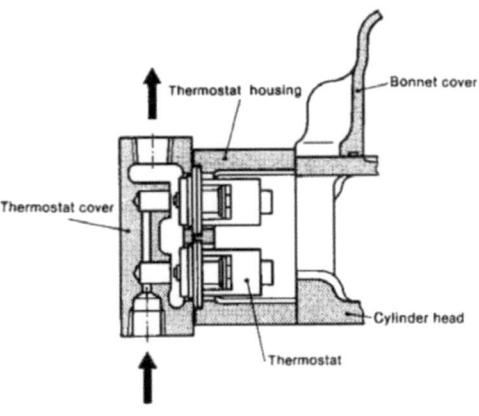

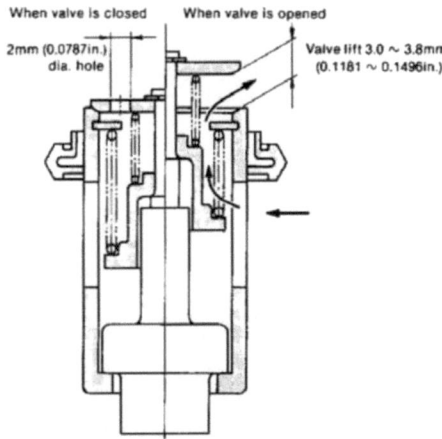

A wax-pellet type thermostat is used for this engine. The "wax-pellet" type is the description given to a quantity of wax in the shape of a small pellet. When the temperature of the cooling water rises, the wax melts and its volume expands. The valve is opened or closed by this variation of volume.

Thermostat operating temperature

Opening temperature	42 ±2°C
Full open temperature	52 ±2°C

Characteristic of Thermostat

Valve opening temperature 42±2°C
Fully opened lift checking temperature 52±2°C

When the seawater temperature is below 42°C, the pumped-up seawater is discharged outside directly from the thermostat section, and circulation of the cooling water into the cylinder is stopped until the water temperature rises. When the water temperature reaches 52°C, the thermostat valve is fully opened.

3-2 Inspection

(1) Remove the water outlet coupling at the top of cylinder body to remove and inspect the thermostat. Remove any dirt or foreign matter that has built up in the thermostat, and check the spring, etc. for damage and corrosion.

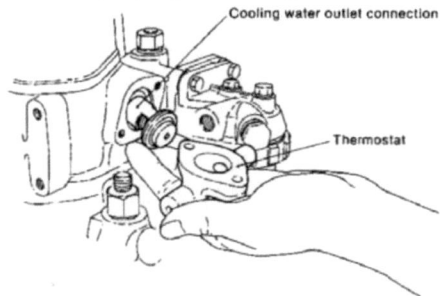

(2) Testing the thermostat
Place the thermostat in a container filled with water. Heat the container with an electric heater. If the thermostat valve begins to open when the water temperature reaches about 42°C and becomes fully open at 52°C, the thermostat may be considered operational. If its behaviour differs much from the above, or if it is found to be broken, replace it.

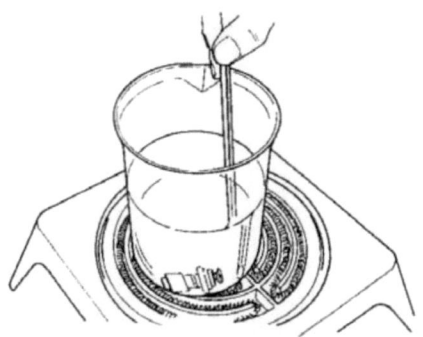

(3) In general, inspect the thermostat after every 300 hours of operation. However, always inspect it when the cooling water temperature rises abnormally and when white smoke is emitted over a long period of time after the engine starting.

(4) Replace the thermostat when it has been in use for a year, or after every 2000 hours of operation.

Part No. code of thermostat	105582—49200

(5) Attaching the thermostat to the cooling water system.
Before attaching the thermostat to the system, be sure to check its packing and make sure there are no leaks.

3-3 Care must be taken when assembling the thermostat

The thermostat cover must be assembled with the arrow mark kept upward.

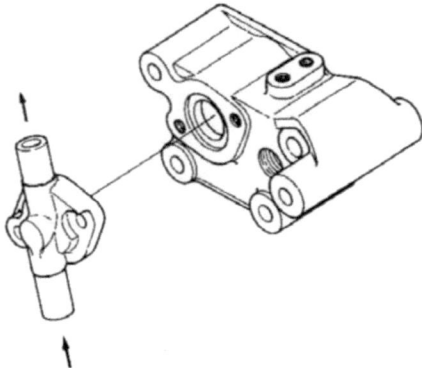

4. Anticorrosion Zinc

4-1 Principles

Anticorrosion zinc is installed to prevent electrolytic corrosion by sea water.
When different metals, i.e., iron and copper, are placed in a highly conductive liquid, such as sea water, the iron gradually rusts. The anticorrosion zinc provides protection against corrosion by being itself corroded in place of the cylinder, cylinder liners and other iron parts.
The anticorrosion zinc is to be put in the following positions.

		1GM10(C)	2GM20(C)	3GM30(C), 3HM35(C)
Cylinder block	Set position	At the side of the fuel valve	At exhaust side	At exhaust side
	Number	1	1	2
Cylinder head	Set position	—	At side cover of cylinder head (rear)	At side cover of cylinder head (rear)
	Number	—	1	1
Type-Size		Flange type 20mm dia × 20mm (0.7874 × 0.7874in.)	Plug type 20mm dia × 30mm (0.7874 × 1.9811in.)	
Part No. of anticorrosion zinc		27210—200200	27210—200300	

Model 1GM10(C)

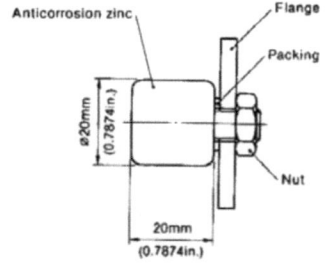

Models 2GM20(C), 3GM30(C) and 3HM35(C)

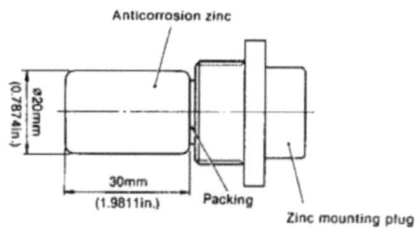

Mounting positon for model 1GM10(C)

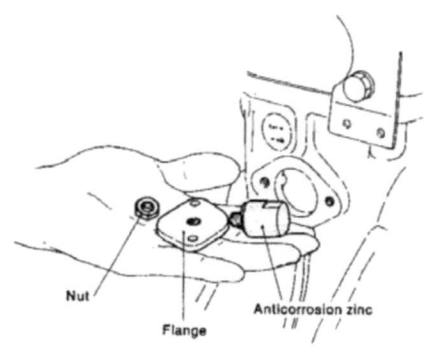

Mounting position for model 2GM20(C)

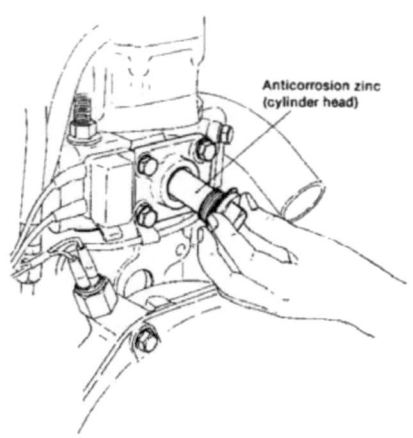

Chapter 7 Direct Sea-Water Cooling System
4. Anticorrosion Zinc

4-2 Inspection

Generally, replace the anticorrosion zinc after every 500 hours of operation. However, since this period depends on the properties of the sea water and operating conditions, periodically inspect the anticorrosion zinc and remove the oxidized film on its surface.
Replace the anticorrosion zinc after 50% corrosion.
Replace the anticorrosion zinc by pulling the old zinc from the zinc mounting plug and screwing in the new zinc.

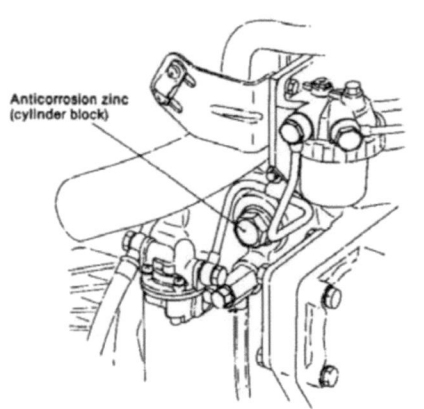
Anticorrosion zinc (cylinder block)

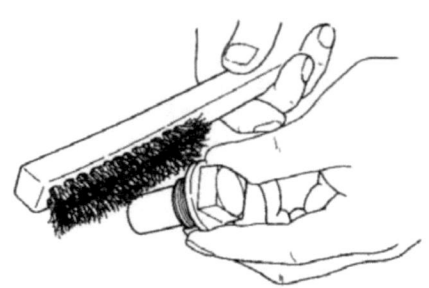

Mounting position for models 3GM30(C) and 3HM35(C)

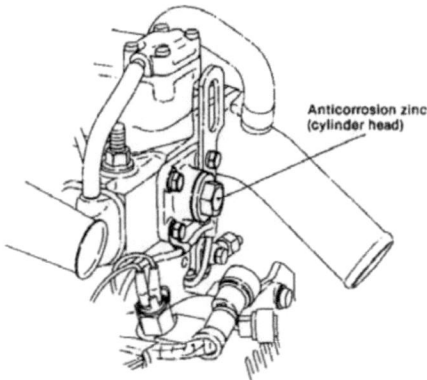
Anticorrosion zinc (cylinder head)

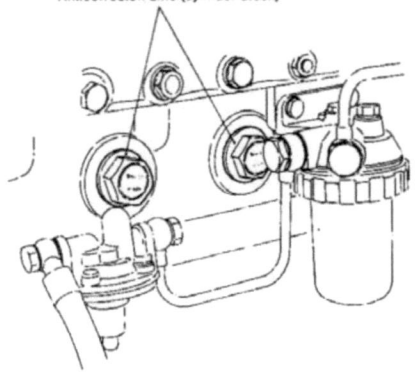
Anticorrosion zinc (cylinder block)

5. Kingston Cock (Optional)

5-1 Construction

The Kingston cock, installed on the bottom of the hull, controls the intake of cooling water into the boat. The Kingston cock also serves to filter the water so that mud, sand, and other foreign matter in the water does not enter the water pump. Numerous holes are drilled in the water side of the Kingston cock, and a scoop strainer is installed to prevent the sucking in of vinyl, etc.

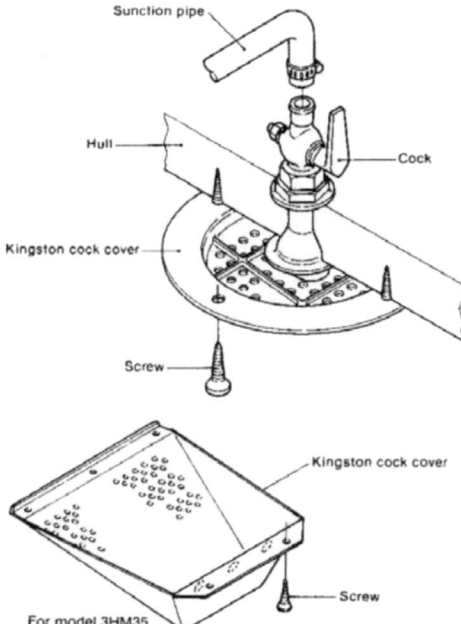

For model 3HM35

5-2 Handling precautions

Caution the user always to close the Kingston cock after each day of use and to confirm that it is open before beginning operation.
If the Kingston cock is left open, water will flow in reverse and the vessel will sink if trouble occurs with the water pump.
On the other hand, if the engine is operated with the Kingston cock closed, cooling water will not be able to get in, resulting in engine and pump trouble.

5-3 Inspection

When the cooling water volume has dropped and the pump is normal, remove the vessel from the water and check for clogging of the Kingston cock.
Moreover, when water leaks from the cock, disassemble the cock and inspect it for wear, and repair or replace it.

6. Bilge Pump and Bilge Strainer (Optional)

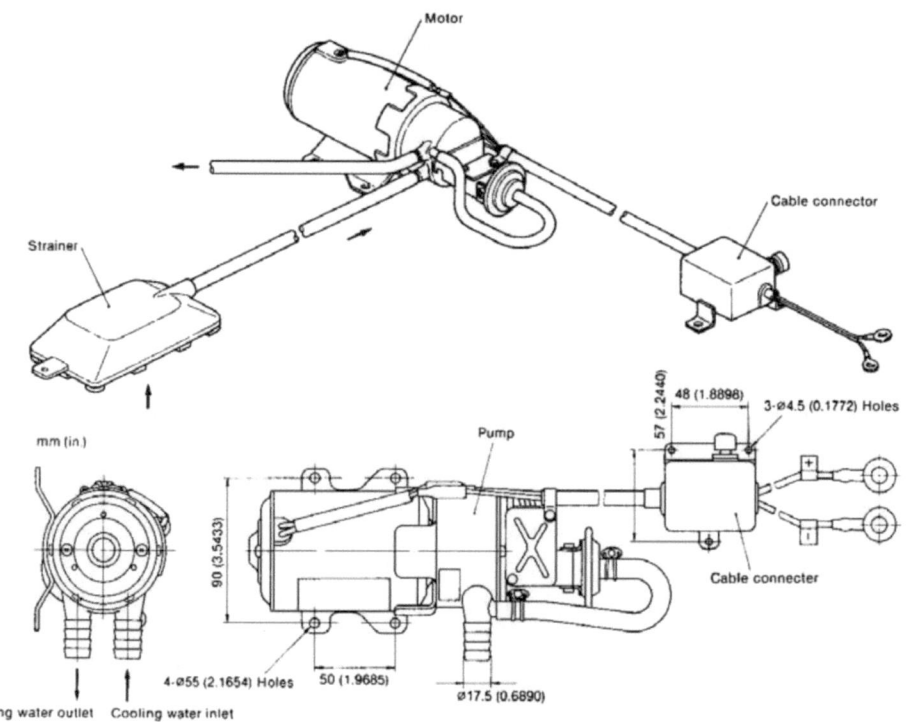

6-1 Bilge pump

6-1.1 Specifications

Code No.	120345-46010 (with strainer)
Model No.	BP190-10
Rating	60 min.
Voltage	12V
Output	90W
Weight	3.0kg (6.6 lb)

6-1.2 Performance of pump (in pure water)

Self-suction performance	Voltage	11.5V
	Max. self-suction lift	1.2m (3.94 ft)
	Self-suction time	4 sec.
Pumping lift performance	Voltage	11.5V
	Current	8A
	Total lift	1m (3.28 ft)
	Lifting volume of water	17 ℓ/min

6-2 Bilge strainer

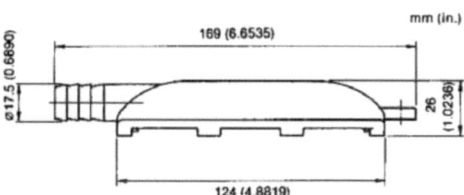

CHAPTER 8
FRESH WATER COOLING SYSTEM

1. Cooling System 8-1
2. Sea Water Pump 8-3
3. Fresh Water Pump 8-4
4. Heat Exchanger 8-7
5. Filler Cap and Subtank 8-11
6. Thermostat 8-13
7. Cooling Water Temperature Switch 8-16
8. Precautions 8-17

1. Cooling System

1-1. System Diagrams

Models 2GM20F, 3GM30F and 3HM35F are constructed from different parts but use the same water flow. The illustration below is of model 3HM35F.

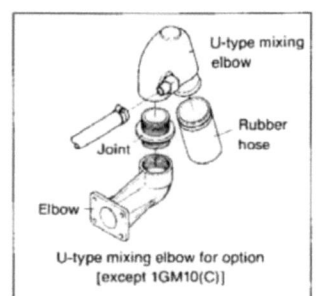

U-type mixing elbow for option
[except 1GM10(C)]

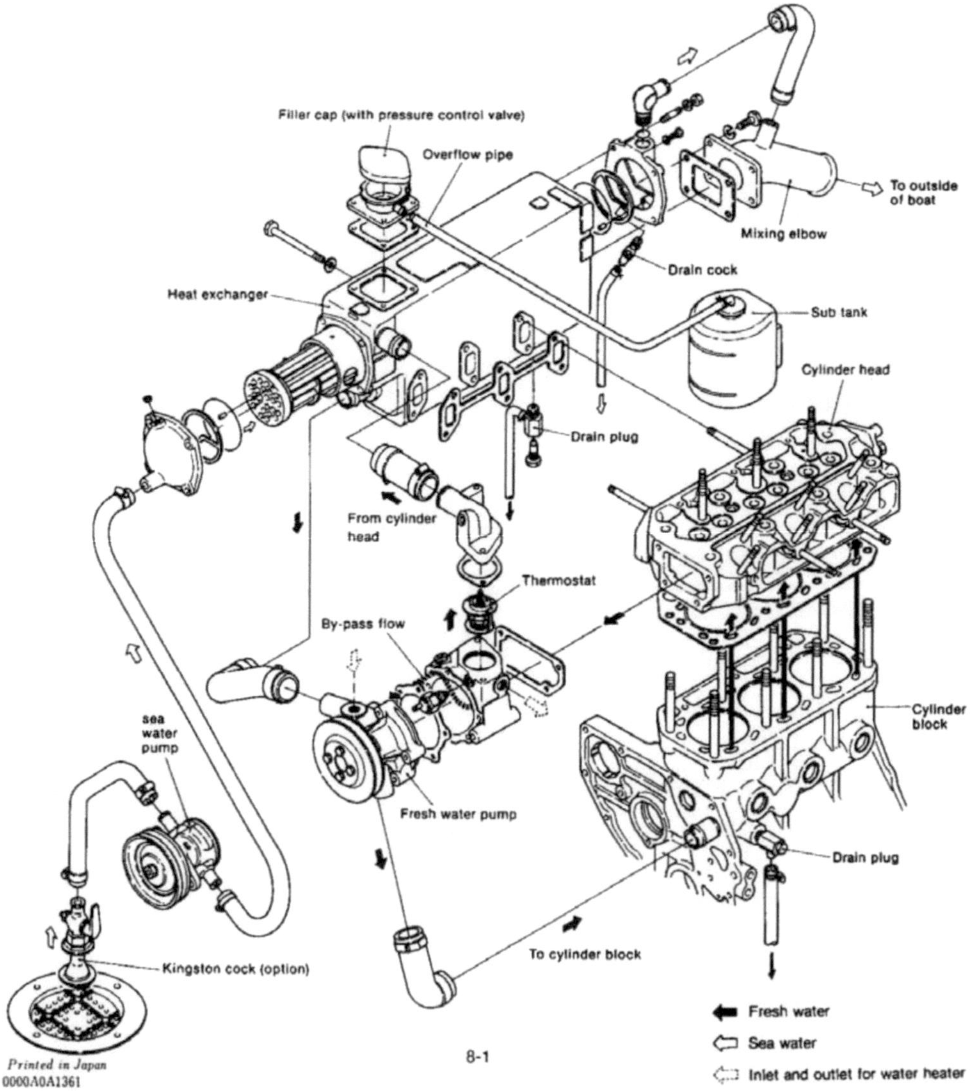

⬅ Fresh water
⬅ Sea water
⬅ Inlet and outlet for water heater

Chapter 8 Fresh Water Cooling System
1. Cooling System

1-2. Cooling system diagram

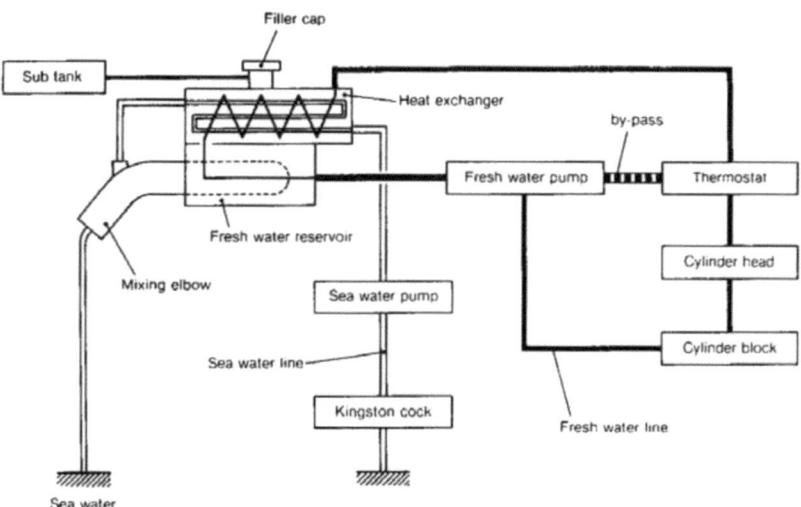

1-3. Cooling system configuration

With fresh water cooled engines, fresh water from the heat exchanger is circulated around the cylinder block and cylinder head. The fresh water itself is cooled by sea water. The fresh water pump forces the fresh water through the cylinder block and cylinder head cooling passages and back to the heat exchanger. The fresh water is kept in constant circulation.

The thermostat is installed at the cylinder head cooling water outlet (fresh water pump mounting bracket). As the thermostat is closed while the fresh water temperature is low—directly after engine starting or when the engine load is light—fresh water flows through the by-pass passage to the suction side of the fresh water pump, and circulates inside the engine without passing through the heat exchanger.

As the fresh water temperature rises the thermostat is opened and fresh water flows into the heat exchanger. The fresh water is cooled in the heat exchanger by sea water in the tube, so the fresh water temperature is always kept at the proper level by the thermostat.

Sea water is delivered by the sea water pump and fed through tubes located inside the cooling pipe to cool the fresh water.

Sea water flows from the heat exchanger into the mixing elbow, and is discharged with the exhaust gas.

2. Sea Water Pump

The sea water pump used for the fresh water-cooled engine is the rubber impeller pump; it is the same type as used for the sea water-cooled engine.
The same sea water pumps are used for models 2GM20F, 3GM30F and 3HM35F; these are also the same types as used for the model 3HM35(C) sea water-cooled engine. However, in the 3HM35F model, the pulley ratio is changed to increase the discharge volume.

	2GM20F, 3GM30F	3HM35F
Engine speed (Max.)	3600rpm	3400rpm
Pulley ratio Crank shaft/Pump shaft	PCø65/PCø85	PCø73/PCø85
Pump shaft speed	2700rpm	2900rpm
Suction head	1m (3.28ft)	1m (3.28ft)
Total head	4m (13.12ft)	4m (13.12ft)
Delivery capacity	1600ℓ/h	1700ℓ/h

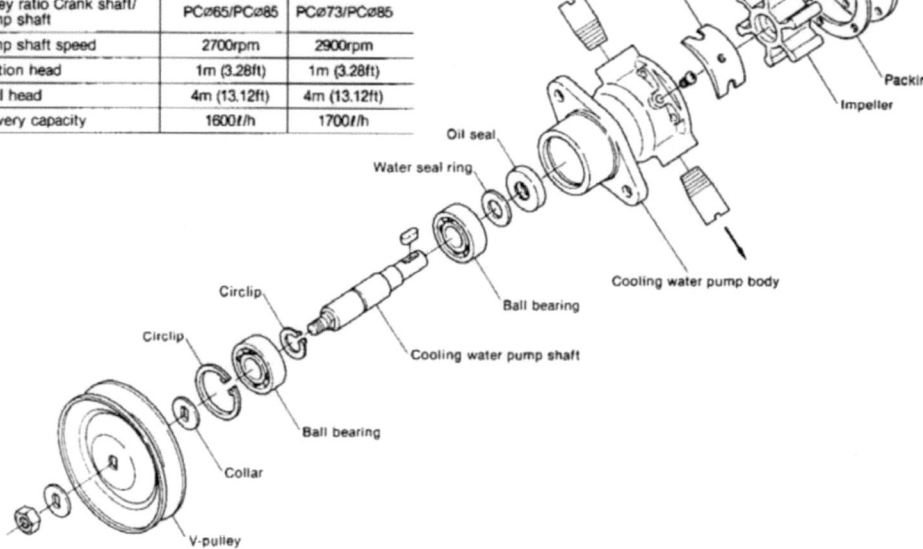

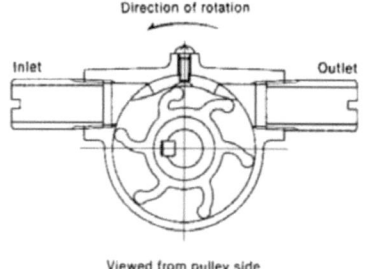

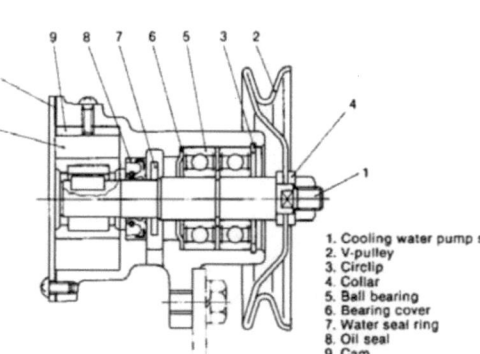

1. Cooling water pump shaft
2. V-pulley
3. Circlip
4. Collar
5. Ball bearing
6. Bearing cover
7. Water seal ring
8. Oil seal
9. Cam
10. Packing
11. Impeller

NOTE: For details on disassembly and reassembly, handling precautions and inspection, refer to "Chapter 7, Section 2. Water pump (P.7-5)".

3. Fresh Water Pump

3-1. Pump construction

The fresh water pump is a centrifugal type pump and is used to move fresh water from the fresh water tank, through the cooling passages in the cylinder block and cylinder head, and then back to the fresh water tank.

The fresh water pump is composed of a pump body, impeller, pump shaft, bearing unit and seals. It is driven by a belt and pulley arrangement at the end of the pump shaft.

The packed bearing unit supports the shaft with roller bearings. It cannot be disassembled.

The impeller is equipped with multiple blades and is mounted on the pump shaft.

The mechanical seal prevents water entering from around the pump shaft. The impeller seal is fixed to the impeller side with spring pressure applied from the pump body side.

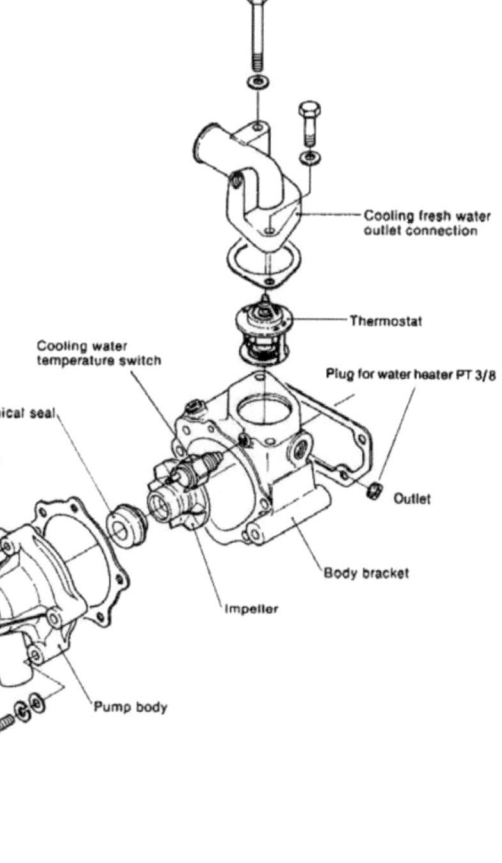

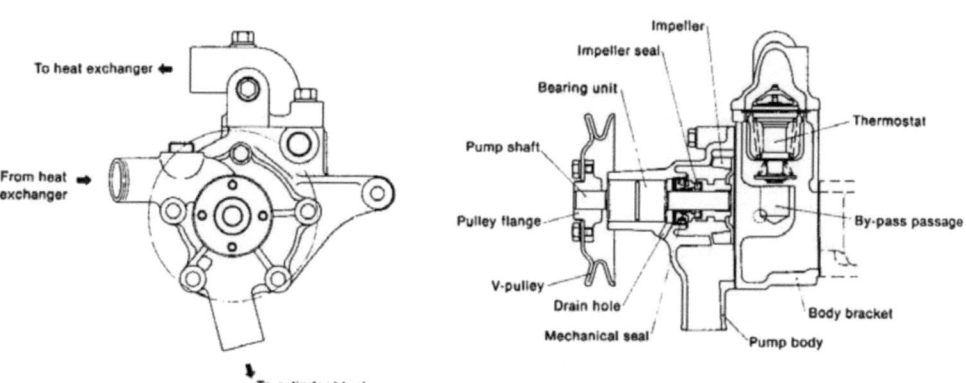

Chapter 8 Fresh Water Cooling System
3. Fresh Water Pump

3-2. Pump capacity and characteristic

	2GM20F, 3GM30F	3HM35F
Crank shaft speed	3600rpm	3400rpm
Pulley ratio Crankshaft/ Pump shaft	PCØ127/PCØ103	PCØ138/PCØ103
Pump shaft speed	4400rpm	4500rpm
Delivery capacity	4000ℓ/h	4200ℓ/h
Total head	3m (9.84ft)	3m (9.84ft)

NOTE: The same type of fresh water pump is used for models 2GM20F, 3GM30F and 3HM35F.

3-3. Pump disassembly

Disassembly of the fresh water pump is difficult and should not be attempted. Faulty units should be replaced. The pump assembly should not be disassembled from the pump body brackets, unless absolutely necessary.

kgf-cm(ft-lb)

Tightening torque for pump setting bolts	40—80 (2.89 ~ 5.79)

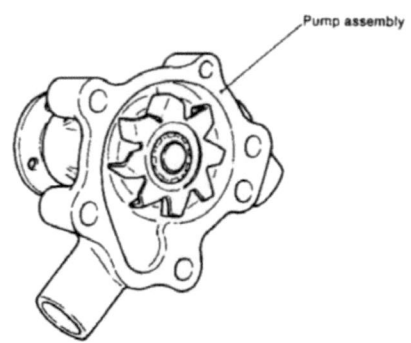

Pump assembly

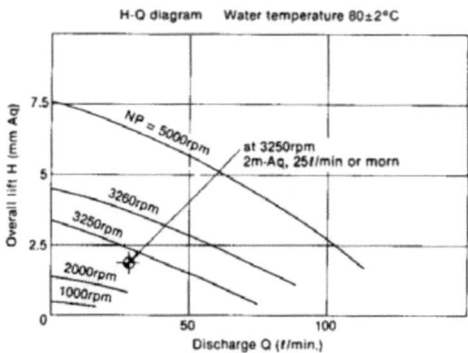

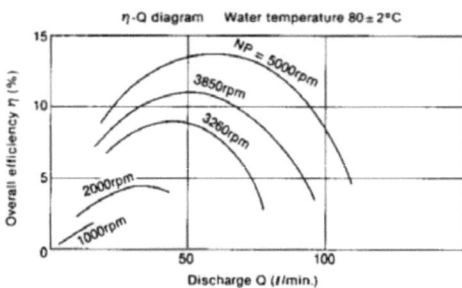

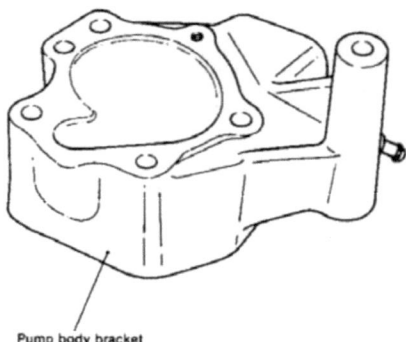

Pump body bracket

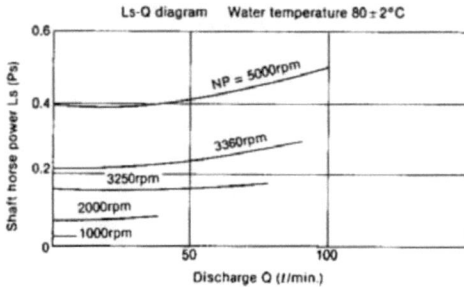

3-4. Inspection and measurement

(1) Confirm smooth rotation by rotating the impeller by hand.
   When the rotation is not smooth, due to bearing play or friction, or abnormal noise is heard, replace the entire pump assembly.
(2) Impeller inspection
   Check impeller for damage, corrosion and water. Replace if required.
(3) Check the holes drilled in the cooling water passage or by-pass passage, and clean or unblock where necessary.

Chapter 8 Fresh Water Cooling System
3. Fresh Water Pump

SM/GM(F)(C)·HM(F)(C)

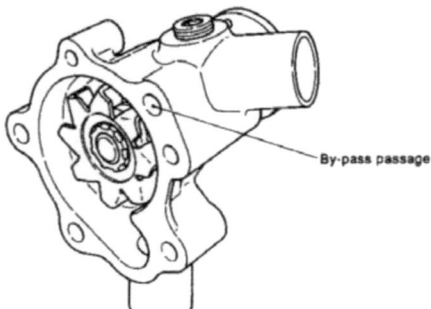
By-pass passage

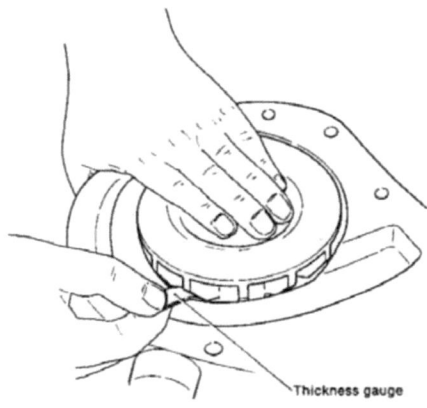
Measuring clearance between impeller and pump body

Thickness gauge

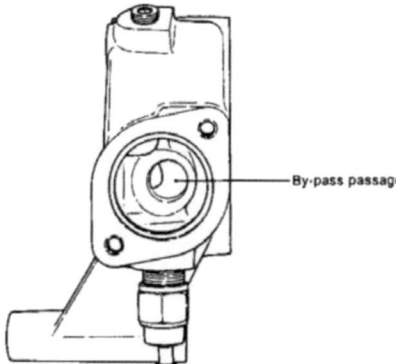
By-pass passage

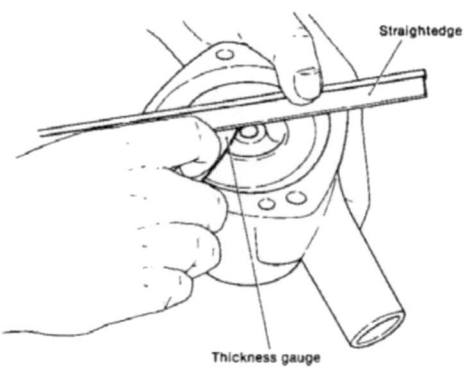
Measuring clearance between impeller and pump body bracket

Straightedge

Thickness gauge

(4) Where water leakage is heavy, due to wear or a damaged mechanical seal and impeller seal, replace the pump assembly with a new one.

(5) Pump body and pump bracket inspection
Clean deposits and rust from body and bracket. Replace if heavily worn or corroded.

(6) Impeller clearances.

mm (in)

	Maintenance standard
Clearance between impeller and body	0.3 ~ 1.1 (0.0118 ~ 0.0433)
Clearance between impeller and bracket	0.5 (0.0197)

To measure clearance between impeller and body, insert a thickness gauge between the two parts at an oblique angle between the two parts.

To measure clearance between impeller and bracket, place a straightedge on the pump body surface and insert a thickness gauge between the straightedge and impeller.

4. Heat Exchanger

4-1. Construction

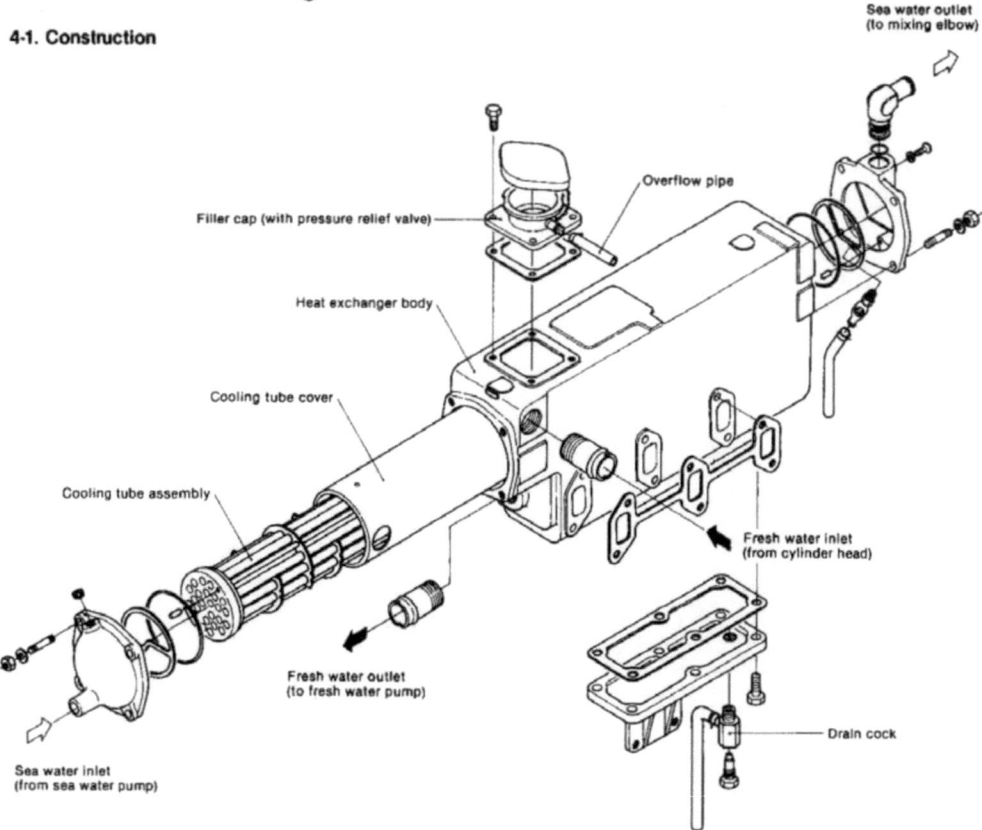

The heat exchanger uses sea water to cool the fresh water, which has reached a high temperature, while being circulated in the cylinder block.
The heat exchanger is a cooling tube which consists of 24 slender tubes and baffle plates, and a cooling tube cover. Sea water passes through the slender tubes, and fresh water passes through the flow path formed between the tubes and baffle plates inside the cooling tube cover.
The lower part of the heat exchanger stores the fresh water, acting as a fresh water tank. An exhaust gas passage, leading out of the storage position, is integrated with the water-cooled exhaust manifold.

Chapter 8 Fresh Water Cooling System
4. Heat Exchanger

SM/GM(F)(C)·HM(F)(C)

The filler cap on top of the heat exchanger is equipped with a pressure relief valve. When pressure exceeds the specified limit, this valve opens to release pressure through the overflow pipe.

On the other hand, when the cooling system pressure becomes negative in relation to the atmospheric pressure, air enters from the overflow pipe.

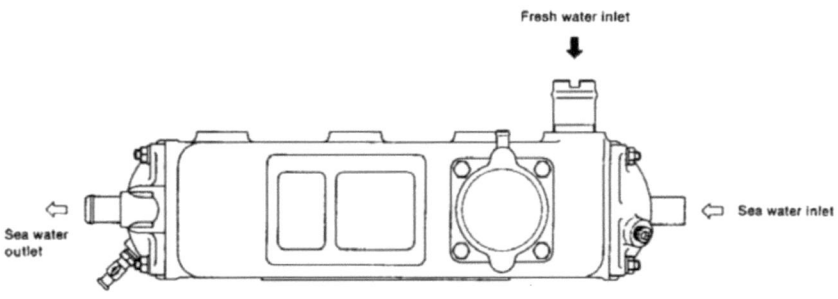

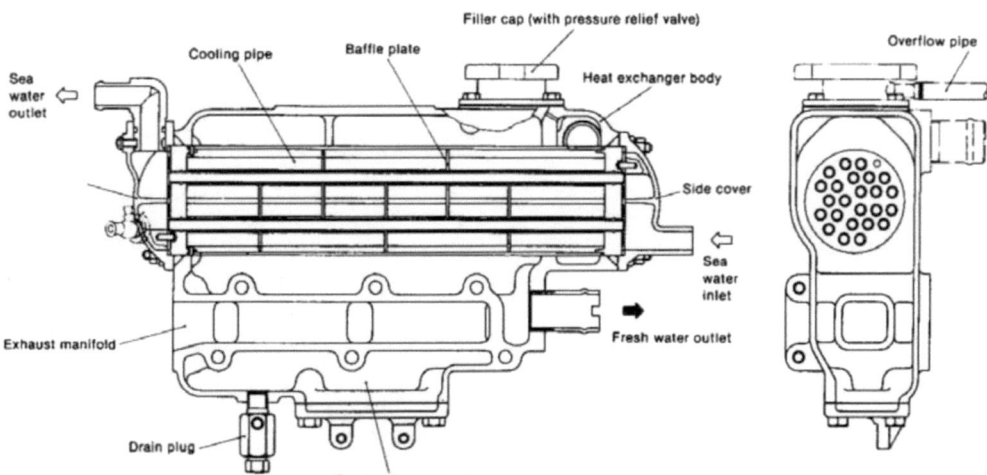

Chapter 8 Fresh Water Cooling System
4. Heat Exchanger

SM/GM(F)(C)·HM(F)(C)

4-2. Water flow in water cooling tube

Fresh water enters the cooling tube from a hole drilled at one end of the tube. It then passes through the flow passage formed by the baffle plates and the tube cover and into the water storage through a hole of the other end.

Sea water enters the side cover at the sea water inlet side, passes through 8 tubes guided by the side cover ribs and then leaves the side cover at the sea water outlet side. Here it passes through another 8 tubes guided by the side cover ribs, and returns to the side cover at the inlet side.
At the inlet side, it is guided by the remaining 8 tubes as at the outlet side, and then flows out to the mixing elbow from the outlet connection via the side cover at the outlet side.

4-3. Specifications

Model of engine		2GM20F	3GM30F	3HM35F
Output (DIN 6270 B rating)	kw/rpm	18.2/3600	27.3/3600	25.4/3400
Pipe dia. X pieces	mm	ø6/ø8 × 24	ø6/ø8 × 24	ø6/ø8 × 24
Radiation area	m^2	0.119	0.163	0.208
Radiation area/HP	m^2/HP	0.0066	0.0060	0.0061
Fresh water capacity	l (cu. in)	2.9 (177.0)	3.4 (207.5)	4.9 (299.0)

Chapter 8 Fresh Water Cooling System
4. Heat Exchanger

SM/GM(F)(C)·HM(F)(C)

4-4. Disassembly

(1) Remove the side covers and pull out the cooling pipe and rubber packings.
NOTE: After the cooling pipe is removed, always replace the rubber packings on both side covers.
(2) Remove filler cap and port.
(3) Remove lower cover and packing.

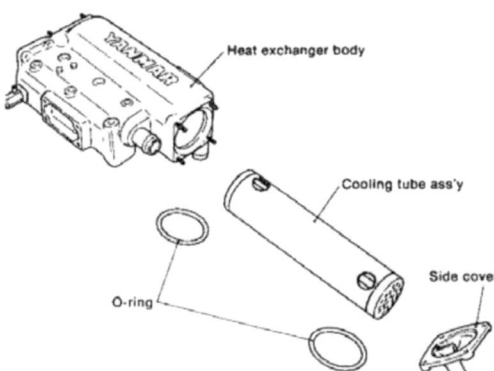

4-5. Inspection and cleaning
4-5.1 Cooling pipe

(1) Inspect for dirt and deposits in the tubes. Clean as required.

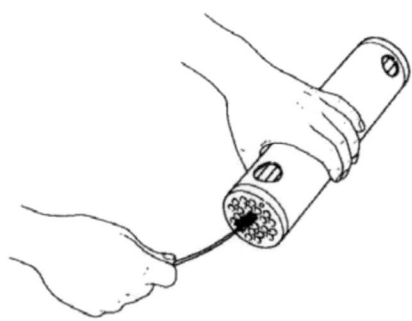

(2) Inspect caulked portions of tubes and flanges for damage. Repair or replace as required.
(3) Inspect the cooling pipe and tubes for leaks. Repair as required.
(4) Check for clogged water passages. Clean as required.

4-5.2 Heat exchanger body

(1) Check for dirt and corrosion build-up inside body and on side covers. Replace if corroded, broken or otherwise damaged.
(2) Check joints at sea water inlet and outlet ports and fresh water inlet and outlet ports. Retighten any loose screws and clean pipes as required.
(3) Check drain cock for clogging. If clogged, clean or repair as required. Retighten screws if necessary.
(4) For inspection of filler cap, anticorrosion zinc, and thermostat, see below.

4-5.3 Leakage test

(1) Test with compressed air and test tank. Seal fresh and sea water ports with rubber caps and immerse tank in a test tank filled with water. Inject compressed air through the overflow pipe and check for air bubbles.
NOTE: Air pressure should be $0.5 \sim 2.0 kgf/cm^2 (7.11 \sim 28.45 lb/in^3)$.

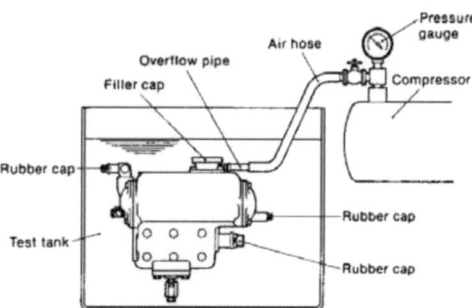

Leakage test using compressed air and test tank

(2) Test using pressure tester
Seal fresh and sea water ports with rubber caps and fill the tank completely with water. Replace the filter cap with a pressure tester and pressurize the tank. If there is a leak, the tank cannot be pressurized or it will only be able to retain pressure for a short time.

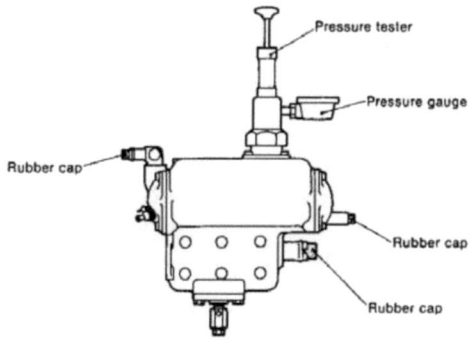

Leakage test using a pressure tester

8-10

5. Filler Cap and Subtank

5-1. Filler cap construction

The filler cap is placed on the fresh water inlet port and is equipped with a pressure control valve.
To attach, place the rocking tab (extension on the attachment section) on the flyneck cam. Then, turn and tighten.
The top seal touches the flyneck tap seat while the pressure valve touches the lower seat.

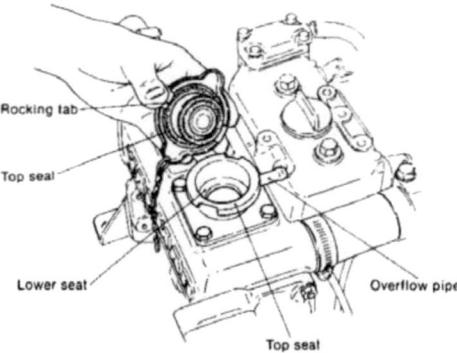

5-2. Filler cap pressure control

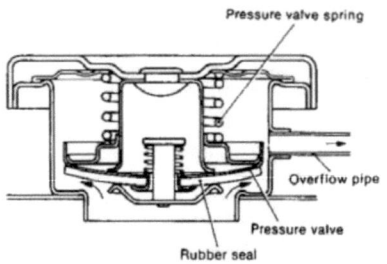

Pressure valve operation

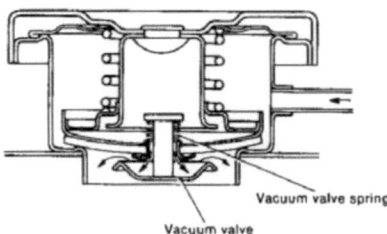

Vacuum valve operation

When the cooling system pressure is within the specified range $0.9 kgf/cm^2$ (12.80 lb/in.^{2}), the pressure valve and vacuum valve are tightly closed on their valve seats. When pressure rises, the pressure valve opens and vapor is discharged from the overflow pipe. When the water cools down and the pressure in the system is lower than atmospheric pressure, the vacuum valve opens and air enters the system through the overflow pipe.
To prevent the pressure valve from opening and resulting water loss, the cooling system can be equipped with a subtank, described below.

Action of Pressure control Valve

Pressure Valve	Opens at $0.9 kgf/cm^2 G$ (12.80 lb/in^2)
Vacuum Valve	Opens at $0.05 kgf/cm^2 G$ (0.71 lb/in^2) or below

5-3. Filler cap inspection

(1) Remove all deposits and rust, check for damage and wear on the seat contacting surfaces, and check spring for proper functioning. Repair or replace as required.
(2) Tester inspection
Attach adaptor and filler cap to tester.
Increase pressure and if pressure remains constant for six seconds, the cap is normal. If pressure does not increase or does not remain constant for six seconds, check for defects. Repair or replace as required.

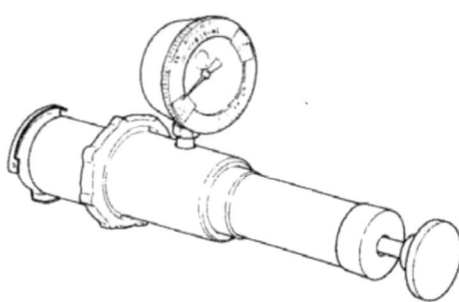

5-4. Subtank function

When the cooling system pressure rises above $0.9 kgf/cm^2$ (12.80 lb/in^2.), the pressure valve opens and vapor is released, reducing the amount of water in the cooling system. The subtank collects this vapor where it condenses. Then, when cooling system pressure falls below atmospheric pressure, the water in the subtank is siphoned back to the main tank.
Use of a subtank is highly recommended, since this allows the engine to be run for longer periods between water replenishment, and the need to open the filler cap is eliminated, thereby removing one possible cause of accidents.

Chapter 8 Fresh Water Cooling System
5. Filler Cap and Subtank

SM/GM(F)(C)-HM(F)(C)

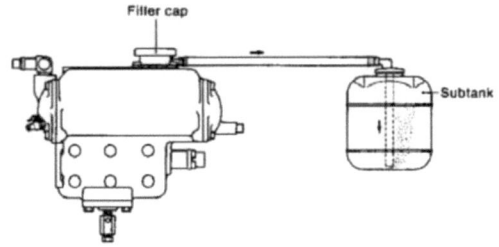

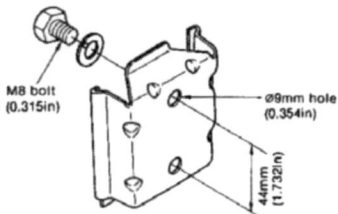

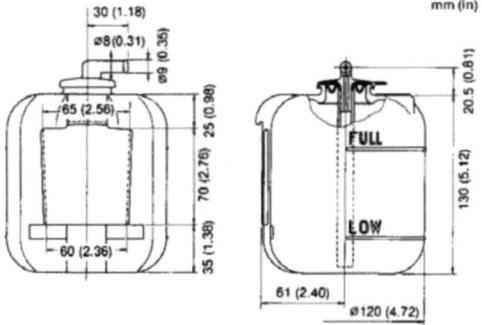

Subtank capacity	Over all capacity	1.25 (76.28)
	Full scale position	about 0.8 (48.82)
	Low scale position	about 0.2 (12.20)
Part No.		120445-44530

(1) Mount the subtank at the same height as the fresh water tank.
(2) Ensure that the length of the overflow pipe is no more than 1m (39.37 in.), and that it does not break.

NOTE: If a subtank is not used, be careful not to immerse the overflow pipe in the bilge, since this can cause bilge water to be siphoned into the cooling system.

5-6. Maintenance during use
(1) Check that when the cooling water is cold the level is within the specified range.
(2) Check that the overflow pipe is not broken, and also that the holes are not blocked up.

5-5. Installation of subtank

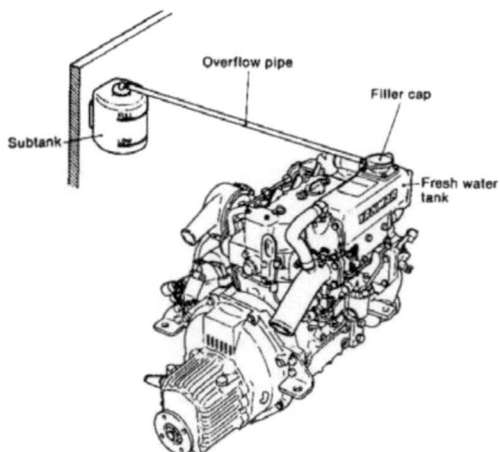

8-12

Printed in Japan
0000A0A1361

6. Thermostat

6-1. Operation

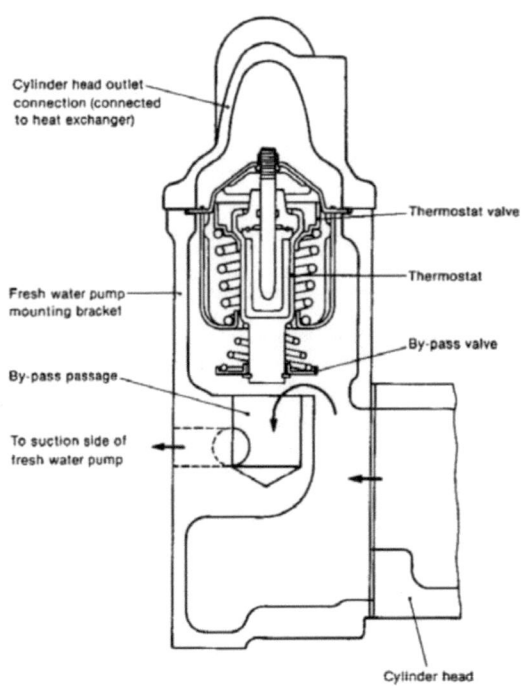

When valve is closed (by-pass passage is opened)

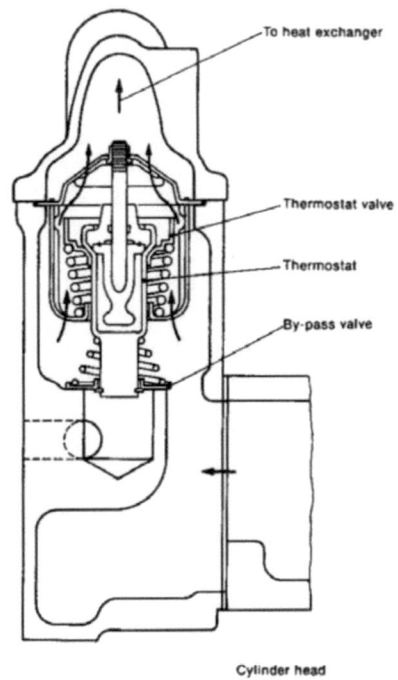

When valve is opened (by-pass passage is closed)

The thermostat opens and closes the by-pass valve and thermostat valve according to the temperature changes of the fresh water in the engine, adjusts the flow of fresh water to the heat exchanger and keeps the fresh water temperature in the engine at the correct level.
The thermostat in the fresh water-cooled engine is a bottom-by-pass type, as shown in the figure, and is installed inside the fresh water pump bracket which combines with the cylinder head cooling water outlet passage.
The thermostat valve is closed while the fresh water temperature is low, and fresh water is fed to the fresh water pump inlet through the drilled hole in the by-pass passage, to be circulated inside the engine.
When the fresh water temperature rises over the valve opening temperature, the thermostat valve opens, and fresh water is fed to the heat exchanger and where it is cooled and then fed to the fresh water pump. With the thermostat valve open, the by-pass passage is throttled. The by-pass passage is completely closed as the temperature rises.

Chapter 8 Fresh Water Cooling System
6. Thermostat

SM/GM(F)(C)·HM(F)(C)

6-2. Construction

A wax-pellet type thermostat is used for this engine. The "wax-pellet" type is the description given to a quantity of wax in the shape of a small pellet. When the temperature of the cooling water rises, the wax melts and its volume expands. The valve is opened or closed by these variations in volume.

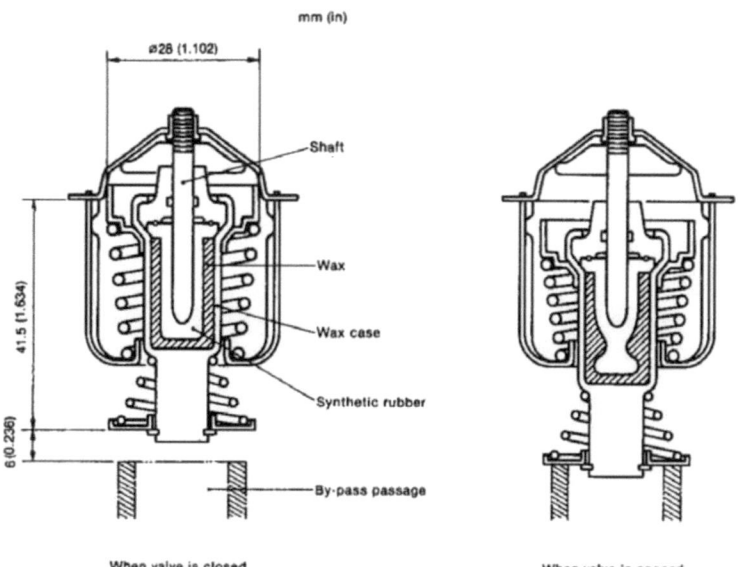

When valve is closed   When valve is opened

Thermostat operating temperature   °C (°F)

Opening temperature	71° (159.8)
Full open temperature (Temperature corresponding to 8mm or more valve lift)	85° (185)

Characteristic of Thermostat

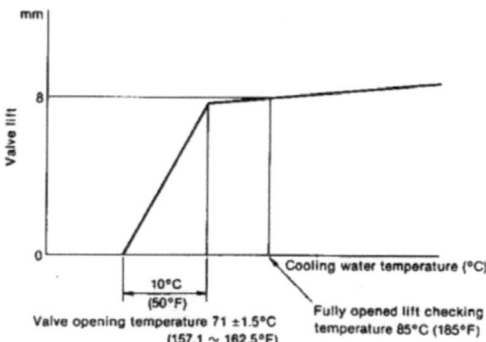

Valve opening temperature 71 ±1.5°C (157.1 ~ 162.5°F)
Fully opened lift checking temperature 85°C (185°F)

8-14

Chapter 8 Fresh Water Cooling System
6. Thermostat

SM/GM(F)(C)·HM(F)(C)

6-3. Inspection

(1) Remove the cooling water outlet connection at the top of the fresh water pump mounting bracket and take out the thermostat.
Remove all deposits and rust, check functioning and inspect parts. Replace if performance has deteriorated or if the spring or other parts are excessively corroded, deformed or otherwise unsuitable.

(3) In general, inspect the thermostat after every 500 hours of operation. However, always inspect it when the cooling water temperature has risen abnormally and when white smoke is emitted for a long period of time after the engine starts.

(4) Replace the thermostat when it has been in use for a year, or after every 2000 hours of operation.

Part No. code of thermostat	121750-49800

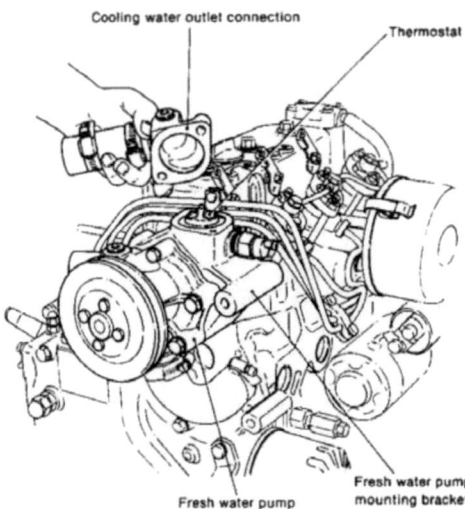

(2) Testing the thermostat
Place the thermostat in a container filled with water. Heat the container with an electric heater. If the thermostat valve begins to open when the water temperature reaches about 71°C and becomes fully open at 85°C, the thermostat may be considered all right. If its behaviour differs much from the above, or if it is found to be broken, replace it.

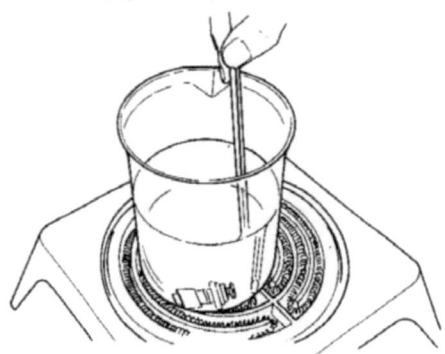

7. Cooling Water Temperature Switch

The cooling water temperature switch is identical to that for the sea water-cooled engine in shape and dimension, but care must be taken when parts are replaced as the operating temperature is different.
This can be checked by the seal color.

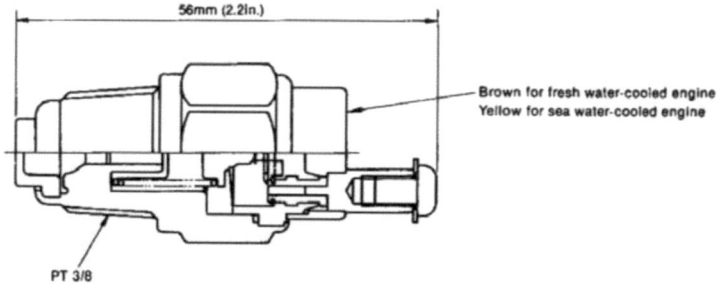

Operating temperature		Current capacity	Response time	Indication color	Parts code
ON	OFF				
95°C(202~193°F)	88°C(187°F) or higher	DC 12V 1A	Within 60 sec.	Green	127610-91350

8. Precautions

8-1. Ventilator

The surface temperature of fresh water cooled engines is higher than sea water cooled engines. Therefore, if the engine room is not well ventilated, engine room temperatures can rise to a point where they will adversely influence engine performance.

8-2. Cooling water

(1) Fresh water
Use clean soft water as cooling water. Hard water will cause calcium build-up, poor heat transmission and a drop in the cooling affect, resulting in overheating.

(2) Fresh water tank capacity ℓ (cu. in)

Model	Capacity
2GM20F	2.9 (177.0)
3GM30F	3.4 (207.5)
3HM35F	4.9 (299.0)

Remove the cap from the fresh water cooler, and check the water level. If the water level is below the top of the cooling pipe, add clean soft water up to the iron plate at the bottom of the filler.

If water is added up to the mouth of the fresh water tank, about 50cc of water will overflow from the filler immediately after the engine is started. This is normal, and is caused by the increase in the volume of the water as its temperature rises. If the water filler cap is removed after the engine has been stopped and allowed to cool, the water level will be 2—3cm from the top of the filler. This is also normal, and is caused by the overflow of the unnecessary water as the temperature of the water rises.

(3) Cooling water (fresh water) level check
Check the level of the cooling water (fresh water) before daily operation. A low cooling water level can cause insufficient pump discharge and the accumulation of scale in the heat exchanger.

(4) Cooling water leakage check during operation
Although checking for water and oil leakage during operation is generally necessary, check for fresh water leakage with special care.
Fresh water leakage is directly related to seizing of the engine.

(5) Fresh water replacement
Replace water every 500 hours. Always use an anti-rust agent.
To drain the water, open the cooling water drain cock and remove the water filler cap. If the filler cap is not removed, a vacuum will be created in the water jacket and not all the water will be drained.

(6) Removing the filler cap
Do not attempt to remove the water filler cap at the top of the fresh water tank while the engine is running, or while the engine is still hot after it has been stopped; steam will escape and may cause serious injury. If removal of the filler cap is unavoidable, place a piece of cloth over the cap and turn the cap slowly, making sure you are in a safe position even if steam escapes.

8-3. Antifreeze

(1) Use permanent type antifreeze in the winter. Freezing of the fresh water will damage the heat exchanger, cylinder head and water jacket.

(2) Antifreeze use
1) Before adding antifreeze, clean the cooling system and check for leaks.
2) Select mixing ratio according to the following table.

ℓ (cu. in)

Temperature	−5°C	−10°C	−15°C	−20°C	−25°C	−30°C
Mixing ratio	12%	22%	29%	35%	40%	44%
2GM20F	0.35 / 21.40	0.64 / 39.10	0.84 / 51.30	1.02 / 62.20	1.16 / 70.80	1.28 / 78.10
3GM30F	0.41 / 25.00	0.75 / 45.80	0.99 / 60.40	1.19 / 72.60	1.36 / 83.00	1.50 / 91.50
3HM35F	0.59 / 36.00	1.08 / 65.90	1.42 / 86.70	1.72 / 105.00	1.96 / 119.60	2.21 / 129.40

NOTE: The temperature selected in the above table should be 5°C lower than the lowest expected temperature in the area.
NOTE: Check the mixing ratio carefully, especially when using premixed coolant.

3) Tighten the drain cock and fill the cooling system. Then, run the engine for approx. 5 to 30 minutes to make sure the solution is well mixed.

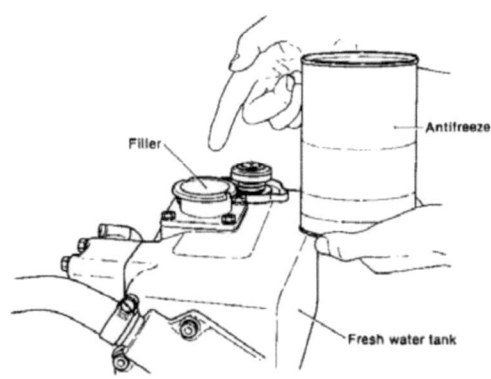

NOTE: Some antifreeze solutions will corrode aluminum. Check carefully before use.
NOTE: When antifreeze protection is no longer necessary, drain water, flush cooling system and refill with fresh water.

8-4. Rust inhibitor

When the fresh water is changed, a rust inhibitor must be added to the new water to prevent rusting.
   Rust inhibitor : Fresh water = 1 : 10
Flush cooling system with fresh water, fill with proper rust inhibitor and then top-up cooling system with fresh water.

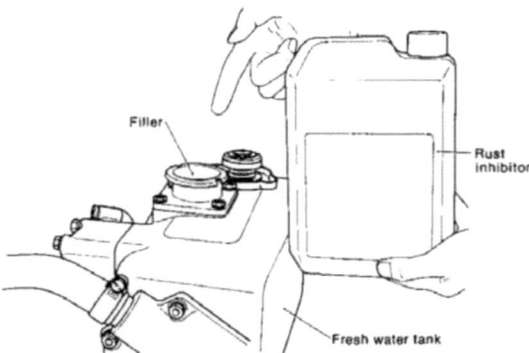

8-5. Idling the engine when stopping

Always idle the engine for ten minutes immediately after starting and prior to stopping. Be sure to idle the engine adequately, especially before stopping. Stop the engine only after its temperature has dropped sufficiently. If the engine is stopped while hot, the hot fresh water will cause the temperature of the water in the heat exchanger pipe to rise, causing a build-up of calcium deposits in the pipe and a drop in the cooling affect.

8-6. Cleaning the heat exchanger tube

If the heat exchanger tube through which the fresh water flows becomes extremely dirty, the cooling effect will deteriorate.
If the C.W. warning lamp lights periodically when the engine is run at the rated output, clean the tube in the fresh water tank with a cleaning agent and then flush the accumulated scale produced by cooling the fresh water from the tube.

CHAPTER 9
MODIFYING THE COOLING SYSTEM

1. General 9-1
2. Disassembly of Sea Water-Cooled Engine 9-2
3. Assembling modified parts
   to the Fresh Water-Cooled Engine 9-7
4. Cautions When the Engine is Installed Inboard 9-12

1. General

1-1. Direct sea water-cooled engine and fresh water-cooled engine

Engine models 2GM20, 3GM30 and 3HM35 are sea water-cooled, and models 2GM20F, 3GM30F and 3HM35F are fresh water-cooled.
The main parts of both sea water-cooled and fresh water-cooled engines are the same; only the cooling systems are different. Sea water-cooled engines can therefore be modified into fresh water cooling by the special parts kit prepared by YANMAR for this modification.

1-2. Modification method

When modifying a seawater-cooled engine into a fresh water-cooled engine, follow the sequence described in Section 2.

1-3. Testing a modified engine

Any engine modified as a fresh water-cooled engine must be given an operating test (running) to check for leakage. This test shall be made before delivery.

1-4. Warranty

Engines modified as fresh water-cooled engines are not covered by the general warranty.

1-5. Kit for modification into a fresh water-cooled engine

The kits for modification into a fresh water-cooled engine differ according to the engine model.
When ordering the modification kit state the following code number.

Applicable Engine Model	2GM20 → 2GM20F	3GM30 → 3GM30F	3HM35 → 3HM35F
Fresh water cooling kit	728271-99510	728374-99510	728671-99510
Mixing elbow Ass'y	O	—	—
Fresh water pump Ass'y		O	
Sea water pump Ass'y	O		—
Heat exchanger Ass'y	*O	*O	*O
Subtank Ass'y		O	
Thermostat Ass'y		O	
Cooling water pipe Ass'y	O		*O
Fuel oil pipe Ass'y	*O	*O	*O
Speed control cable bracket Ass'y	O	—	—
V-belt and other parts Ass'y		O	*O

NOTES: O parts marked are those included in the modification kit (necessary for modification).
— parts marked are those not included in the kit (unnecessary for modification).
*O parts marked are those which differ according to the engine model (not interchangeable).

2. Disassembly of sea water-cooled engine

2-1. Drain the cooling sea water

Locations of Cooling Water Drain Plugs

	2GM20	3GM30	3HM35
Cylinder block	O (Intake side)	O (Exhaust side)	O (Exhaust side)
Cooling water pump	O	O	O
Exhaust manifold	—	O	O

Note: CSW = Cooling Sea Water
CFW = Cooling Fresh Water

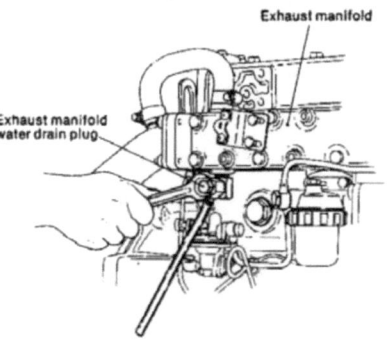

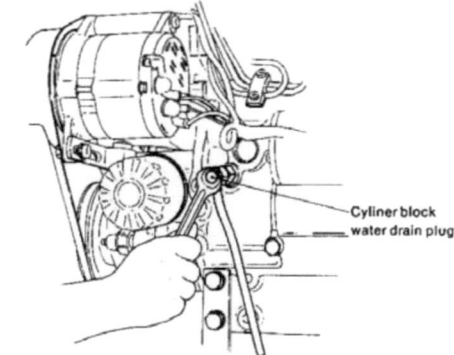

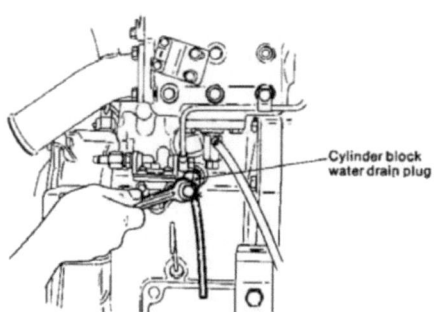

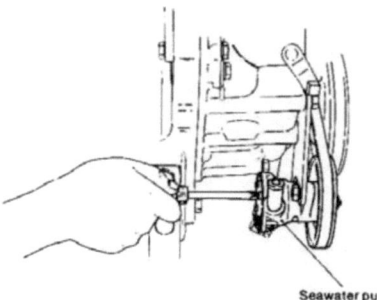

Seawater pump

2-2. Remove the cooling water pipe

(1) For model 2GM20, remove the CSW hose between the thermostat and mixing elbow.
(2) For models 3GM30 and 3HM35, remove the CSW hose between the thermostat and exhaust manifold.

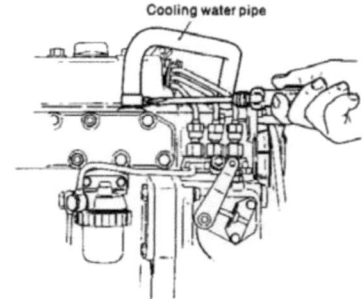

2-3. Remove the fuel oil pipe

(1) Remove the fuel pipe between the oil filter and fuel pump.
(2) Remove the fuel pipe between the fuel feed pump and fuel filter.

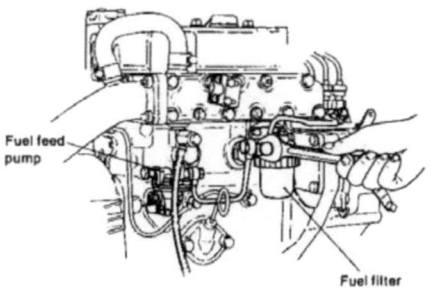

Chapter 9 Modifying The Cooling System
2. Disassembly of Sea Water-Cooled Engine

SM/GM(F)(C)·HM(F)(C)

2-4. Remove the fuel filter (2GM20)
For models 3GM30 and 3HM35, the filter may be removed as assembled to the exhaust manifold.

2-5. Remove the remote control bracket (2GM20)
For models 3GM30 and 3HM35, the bracket may be removed as assebled on the exhaust manifold.

2-6. Remove the mixing elbow (2GM20)
For models 3GM30 and 3HM35 the elbow may be removed as assembled on the exhaust manifold.

2.7 Remove the exhaust manifold (3GM30, 3HM35)
(1) For models 3GM30 and 3HM35, the exhaust manifold may be removed with the fuel filter, remote control bracket and mixing elbow assembled on the exhaust manifold.
(2) Remove the exhaust manifold fixing studs.

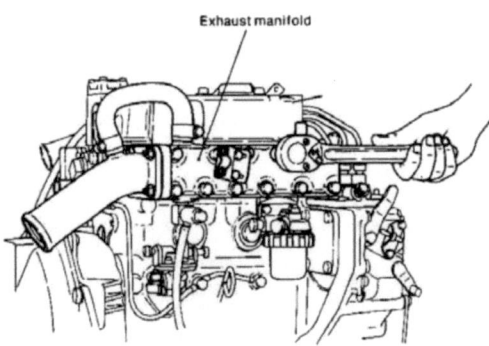

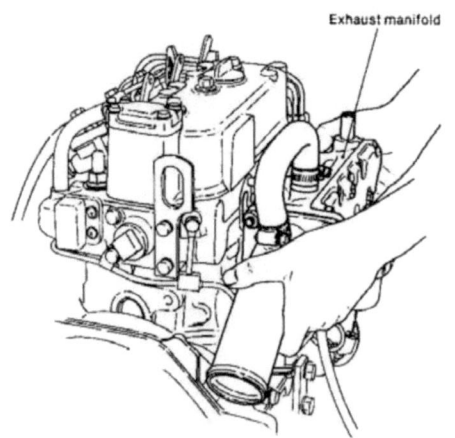

2-8. Remove the cooling water pipe
Remove the CSW hose between the CSW pump and by-pass metal fitting.

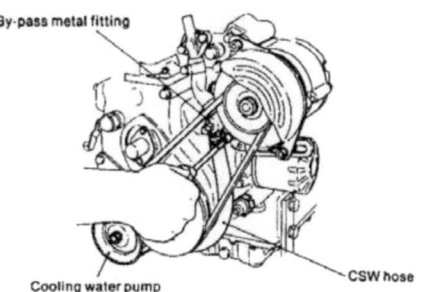

2-9. Remove the electrical wiring
Remove the wiring connected to the alternator and cooling water temperature sender.

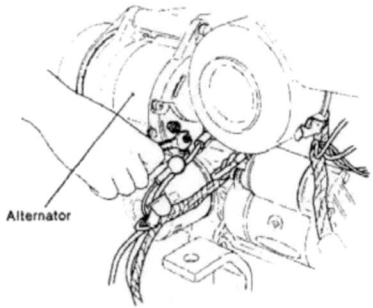

2-10. Remove the alternator
Remove the alternator cover and V belt after loosening the alternator adjusting bolt.
The alternator can be more easily removed when removed as assembled on the thermostat bracket.

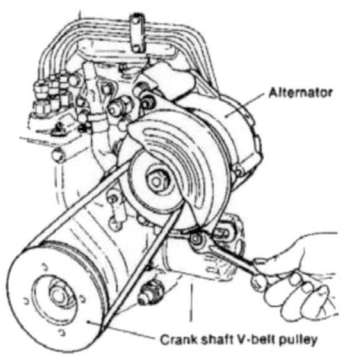

Chapter 9 Modifying The Cooling System
2. Disassembly of Sea Water-Cooled Engine

2-11. Remove the thermostat cover
(1) Loosen the cramp of the cooling water hose between the by-pass metal fitting and thermostat at the by-pass metal fitting side.

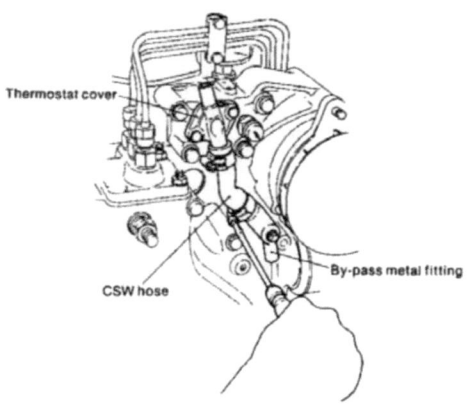

(2) Remove the thermostat cover.

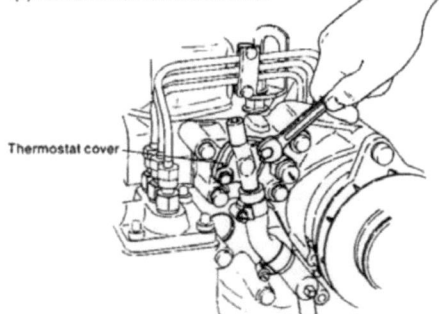

2-12. Remove the high pressure pipe anti-swing metal fitting from the thermostat bracket.

2-13. Remove the thermostat bracket
Remove the cooling water temperature sensor and alternator as assembled on the thermostat bracket.

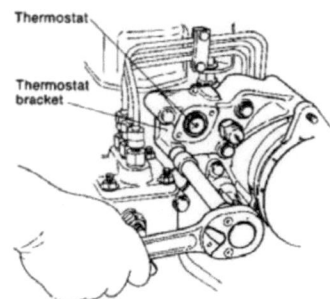

2-14. Remove the cooling water by-pass connection
(1) Remove the cooling water by-pass metal fitting (L-type joint).

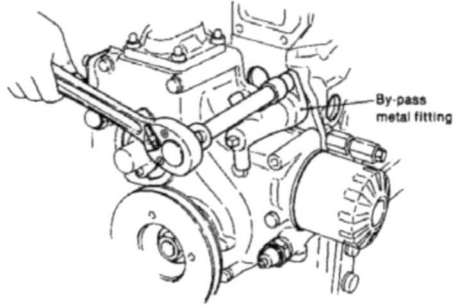

(2) Extract the by-pass connection screwed into the cylinder block.

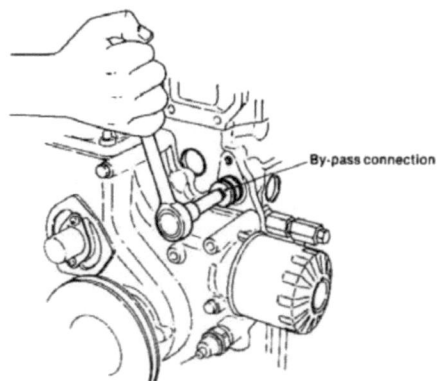

2-15. Remove the CSW pump

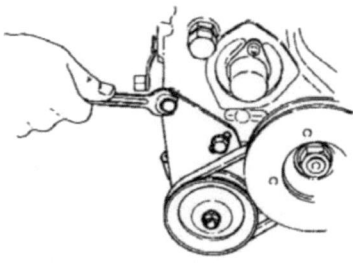

Chapter 9 Modifying The Cooling System
2. Disassembly of Sea Water-Cooled Engine

SM/GM(F)(C)-HM(F)(C)

2-16. Remove the intake silencer

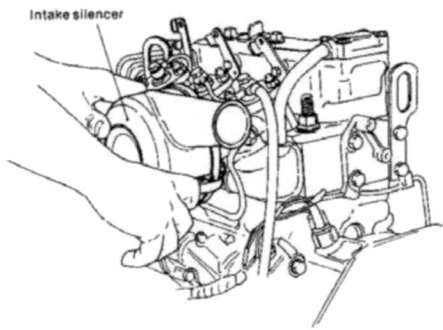

2-17. Remove the fuel high pressure pipe

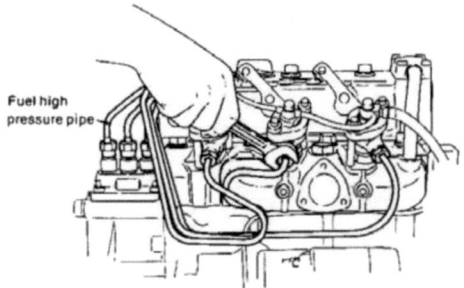

2-18. Remove the lifting hook at the front of the engine (3GM30, 3HM35)

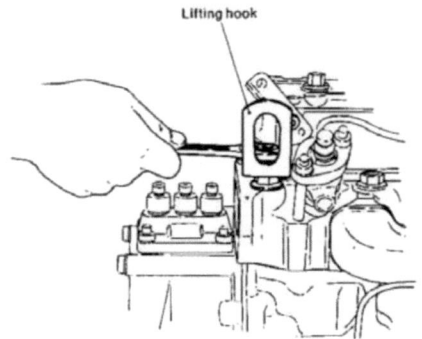

2-19. Remove the lifting hook at the rear or the engine together with the rear cylinder head cover

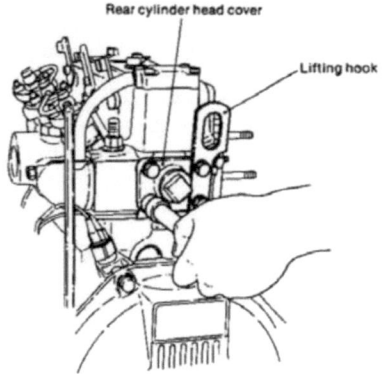

Chapter 9 Modifying The Cooling System
2. Disassembly of Sea Water-Cooled Engine

The disassembly necessary to modify a sea water-cooled engine into a fresh water-cooled engine is completed with this step. The removed parts, and the appearance of the engine after disassembly are shown below:

Removed parts

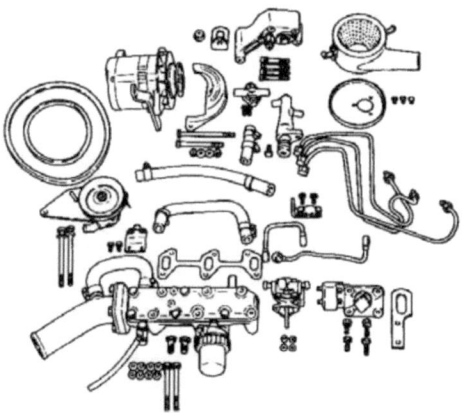

Appearance of engine after disassembly (Example Model, 3GM30)

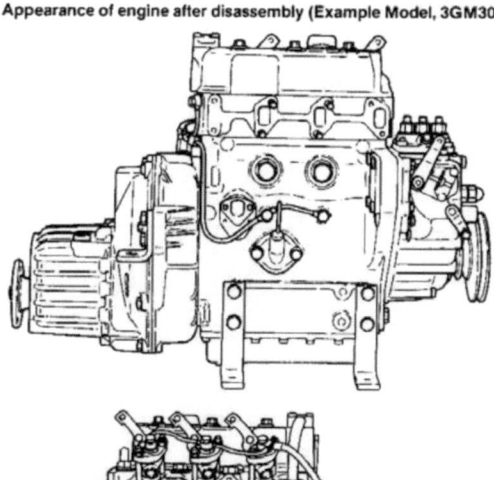

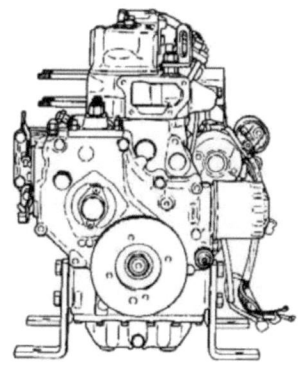

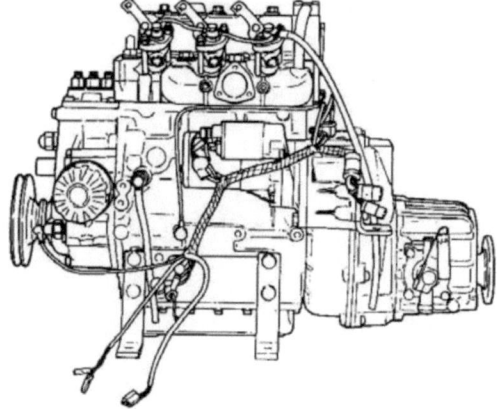

3. Assembling modified parts to the fresh water-cooled engine

The parts required to modify a sea water-cooled engine to a fresh water-cooled engine are as shown below.

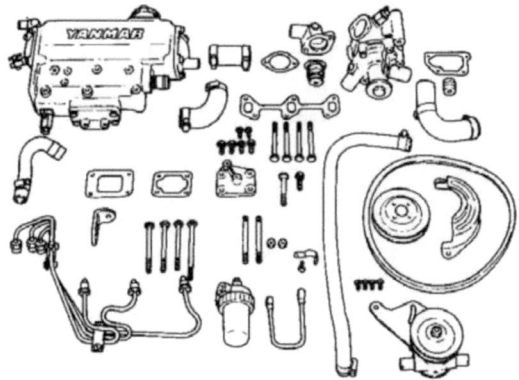

3-1. Assemble the rear cylinder head cover together with the rear lifting hook.

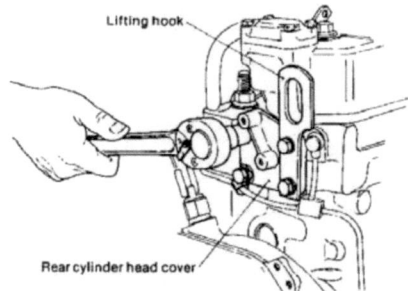

NOTE: New packing should be used.
Apply Threebond No.4 on both surfaces of packing.

3-2. Assemble the front lifting hook (3GM30, 3HM35)
The hook on the model 2GM20 is in a position not affected by the modification.

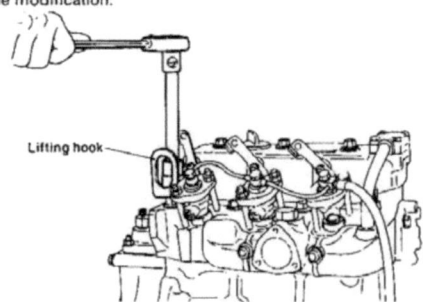

NOTE: Use the special lifting hook for the fresh water-cooled engine.

3-3. Assemble the fuel injection tube

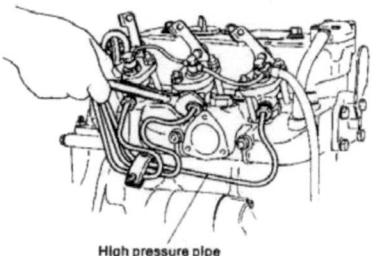

NOTE: Use the special high pressure pipe for a fresh water-cooled engine. The shape and dimensions are different from those for a sea water-cooled engine.

3-4. Assemble the intake silencer

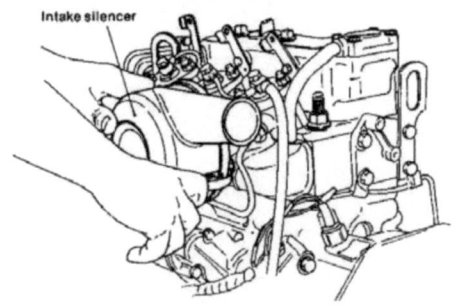

NOTE: The intake silencer is the same for both the fresh water and sea water-cooled engines.

Chapter 9 Modifying The Cooling System
3. Assembling Modified Parts to the Fresh Water-Cooled Engine

3-5. Assemble the CFW joint to the cylinder block
Apply Threebond No.20 to the threads and screw.

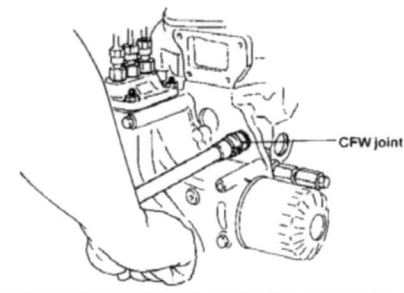

| Tightening torque | 2.5 ~ 3.5 kgf-m(18 ~ 25 ft-lb) |

3-6. Assemble the CFW pump assembly
Assemble after applying Threebond No.4 to both surfaces of the packing.

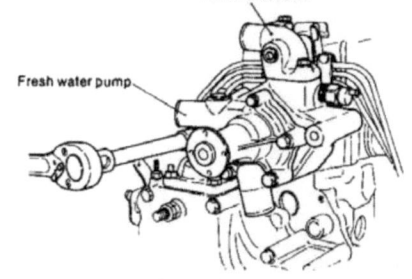

| Tightening torque | 2 ~ 2.5 kgf-m(14.5 ~ 18 ft-lb) |

3-7. Assemble the thermostat and thermostat cover
NOTE: Apply Threebond No.4 to both surfaces of the packing.

3-8. Assemble the CFW hose
(1) Connect the CFW hose between the CFW pump and cylinder block and tighten the hose clamp.

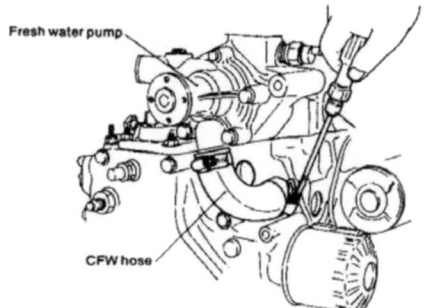

(2) Connect the CFW hose between the CFW pump and heat exchanger by connecting it to the CFW pump; tighten the hose clamp slightly. The hose clamp will be securely tightened after the heat exchanger is assembled.

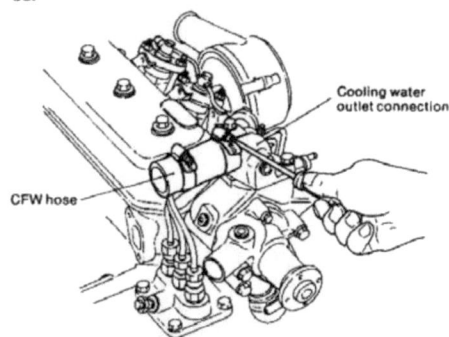

3-9. Assemble the CSW pump

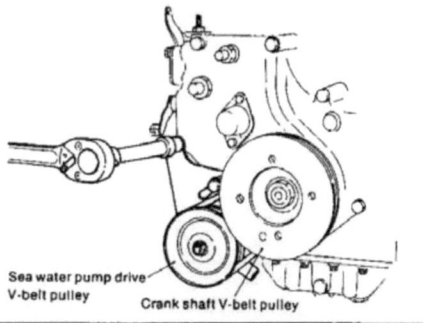

| V-belt tension Pushed with a force of 10kg (22 lb.) | 5 ~ 7mm (0.1969 ~ 0.2756 in.) |

3-10. Insert the stud bolt for fitting the heat exchanger
Apply Three bond 203M to the threads.

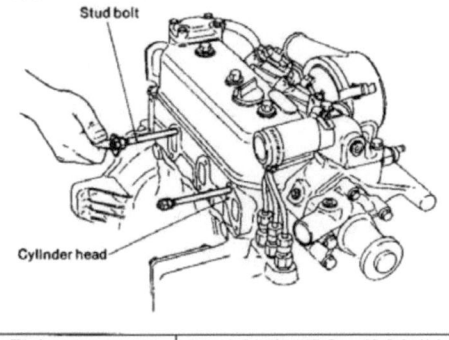

| Tightening torque | 1 ~ 1.5 kgf-m(7.3 ~ 10.9 ft-lb) |

Chapter 9 Modifying The Cooling System
3. Assembling Modified Parts to the Fresh Water-Cooled Engine

3-11. Assemble the heat exchanger

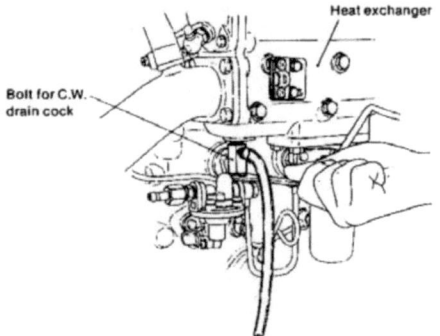

For model 3GMD, connect the pipes after removing the CW drain cock bolt at the bottom of the heat exchanger to prevent the pipe from jamming against the fuel feed pump.

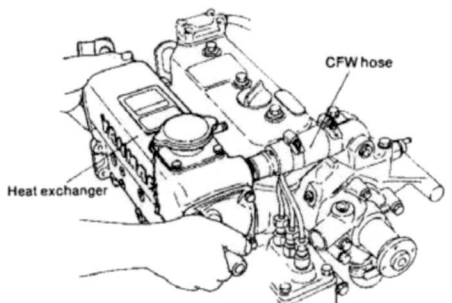

NOTE: New gasket packing must be used.

| Tightening torque | 2 ~ 2.5 kgf-m (14.5 ~ 18 ft-lb) |

3-12. Assemble the fuel filter

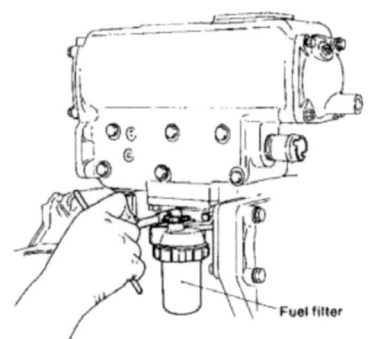

NOTE: The same fuel filter is used as for a sea water-cooled engine.

3-13. Assemble the fuel oil pipe

(1) Connect the fuel oil pipe between the fuel feed pump and fuel injection pump.
(2) Connect the fuel oil pipe between the fuel filter and fuel injection pump.

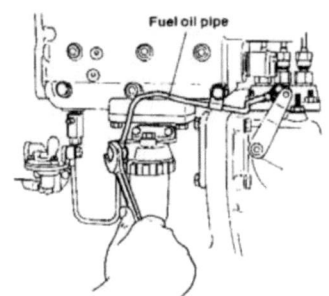

3-14. Assemble the CFW hose

Connect the CFW hose between the CFW pump and heat exchanger and tighten the hose clamp.

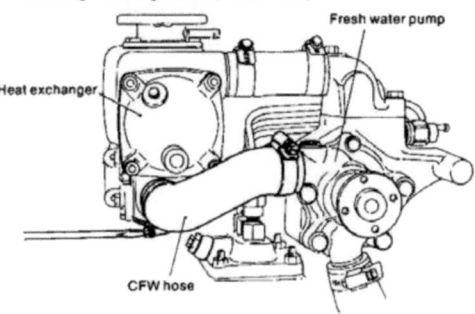

3-15. Assemble the CSW hose

Connect the CSW hose between the CSW pump and heat exchanger and tighten the hose clamp.

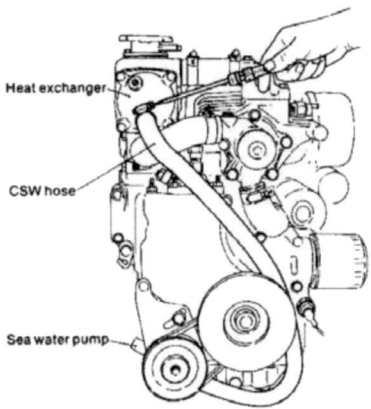

3-16. Assemble the mixing elbow

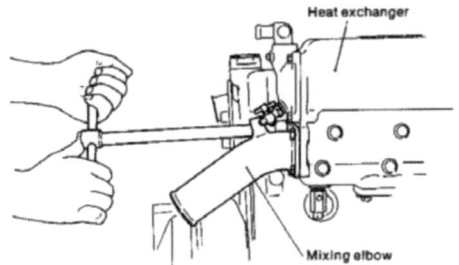

NOTE: New gasket packing must be used.

3-17 Connect the CSW hose between the head exchanger and mixing elbow and tighten the hose clamp.

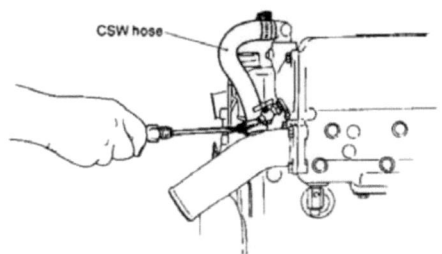

3-18. Assemble the CFW pump V-belt pulley

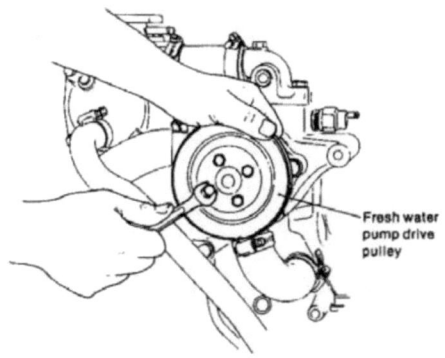

3-19. Assemble the alternator

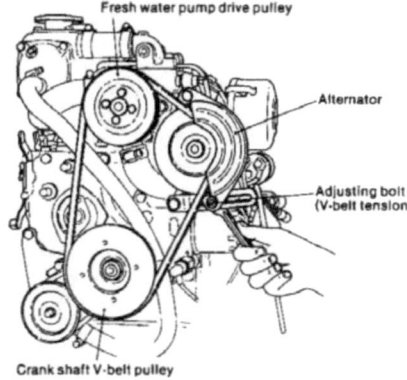

V-belt tension Depressed with a force of 10kg (22 lb.)	Approx. 10mm (Approx. 0.3937 in.)

3-20. Assemble the cooling water temperature sensor

First fit the cooling water temperature sender to the CFW pump and then assemble both units together.

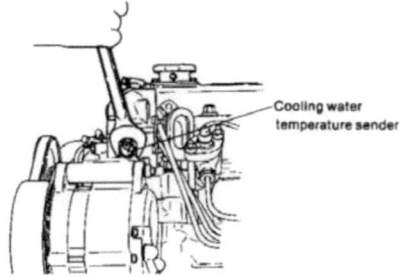

NOTE: Apply Threebond No.4 to the threads.

3-21. Connect electrical wiring

Connect the electrical wirings to the alternator and cooling water temperature sender.

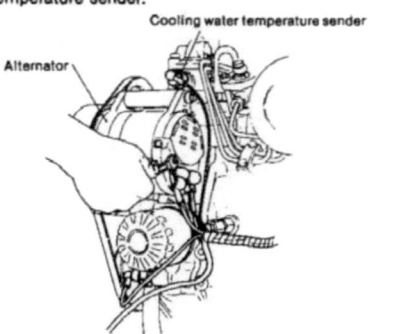

3-22. Assemble the remote control bracket.

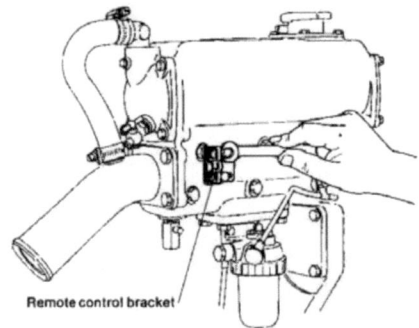
Remote control bracket

The sea water-cooled engine has now been modified as a fresh water-cooled engine.

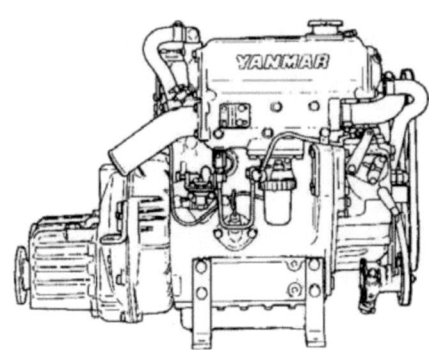

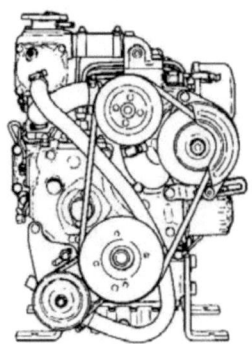

4. Cautions when the engine is installed inboard

(1) In the case of a fresh water-cooled engine, a fresh water subtank must be installed. For the installation method, refer to the "Installation of the subtank" section.
(2) A seawater drain cock and the fresh water drain plug are provided in the heat exchanger; a drain hose should be fitted to each plug.

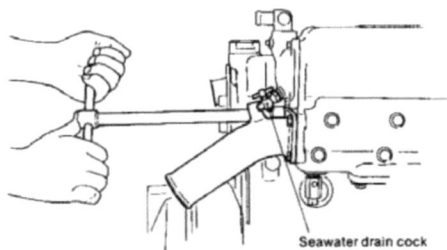

Seawater drain cock

(3) There is no problem when the engine is installed in a newly built ship, but when an engine in use is modified, care must be taken because the cooling water piping is different.

	2GM20 → 2GM20F	3GM30 → 3GM30F	3HM35 → 3HM35F
Hose at CSW pump inlet (Kingston cock—CSW pump) outer dia/inner dia	φ20/13 → φ24/17	φ20/13 → φ24/17	φ24/17 = φ24/17
Kingston cock to be used	10A → 15A	10A → 15A	15A

NOTE: Kingston cocks are optional.

10A	Part No.43662 — 010030	1GM10, 2GM20, 3GM30
15A	Part No.43662 — 015020	2GM20F, 3GM30F, 3HM35F

Pin diameter of mixing elbow is different
2GM20 → 2GM20F
(44mm)   (51mm)
(1.7323in.) (2.0079in.)

CHAPTER 10
REDUCTION AND REVERSING GEAR

[A] For Engine Models 1GM10, 2GM20(F) and 3GM30(F)
1. Construction.................................... 10-1
2. Shifting Device................................ 10-7
3. Inspection and Servicing..................... 10-14
4. Disassembly................................... 10-19
5. Reassembly................................... 10-24

[B] For Model 3GM35(F)
1. Construction.................................. 10-29
2. Installation................................... 10-33
3. Operation and Maintenance.................. 10-34
4. Inspection and Servicing..................... 10-35
5. Disassembly................................... 10-40
6. Reassembly................................... 10-44

[C] Marine Gear Models KM2P, KM3P and KM3V
for Engine Models 1GM10, 2GM20(F) and 3GM30(F)
1. Construction.................................. 10-50
2. Shifting Device................................ 10-56
3. Inspection and Servicing..................... 10-61
4. Disassembly................................... 10-68
5. Reassembly................................... 10-73

[D] V-drive Gear, Model KM3V
1. Construction.................................. 10-77
2. Specifications................................. 10-80
3. Power Transmission System.................. 10-81
4. Cooling System (Sea-water Cooling Engine)......... 10-82
5. Piping Diagrams............................... 10-85
6. Inspection and Servicing..................... 10-90
7. Shim Adjustment for V-drive Gear Shaft,
   and Backlash Adjustment for V-drive Gear Shaft and
   Drive Gear.................................... 10-92
8. Disassembly................................... 10-94
9. Reassembly................................... 10-97

Printed in Japan
0000 A 0 A 1361

Chapter 10 Reduction and Reversing Gear
1. Construction

SM/GM(F)(C)·HM(F)(C)

[A] For engine models 1GM10, 2GM20(F) and 3GM30(F)

1. Construction

1-1 Construction

This clutch is a cone-type, mechanically operated clutch. When the drive cone (which is connected to the output shaft by the lead spline) is moved forward or backward, its taper contacts with the large gear and transfers power to the output shaft.

The construciton is simple when compared with other types of clutch and it serves to reduce the number of components, making for a lighter, more compact unit which can be operated smoothly. Although it is small, the power transmission efficiency is high even under a heavy load. Its durability is high and it is reliable as high grade materials are used for the shaft and gear, and a taper roller bearing is incorporated. Power transmission is smooth as connection with the engine is made through the damper disc.

- The drive cone is made from special aluminum bronze which has both higher wear-resistance and durability. The drive cone is connected with the output shaft through the thread spline. The taper angle, diameter of the drive cone, twist angle, and diameter of the thread spline, are designed to give the greatest efficiency, thus ensuring that the drive cone can be readily engaged or disengaged.
- Helical gears are used for greater strength. The intermediate shaft is supported at 2 points to reduce deflection and gear noise.
- The clutch case, mounting flange and side cover are made from an aluminum alloy of special composition to reduce weight. It is also anticorrosive against seawater.
- As the damper disc is fitted to the output shaft, power can be transmitted smoothly. For the damper disc, springs of different strengths are used so that two stages of torque and twist angle are applied. That is, in the first stage, only the weak spring is used, and the strong spring comes into action for a torque higher than a predetermined value.

This prevents gear noise due to torsional vibration as well as absorbing shock when engaging.

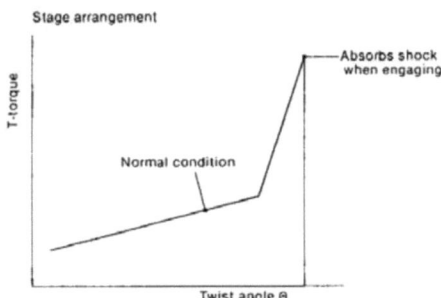

Stage arrangement

- The oil level dipstick hole doubles as a breather in addition to being the oil supply port. There is a small clearance between the dipstick and the inside of the dipstick tube which functions as a breather.
- The engagement between the cone and the large gear can be maintained even when the load on the propeller is zero. This is done by the action of the notch and spring joint on the operation lever in the operation device.

The operation device can still be used without adjusting the remote operation device when the cone is internally worn, because it is compensated for by the spring joint.

- In order to reduce friction on the operation lever shaft, a needle bearing is used to allow smooth operation.

10-1

Chapter 10 Reduction and Reversing Gear
1. Construction

SM/GM(F)(C)·HM(F)(C)

1·2 Specifications

Model			KM2-C			KM3-A		
For engine models			1GM10, 2GM20(F)			3GM30(F)		
Clutch			Constant mesh gear with servo cone clutch (wet type)					
Reduction ratio	Forward		2.21	2.62	3.22	2.36	2.61	3.20
	Reverse		3.06	3.06	3.06	3.16	3.16	3.16
Propeller shaft rpm (Forward) rpm			1540	1298	1055	1441	1303	1063
Direction of rotation	Input shaft		Counter-clockwise, viewed from stern					
	Output shaft	Forward	Clockwise, viewed from stern					
		Reverse	Counter-clockwise, viewed from stern					
Remote control	Control head		Single lever control					
	Cable		Morse, 33-C					
	Clamp		YANMAR made, standard accessory					
	Spring joint		YANMAR made, standard accessory					
Output shaft coupling	Outer diameter		ø100mm (3.93")					
	Pitch circle diameter		ø78mm (3.07")					
	Connecting bolt holes		4—ø10.5mm (4—ø0.41")					
Position of shift lever			Left side, viewed from stern					
Lubricating oil			SAE #10W-30, CC class					
Lubricating oil capacity			0.25ℓ			0.3ℓ		
Dry weight			9.5kg (20.9 lbs)			11.0kg (24.3 lbs)		

Models KM2C and KM3A reduction and reverse gear boxes, shafts and gears are the same except for the following items:
• No. of gear teeth (derives different gear ratios).
• Distance between bearings for input and output shafts.
• Clutch case, mounting flange.

Chapter 10 Reduction and Reversing Gear
1. Construction

SM/GM(F)(C)·HM(F)(C)

1-3 Power transmission system
1-3.1 Arrangement of shafts and gears

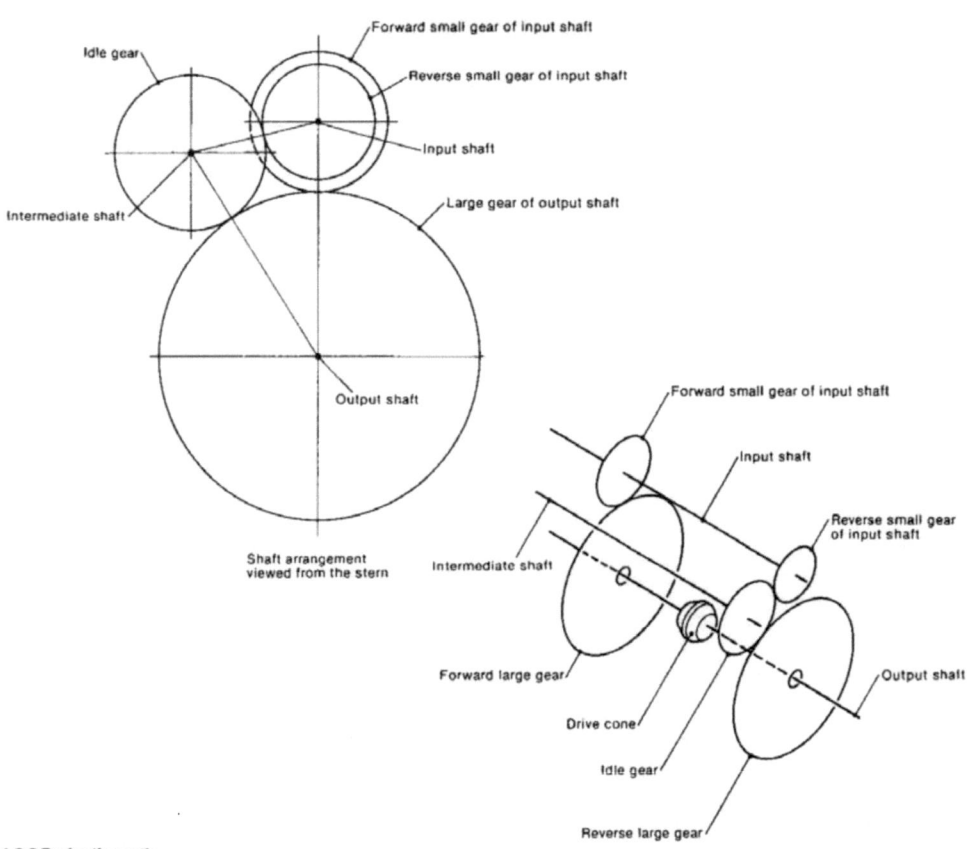

1-3.2 Reduction ratio
Forward

Model	No. of teeth of forward small gear Zif	No. of teeth of forward large gear Zof	Reduction ratio Zof/Zif
KM2-C	24	53	53/24 = 2.21
	21	55	55/21 = 2.62
	18	58	58/18 = 3.22
KM3-A	25	59	59/25 = 2.36
	23	60	60/23 = 2.61
	20	64	64/20 = 3.20

Reverse

Model	No. of teeth of reverse small gear Zir	No. of teeth of intermediate shaft gear Zi	No. of teeth of reverse large gear Zdr	Reduction ratio Zi/Zir·Zdr/Zi
KM2-C	18	26	55	55/18 = 3.06
KM3-A	19	26	60	60/19 = 3.16

10-3

Chapter 10 Reduction and Reversing Gear
1. Construction

SM/GM(F)(C)·HM(F)(C)

1-3.3 Power transmission routine — Forward

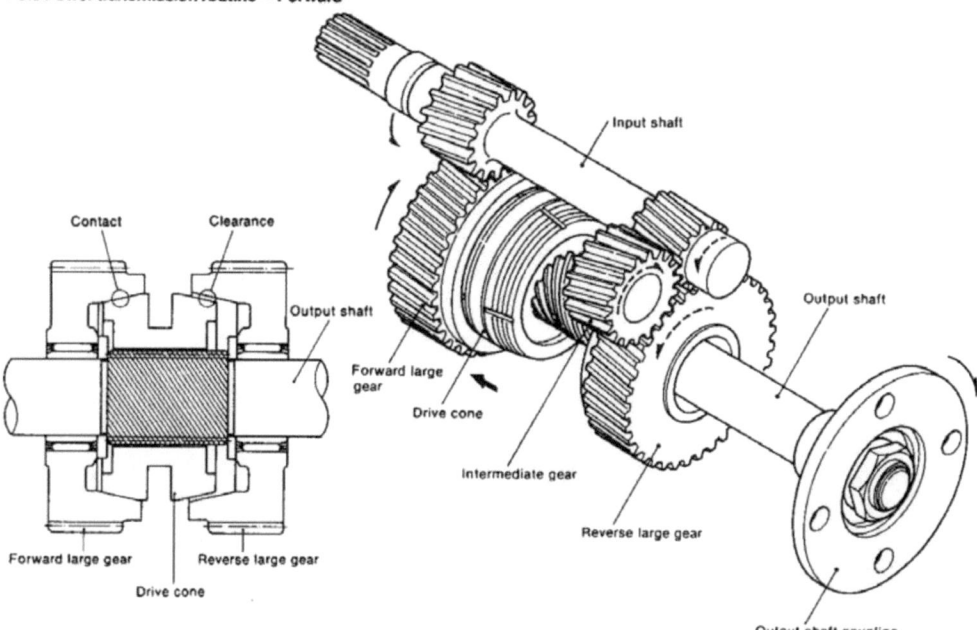

1-3.4 Power transmission routine — Reverse

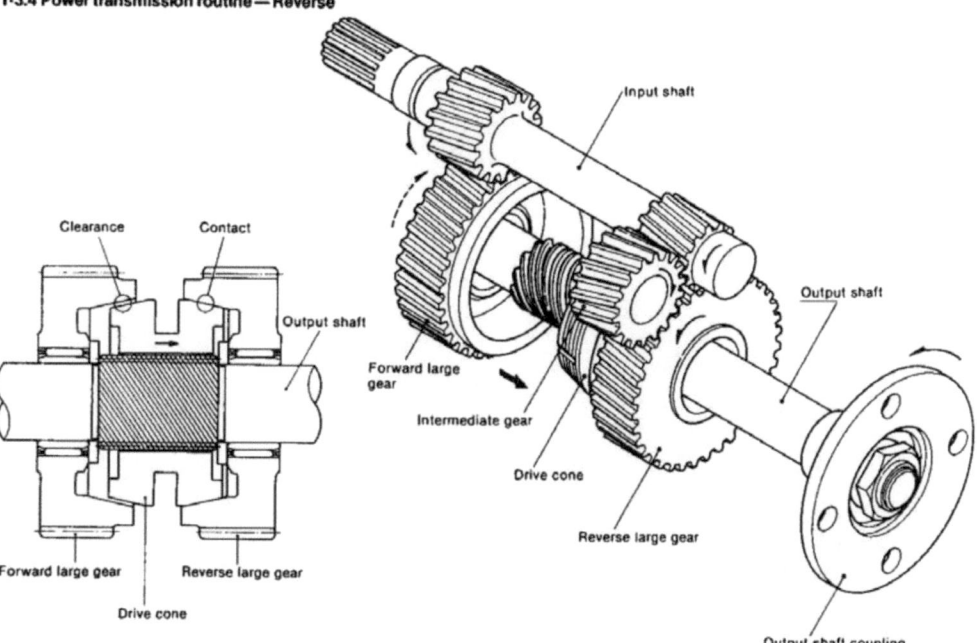

Chapter 10 Reduction and Reversing Gear
1. Construction

SM/GM(F)(C)·HM(F)(C)

1-4 Drawing

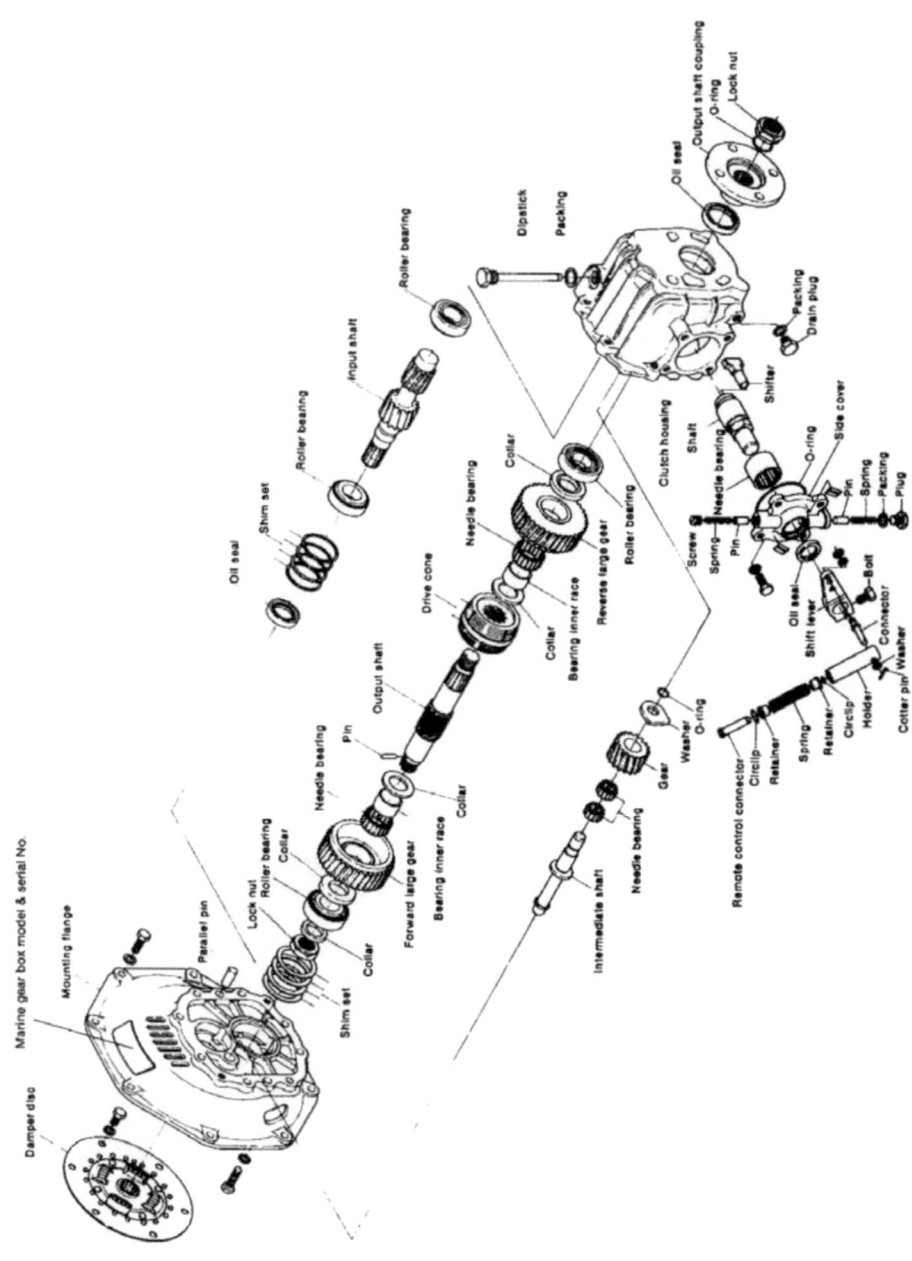

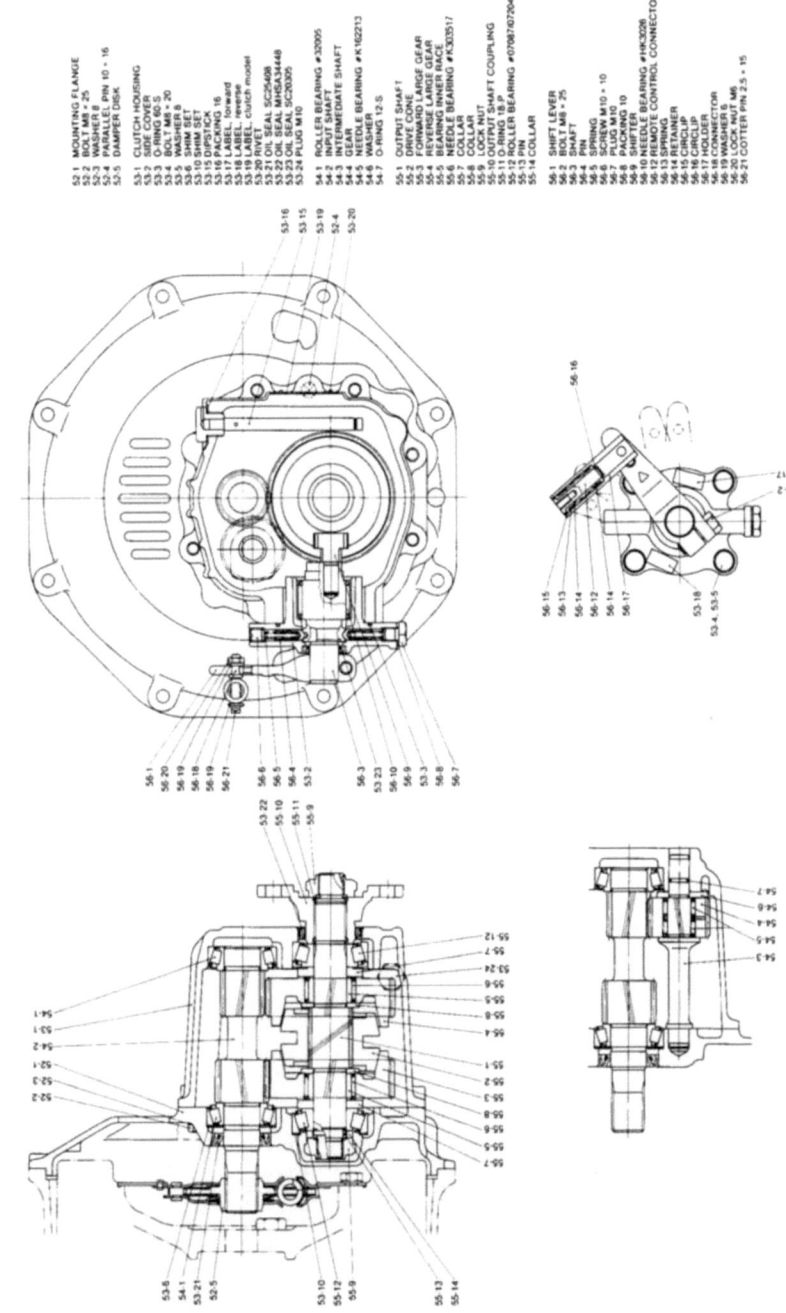

2. Shifting Device

2-1 Construction of shifting device

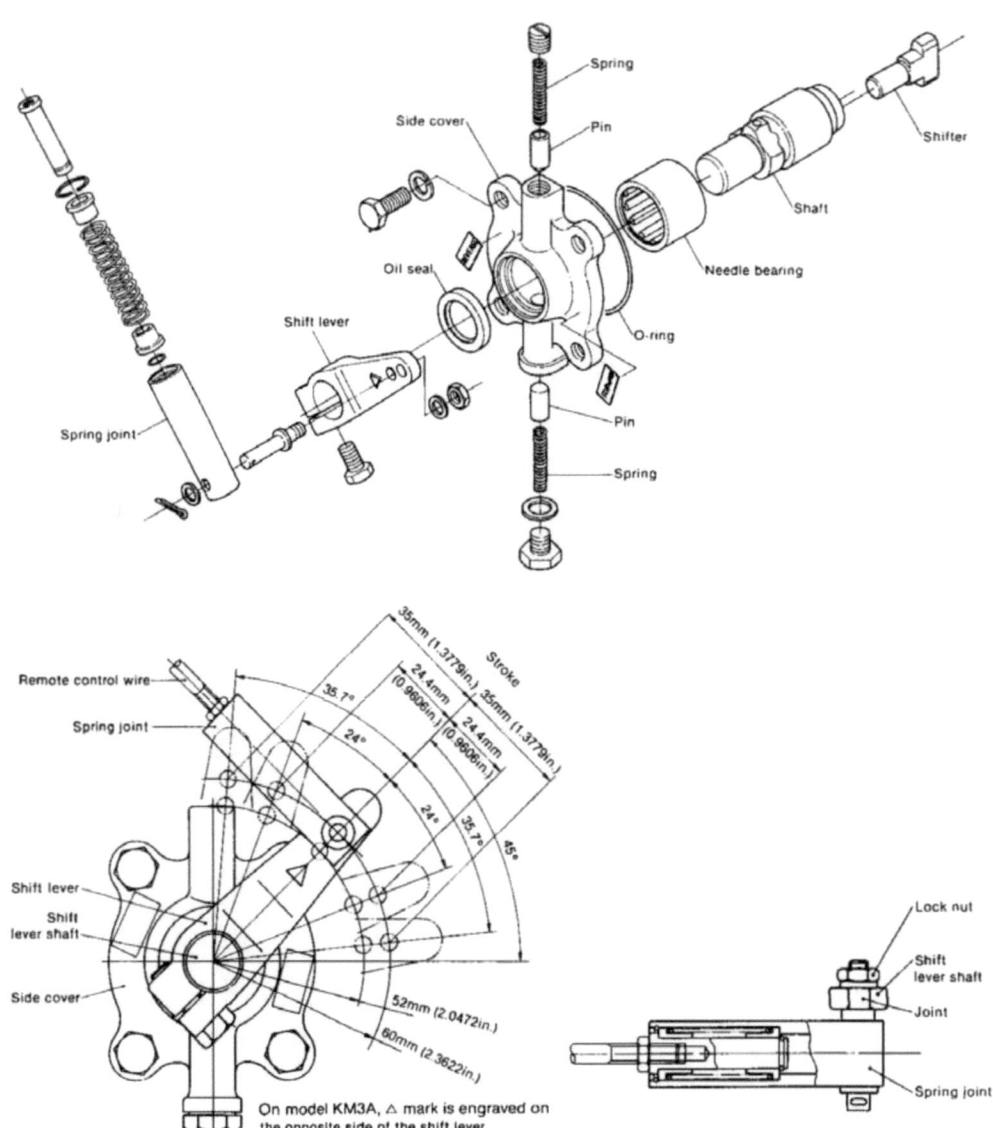

On model KM3A, △ mark is engraved on the opposite side of the shift lever.

Chapter 10 Reduction and Reversing Gear
2. Shifting Device

SM/GM(F)(C)·HM(F)(C)

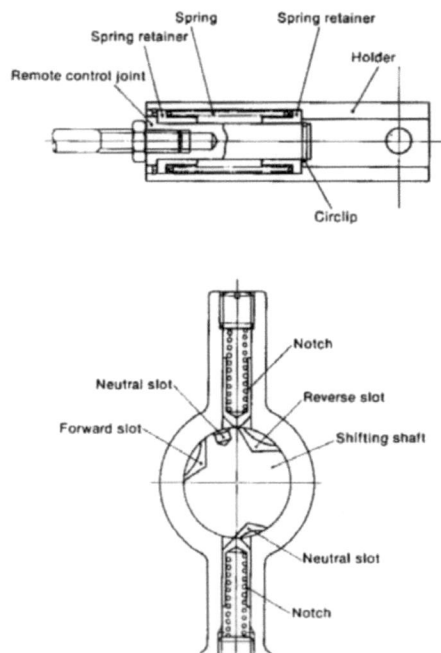

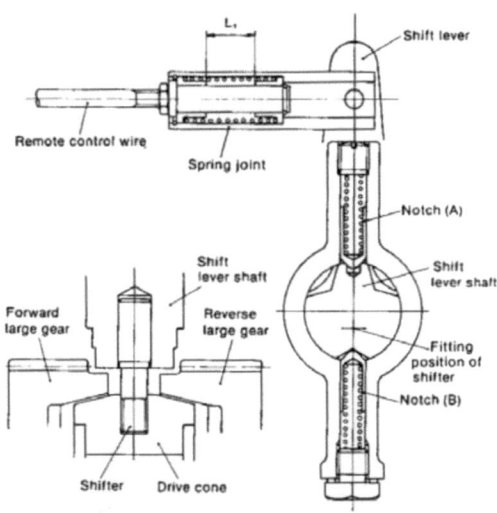

The shift lever shaft is supported by the side cover in which it rotates. Around the shift lever shaft, there are slots which engage the notch in order to control transmission of rotary power either forward or reverse, or to keep it in neutral. The notch engages each slot by the force of the notch spring.
The shifter is set at the end of the shift lever shaft eccentric to the shaft center line and the angular movement of the shift shaft (i.e. rotation). The shifter is moved forward or backward along the line of the output shaft and this in turn moves the drive cone forward or backward.
The spring joint contains a spring and 2 spring retainers in the holder, and the remote control joint is connected to the spring retainers so that it can slide a fixed distance. By pushing or pulling the remote control joint with the holder fixed, the remote control joint moves to a position where the two spring retainers touch.

2-2 Action of the shifting device
2-2.1 Changing from neutral to forward
The relationship between the spring joint and the notch is as shown in the following figure, and the two spring retainers are the maximum distance apart.

Neutral position
The shift lever is kept securely in the neutral position by notches (A) and (B).

Changing the power transmitting direction to forward is explained below.
When pushed forward, the remote control joint moves the spring retainers. The spring is compressed until the two spring retainers touch.

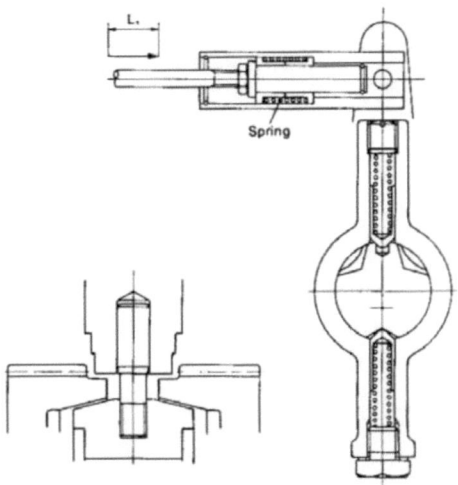

L_1 position of remote operation stroke

The spring in the spring joint is compressed, but the shift lever does not move.

10-8

Chapter 10 Reduction and Reversing Gear
2. Shifting Device

SM/GM(F)(C)·HM(F)(C)

By pushing the remote control joint the holder moves, and the shift lever and the shift lever shaft also move to disengage the notch from the neutral position.

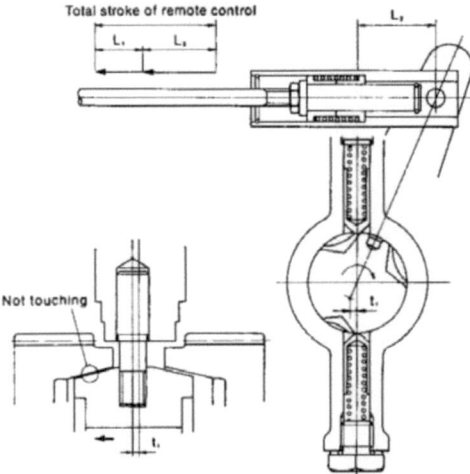

Forced moving position
When the shift lever is forcibly moved through distance L_2 the shifter moves distance t_1. In this position, the drive cone has not yet made contact. However, notches (A) and (B) are disengaged from the neutral notch slot, and notch (A) is positioned on the tapered surface.

The shift lever shaft is turned by the movement of the remote control joint. When the notch touches the tapered part of the forward setting slot, it is pushed by the notch spring force and turns the shift lever forward. At the same time, as the remote control joint is fixed by the two retainers of the spring joint being in contact with each other,

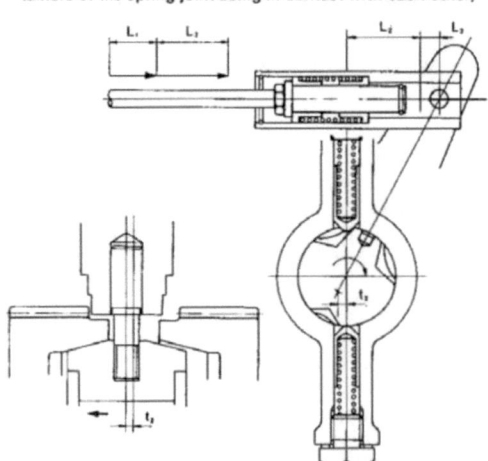

the holder is moved by the spring reaction so that the shift lever is pushed forward.
By the actions of the notch spring and spring joint, the shifter maintains pressure on the drive cone.

Engaging position for forward
By means of the shift lever shaft turning force which is caused by the spring in the joint and the notch (A), the shifter is moved distance L_3 and engagement is complete. Pressure is maintained on the drive cone after engagement.

2-2.2 Engagement from forward to neutral
Engagement for reverse is the same as for forward, that is, return to the neutral position and move the remote control joint forward.

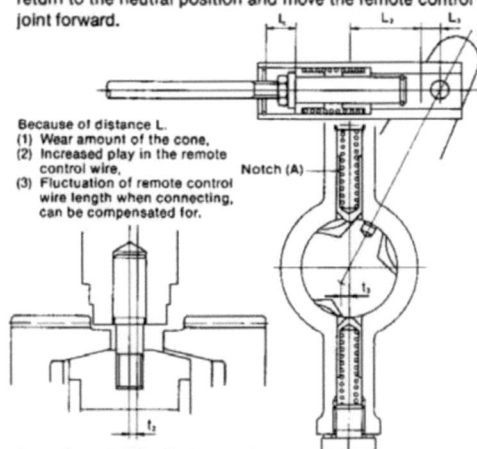

Because of distance L,
(1) Wear amount of the cone,
(2) Increased play in the remote control wire,
(3) Fluctuation of remote control wire length when connecting, can be compensated for.

Notch (A)

Engaging position for forward
The drive cone, which is moved by the spring in the joint and notch (A), is kept under force until distance L becomes zero even when the cone is worn.

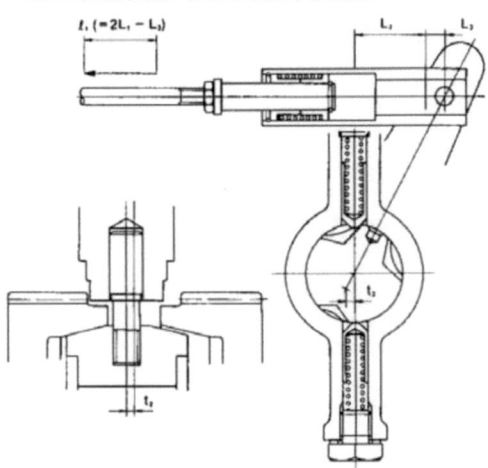

Chapter 10 Reduction and Reversing Gear
2. Shifting Device

Position of remote control stroke l_1

The shift lever does not move although the spring in the joint is compressed. The cone is kept in contact due to the transmission of torque when idling.

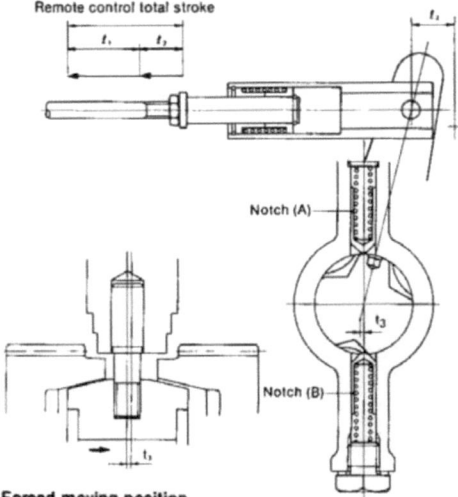

Forced moving position

The shift lever is forcibly moved through distance l_2, overcoming light friction due to the transmitting torque and the drive cone separates. Notch (A) disengages and notch (B) engages.

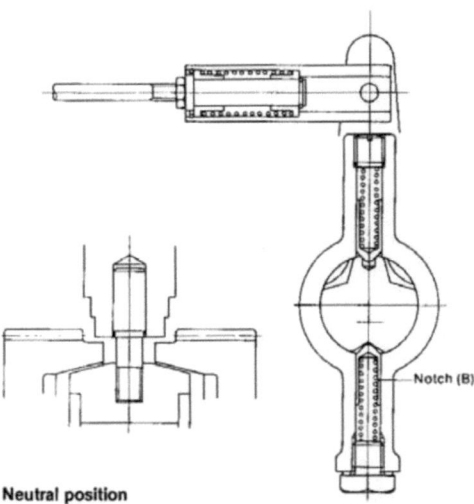

Neutral position

The shift lever is returned to neutral by the turning force generated on the shift lever shaft by the spring in the joint and notch (B).

2-3 Clutch shifting force
(reference value) [Engine at 1000rpm]

Shifting Position Shifting Direction	Shift lever position at 60mm	Remote control handle position at 170mm (cable length, 5m)
Engaging stroke	Approx. 3kg (6.6 lbs)	3 ~ 4kg (6.6 ~ 8.8 lbs)
Disengaging stroke	—	6 ~ 8kg (13.2 ~ 17.6 lbs)

Disengaging stroke:
(1) At the initial stage of usage, the stroke may be heavier than the above value, but the stroke gets light when adopted.
(2) It varies according to the idling speed of the engine. The lower the rotation becomes, the lighter the stroke becomes.
(3) The longer the remote control cable, the more bent it becomes, and the smaller the bending radius, the heavier the disengaging stroke.
[33-C minimum bending radius 203.2mm(8")]
(4) When the spring joint is attached to the shift lever at 52mm distance from the center of the lever shaft, the disengaging stroke will be 15% heavier then when attached at a distance of 60mm.

2-4 Adjustment
When the clutch side cover is removed, make the following adjustments at the time of reassembly.

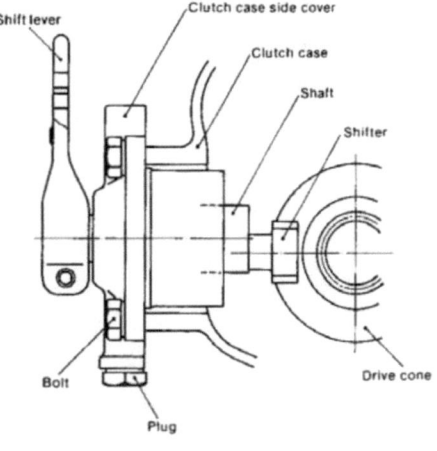

Chapter 10 Reduction and Reversing Gear
2. Shifting Device

(1) Shift the slot into the drive cone so that it extends as far as the center of the two large gears.

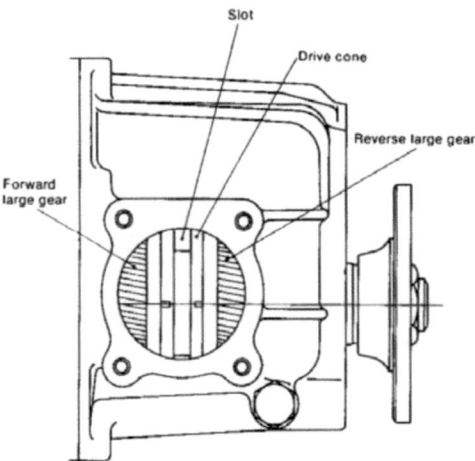

(2) Set the shift lever at neutral position. (Note that the shift lever can be rotated 360° when it is removed from case.) The neutral position is the position where the shifter comes downwards when the plug is below. When the plug is at the bottom, in the neutral position the shifter points downwards.

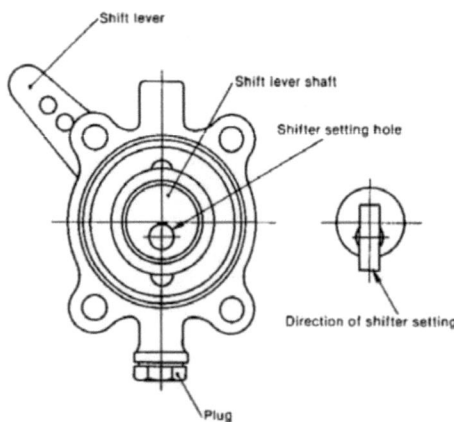

(3) Put the shifter of the side cover at bottom, and set the shifter to the ditch in the drive cone at the center of the forward and reverse gears. Do not move the drive cone from the center of the two gears at the time of the reassembly. (Note that 2mm diameter clearance is provided in the holes of the side cover, and the gear case.) This is for adjusting the difference between the engaging, and disengaging strokes.)

(4) Fit the shift lever locating jig into the holes of the side cover through the 15mm dia. holes as shown.

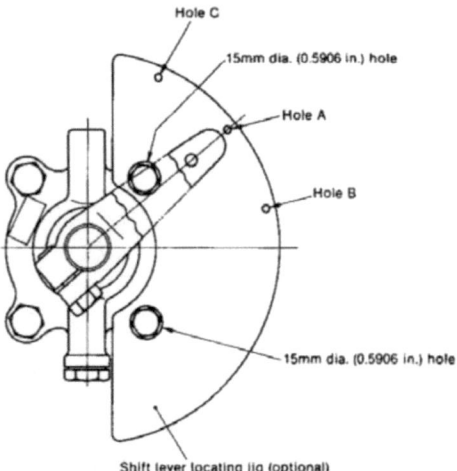

(5) Put the shift lever in neutral and check that the tip of the lever is aligned with hole A of the jig. It is not, loosen the fixing bolt on the shift lever, align it, then tighten the bolt. Take care to leave approximately 0.5mm (0.0197in.) clearance between the shift lever and the side cover.

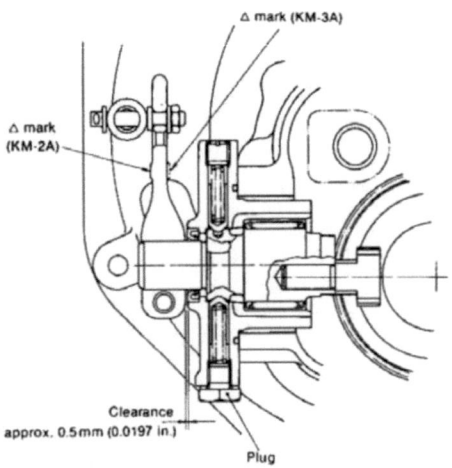

10-11

Chapter 10 Reduction and Reversing Gear
2. Shifting Device

SM/GM(F)(C)·HM(F)(C)

(6) Move the shift lever forward or back, and visually check the respective distances between the tip of the shift lever and holes B and C. Also check the difference between these distances.

(7) When these two distances are not equal, slightly loosen the four setting bolts of the side cover so that it can be moved a little in the shaft direction.

(8) When the distance is larger than normal in the forward setting, move the side cover slightly to the engine side.

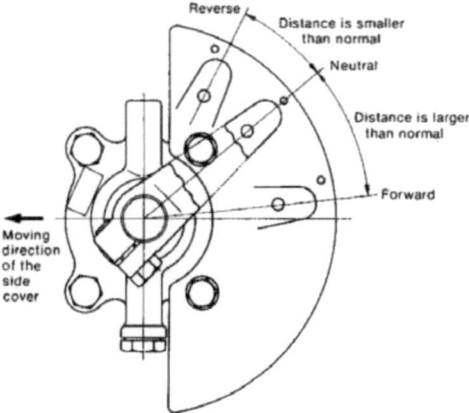

(9) When the distance is larger than normal in the reverse setting, move the side cover slightly to the propeller side.

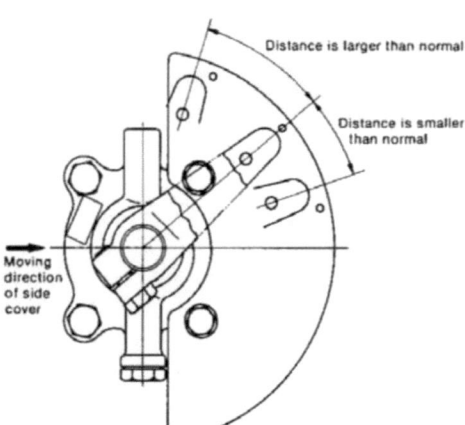

(10) When the distances are equal between neutral and forward and neutral and reverse tighten the setting bolts of the side cover.

(11) Although these distances may be equal both for forward and reverse, there might be some discrepancy between holes B and C due to difference in machining. However, if the discrepancy is the same for forward and reverse there is no problem.

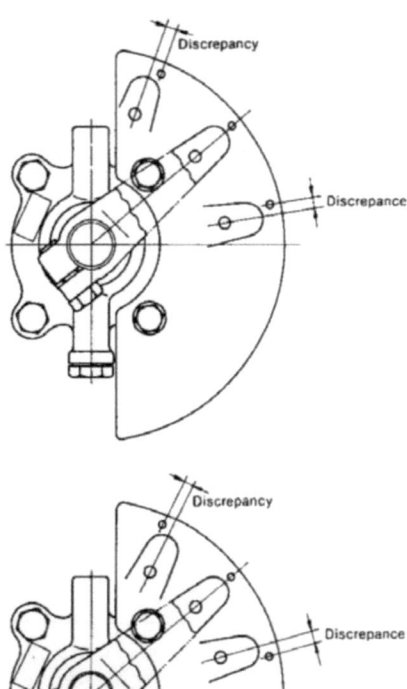

(12) Install the spring joint on the shift lever.
(Only when it is dismantled in the boat).
NOTE: *When the shift device is removed in the boat, the engine must always be stopped.*

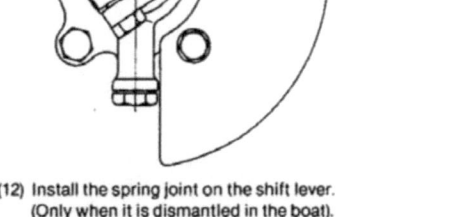

10-12

Printed in Japan
0000A0A1361

2-5 Inspect for the following points (to be inspected every 2-3 months)

(1) Looseness at the connection of the spring joint and the remote control cable.
(2) Looseness of the attaching nut of the spring joint and the shift lever.
(3) To make sure that the value of A, and B is not "Zero" at the engaging position of the remote control lever. If the value is "Zero", untighten the bolt of the side cover, and adjust according to the steps described in 2-4.

When the cone for the forward side gets worn, the value of B is decreased, and for reverse side, the value of A is decreased. When the play in the remote control system is increased, both values of A and B are decreased.

2-6 Cautions

(1) Always stop the engine when attaching, adjusting, and inspecting.
(2) When conducting inspection immediately after stopping the engine, do not touch the clutch. The oil temperature is often raised to around 90°C (194°F).
(3) Half-clutch operation is not possible with this design and construction. Do not use with the shift lever halfway to the engaged position.
(4) Set the idling engine speed at between 750 and 800 rpm.

NOTE: The dual (Two) lever remote control device cannot be used.

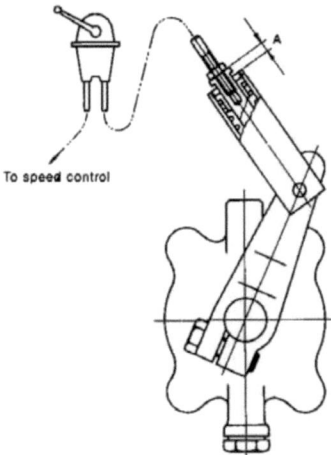

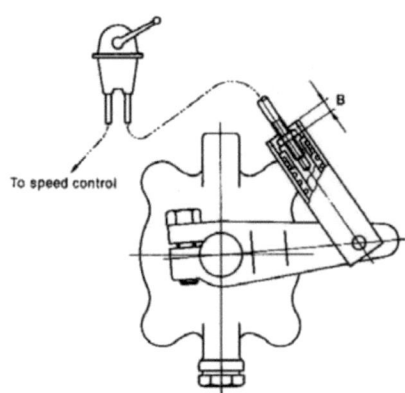

3. Inspection and Servicing

3-1 Clutch case
(1) Check the clutch case with a test hammer for cracking. Perform a color check when required.
If the case is cracked, replace it.
(2) Check for staining on the inside surface of the bearing section.
Also, measure the inside diameter of the case.
Replace the case if it is worn beyond the wear limit.

3-2 Bearing
(1) Rusting and damage.
If the bearing is rusted or the taper roller retainer is damaged, replace the bearing.
(2) Make sure that the baarings rotate smoothly.
If rotation is not smooth, if there is any binding, or if any abnormal sound is evident, replace the bearing.

3-3 Gear
Check the surface, tooth face conditions and backlash of each gear. Replace any defective part.
(1) Tooth surface wear.
Check the tooth surface for pitching, abnormal wear, dents, and cracks. Repair lightly damaged gears and replace heavily damaged gears.
(2) Tooth surface contact.
Check the tooth surface contact. The amount of tooth surface contact between the tooth crest and tooth flank must be at least 70% of the tooth width.
(3) Backlash.
Measure the backlash of each gear, and replace the gear when it is worn beyond the wear limit.

mm (in.)

	Maintenance standard	Wear limit
Input shaft forward gear and output shaft forward gear	0.06 ~ 0.12 (0.0024 ~ 0.0047)	0.2 (0.0079)
Input shaft reverse gear and intermediate gear	0.06 ~ 0.12 (0.0024 ~ 0.0047)	0.2 (0.0079)
Intermediate gear and output shaft reverse gear	0.06 ~ 0.12 (0.0024 ~ 0.0047)	0.2 (0.0079)

(The same dimensions apply to both KM2-C and KM3-A)

3-4 Forward and reverse large gears
(1) Contact surface with drive cone.
Visually inspect the tapered surface of the forward and reverse large gears where they make contact with the drive cone to check if any abnormal condition or sign of overheating exists.
If any defect is found, replace the gear.

(2) Forward/reverse gear needle bearing.
When an abnormal sound is produced at the needle bearing, visually inspect the rollers; replace the bearing if the rollers are faulty.

Rollers

3-5 Drive cone
(1) Visually inspect that part of the surface that comes into contact with the circumferential triangular slot to check for signs of scoring, overheating or wear. If deep scoring or signs of overheating are found, replace the cone.

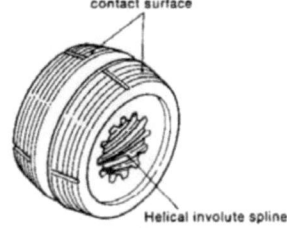

contact surface

Helical involute spline

(2) Check the helical involute spline for any abnormal condition on the tooth surface, and repair or replace the part should any be found.
(3) Measure the amount of wear on the tapered contact surface of the drive cone, and replace the cone when the wear exceeds the specified limit.

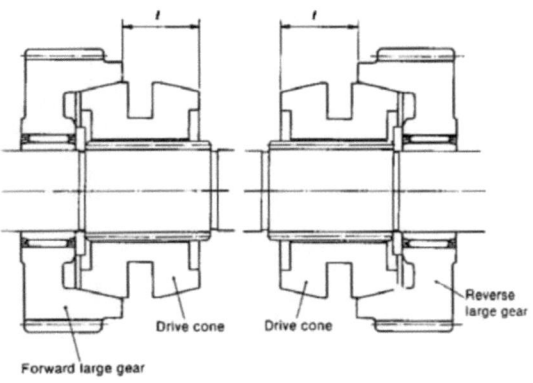

Forward large gear   Drive cone   Drive cone   Reverse large gear

Tapered surface

Chapter 10 Reduction and Reversing Gear
3. Inspection and Servicing

SM/GM(F)(C)·HM(F)(C)

mm (in.)

Dimensions l		Standard dimensions	Limited dimensions
	KM2-C	24.4 ~ 24.7 (0.9606 ~ 0.9724)	24.1 (0.9488)
	KM3-A	29.9 ~ 30.2 (1.1772 ~ 1.1890)	29.6 (1.1654)

NOTE: When dismantled, the forward or reverse direction of the drive cone must be clearly identified.

(4) Measure the dimension of the slot width of the drive cone, and replace the cone when the dimension is over the specified limit.

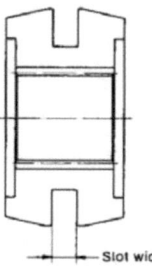

Slot width

mm (in.)

	Standard dimensions	Standard clearance	Allowable clearance	Limited clearance
Slot width of drive cone	$8^{+0.1}_{0}$ (0.3150 ~ 0.3189)	0.15 ~ 0.3 (0.0059 ~ 0.0118)	0.6 (0.0236)	8.3 (0.3268)
Shifter width	$8^{-0.15}_{-0.20}$ (0.3071 ~ 0.3090)			7.7 (0.3031)

3-6 Thrust collar

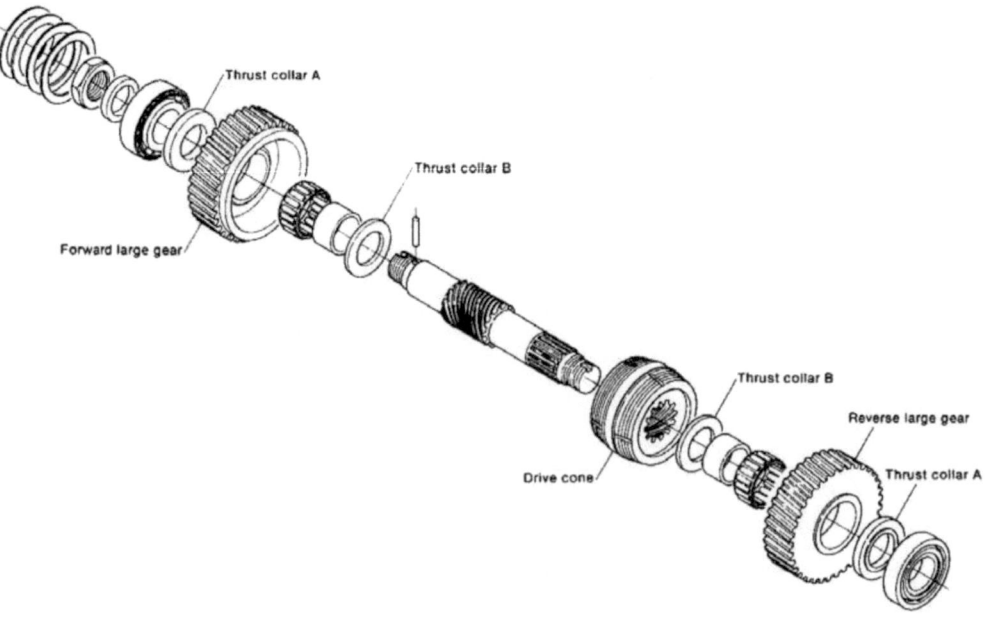

(1) Visually inspect the sliding surface of thrust collar A or B to check for signs of overheating, scoring, or cracks. Replace the collar if any abnormal condition is found.
(2) Measure the thickness of thrust collar A or B, and replace it when the dimension exceeds the specified limit.

3-9 Output shaft

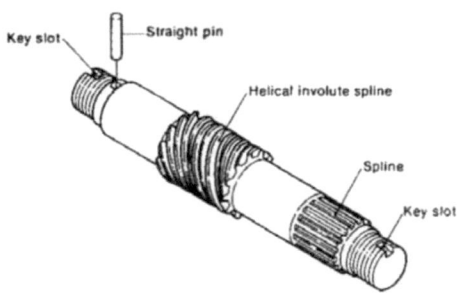

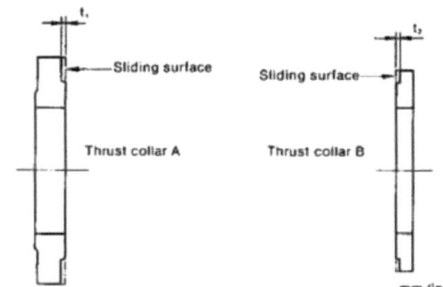

mm (in.)

Stepped wear	Limit for use
Thrust collar A, t_1	0.05 (0.0020)
Thrust collar B, t_2	0.20 (0.0079)

(1) Visually inspect the spline and the helical involute spline, and repair or replace a part when any abnormal condition is found on its surface.

3-10 Intermediate shaft

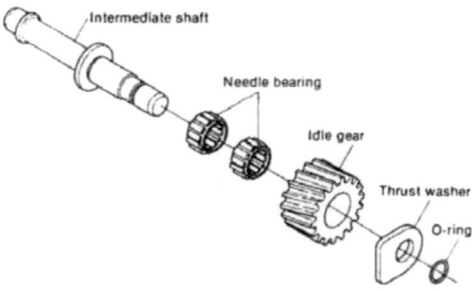

3-7 Oil seal of output shaft
Visually inspect the oil seal of the output shaft to check if there is any damage or oil leakage; replace the seal when any abnormal condition is found.

3-8 Input shaft

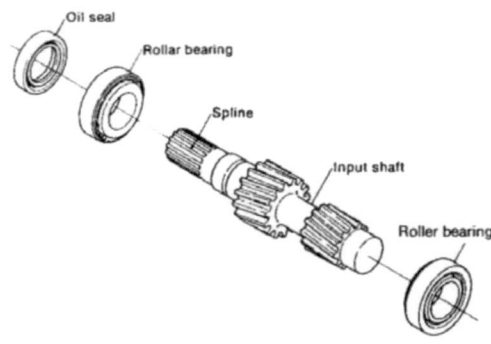

(1) Needle bearing dimensions, staining.
Check the surface of the roller to see whether the needle bearing sticks or is damaged. Replace if necessary.

3-11 Shifting device
3-11.1 Shifter

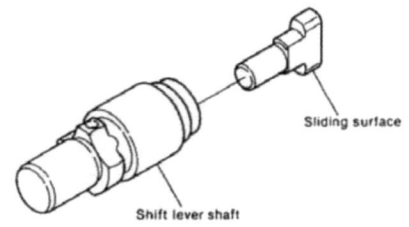

(1) Spline part.
Whenever uneven wear and/or scratches are found, replace with a new part.
(2) Surface of oil seal.
If the sealing surface of the oil seal is worn or scratched, replace.

Chapter 10 Reduction and Reversing Gear
3. Inspection and Servicing

SM/GM(F)(C)·HM(F)(C)

(1) Visually inspect the surface in contact with the drive cone, and replace the shifter when signs of overheating, damage or wear are found.
(2) Measure the width of the shifter, and replace it when the wear exceeds the specified limit. Also measure the diameter of the shifter shaft, and replace it when the wear exceeds the specified limit.

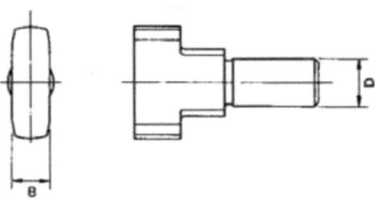

mm (in.)

	Standard dimensions	Clearance	Allowable clearance	Specified limit
Slot width of drive cone	$8^{+0.1}_{0}$ (0.3150 ~ 0.3189)	0.15 ~ 0.3 (0.0059 ~ 0.0118)	0.6 (0.0236)	8.3 (0.3268)
Shifter width	$8^{-0.15}_{-0.20}$ (0.3070 ~ 0.3091)			7.7 (0.3031)
Shifter shaft diameter	$10^{-0.005}_{-0.014}$ (0.3931 ~ 0.3935)	0.005 ~ 0.029 (0.0002 ~ 0.0011)	0.05 (0.0020)	9.95 (0.3917)
Shift lever shaft diameter	$10^{+0.015}_{0}$ (0.3937 ~ 0.3943)			10.05 (0.3957)

3-11.2 Notch slot of shift lever shaft

Visually inspect the notch slot of the shift lever shaft to check for any abnormal wear or cracking, replace any defective part if found.

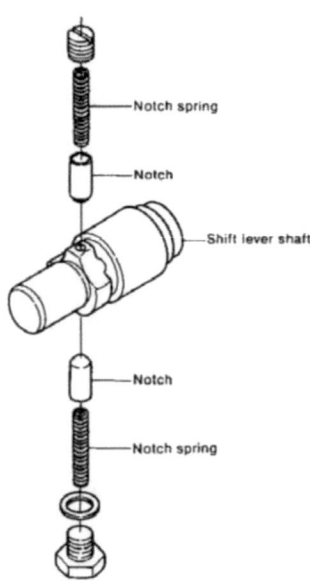

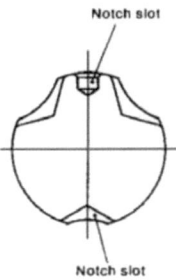

Notch slot

Notch slot

3-11.3 Notch

Visually inspect the tip of the notch to check for wear, damage or deformation. Replace the notch if it is found to be defective in any way.

3-11.4 Notch spring

Visually inspect the notch spring to check for any damage, corrosion or permanent set; replace the spring when it is found to be defective.

Free length	34mm (1.3386in.)
Spring coefficient	0.459kg (0.992 lb)
Set length	25.5mm (1.0039in.)
Set load	3.90kg (8.598 lb)

3-12 Spring joint

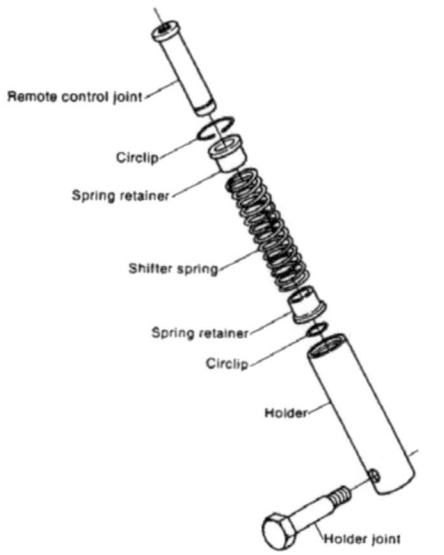

3-13 Damper disc

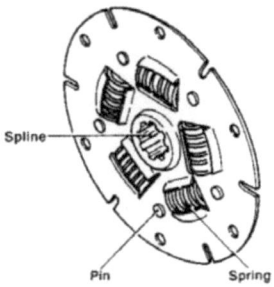

(1) Spline part.
  Whenever uneven wear and/or scratches are found, replace with a new part.
(2) Spring.
  Whenever uneven wear and/or scratches are found, replace with a new part.
(3) Pin wear.
  Whenever uneven wear and/or scratches are found, replace with a new part.
(4) Whenever a crack or damage to the spring slot is found replace the defective part with a new one.

(1) Check each part for abnormal play, and replace if play is excessive.
(2) When the movement of each part is not smooth, measure the tension and replace as a complete unit when it exceeds the specified limit.

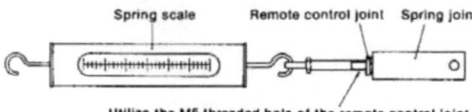

Utilize the M5 threaded hole of the remote control joint.

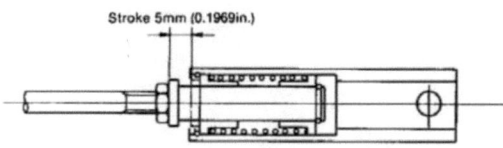

	Standard value	Limit value
		kg (lb)
Tension (at the position of 5mm stroke)	2.8 (6.17)	2.5 (5.51)

4. Disassembly

4-1 Dismantling the clutch
(1) Remove the remote control cable.
(2) Remove the clutch assembly from the engine mounting flange.

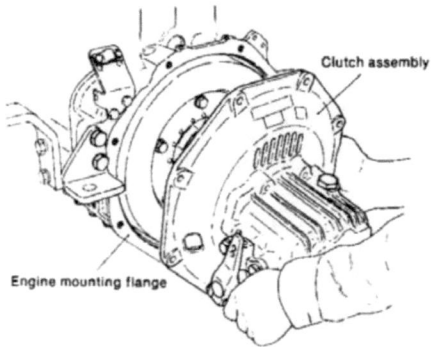

(3) Drain the lubricating oil.
Drain the lubricating oil by loosening the plug at the bottom of the clutch case.

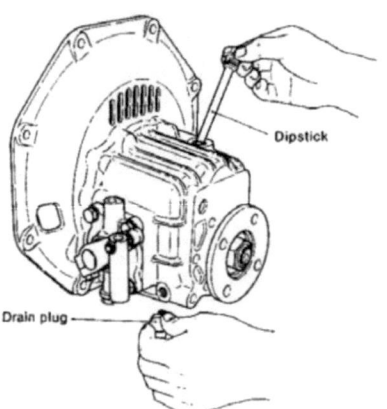

(4) Remove the end nut and output shaft coupling.

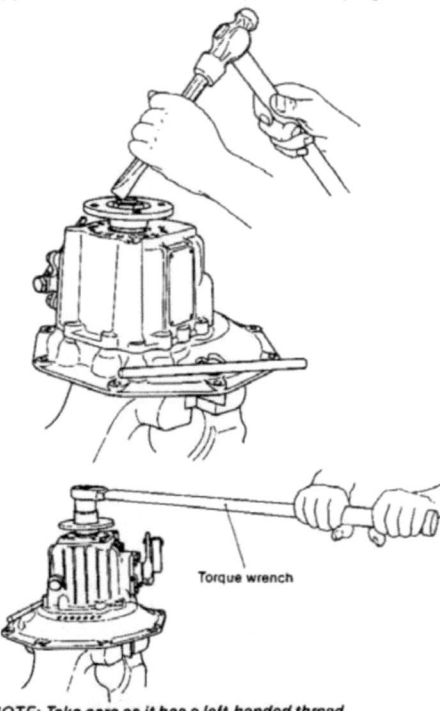

NOTE: Take care as it has a left-handed thread.

(5) Remove the oil dip stick and packing.
(6) Remove the fixing bolts on the side cover, and also remove the shift lever shaft, shift lever and shifter.

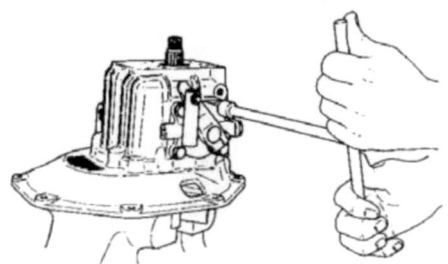

Chapter 10 Reduction and Reversing Gear
4. Disassembly

SM/GM(F)(C)-HM(F)(C)

(7) Remove the bolts which secure the mounting flange to the case body, give light taps to the left and right with a plastic headed hammer while supporting the clutch case with your hand, then remove the mounting flange.

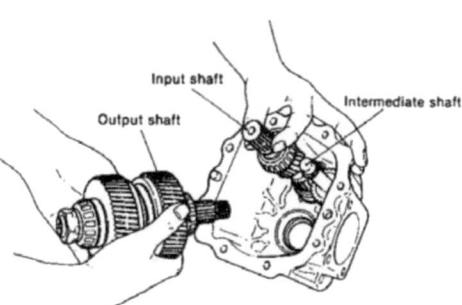

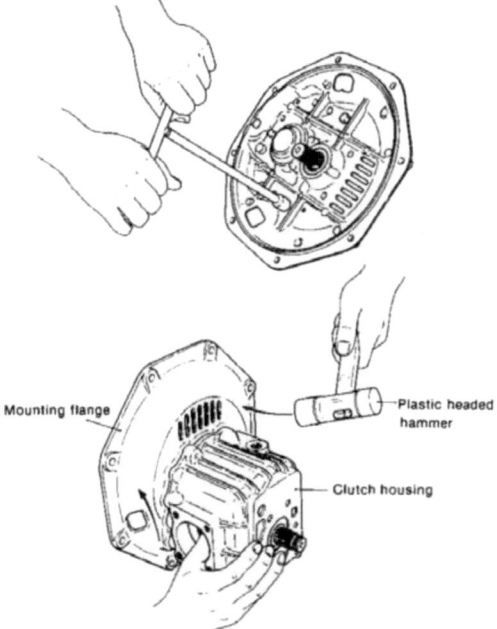

(9) Take out the intermediate shaft and input shaft. When taking out the intermediate shaft, place a bolt or spacer on the shaft hole of the case, and drive the shaft out by tapping it lightly.

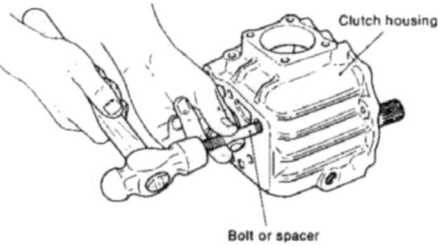

(8) Withdraw the output shaft assembly.

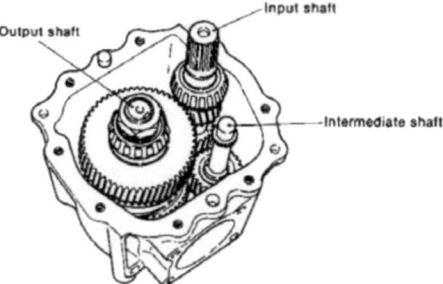

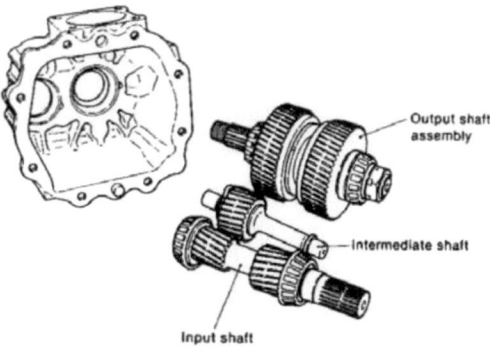

10-20

Printed in Japan
0000A0A1361

Chapter 10 Reduction and Reversing Gear
4. Disassembly

SM/GM(F)(C)·HM(F)(C)

(10) Remove the oil seal of the output shaft from the case body.

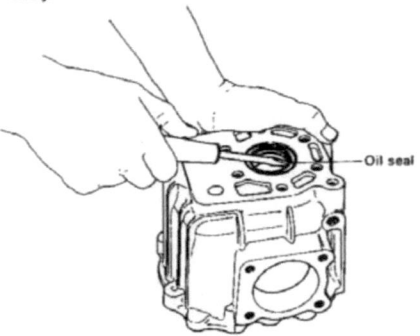

(11) Remove the outer bearing race from the case body by using the special tool.

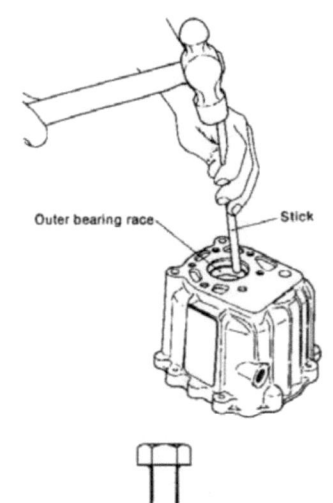

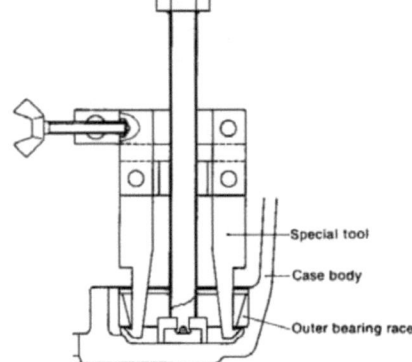

(12) Remove the oil seal of the input shaft from the mounting flange.
(13) Remove the outer bearing race from the mounting flange in the same way as with the case body.
(14) Remove each adjusting plate from the input or output shaft.

NOTE: *The same adjusting plates can be reused when the following parts are not replaced. When any part is replaced however, re-adjustment is necessary.*

4-2 Removal of the output shaft

(1) Take out the reverse large gear, thrust collar A and inner bearing race.
The reverse large gear must be withdrawn using a pulley extracter, by fixing the nut at the forward end in a vice.

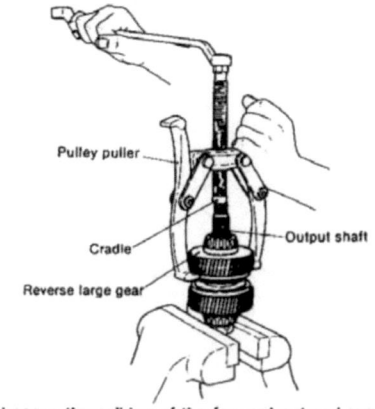

(2) Loosen the calking of the forward nut and remove the nut and spacer.
Remove the nut by using a torque wrench after setting the output shaft coupling and fixing the coupling bolt in a vice.

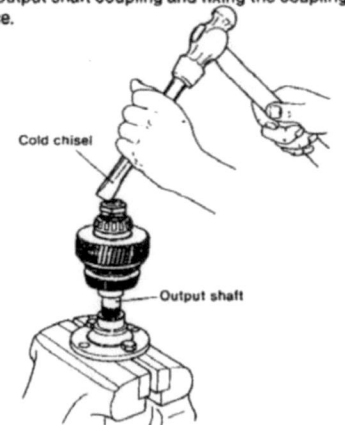

Chapter 10 Reduction and Reversing Gear
4. Disassembly

SM/GM(F)(C)-HM(F)(C)

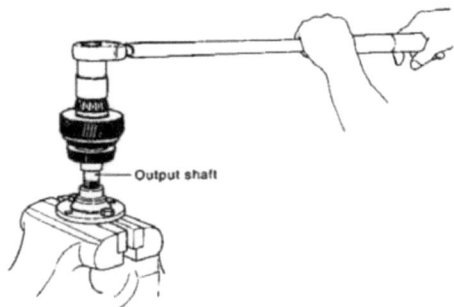

(3) Place the pulley extractor against the end surface of the forward large gear, and withdraw the forward large gear, thrust collar A and inner bearing race.

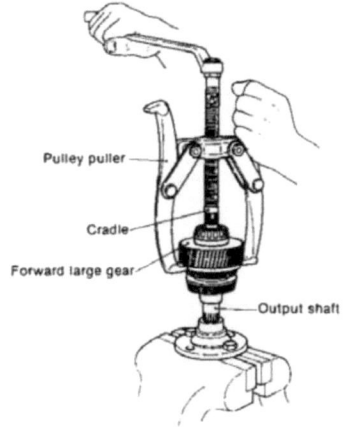

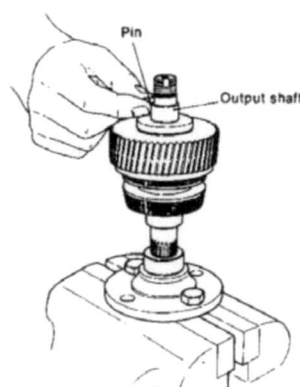

NOTE: Take care as the nut has left-handed thread.

(4) While gripping the drive cone, tap the end of the shaft with a plastic beaded hammer, and withdraw the thrust collar B and inner needle bearing race. A pulley extractor may be used.

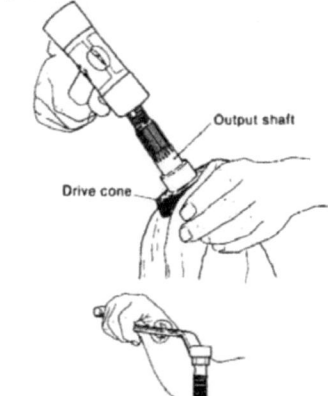

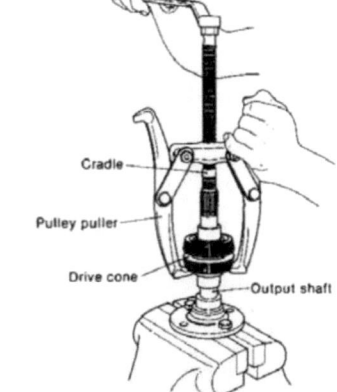

4-3 Removal of the intermediate shaft
(1) Remove the "O" ring.
(2) Remove the thrust washer.
(3) Remove the intermediate gear and needle bearing.

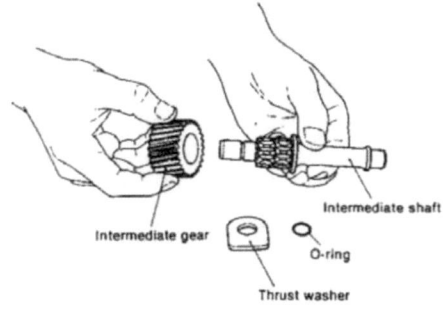

Chapter 10 Reduction and Reversing Gear
4. Disassembly

4-4 Dismantling the side cover assembly (Shifting device)

(1) Loosen the bolt of the shift lever, and remove the shift lever from the shaft.

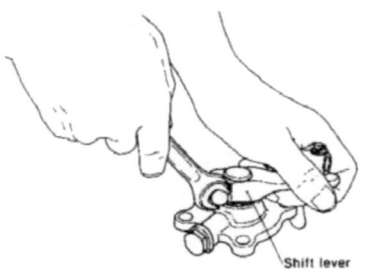

(2) Remove the stop screw for the notch and plug, and take out the notch and spring.

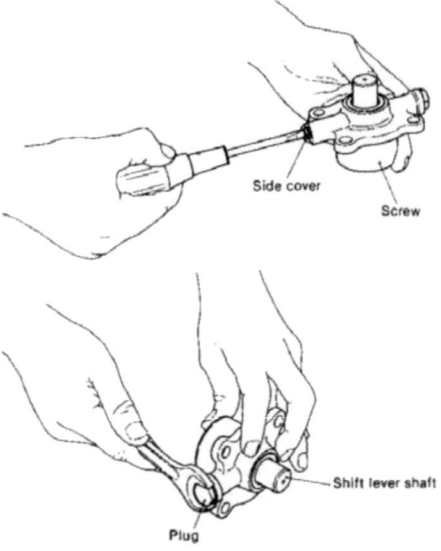

(3) Take out the shifter.

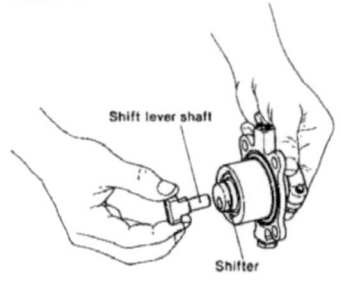

(4) Withdraw the shift lever shaft.

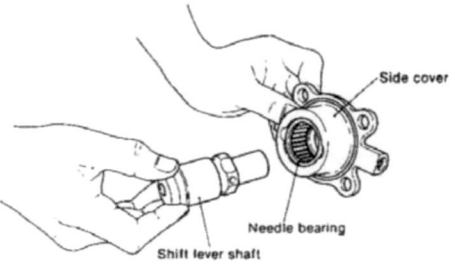

(5) Remove the oil seal.
(6) After removing the calking for locking, heat the needle bearing portion up to about 100°C, and extract the needle bearing from the side cover.

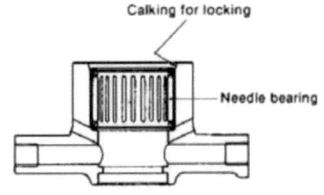

5. Reassembly

5-1 Reassembly of output shaft

(1) Fit the forward side thrust collar B onto the shaft.
(2) Drive in the forward end inner needle bearing race using a jig.

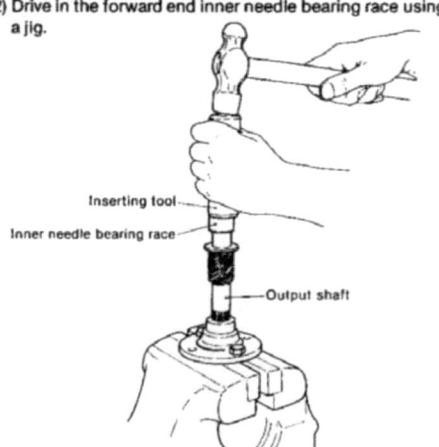

(3) Assemble the needle bearing and forward large gear.
NOTE: Check that the forward large gear rotates smoothly.
(4) Fit the thrust collar A and pin, and drive in the inner bearing race using a jig.

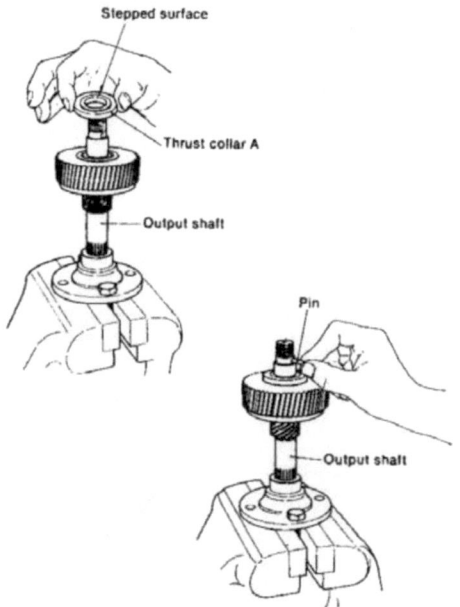

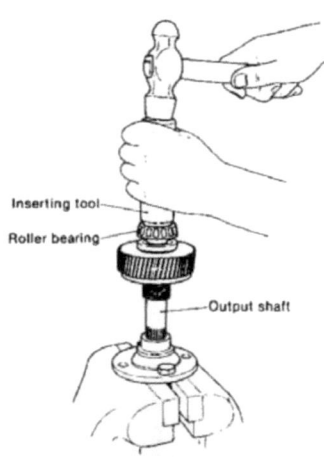

NOTES: 1) Drive in with a plastic headed hammer. Do not hit it hard.
2) When fitting the thrust collar A, note the fitting direction. Fit it keeping the stepped surface toward the bearing side.
3) Note that the pin cannot be fitted after the inner bearing race has been driven in.

(5) Assemble the collar and pin so that the pin is in the groove of the collar.
(6) Set and tighten the forward end nut. Insert the bolt into the coupling, and fix it in a vice, keeping the spline part upward.
Insert the shaft into the spline of the coupling, fit the spacer, and tighten the nut with a torque wrench.

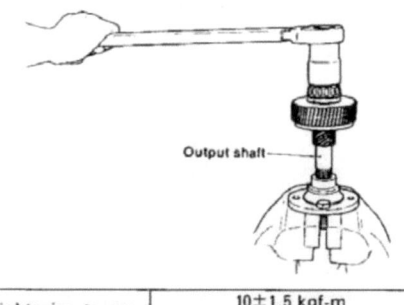

Tightening torque	10±1.5 kgf-m (61.5 ~ 83.2 ft-lb)

(The same torque applies to both models KM2-C and KM3-A).

NOTES: 1) Take care as it is a left-handed thread.
2) Use the reverse side nut used before dismantling as the forward end nut. This is so as not to match the calked portion to the same point.

Chapter 10 Reduction and Reversing Gear
5. Reassembly

5-2 Reassembly of the clutch

(1) Fit the oil seal and bearing outer race in the clutch case.
(2) Insert the input shaft into the clutch case.
(3) Drive the intermediate shaft into the clutch case.

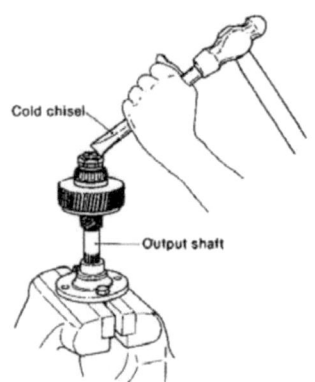

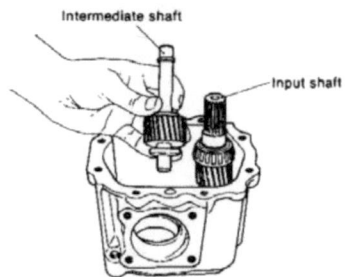

(7) Insert the drive cone while keeping the output shaft set for reverse.

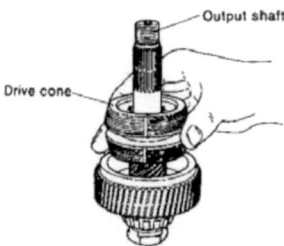

NOTES: 1) If the output shaft is not fitted into the clutch case before driving-in the intermediate shaft, it cannot be assembled.
2) Note the assembly direction of the thrust washer.

(4) Insert the output shaft into the clutch case.

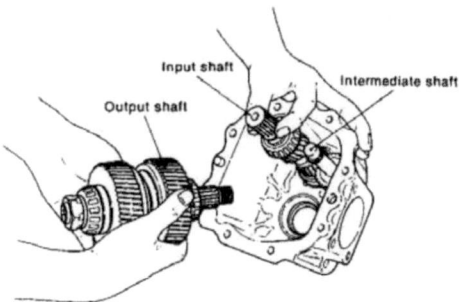

(8) Apply procedures 1 through 4 to the forward end.

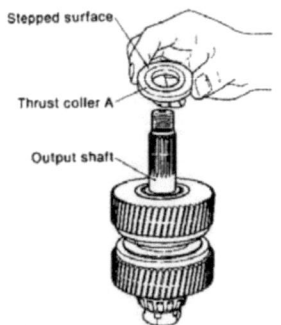

NOTE: Fit thrust collar A so that the stepped surface faces the bearing side.

(5) Check the thickness of shims for both input and output shafts. When the component parts are not replaced after dismantling, the same shims can be reused. When the clutch case flange or any one of the following parts is replaced, the thickness of shim must be determined in the following manner.

For input shaft parts: input shaft, bearing.
For output shaft parts: output shaft, thrust collar A, thrust collar B, gear, bearing.
1) Measure the distance between the clutch case body and the mounting flange, A or D for each shaft.
2) Fit the outer bearing race to each shaft, and measure the distance (B or C) between bearings.

Chapter 10 Reduction and Reversing Gear
5. Reassembly

SM/GM(F)(C)·HM(F)(C)

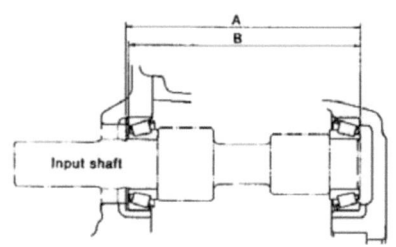
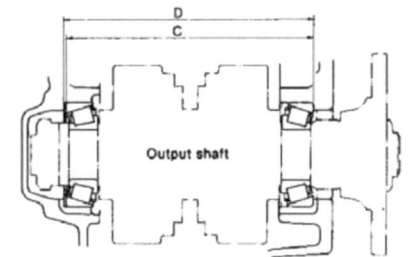

mm (in.)

	A	B	C	D
KM2-C	116.40 ~ 116.75 (4.5827 ~ 4.5964)	115.2 ~ 116.1 (4.5354 ~ 4.5709)	121.48 ~ 122.53 (4.7827 ~ 4.8240)	122.60 ~ 122.95 (4.8268 ~ 4.8406)
KM3-A	127.4 ~ 127.75 (5.0157 ~ 5.0295)	126.2 ~ 127.1 (4.9685 ~ 5.0039)	134.56 ~ 136.0 (5.2976 ~ 5.3543)	136.0 ~ 136.35 (5.3543 ~ 5.3681)

3) Determine the thickness of shim so that the values of clearance and interference after fitting comply with the values in the following table.

Clearance (or interference) for each shaft   mm (in.)

Input shaft	±0.05 (±0.0020)
Output shaft	0 ~ −0.1 (0 ~ −0.0039)

NOTE: Negative value shows interference.

Adjusting plate

	Part No.	Thickness mm (in.)	No. of shims
Input shaft	177088-02350	0.5 (0.0197)	1
		0.4 (0.0157)	1
		0.3 (0.0118)	2
Output shaft	177090-02250	1.0 (0.0394)	1
		0.5 (0.0197)	1
		0.3 (0.0118)	1
		0.1 (0.0039)	2

(6) Fit the adjusting plate to the mounting flange, and drive in the outer bearing race.

NOTE: The outer bearing race can be easily driven in by heating the mounting flange to about 100°C, or by cooling the outer race with liquid hydrogen.

(7) Apply non-drying liquid packing around the outer surface of the oil seal, and insert the oil seal into the mounting flange while keeping the spring part of the oil seal facing the inside of the case.

(8) Apply non-drying liquid packing to the matching surfaces of the mounting flange and the case body.

(9) Insert the input shaft and output shaft into the shaft holes of the mounting flange, assemble the mounting flange on the case body, and tighten the bolt.

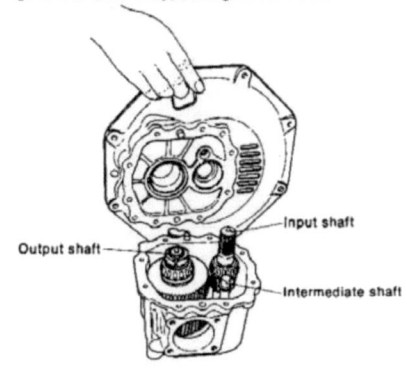

10-26

Chapter 10 Reduction and Reversing Gear
5. Reassembly

NOTE: Apply non-drying liquid packing to either the mounting flange or the case body.

(10) Assemble the output shaft coupling on the output shaft, and fit the O-ring.
(11) Tighten the end nut by using a torque wrench, then calk it.

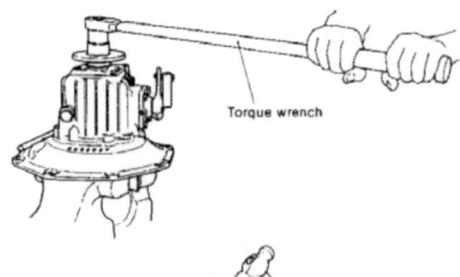

Torque wrench

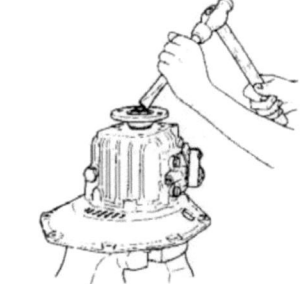

NOTE: Take care as it is a left-handed thread.

Tightening torque	10±1.5 kgf-m (61.5 ~ 83.2 ft-lb)

(The same torque applies to both models KM2-C and KM3-A).

5-3 Reassembly of the shifting device
(1) Fit the oil seal and needle bearing to the side cover.

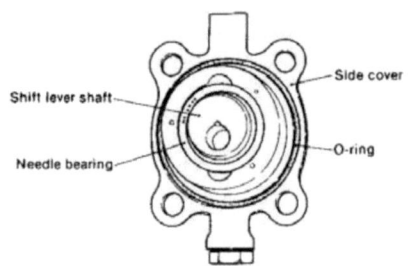

Shift lever shaft — Side cover
Needle bearing — O-ring

(2) Fit the shift lever.

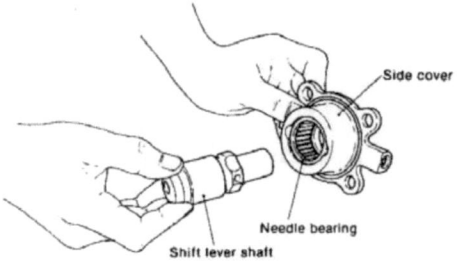

Side cover
Needle bearing
Shift lever shaft

(3) Fit the notch and spring, and screw in the plug and stop screw.

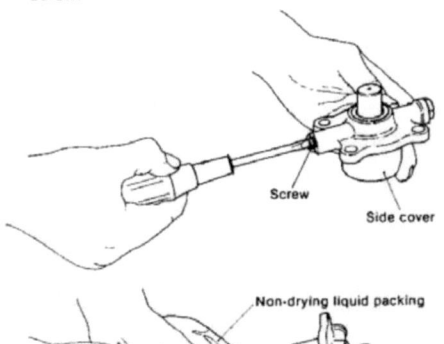

Screw
Side cover

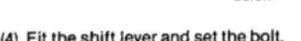

Non-drying liquid packing

Screw

(4) Fit the shift lever and set the bolt.

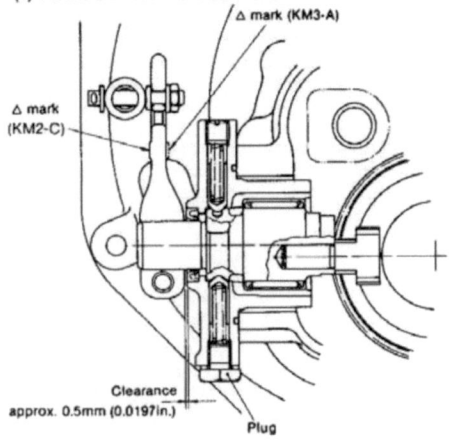

△ mark (KM3-A)
△ mark (KM2-C)

Clearance approx. 0.5mm (0.0197in.)
Plug

NOTE: The clearance between the surface of the side cover and the operation lever is to be 0 ~ 0.5mm (0 ~ 0.0197in.)

(5) Fit the shifter to the shift lever shaft.
(6) Fit the side cover to the clutch case. Ensure that the shifter engages the groove of the drive cone.

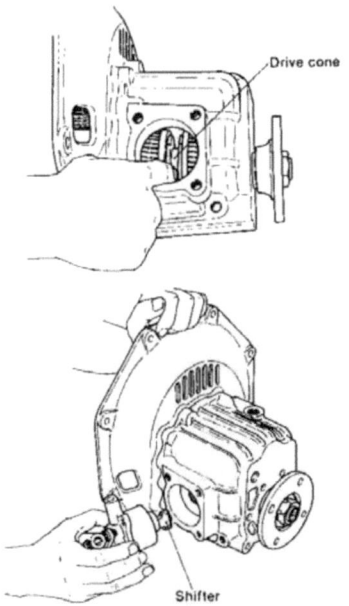

Drive cone

Shifter

(7) Check that the lever turns smoothly.
NOTE: The lever may not turn smoothly if the housing is not filled with lubricating oil.
(8) Fit the spring joint, and set the remote control cable after adjusting.
For fitting and adjustment refer to the detailed explanation in the appropriate section.

Chapter 10 Reduction and Reversing Gear
1. Construction

SM/GM(F)(C)-HM(F)(C)

For model 3GM35(F)

1. Construction

1-1 Construction

The Kanzaki-Carl Hurth KBW10 reduction reversing gear was developed jointly by Kanzaki Precision Machine Co., Ltd., a subsidiary of Yanmar and one of Japan's leading gear manufacturers, and Carl Hurth Co.

The KBW10 consists of a multi-disc clutch and reduction gear housed in a single case. It is small, light, simply constructed and extremely reliable.

*The force required to shift between forward and reverse can be controlled by a cable type remote control system much smaller and simpler than other types of reduction reversing gears.

*The friction discs are durable sinter plates, and the surface of the steel plates are corrugated in a sine curve shape to ensure positive engagement and disengagement and minimum loss of transmission force.

*Because of the special construction of this gear, the optimum pressure is automatically applied to the clutch plate in direct proportion to the input shaft torque.

1-2 Specifications

Engine model			3HM35(F)	
Nomenclature			KBW10E	
Reduction system			One-stage reduction, helical gear	
Reversing system			Constant mesh gear	
Clutch			Wet type multi-disc, mechanically operated	
Reduction ratio	Forward		2.14	2.83
	Reverse		2.50	
Direction of rotation	Input shaft		Counterclockwise as viewed from stern	
	Output shaft	Forward	Clockwise as viewed from stern	
		Reverse	Counterclockwise as viewed from stern	
Lubricating oil			DEXRON-ATF	
Lubricating oil capacity			0.7ℓ	

Chapter 10 Reduction and Reversing Gear
1. Construction

SM/GM(F)(C)-HM(F)(C)

1-3 Power transmission system

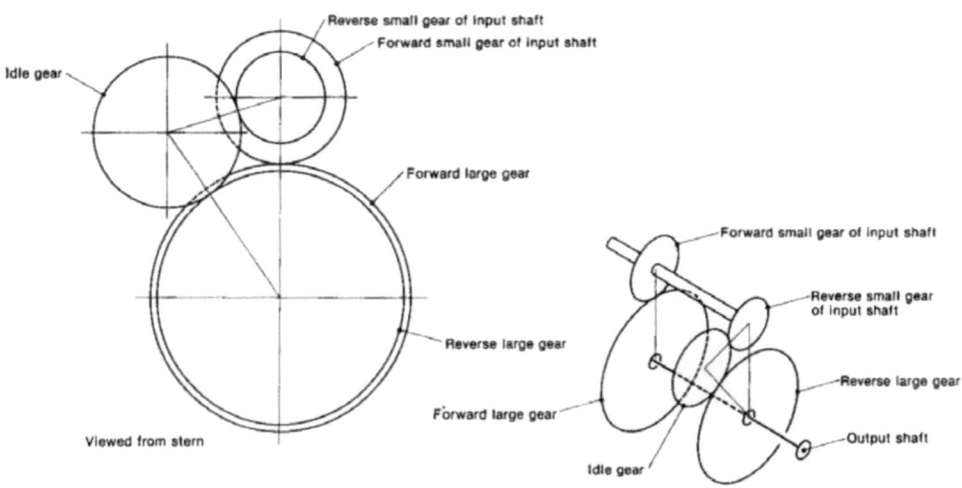

Forward			Reverse			
Number of teeth		Reduction ratio	Number of teeth			Reduction ratio
Forward small gear of input shaft	Forward large gear		Reverse small gear of input shaft	Idle gear	Reverse large gear	
22	47	47/22 = 2.14	18	25	45	45/18 = 2.50
18	51	51/18 = 2.83				

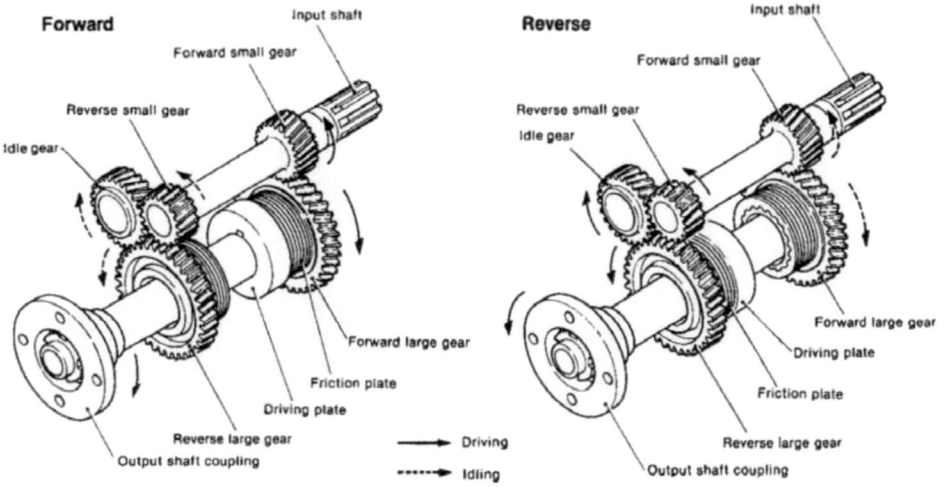

10-30

Printed in Japan
0000A0A1361

Chapter 10 Reduction and Reversing Gear
1. Construction

SM/GM(F)(C)·HM(F)(C)

1-4 Drawing

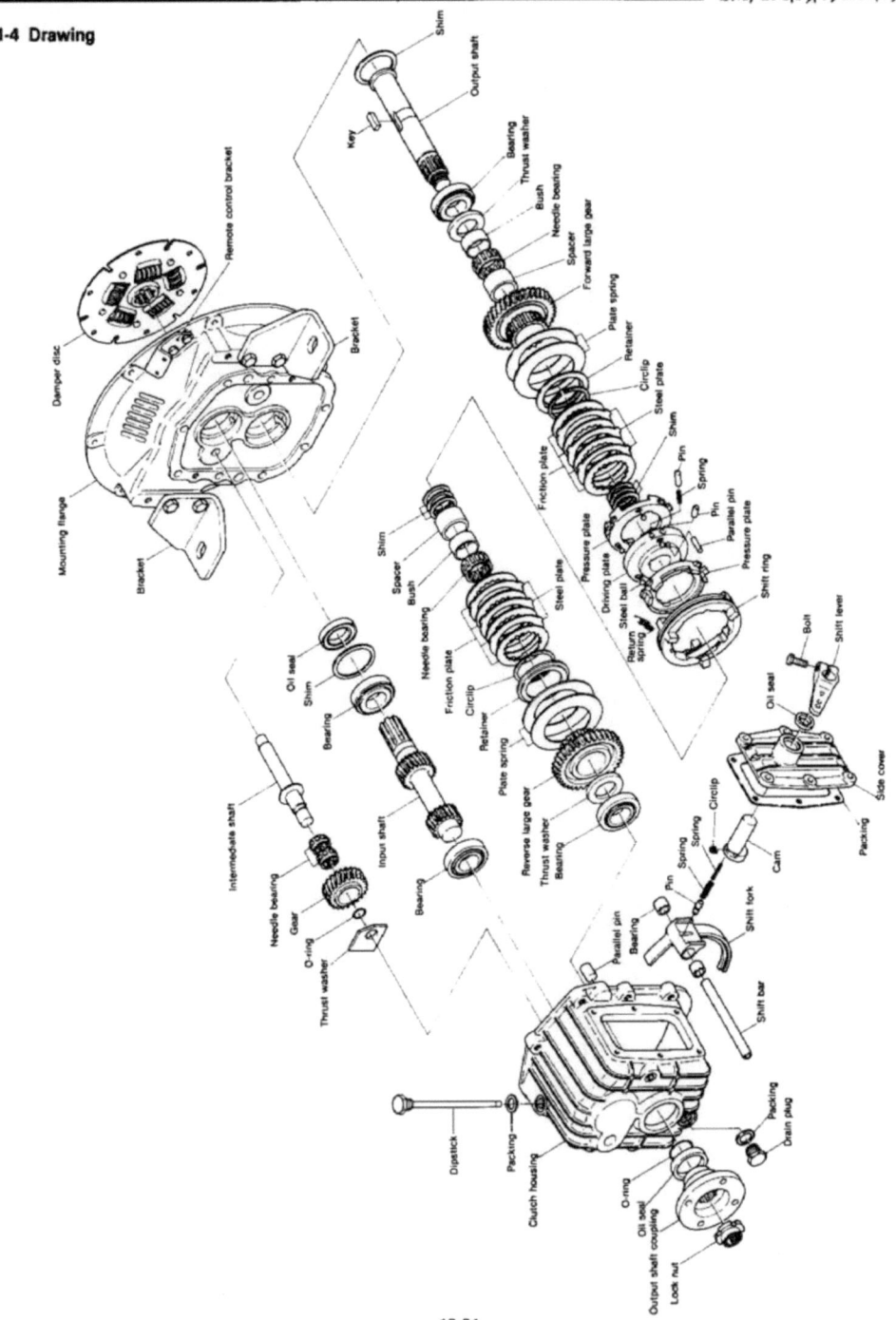

Chapter 10 Reduction and Reversing Gear
1. Construction

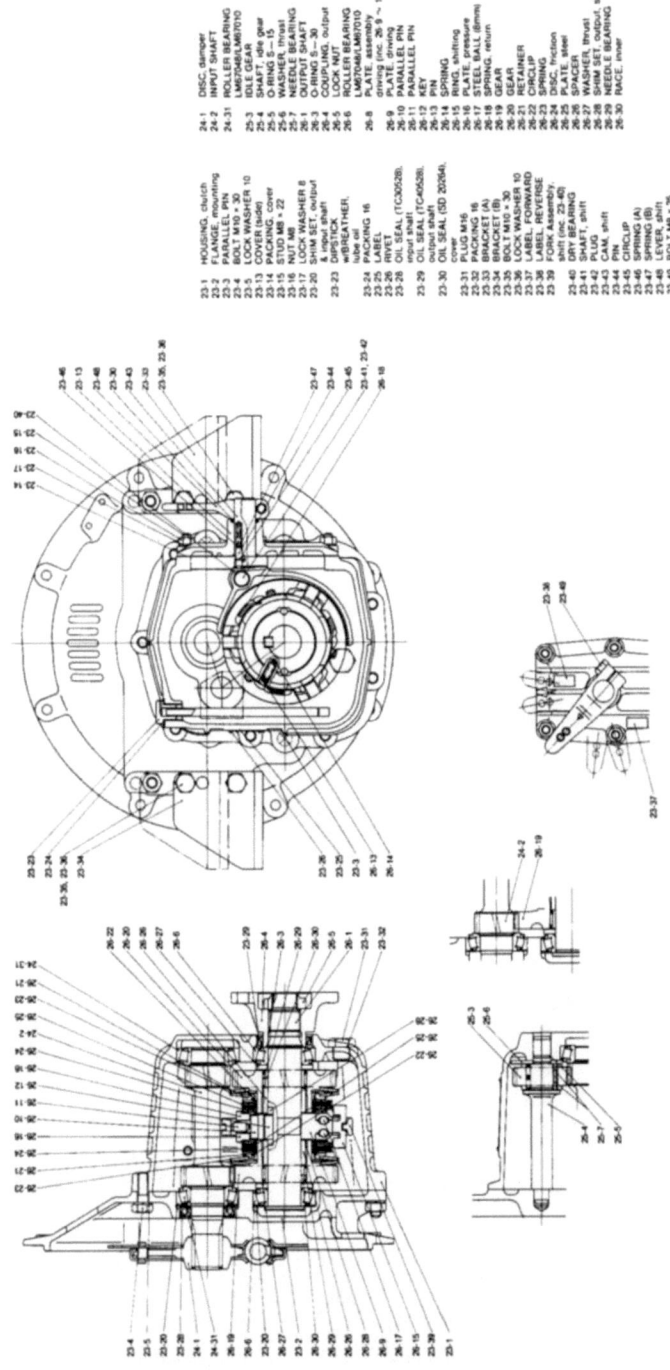

2. Installation

2-1 Installation angle
During operation the angular inclination of the gearbox in the longitudinal direction must be less than 20° relative to the water line.

2-2 Remote control unit
This marine gearbox is designed for single lever control to permit reversing at full engine speed (e.g. to avoid danger, etc.). Normally, Morse or Teleflex single lever control is employed. During installation, make sure that the remote control lever and shift lever on the marine gearbox are coordinated. Shifting the lever toward the propeller side produces forward movement, while moving the lever toward the engine side causes the vessel to move in the reverse direction.

To connect the linkage, the operating cable must be positioned at right angles to the shift lever when the shift lever is in the neutral position.

The shift play, measured at the pivot point of the shift lever, must be at least 35mm on each side (reverse and forward) of the neutral position. Greater shift play has no adverse effect on the marine gearbox. After connecting the linkage, confirm that the remote control and the shift lever on the marine gearbox work properly.

A typical linkage arrangement is illustrated in the figure below.

When the cable is attached to the hole 52mm (2.0472in.) from the center of the rotation of the shift lever, these strokes must be 30mm (1.1811in.)

mm(in.)

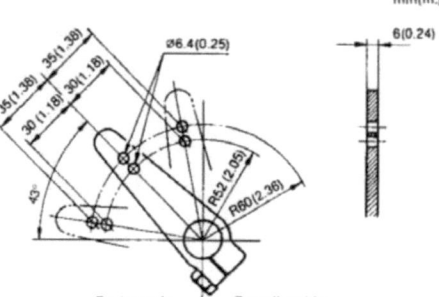

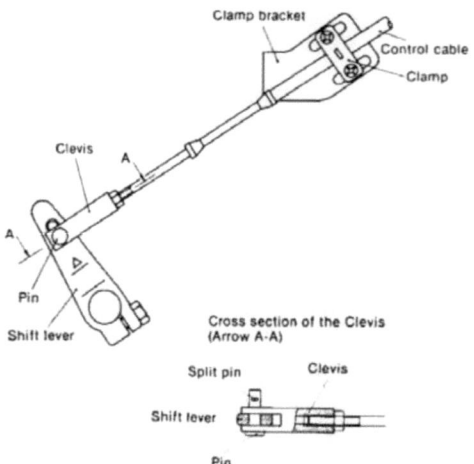

NOTE: Since the cable stroke may be insufficient, two holes are drilled in the shift lever.
When the cable is attached to the hole 60mm (2.3622in.) from the center of the rotation of the shift lever, the strokes from the center to the forward and reverse sides must both be 35mm (1.3780in.).

3. Operation and Maintenance

3-1 Lube oil
(1) Oil level
The oil level should be checked each month and must be maintained between the groove and the end of the dipstick. The groove indicates the maximum oil level and the end of the dipstick is the minimum oil level. When checking the oil level with the dipstick, do not screw in the oil filler screw; it should rest on top of the oil filler hole.

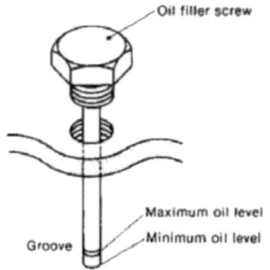

(2) Oil change
Change the oil after the first 100 hours of operation, and every 300 hours of operation thereafter. When adding oil between oil changes, always use the same type of oil that is in the marine gearbox.

(3) Recommended brands of lube oil

Supplier	Brand name
SHELL	SHELL DEXRON
CALTEX	TEXAMATIC FLUID (DEXRON)
ESSO	ESSO ATF
MOBIL	MOBIL ATF220
B.P. (British Petroleum)	B.P. AUTRAN DX

3-2 Precautions
Do not stop the shift lever halfway between the neutral and forward or reverse positions. The lever must be set to the neutral position or shifted into forward or reverse in a single motion.

3-3 Side cover
The internal shifting mechanism has been carefully aligned at the factory. Improper removal of the side cover can cause misalignment. If the side cover must be removed, proceed as follows:
—Before removing the cover, put alignment marks on the side cover and the case to facilitate accurate installation.
—When installing the side cover, put the shift lever in neutral so that the cam lobe on the shift lever engages the groove on the internal shift mechanism. When the cam lobe and groove are engaged properly there will be no clearance between the body and the side cover. Do not use packing or gaskets when installing the side cover.

—After making sure that the cam lobe and notches are aligned properly, securely tighten all the bolts. After tightening the bolts, move the lever back and forth. Positive contact should be felt and a click should be clearly audible as the gears shift; otherwise, the cam and notch are not properly engaged, and the cover must be loosened and readjusted until proper engagement is achieved.

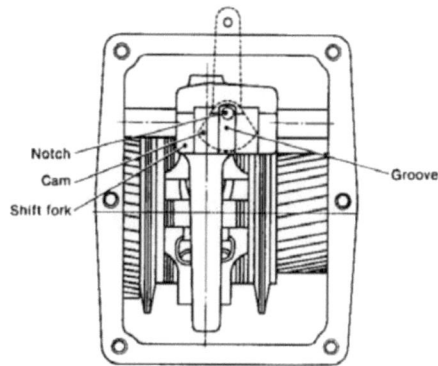

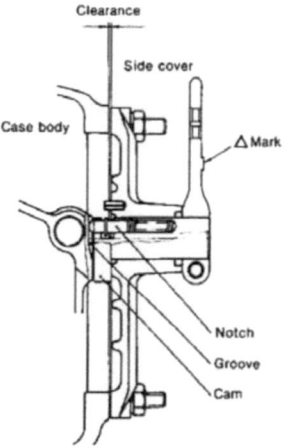

4. Inspection and Servicing

4-1 Clutch case
(1) Check the clutch case with a test hammer for cracking. Perform a color check when required.
If the case is cracked, replace it.
(2) Check for staining on the inside surface of the bearing section.
Also, measure the inside diameter of the case.
Replace the case if it is worn beyond the wear limit.

4-2 Bearing
(1) Rusting and damage
If the bearing is rusted or the taper roller retainer is damaged, replace the bearing.
(2) Make sure that the bearings rotate smoothly.
If rotation is not smooth, if there is any binding, or if an abnormal sound is heard, replace the bearing.

4-3 Gear
(1) Tooth surface wear
Check the tooth surface for pitching, abnormal wear, dents, and cracks. Repair lightly damaged gears and replace heavily damaged gears.
(2) Tooth surface contact
Check the tooth surface contact. The amount of tooth surface contact between the tooth crest and tooth flank must be at least 70% of the tooth width.
(3) Backlash
Measure the backlash of each gear, and replace the gear when it is worn beyond the wear limit.

mm (in.)

	Maintenance standard	Wear limit
Input shaft forward gear and output shaft forward gear	$0.1 \sim 0.2$ (0.0040 $\sim$ 0.0079)	0.3 (0.0118)
Input shaft reverse gear and intermediate gear	$0.1 \sim 0.2$ (0.0040 $\sim$ 0.0079)	0.3 (0.0118)
Intermediate gear and output shaft reverse gear	$0.1 \sim 0.2$ (0.0040 $\sim$ 0.0079)	0.3 (0.0118)

(4) Forward/reverse gear spline
1) Check the spline for damage and cracking.
2) Step wear of spline
Step wear depth limit: 0.1mm (0.0040in.)

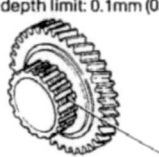
Step wear of spline

(5) Forward/reverse gear needle bearing
When an abnormal sound is produced at the needle bearing, visually inspect the rollers; replace the bearing if the rollers are faulty.
Rollers

4-4 Steel plate
(1) Burning, scratching, cracking
Replace any steel plates that are discolored or cracked.
(2) Warping measurement

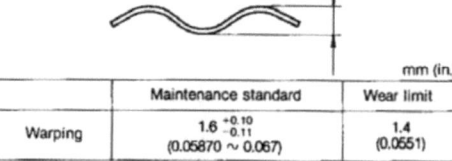

mm (in.)

	Maintenance standard	Wear limit
Warping	$1.6 ^{+0.10}_{-0.11}$ (0.05870 $\sim$ 0.067)	1.4 (0.0551)

(3) Steel plate pawl width measurement

Pawl

Measure the width of the steel plate pawl and the width of the pressure plate; replace the plate when the clearance exceeds the wear limit.

Steel plate width
Wear must be under 0.2mm (0.0079 in.)

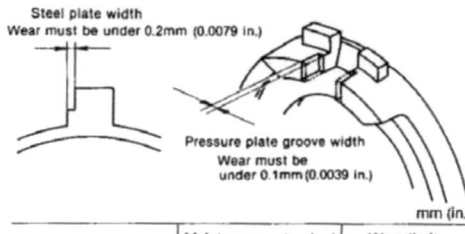

Pressure plate groove width
Wear must be under 0.1mm (0.0039 in.)

mm (in.)

	Maintenance standard	Wear limit
Steel plate width	$12 ^{0}_{-0.2}$ (0.4646 $\sim$ 0.4724)	Worn 0.2 (0.0079)
Pressure plate groove	$12 ^{+0.1}_{0}$ (0.4724 $\sim$ 0.4764)	Worn 0.1 (0.0039)
Clearance	$0 \sim 0.3$ (0 $\sim$ 0.0118)	$0.3 \sim 0.6$ (0.0118 $\sim$ 0.0236)

Chapter 10 Reduction and Reversing Gear
4. Inspection and Servicing

SM/GM(F)(C)·HM(F)(C)

4-5 Friction plate
(1) Check the friction plate for burning, scoring, or cracking. Repair the plate when the damage is light and replace the plate if the damage is heavy.
(2) Friction surface wear
Measure the thickness of the friction plate, and replace the plate when it is worn beyond the wear limit.

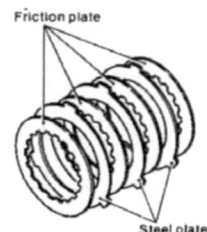

4-6 Pressure plate

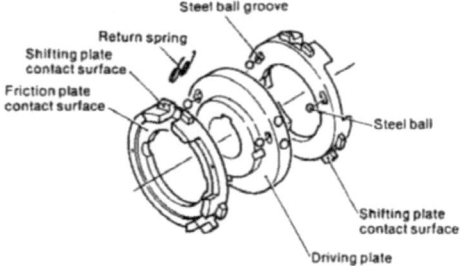

(1) Steel ball groove
Check the steel ball groove for stains and wear. Replace the pressure plate if the groove is noticeably worn.
(2) Friction plate contact surface
Check the contact face for stains and damage.
(3) Shifting plate contact surface
(4) Worn parts measurement

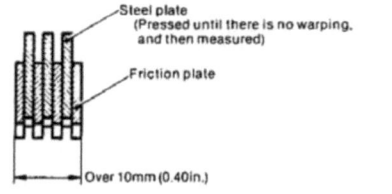

mm (in.)

	Maintenance standard	Wear limit
Friction plate thickness	$1.7 \,^{0}_{-0.05}$ (0.0650 ~ 0.0670)	1.5 (0.0591)

The assembled friction plate and steel plate dimensions must be over 10mm (0.0040in.).

Both sides of the friction plate have a 0.35mm copper sintered layer. Replace the friction plate when this layer is worn more than 0.2mm on one side (standard thickness $1.7 \,^{0}_{-0.05}$ mm). However, the sum of the wear of the four friction plates must not exceed 0.8mm. When this value is exceeded, replace all friction plates. In unavoidable circumstances, it is permissible to replace only the friction plate with the greatest amount of wear.

(3) Friction plate and gear spline back clearance
Measure the clearance between the friction plate spline collar and the output shaft gear spline, and replace the plate or spline when they are worn beyond the wear limit.

mm (in.)

	Maintenance standard	Wear limit
Standard backlash	0.20 ~ 0.61 (0.0079 ~ 0.0240)	0.9 (0.0354)

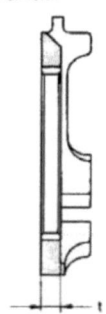

mm (in.)

	Maintenance standard	Wear limit
Thickness: t	$6.6 \,^{0}_{-0.2}$ (0.2520 ~ 0.2598)	6.3 (0.2480)

(5) Return spring permanent strain.
Make sure the length (free length) is within the values specified in the figure.

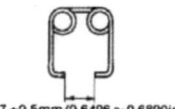

17 ±0.5mm (0.6496 ~ 0.6890in.)

10-36

Printed in Japan
0000A0A1361

4-7 Driving plate

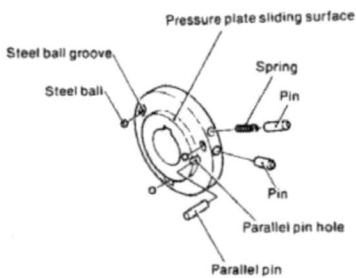

(1) Check the key groove for scoring and cracking, and the output shaft fitting section for burning. Repair if the damage is light and replace the driving plate if the damage is heavy.
(2) Outside diameter of pressure plate sliding part; others

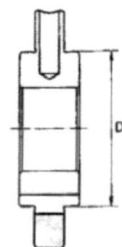

mm (in.)

	Maintenance standard	Wear limit
Outside diameter: D	ø59 $_{-0.134}^{-0.060}$ (2.3176 ~ 2.3205)	ø58.8 (2.3150)

(3) Steel ball groove wear and stains.
(4) Determine the amount of wear and play of both the axial and circumferential direction pins.
(5) Permanent spring strain.

mm (in.)

	Maintenance standard	Wear limit
Spring free length	32.85 (1.2933)	32 (1.2598)

(6) Pin end wear.

4-8 Retainer

(1) Check for stains and damage on the friction plate contact surface.
(2) Check for wear and cracking on the plate spring contact surface.
(3) Measurement of dimensions

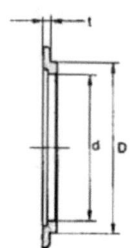

mm (in.)

	Maintenance standard	Wear limit
d	ø57.5 $_{+0.060}^{+0.106}$ (2.2661 ~ 2.2680)	ø57.8 (2.2756)
D	ø66 $_{-0.1}^{0}$ (2.5945 ~ 2.5984)	ø65.7 (2.5866)
t	2.8 $_{-0.08}^{0}$ (0.1071 ~ 0.1102)	2.6 (0.1024)

4-9 Plate spring

(1) Permanent strain

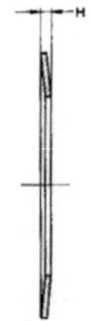

mm (in.)

	Maintenance standard	Wear limit
H: when plate spring is free	6.25 ±0.1 (0.2421 ~ 0.2500)	6.0 (0.2362)

4-10 Thrust collar

The gear side of the thrust washer has a 0.3mm copper sintered layer. Replace the thrust collar when the thickness is less than 4.75mm (standard thickness: 5 $_{-0.1}^{0}$ mm).

4-11 Shift ring

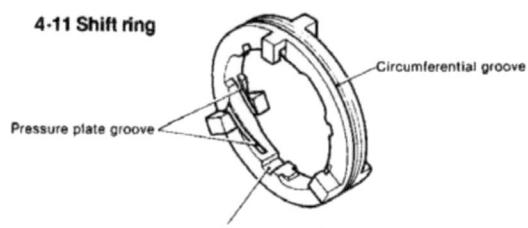

(1) Circumferential groove wear.

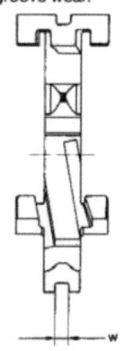

	Maintenance standard	Wear limit
Shifting groove: w	$6^{+0.1}_{0}$ (0.2362 ~ 0.2402)	6.3 (0.2480)

mm (in.)

(2) Pressure plate groove wear.
Whenever uneven wear and/or scratches are found, replace with a new part.
(3) Parallel pin contact part wear.
Whenever uneven wear and/or scratches are found, replace with a new part.

4-12 Shift fork and shift lever

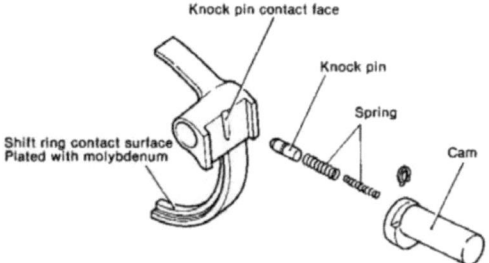

(1) End wear.
The shift ring contact surface of the shift fork is plated with molybdenum (thickness: 0.04—0.05mm). If this plating is peeled or worn to such an extent that the base metal of the shift fork is exposed, replace the shift fork.
(2) Cam surface wear and stains.
Whenever uneven wear and/or scratches are found, replace with a new part.
(3) Pin part play.
Whenever uneven wear and/or scratches are found, replace with a new part.
(4) Notch end wear.
Whenever uneven wear and/or scratches are found, replace with a new part.

4-13 Output shaft

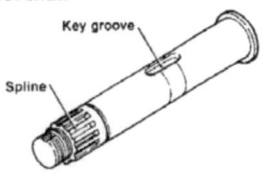

(1) Key groove.
Whenever uneven cracks and/or stains are found, replace with a new part.

4-14 Damper disc

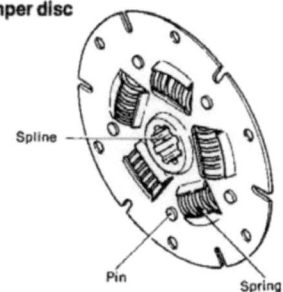

(1) Spline part
Whenever uneven wear and/or scratches are found, replace with a new part.
(2) Spring.
Whenever uneven wear and/or scratches are found, replace with a new part.
(3) Pin wear.
Whenever uneven wear and/or scratches are found, replace with a new part.

4-15 Input shaft

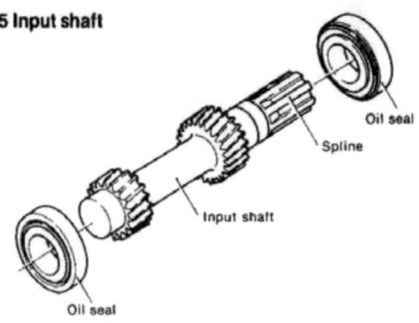

(1) Spline part
Whenever uneven wear and/or scratches are found, replace with a new part.
(2) Surface of oil seal.
If the sealing surface of the oil seal is worn or scratched, replace.

4-16 Intermediate shaft

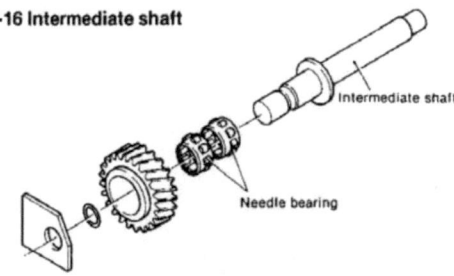

(1) Needle bearing dimensions, staining.
Check the surface of the roller to see whether the needle bearing sticks or is damaged. Replace if necessary.

5. Disassembly

5-1 Disassembling the clutch and accessories
(1) Remove the drain plug and packing, and drain the oil from the clutch.
(2) Uncaulk the output shaft lock nut, and remove the nut using a disassembly tool.

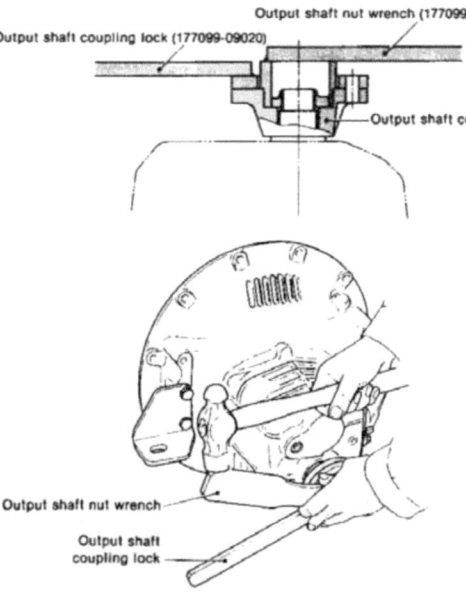

(3) Remove the output coupling.
(4) Remove the dipstick and packing.
(5) Remove the case cover M8 nut super lock washer; remove the case cover, with the operating lever, shift cam, etc. in position.
(6) Remove the shift bar plug with a hexagonal bar spanner (width across flats: 8mm (0.0394in.), and pull the shift bar from the case, using the M10 pulling bolt at the end of the shift bar.

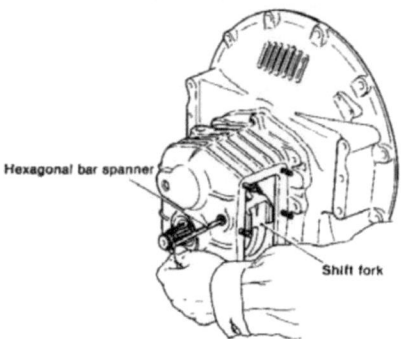

(7) Remove the shift fork.

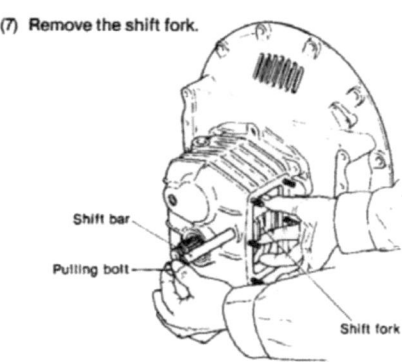

(8) Remove the M10 bolt and super lock washer on the mounting flange.
(9) Screw the M10 bolt into the M10 pulling bolt hole of the mounting flange, and remove the mounting flange. Do not remove the parallel pin.

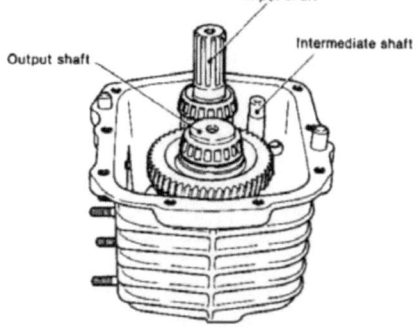

(10) Remove the output shaft, intermediate shaft, and input shaft from the case, in that order.

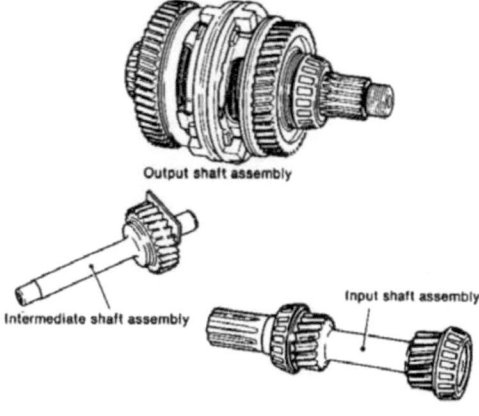

Chapter 10 Reduction and Reversing Gear
5. Disassembly

(11) Heat the case body to about 100°C and remove the outer race of the input shaft and output shaft bearings. If the outer races are difficult to remove, tap them out with a plastic hammer from the rear of the case, or pull them by using the pulling groove in the case at the rear of the races.

(12) Remove the outer race of the bearing from the mounting flange as described in step (11) above.

(13) Remove the input shaft and output shaft adjusting plates.

NOTE: If the following parts are not replaced, the adjusting plates may be reused without readjustment. However, if even one part is replaced, readjustment is necessary.
Input shaft part: 24-2, 24-31
Output shaft part: 26-6, 26-9, 26-26, 26-27, 26-28, 26-30

(14) Pull the oil seal from the case.
(15) Pull the oil seal from the mounting flange.

5-2 Disassembling the input shaft
Pull the bearing from the input shaft.
NOTE: Do not disassemble unless the input shaft parts are damaged.

5-3 Disassembling the output shaft
(1) Remove the O-ring.

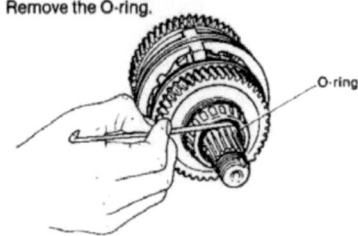

(2) Remove the output shaft by pressing the threaded end of the output shaft with a press, or tapping it with a hammer.

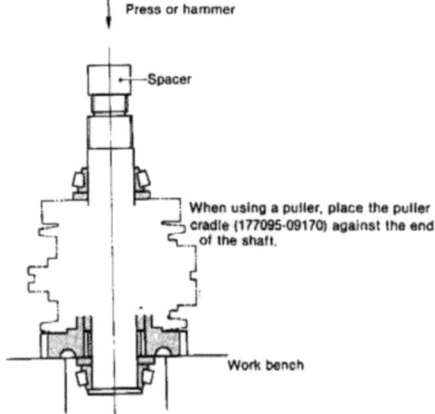

When using a puller, place the puller cradle (177095-09170) against the end of the shaft.

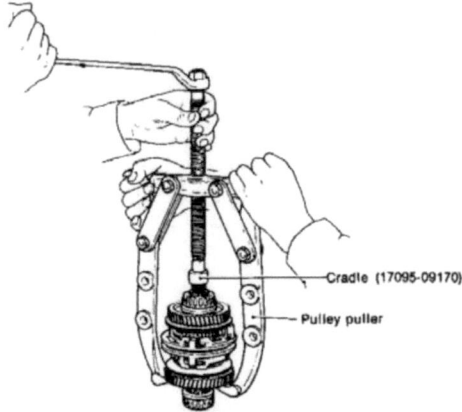

Cradle (17095-09170)
Pulley puller

NOTE 1: When removing the shaft, place spacers between the shaft and the press to prevent damage.

NOTE 2: Make sure that the forward large gear parts and reverse large gear parts are not mixed together once they are removed.

(3) Remove the adjusting plate.
NOTE: Record the thickness of the adjusting plate to facilitate reassembly.
If the parts are not replaced, the adjusting plate may be reused without readjustment. However, if even one part is replaced, readjustment is required.

(4) Remove the key.
To facilitate removal, clamp the key with a vice.

(5) Remove the adjusting plate.
NOTE: Record the thickness of the adjusting plate to facilitate reassembly.
If the parts are not replaced, the adjusting plate may be reused without adjustment. However, if even one part is replaced, readjustment is required.

(6) Remove the spacer and needle bearing.
(7) Cover the outer race of the forward bearing, and pull out the output shaft about 10mm (0.3937in.) by pressing the threaded end of the output shaft with a press, or tapping it with a hammer.
NOTE: Do not pull it out more than 10mm (0.3937in.); otherwise damage may result.

10-41

Chapter 10 Reduction and Reversing Gear
5. Disassembly

(8) Insert the disassembly tool between the collar of the output shaft and the bearing; next remove the bearing inner race, thrust collar, and bearing from the output shaft with a press or hammer.

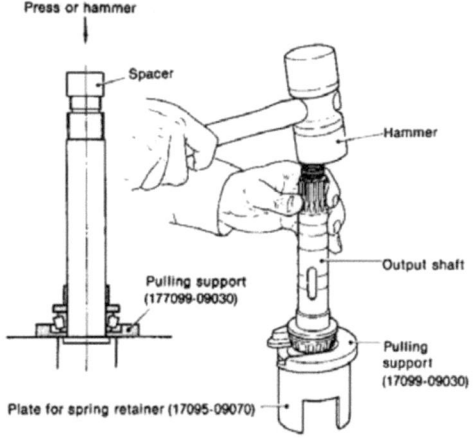

(9) Remove the friction plates and steel plates from the forward large gear.
(10) Using a disassembly tool, compress the plate spring and remove the circlip from the forward large gear.

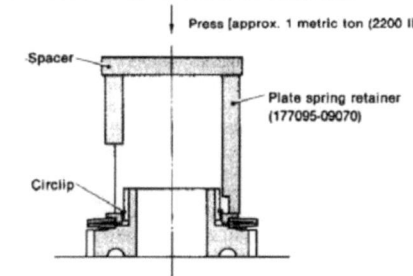

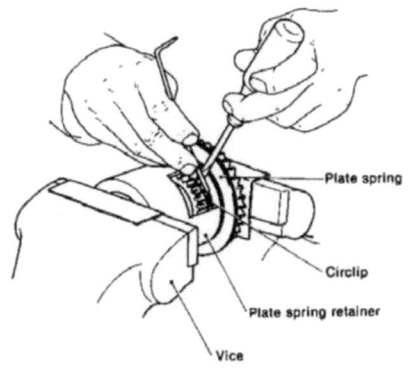

(11) Remove the retainer and plate spring.
(12) Remove the parts from the reverse large gear as described in steps (9)—(11) above.
(13) Remove the pressure plate return spring; remove the pressure plate and steel ball.

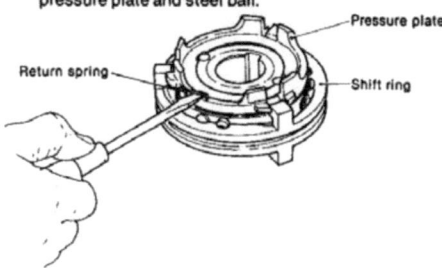

(14) Remove the shift ring.
To disassemble, remove the three knock pins. When disassembling the shift ring, cover it with a cloth to prevent it being lost.
(15) Remove the knock pin and spring from the driving plate.

5-4 Disassembling the intermediate shaft
(1) Place a spacer against the case side end of the intermediate shaft and remove the shaft from the case by tapping the spacer with a hammer.

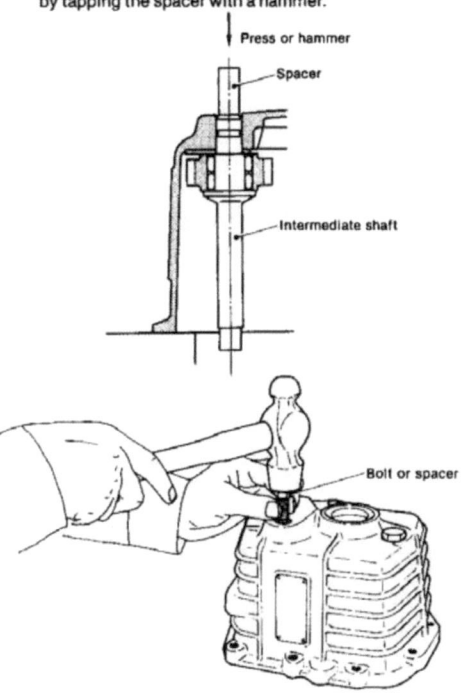

Chapter 10 Reduction and Reversing Gear
5. Disassembly

(2) Remove the O-ring.
(3) Remove the idle gear, needle bearing, and thrust washer.

5-5 Disassembling the operating system
(1) Loosen the M8 bolt of the shift lever; remove the shift lever.
(2) Pull the shift cam.
(3) Push in the knock pin and remove the circlip.
(4) Remove the knock pin and spring.
(5) Pull the oil seal from the case side cover.

6. Reassembly

6-1 Reassembly precautions
(1) Before reassembling, clean all parts in washing oil, and replace any damaged or worn parts.
Remove non-dry packing agent from the mating surface with a blunt knife.
(2) Pack the oil seal and O-ring parts with grease.
(3) Coat the mating surfaces of the case with wet packing.

6-2 Reassembling the output shaft
(1) Reassembling forward large gear and plate spring
  1) Insert the two plate springs of the forward large gear so that their large diameter sides are opposite each other.
  2) Insert the retainer and install the circlip.
  3) Compress the plate spring, using the disassembly tool, and snap the circlip into the groove on the outside of the spline of the forward large gear.

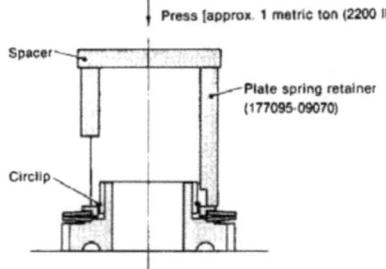

(2) Reassemble the reverse large gear and plate spring, retainer, and circlip as described in step (1) above.
(3) Determining the forward adjusting plate thickness

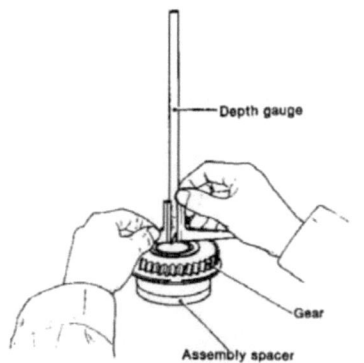

NOTE: As mentioned in section 5-3. (5), if no parts need to be replaced, the adjusting plate can be reused without adjustment.

1) Position the assembled large gear on the assembly tool so that the spline part is on the bottom; insert the spacer and bearing inner race into the gear.

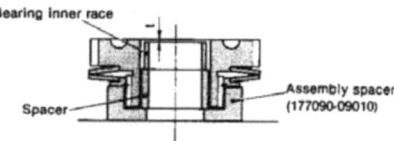

2) Adjust the thickness of the adjusting plate until it conforms to the dimension shown in the figure.
3) Two adjustment plates of 0.5mm (0.0197in.) and 0.3mm (0.0118in.) are available.
Combine these plates to obtain the "t" dimension.
(4) Determine the thickness of the reverse adjusting plate by following the procedure described in step (3) above.
(5) First, insert a friction plate into the spline part of the forward large gear; next insert steel plates and friction plates alternately. Finally, insert a friction plate (four friction plates and three steel plates).
(6) Insert the friction plates and steel plates into the spline part of the reverse large gear in the same manner as described in step (5) above (four friction plates and three steel plates).
(7) Press the inner race of the bearing onto the output shaft up to the collar, using an assembly tool.
NOTE: The inner race can be installed easily by preheating it to approximately 100°C.

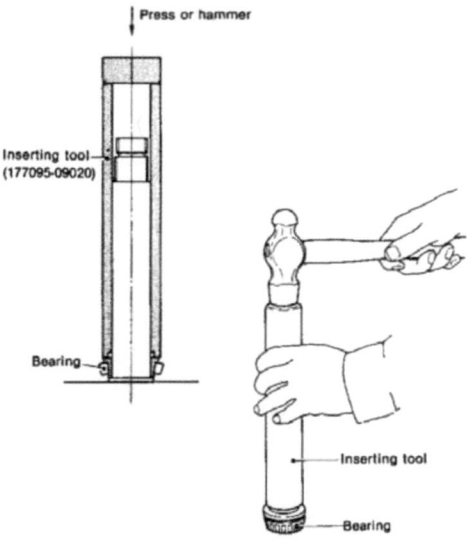

10-44

Chapter 10 Reduction and Reversing Gear
6. Reassembly

(8) Insert the thrust collar, with the sintered surface (brown surface) facing the gear side.
(9) Press the bearing inner race onto the output shaft, using an assembly tool.

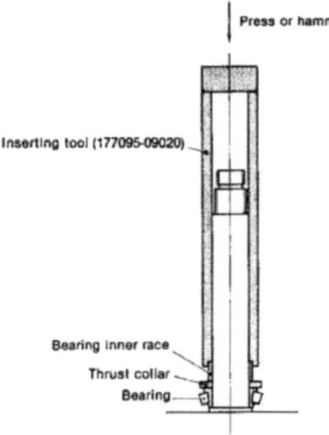

(10) Insert the needle bearing.
(11) Insert the spacer and adjusting plate.
(12) Fit the key so that the fillet side is facing the threaded part of the output shaft.

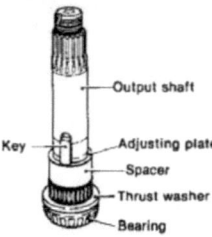

(13) Insert the forward large gear, together with the friction plates and steel plates. At this time, align the three pawls on the outside of the steel plates.

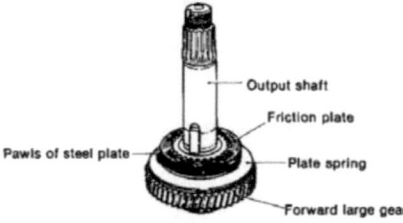

(14) Cover the friction plates and steel plates with the pressure plate so that the pawls of the steel plate fit into the three notches on the pressure plate.
(15) Insert the three steel balls into the three grooves in the pressure plate.

(16) Insert the drive plate into the output shaft so that the side with the identification groove faces the forward large gear side.
NOTE: Make sure that the three steel balls are in the three grooves of the driving plate.
At the same time, make sure that the pin for the driving plate fits into the groove of the torque limiter for the pressure plate.

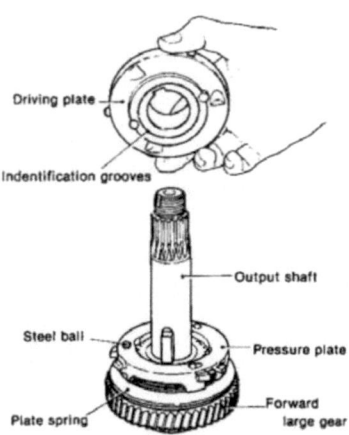

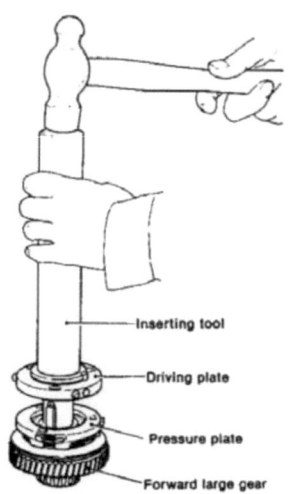

Chapter 10 Reduction and Reversing Gear
6. Reassembly

(17) Insert the adjusting plate and spacer.
(18) Press the bearing inner race, using an assembly tool.

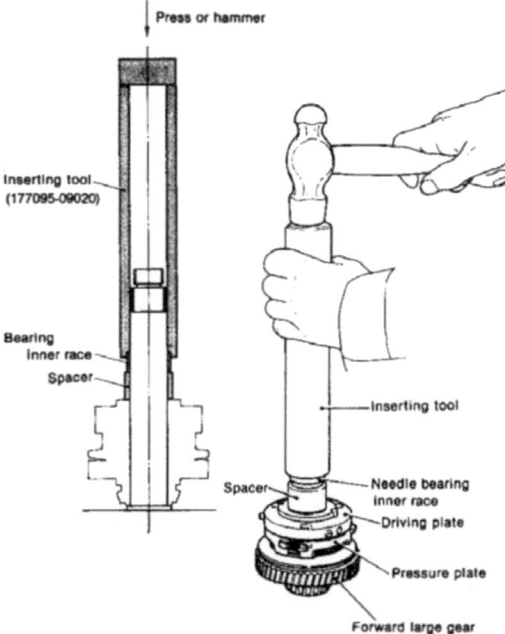

(19) Insert the knock pins and springs into the three holes around the circumference of the driving plate.
(20) Cover the driving plate with the shift ring so that the side with the identification groove faces the forward large gear side; install the ring so that the knock pins are pushed in.

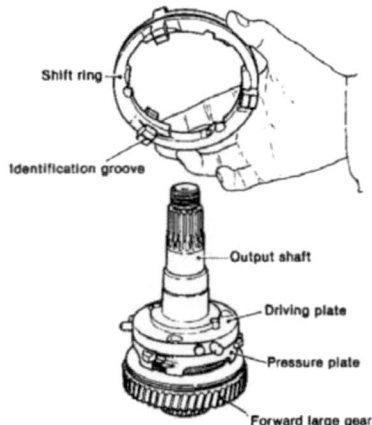

(21) Insert the three steel balls into the three grooves in the driving plate.
(22) Place the pressure plate onto the driving plate so that the steel balls enter the three grooves of the pressure plate.
(23) Insert the three pressure plate return springs between the shift ring and the driving plate, and attach them to the small holes in the side of the pressure plate.
(24) Insert the reverse large gear [see step (6)] so that the three pawls of the steel plates enter the notches around the circumference of the pressure plate.
(25) Insert the needle bearing.
(26) Insert the thrust washer so that the sintered side (brown side) faces the gear side.
(27) Press the inner race of the bearing, using an assembly tool. Make sure that the direction of the bearing is corret.

NOTE: *The bearing inner race can be installed easily by preheating it to approximately 100°C.*

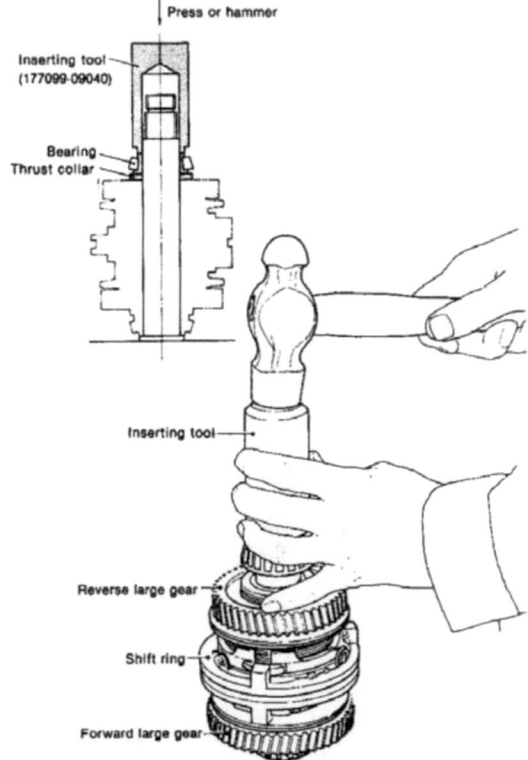

(28) Insert the O-ring.
(29) With the shift ring in the reverse position, check the forward large gear to make sure it rotates smoothly. Next, with the shift ring in the forward position, check the reverse large gear to make sure it rotates smoothly.

Chapter 10 Reduction and Reversing Gear
6. Reassembly

SM/GM(F)(C)·HM(F)(C)

6-3 Reassembling the input shaft
Press the inner race of the bearing onto the input shaft. Make sure that the direction of the bearing is correct.

NOTE: The bearing inner race can be easily installed by preheating it to approximately 100°C.

6-4 Reassembling the intermediate shaft
NOTE: Assemble the intermediate shaft as described in section 6-5. (5).

(1) Insert the needle bearing and idle gear on the intermediate shaft. Then insert the thrust washer.

NOTE: Pay careful attention to the assembling direction of the thrust washer.

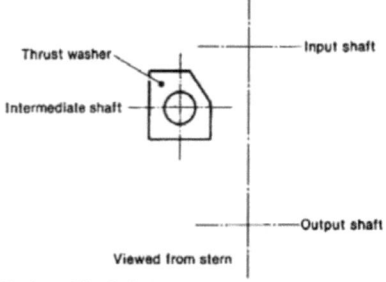

Viewed from stern

(2) Insert the O-ring.
(3) Press the assembled intermediate shaft into the case with a press or hammer.

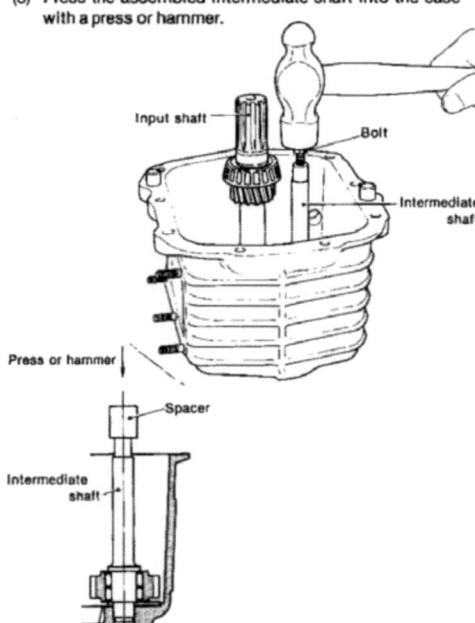

(4) Make sure that the idle gear rotates smoothly.

6-5 Installing the input shaft and output shaft
(1) Determining the thickness of the input shaft adjusting plate and output shaft adjusting plate

NOTE: As mentioned in section 5-1. (13), when none of the parts are replaced, the adjusting plate can be reused without readjustment.

1) Measure length "A" "D" between the cases of each shaft of the case body and mounting flange.
2) Cover each bearing with the bearing outer race, and measure length "B" "C" between the bearings.

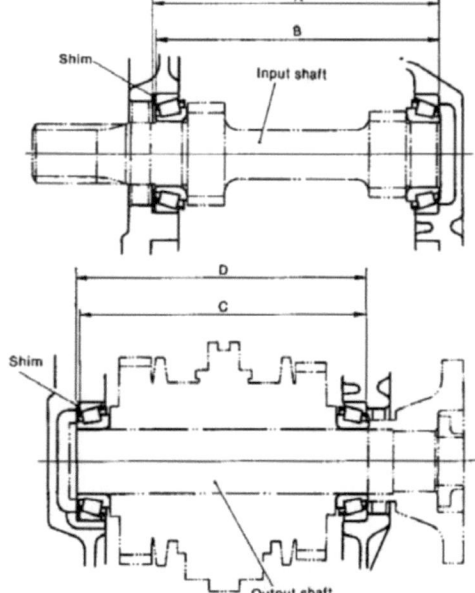

3) Adjust the input shaft adjusting plate thickness so that the clearance or tightening allowance is less than 0.05mm (0.0020in.).
4) Adjust the output shaft adjusting plate thickness so that the tightening allowance is within 0 ~ 0.1mm (0~0.0040in.).
5) Four adjusting plates of 1mm (0.0394in.), 0.5mm (0.0197in.), 0.3mm (0.0118in.) and 0.1mm (0.0040in.)are available.
Combine these plates to obtain the desired adjusting plate measurement.

(2) Insert the adjusting plate into the mounting flange, and press the outer race of the bearing.
Also, press the outer race of the bearing into the case.

NOTE: The outer race can be installed easily by heating the mounting flange and case to approximately 100°C, or by cooling the bearing outer race with liquid nitrogen, etc.

(3) Coat the circumference of the oil seal with a non-dry packing agent, and press it onto the mounting flange and case so that the spring part of the oil seal is inside the case.

Chapter 10 Reduction and Reversing Gear
6. Reassembly

(4) Coat the mating surfaces of the mounting flange and case with a non-dry packing agent.
Wipe off oil and dirt on the mating surface of the case and coat with a thin film of non-dry packing agent.

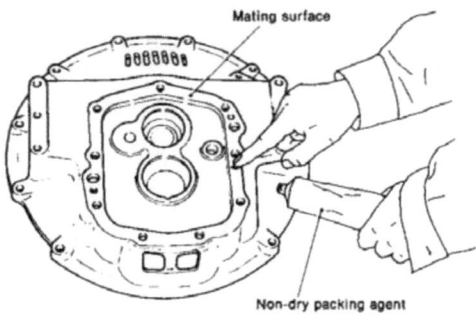

(5) Insert the input shaft into the case, assemble the intermediate shaft as described in section 6-4 and then insert the output shaft into the case.
(6) Align the mounting flange with the case, and insert the parallel pin by tapping the mounting flange with a plastic hammer.
(7) Insert the super lock washer and tighten the M10 bolt.
(8) Install the dipstick and packing.
(9) Install the drain plug and packing.

6-6 Reassembling and installing the operating system

(1) Insert the shift fork into the case from the side, insert the shift bar.
NOTE: Insert the shift bar with the threaded end towards the outside (output shaft coupling side).

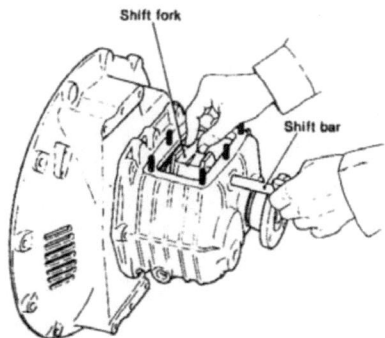

(2) Coat the threaded part of the shift bar plug with a non-dry packing agent and secure it to the case with a hexagonal bar spanner (width across flats: 8mm (0.3150in.).
NOTE: Put the shift fork into neutral before installing.

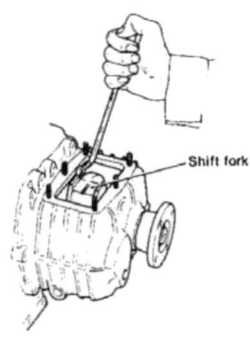

(3) Coat the circumference of the oil seal with a non-dry packing agent and press the seal to the case cover.
(4) Insert the spring into the shift cam.
(5) Insert the knock pin into the shift cam from the front end, and lock with the circlip.
(6) Insert the assembled shift cam into the case cover.
(7) Fit the shift lever to the shift cam, and tighten the M8 bolt.
NOTE: The shift cam must rotate smoothly.
(8) Replace the packing if it is damaged.
(9) Attach the case side cover together with the operating system to the case body.
At this time, make sure that the shift cam is fitted to the shift fork, and that the shift lever is in neutral.
NOTE: Put the shift fork into neutral before installing.
(10) Insert the super lock washer, and tighten the M8 nut.
(11) Shift the shift lever to forward and reverse to make sure that the lever operates normally.
If the lever does not operate normally, loosen the M8 nut, slide the case side cover forward, backward, and to the left and right, then re-tighten with the M8 nut in the position at which the lever operates normally.
NOTE: If the lever operates normally a click will be heard when it is put into forward and reverse.

6-7 Installing the output shaft coupling

(1) Install the output shaft coupling on the output shaft.
(2) Tighten and caulk the output shaft lock nut, using the assembly tool.
   Tightening torque.......... 9.5kgf-m (68.7ft-lb)

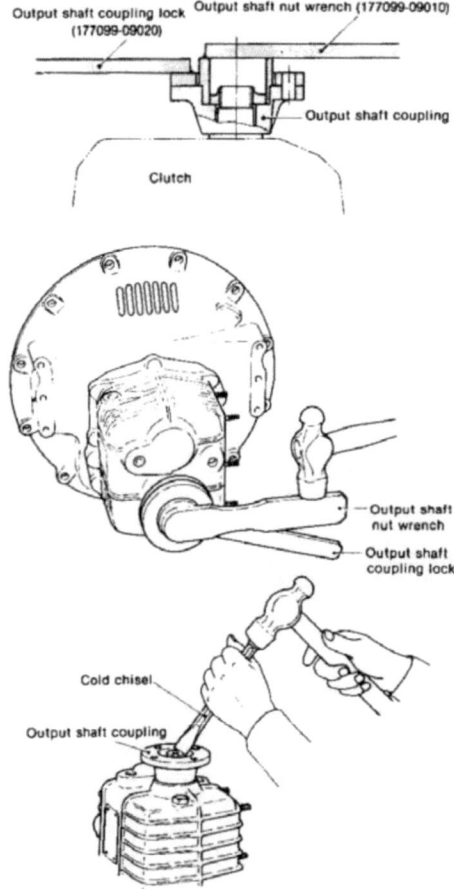

(3) Shift the shift lever to the neutral position and make sure the clutch engages when the shift lever is put into forward and reverse.
The input/output shafts will not rotate smoothly if the side gap of the bearing is too small in relation to the thickness of the adjusting plate.

[C] Marine Gear Models KM2P, KM3P and KM3V

for Engine Models 1GM10, 2GM20(F) and 3GM30(F)

Applicable Engine Models & Serial Nos.

(Effective from:)

KM2P 1GM10	E/# 03413	and after Aug. 1985
KM2P 2GM20(F)	E/# 03567	and after Aug. 1985
KM3P 3GM30(F)	E/# 01888	and after Aug. 1985

1. Construction

1-1 Construction

This clutch is a cone-type, mechanically operated clutch. When the drive cone (which is connected to the output shaft by the lead spline) is moved forward or backward, its taper contacts with the large gear and transfers power to the output shaft.

The construction is simple when compared with other types of clutch and it serves to reduce the number of components, making for a lighter, more compact unit which can be operated smoothly. Although it is small, the power transmission efficiency is high even under a heavy load. Its durability is high and it is reliable as high grade materials are used for the shaft and gear, and a taper roller bearing is incorporated. Power transmission is smooth as connection with the engine is made through the damper disc.

- The drive cone is made from special aluminum bronze which has both higher wear-resistance and durability. The drive cone is connected with the output shaft through the thread spline. The taper angle, diameter of the drive cone, twist angle, and diameter of the thread spline, are designed to give the greatest efficiency, thus ensuring that the drive cone can be readily engaged or disengaged.
- Helical gears are used for greater strength. The intermediate shaft is supported at 2 points to reduce deflection and gear noise.
- The clutch case and mounting flange are made from an aluminum alloy of special composition to reduce weight. It is also anticorrosive against seawater.
- As the damper disc is fitted to the output shaft, power can be transmitted smoothly. For the damper disc, springs of different strengths are used so that two stages of torque and twist angle are applied. That is, in the first stage, only the weak spring is used, and the strong spring comes into action for a torque higher than a predetermined value.

This prevents gear noise due to torsional vibration as well as absorbing shock when engaging.

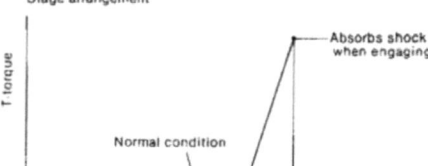
Stage arrangement

- There is a small clearance between the dipstick and the inside of the dipstick tube. A small hole in the dipstick works as a breather.
- When the load on the propeller is removed, the engagement of the drive cone and the large gear is maintained by the shifter and V-groove of the drive cone. Even when the drive cone's tapered area and V-groove are worn, this engagement is maintained by the shift lever device and accordingly no adjustment of the remote control cable is required.
- The cup spring on the rear of the larger gear absorbs rotational fluctuations and stabilizes the engagement of the drive cone and the larger gear. Thus, the durability of the cone against wear is enhanced.

Chapter 10 Reduction and Reversing Gear
1. Construction

SM/GM(F)(C)·HM(F)(C)

1-2 Specifications

Model			KM2P			KM3P		
For engine models			1GM10, 2GM20(F)			3GM30(F)		
Clutch			Constant mesh gear with servo cone clutch (wet type)					
Reduction ratio	Forward		2.21	2.62	3.22	2.36	2.61	3.20
	Reverse		3.06	3.06	3.06	3.16	3.16	3.16
Propeller shaft rpm (Forward)			1540	1298	1055	1441	1303	1063
Direction of rotation	Input shaft		Counter-clockwise, viewed from stern					
	Output shaft	Forward	Clockwise, viewed from stern					
		Reverse	Counter-clockwise, viewed from stern					
Remote control	Control head		Single lever control					
	Cable		Morse, 33-C (cable travel 76.2mm or					
	Clamp		YANMAR made, standard accessory					
	Cable connector		YANMAR made, standard accessory					
Output shaft coupling	Outer diameter		ø100mm (3.93")					
	Pitch circle diameter		ø78mm (3.07")					
	Connecting bolt holes		4—ø10.5mm (4—ø0.41")					
Position of shift lever			Left side, viewed from stern					
Lubricating oil			SAE #10W-30, CC class					
Lubricating oil capacity			0.3 ℓ			0.35 ℓ		
Dry weight			10.3 kg (22.7 lbs)			11.5 kg (25.4 lbs)		

Models KM2P and KM3P reduction and reverse gear boxes, shafts and gears are the same except for the following items:
- No. of gear teeth
- Distance between bearings for input and output shafts.
- Clutch case, mounting flange.

Chapter 10 Reduction and Reversing Gear
1. Construction

SM/GM(F)(C)·HM(F)(C)

1-3 Power transmission system
1-3.1 Arrangement of shafts and gears

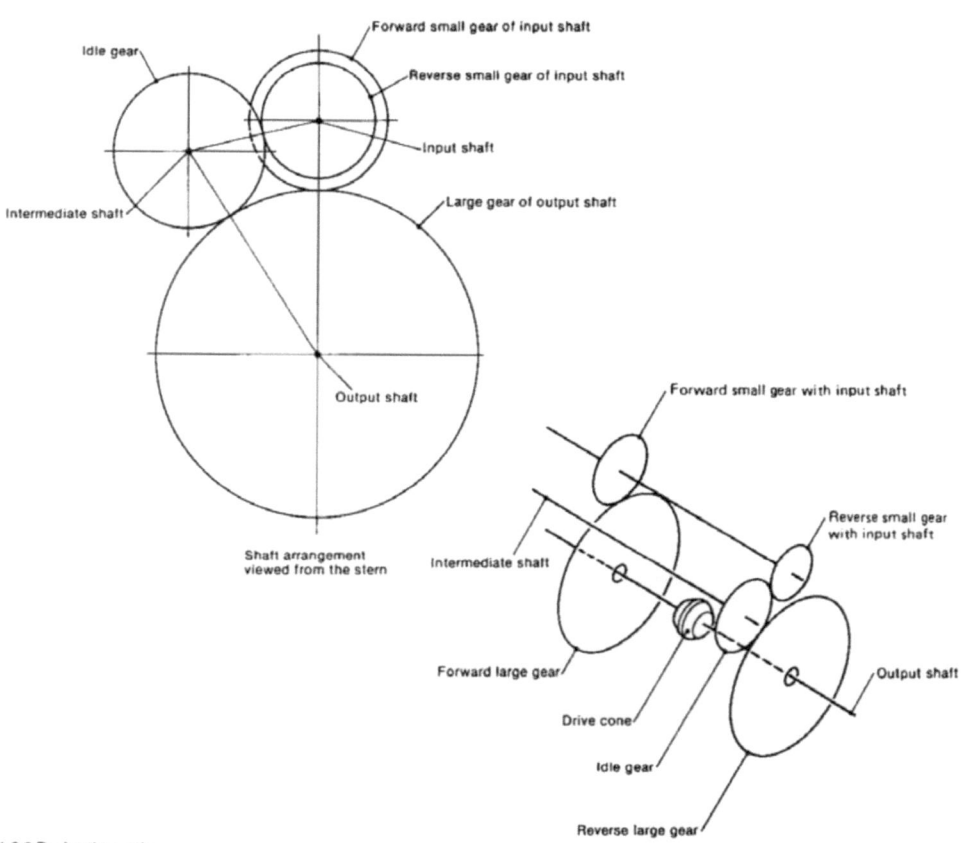

1-3.2 Reduction ratio
Forward

Model	No. of teeth of forward small gear Zif	No. of teeth of forward large gear Zof	Reduction ratio Zof/Zif
KM2P	24	53	53/24 = 2.21
	21	55	55/21 = 2.62
	18	58	58/18 = 3.22
KM3P	25	59	59/25 = 2.36
	23	60	60/23 = 2.61
	20	64	64/20 = 3.20

Reverse

Model	No. of teeth of reverse small gear Zir	No. of teeth of intermediate shaft gear Zi	No. of teeth of reverse large gear Zdr	Reduction ratio Zi/Zir·Zdr/Zi
KM2P	18	26	55	55/18 = 3.06
KM3P	19	26	60	60/19 = 3.16

Chapter 10 Reduction and Reversing Gear
1. Construction

SM/GM(F)(C)-HM(F)(C)

1-3.3 Power transmission routine — Forward

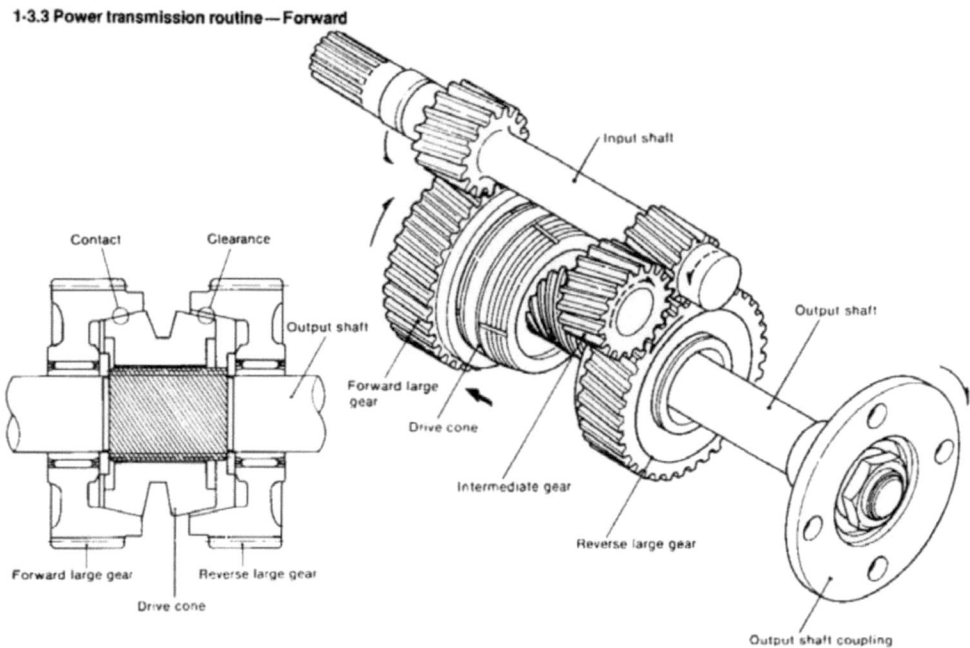

1-3.4 Power transmission routine — Reverse

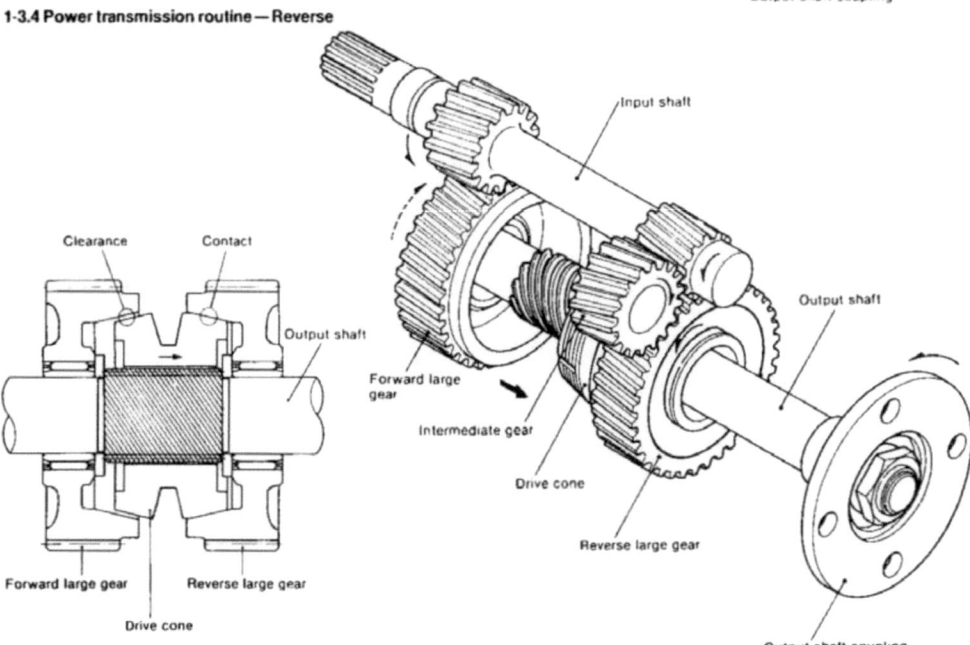

Chapter 10 Reduction and Reversing Gear
1. Construction

SM/GM(F)(C)·HM(F)(C)

1-4 Drawing

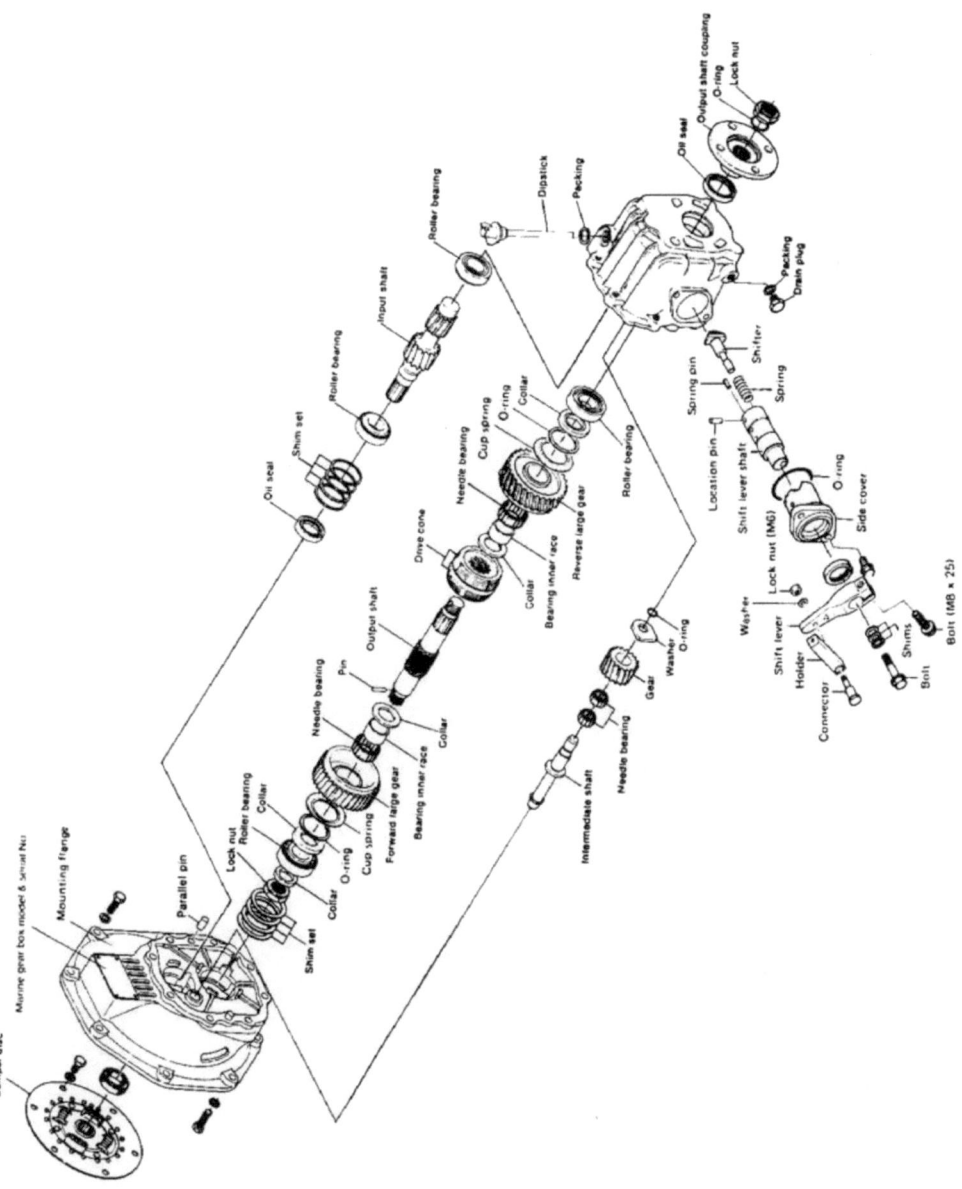

Chapter 10 Reduction and Reversing Gear
1. Construction

SM/GM(F)(C)·HM(F)(C)

1-5 Sectional view

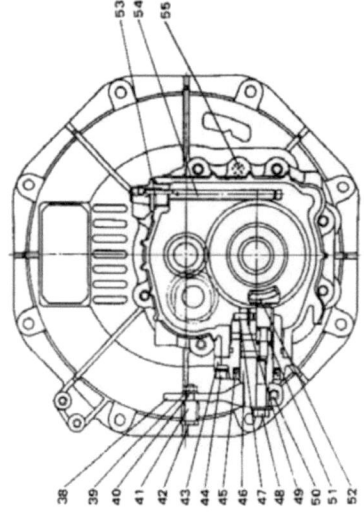
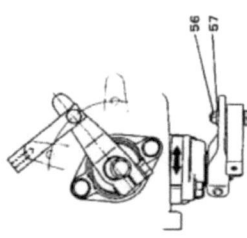
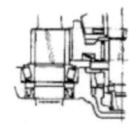
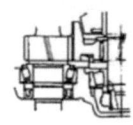
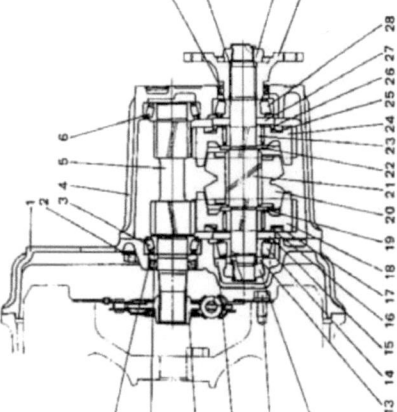
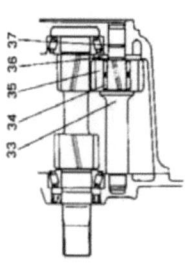

1	Mounting flange
2	Bolt M8 x 25
3	Bearing
4	Clutch case
5	Input shaft
6	Bearing
7	Oil seal
8	Shim
9	Dumper disk
10	Shim
11	Bolt M8 x 14
12	Lock nut
13	Collar
14	Bearing
15	Thrust collar A
16	Spring retainer
17	Cup spring
18	Forward gear
19	Thrust collar B
20	Drive cone
21	Output shaft
22	Thrust collar B
23	Inner race
24	Reverse gear
25	Cup spring
26	Spring retainer
27	Thrust collar A
28	Bearing
29	Oil seal
30	O-ring
31	Lock nut
32	Coupling
33	Idle gear shaft
34	Bearing
35	Idle gear
36	Thrust washer
37	O-ring
38	Shift lever
39	Lock nut
40	Washer
41	Holder
42	Connector
43	Side cover
44	Bolt M8 x 25
45	Oil seal
46	Shift lever shaft
47	Location pin
48	Stopper bolt
49	Spring pin
50	Bolt M8 x 25
51	Spring
52	Shifter
53	Washer
54	Dipstick
55	Parallel pin
56	Lock nut
57	Washer

Printed in Japan
0000 A 0 A 1361

2. Shifting Device

2-1 Construction of shifting mechanism

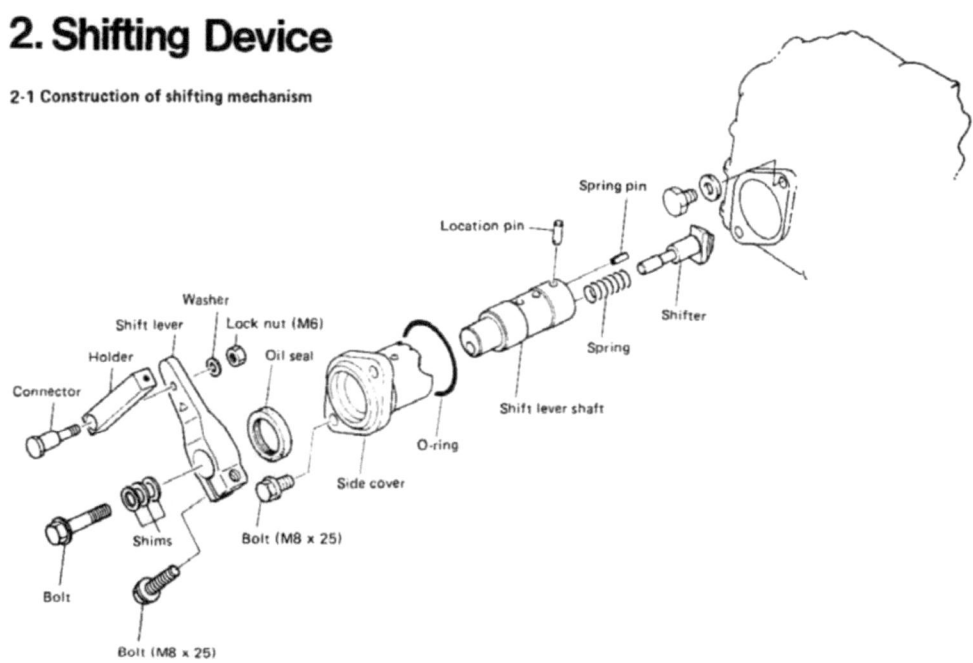

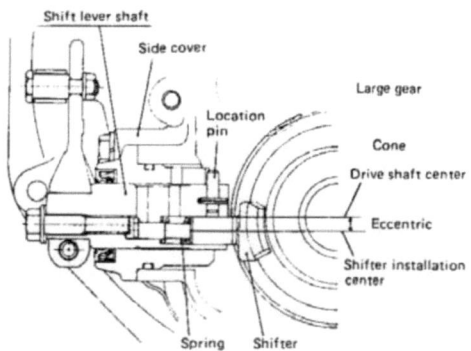

The shift lever shaft is installed on the side cover with neutral, forward and reverse positions provided on this cover. The neutral, forward and reverse location pins of the shift lever shaft are constantly inserted into their respective grooves on the shift lever by the tension of the shifter spring. The shifter is set on the eccentric hole of the shift lever shaft and moves the drive cone in the neutral position either to the forward or reverse positions, and then back to the neutral position. (The shift lever shaft moves slightly to the shift lever or drive cone side when the shift lever is placed in the forward or reverse positions.)

Chapter 10 Reduction and Reversing Gear
2. Shifting Device

2-2 Forward and reverse clutch operation
(Neutral ⇒ Forward; Neutral ⇒ Reverse)

When the shift lever is moved to the forward position from the neutral position, the shift liver shaft starts to revolve, and the location pin disengages from the neutral V-groove position of the side cover. (Shift lever moves approx. 0.5mm to the drive cone side.) At this time the shifter which is set on the eccentric hole of the shift lever shaft, moves the drive cone's V-groove to the forward large gear.

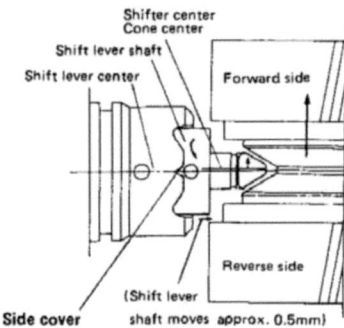

Side cover neutral position (V-groove)

When the location pin of the shift lever shaft falls in the forward position groove of the side cover, (the shift lever shaft moves to the shift lever side approx. 3mm), and the shifter starts to press the drive cone V-groove to the forward large gear side through the spring force.

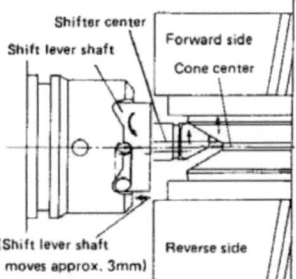

At forward engagement position

2-3 Engagement and disengagement of clutch
(Forward ⇒ Neutral; Reverse ⇒ Neutral)

When the shift lever is moved to the forward position from the neutral position, the shift lever shaft starts to revolve, and the location pin disengages from the forward position groove of the side cover. (The shift lever shaft moves approx. 3mm to the drive cone side.) At this time, the shifter which is set on the eccentric hole of the shift lever shaft is moved to the neutral side (reverse large gear side). The drive cone, however, is engaged with the forward large gear through the torque force produced by the revolving centrifugal force.

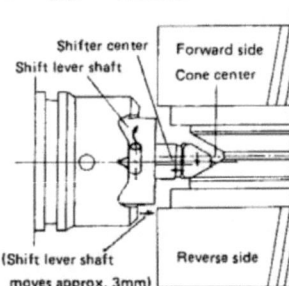

Further, when the shift lever shaft starts to revolve, and the positioning pin falls in to the neutral V-groove position of the side cover (the shift lever shaft travels approx. 5mm to the shift lever side), the shifter moves to the shift lever side (to the spring side) while moving the V-groove of the drive cone to the reverse large gear side. The movement of the shifter to the shift lever side, however, is stopped when the shifter end contacts the stopper bolt. The shifter only works to press the V-groove of the drive cone to the reverse large gear side. Thus, the drive cone is disengaged from the forward large gear. After this disengagement, the transmission torque of the drive cone is decreased to zero and the shift lever is returned to the neutral position by the spring force.

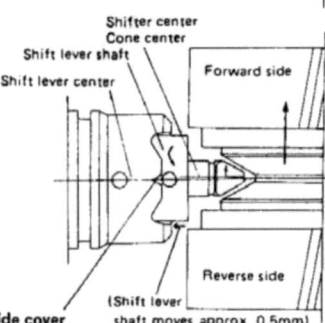

Side cover neutral position (V-groove)

2-4 Clutch shifting force

Shifting position / Shifting direction	Shift lever position at 56mm	Remote control handle position at 170mm (Cable length, 4m)
Engaging force at 1000 rpm	3 ~ 4 kg (6.6 ~ 8.8 lbs)	4 ~ 5 kg (8.8 ~ 11.0 lbs)
Disengaging force at 1000 rpm	3.5 ~ 5 kg (7.7 ~ 11.0 lbs)	4 ~ 6 kg (8.8 ~ 13.2 lbs)

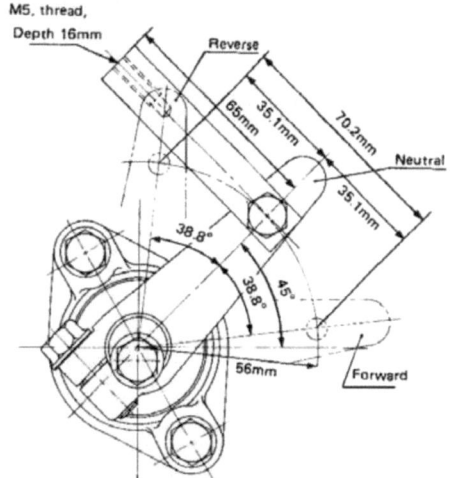

2-5 Adjustment of shifting device

Whenever the side cover, shift lever shaft, shifter, stopper bolt or drive cone is replaced, be sure to adjust the clearance between the shifter end and the stopper bolt by using shims. When the adjustment of this clearance is not proper the drive cone may not be properly fitted when the shift lever is moved to the neutral position either from the forward or reverse position.

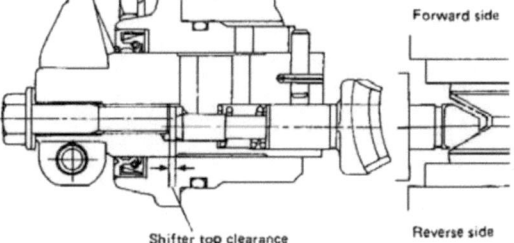

Shifter top clearance

Chapter 10 Reduction and Reversing Gear
2. Shifting Device

SM/GM(F)(C)·HM(F)(C)

2-5.1 Measurement and adjustment of clearance

(a) Assemble the shifting mechanism (without installing the stopper bolt of the shifter) to the marine gear case.

NOTE: Ensure the correct direction of the shifter before assembly.

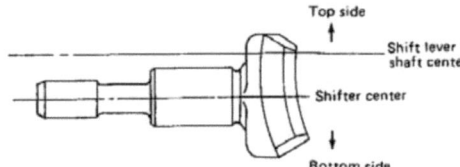

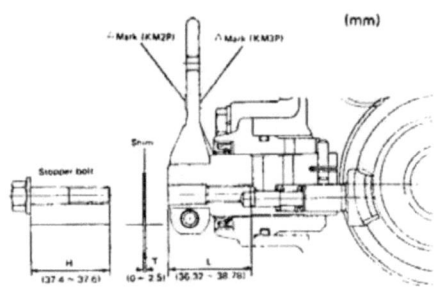

(mm)

(b) Turn the shift lever 10 ~ 15 degrees either to the forward or reverse position from the neutral position.
(c) Measure the L-distance between the shift lever shaft end surface and the shifter's end.
(d) Measure the H-distance (the distance from the neck of the stopper bolt to its end).
(e) Obtain the shim thickness "T" by the following formula.

$$T = (H - L + 1.25) \pm 0.1mm \ (0.004in.)$$

NOTE: Shim set includes one piece each of 1mm, 0.4mm, 0.3mm, 0.25mm shims.
(YANMAR Part No. 177088-06380)

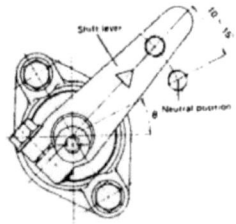

NOTE. Shift lever must be installed in the direction of the △-mark ensuring the specified installation angle (θ).

(f) Insert shim (s) of proper thickness to the stopper bolt side and tighten it to the shift lever shaft.

NOTE: When tightening the stopper bolt, apply either a non-drying type liquid packing (TREE BOND No.1215), or a seal tape around the bolt threads.

$\theta =$	KM2P	KM3P
	40°	45°

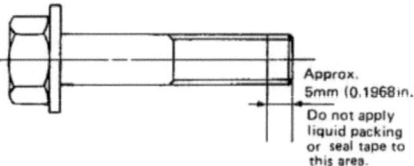

2-5.2 Inspect for the following points
(to be inspected every 2-3 months)

(1) Looseness at the connection of the cable connector and the remote control cable.
(2) Looseness of the attaching nut of the cable connector and the shift lever.

Chapter 10 Reduction and Reversing Gear
2. Shifting Device

2-6 Adjustment of the remote control head Marine gearbox control side

(1) Equal distribution of the control lever stroke.

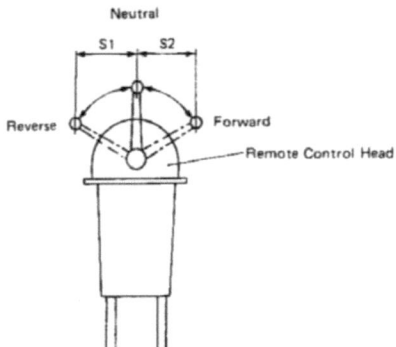

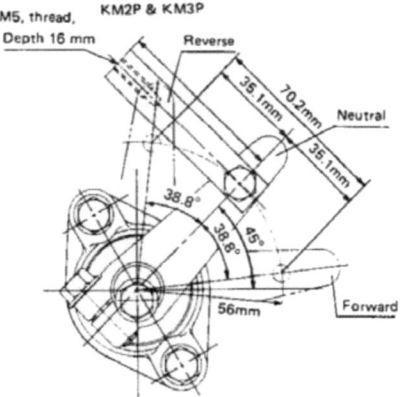

M5, thread, Depth 16 mm KM2P & KM3P

The stroke between the neutral position → forward position (S2), and the neutral position → reverse positon (S1) must be equalized.
When either stroke is too short, clutch engagement becomes faulty.

(2) Equalizing the travel distance of the control cable.

After ensuring the equal distribution of the stroke described in (1), connect the cable to the control head. Adjust that the cable shift travel of the S1 and S2 control lever strokes becomes identical.

2-7 Cautions

(1) Always stop the engine when attaching, adjusting, and inspecting.
(2) When conducting inspection immediately after stopping the engine, do not touch the clutch. The oil temperature is often raised to around 90°C (194°F).
(3) Half-clutch operation is not possible with this design and construction. Do not use with the shift lever halfway to the engaged position.
(4) Set the idling engine speed at between 750 and 800 rpm.

NOTE: The dual (Two) lever remote control device cannot be used.

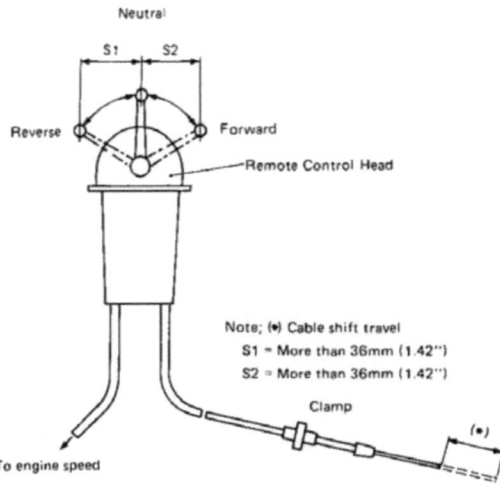

Note; (•) Cable shift travel
S1 = More than 36mm (1.42")
S2 = More than 36mm (1.42")

3. Inspection and Servicing

3-1 Clutch case

(1) Check the clutch case with a test hammer for cracking. Perform a color check when required.
If the case is cracked, replace it.
(2) Check for staining on the inside surface of the bearing section.
Also, measure the inside diameter of the case.
Replace the case if it is worn beyond the wear limit.

3-2 Bearing

(1) Rusting and damage.
If the bearing is rusted or the taper roller retainer is damaged, replace the bearing.
(2) Make sure that the bearings rotate smoothly.
If rotation is not smooth, if there is any binding, or if any abnormal sound is evident, replace the bearing.

3-3 Gear

Check the surface, tooth face conditions and backlash of each gear. Replace any defective part.
(1) Tooth surface wear.
Check the tooth surface for pitting, abnormal wear, dents, and cracks. Repair the lightly damaged gears and replace heavily damaged gears.
(2) Tooth surface contact.
Check the tooth surface contact. The amount of tooth surface contact between the tooth crest and tooth flank must be at least 70% of the tooth width.
(3) Backlash.
Measure the backlash of each gear, and replace the gear when it is worn beyond the wear limit.

mm (in.)

	Maintenance standard	Wear limit
Input shaft forward gear and output shaft forward gear	0.06 ~ 0.12 (0.0024 ~ 0.0047)	0.2 (0.0079)
Input shaft reverse gear and intermediate gear	0.06 ~ 0.12 (0.0024 ~ 0.0047)	0.2 (0.0079)
Intermediate gear and output shaft reverse gear	0.06 ~ 0.12 (0.0024 ~ 0.0047)	0.2 (0.0079)

(The same dimensions apply to both KM2P and KM3P.)

3-4 Forward and reverse large gears

(1) Contact surface with drive cone.
Visually inspect the tapered surface of the forward and reverse large gears where they make contact with the drive cone to check if any abnormal condition or sign of overheating exists.
If any defect is found, replace the gear.

(2) Forward/reverse gear needle bearing.
When an abnormal sound is produced at the needle bearing, visually inspect the rollers; replace the bearing if the rollers are faulty.

Rollers

3-5 Drive cone

(1) Visually inspect that part of the surface that comes into contact with the circumferential triangular slot to check for signs of scoring, overheating or wear. If deep scoring or signs of overheating are found, replace the cone.

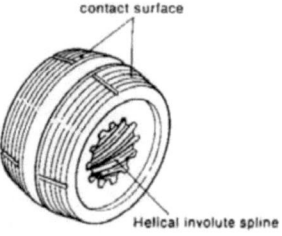

(2) Check the helical involute spline for any abnormal condition on the tooth surface, and repair or replace the part should any defect be found.
(3) Measure the amount of wear on the tapered contact surface of the drive cone, and replace the cone when the wear exceeds the specified limit.

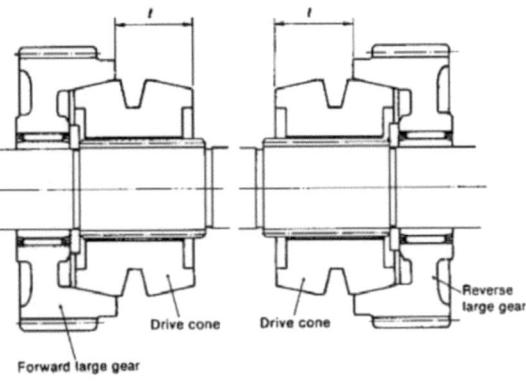

Chapter 10 Reduction and Reversing Gear
3. Inspection and Servicing

SM/GM(F)(C)·HM(F)(C)

mm (in.)

Dimensions *l*		Standard dimensions	Limited dimensions
	KM2P	29.2 ~ 29.8 (1.1496 ~ 1.1732)	28.1 (1.1063)
	KM3P	32.7 ~ 33.3 (1.2874 ~ 1.3110)	32.4 (1.2756)

NOTE: *When dismantled, the forward or reverse direction of the drive cone must be clearly identified.*

(4) If the wear of the V-groove of the drive cone is excessive, replace the part.

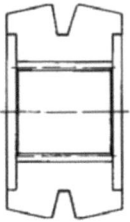

3-6 Thrust collar

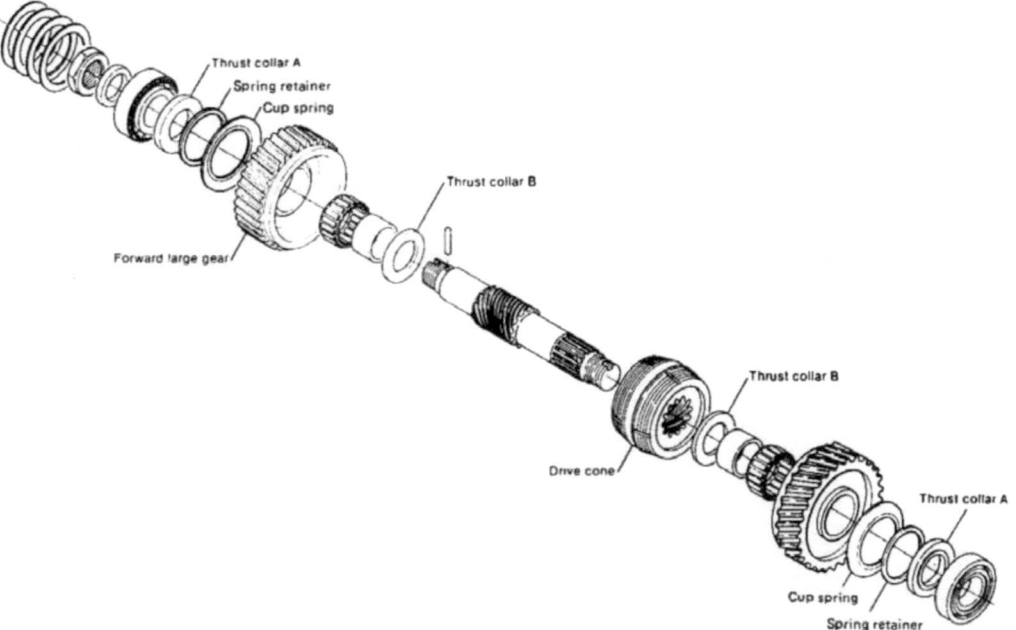

10-62

Chapter 10 Reduction and Reversing Gear
3. Inspection and Servicing

SM/GM(F)(C)·HM(F)(C)

(1) Visually inspect the sliding surface of thrust collar A or B to check for signs of overheating, scoring, or cracks. Replace the collar if any abnormal condition is found.
(2) Measure the thickness of thrust collar A or B, and replace it when the dimension exceeds the specified limit.

3-8 Oil seal of output shaft
Visually inspect the oil seal of the output shaft to check if there is any damage or oil leakage; replace the seal when any abnormal condition is found.

3-9 Input shaft

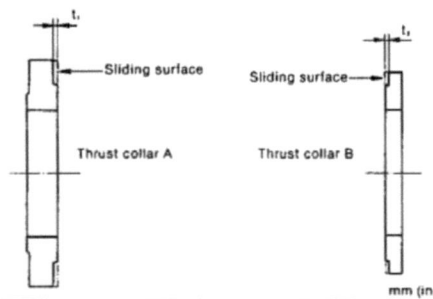

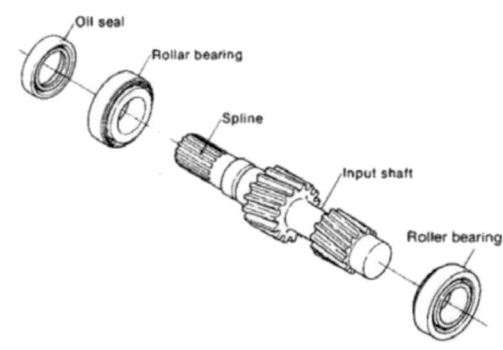

Stepped wear	Limit for use mm (in.)
Thrust collar A, t_1	0.05 (0.0020)
Thrust collar B, t_2	0.20 (0.0079)

3-7 Cup spring and spring retainer
(1) Check for cracks and damage to the cup spring and spring retainer. Replace the part if defective.
(2) Measure the free length of the cup spring and the thickness of the spring retainer. If the length or the thickness deviates from the standard size, replace the part.

(1) Spline part.
Whenever uneven wear and/or scratches are found, replace with a new part.
(2) Surface of oil seal.
If the sealing surface of the oil seal is worn or scratched, replace.

3-10 Output shaft

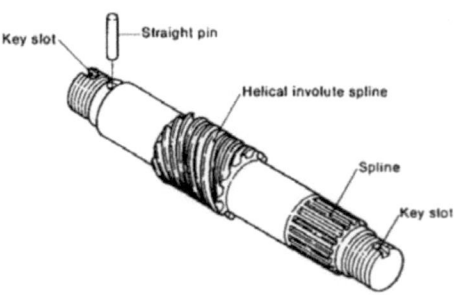

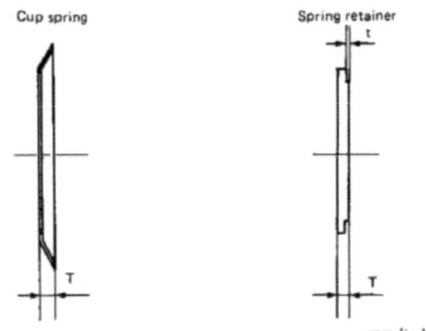

	Standard	Limit mm (in.)
Cup spring, T	2.8 ~ 3.1 (0.1102 ~ 0.1220)	2.6 (0.1024)
Spring retainer, T	2.92 ~ 3.08 (0.1150 ~ 0.1213)	2.8 (0.1102)
Spring retainer, t	—	0.1 (0.0040)

(1) Visually inspect the spline and the helical involute spline, and repair or replace a part when any abnormal condition is found on its surface.

Chapter 10 Reduction and Reversing Gear
3. Inspection and Servicing

SM/GM(F)(C)·HM(F)(C)

3-11 Intermediate shaft

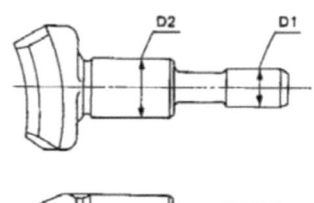

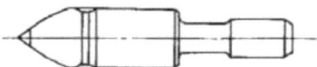

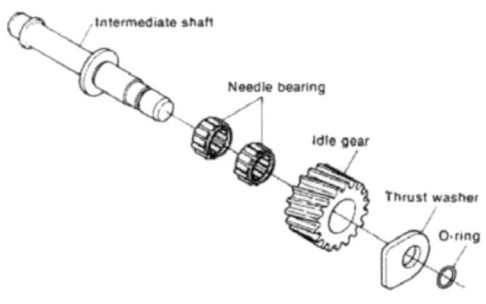

(1) Needle bearing dimensions, staining.
Check the surface of the roller to see whether the needle bearing sticks or is damaged. Replace if necessary.

	Standard	Limit
D1	66.9 ~ 67.0	65
	(2.6338 ~ 2.6378)	(2.5591)
D2	11.966 ~ 11.984	11.95
	(0.4711 ~ 0.4718)	(0.4705)
Shift lever shaft, Shifter insert hole	12.0 ~ 12.018	12.05
	(0.4724 ~ 0.4731)	(0.4744)

mm (in.)

3-12 Shifting device
3-12.1 Shifter

3-12.2 Shift lever shaft and location pin
(1) Check the shift lever shaft and location pin for damage or distortion, and replace defective parts. If the location pin must be replaced, replace it together with the shift lever shaft.
(2) Measure the diameter of the shift lever shaft and the shifter insertion hole. Replace the part if the size deviates from the standard value.

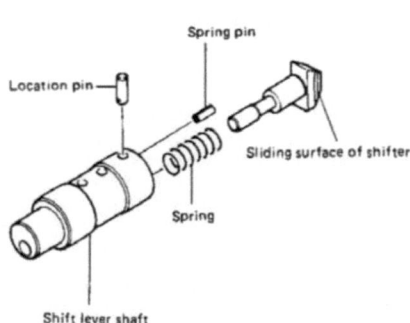

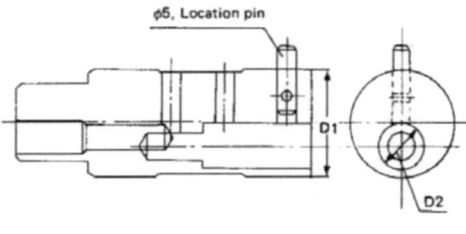

(1) Visually inspect the surface in contact with the drive cone, and replace the shifter when signs of overheating, damage or wear are found.
(2) Measure the shaft diameter of the shifter. Replace the shaft if the size deviates from the standard.

	Standard	Limit
D1	27.959 ~ 27.98	27.90
	(1.1001 ~ 1.1016)	(1.0984)
D2	12.0 ~ 12.018	12.05
	(0.4724 ~ 0.4731)	(0.4744)
Side cover, Shift insert hole	28.0 ~ 28.021	28.08
	(1.1024 ~ 1.1032)	(1.1055)

mm (in.)

Chapter 10 Reduction and Reversing Gear
3. Inspection and Servicing

SM/GM(F)(C)·HM(F)(C)

3-12.3 Shifter spring
(1) Check the spring for scratches or corrosion.
(2) Measure the free length of the spring.

Shifter spring	Standard		Limit
Free length	22.6 mm	(0.890in.)	19.8 mm (0.780in.)
Spring constant	0.854 kgf/mm (1.88 lbs/0.04in.)		—
Length when attached	14.35 mm	(0.5650 in.)	—
Load when attached	7.046 kg	(15.54 lbs)	6.08 kg (13.41 lbs)

3-12.4 Stopper bolt
Check the stopper bolt. If it is worn or stepped, replace.

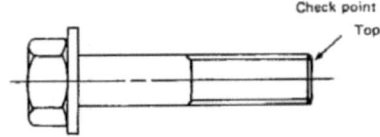

3-13 Damper disc

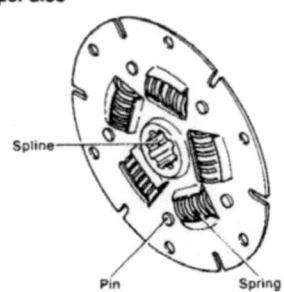

(1) Spline part.
　Whenever uneven wear and/or scratches are found, replace with a new part.
(2) Spring.
　Whenever uneven wear and/or scratches are found, replace with a new part.
(3) Pin wear.
　Whenever uneven wear and/or scratches are found, replace with a new part.
(4) Whenever a crack or damage to the spring slot is found replace the defective part with a new one.

3-12.5 Side cover and oil seal
(1) Check the neutral, forward and reverse position grooves. Replace if the grooves are worn.
(2) Measure the insertion hole of the shift lever shaft. Replace if the size deviates from the standard value.
(3) Check the oil seal and the O-ring for damage. Replace if the part is defective.

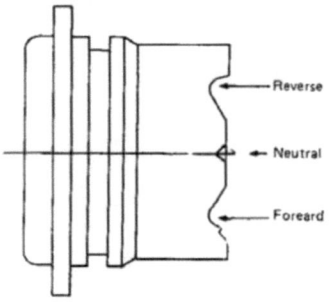

3-14 Shim adjustment for output and input shafts
Check the thickness of shims for both input and output shafts. When the component parts are not replaced after dismantling, the same shims can be reused. When the clutch case and flange or any one of the following parts is replaced the thickness of shim must be determined in the following manner.

For input shaft parts: input shaft, bearing.
For output shaft parts: output shaft, thrust collar A, thrust collar B, gear, bearing.

Chapter 10 Reduction and Reversing Gear
3. Inspection and Servicing

SM/GM(F)(C)·HM(F)(C)

(1) Shim thickness (T1) measurement of input shaft
  (a) Measure the bearing insertion hole depth (A) of the mounting flange, and the bearing insertion hole depth (A') of the clutch case.
  (b) Measure the length (B) between the bearing outer races of the input shaft assembly.
  (c) Obtain the (T1) thickness by the following formula:

  $T_1 = A + A' - B$ (T1: Clearance ± 0.05mm)

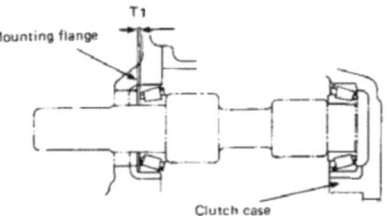

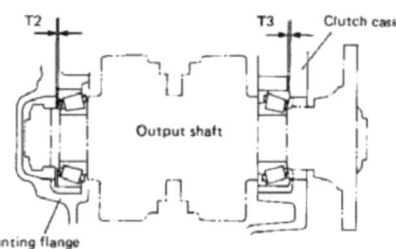

(c) Measure the (F) and (E) length from the outer race end of the clutch case bearing included in the output shaft assembly.

NOTE. Before measuring the (F) and (E) length, press the forward large gear and the reverse large gear to the drive cone until there is no clearance among them.

(d) Obtain the (T2) and (T3) thickness by the following formulas:

$T_2 = C + C' - D - T_3$ (Clearance $\pm^{0.1mm}_{0}$)

T_3 (KM2P) = $C' - 48.3 - \frac{E}{2} - F$ (Clearance ±0.05mm)

T_3 (KM3P) = $C' - 47.3 - \frac{E}{2} - F$ (Clearance ±0.05mm)

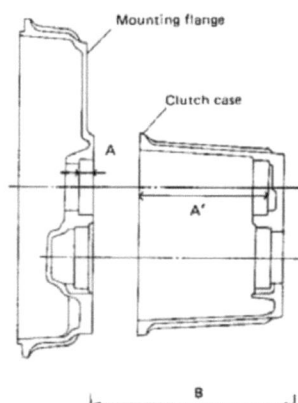

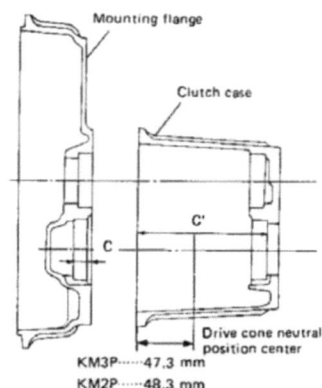

KM3P······47.3 mm
KM2P······48.3 mm

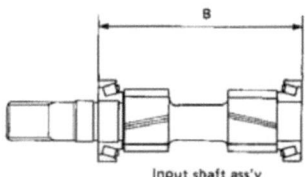

Input shaft ass'y

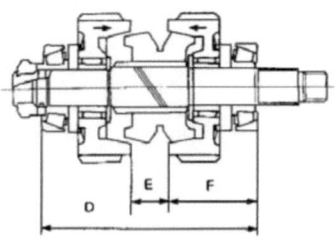

Output shaft ass'y

(2) Shim thickness (T2, T3) measurement of output shaft

  (a) Measure the bearing insertion hole depth (C) of the mounting flange, and the bearing insertion hole depth (C') of the clutch case.

  (b) Measure the length (D) between the bearing outer races.

NOTE: Tighten the mounting flange nut of the output shaft assembly with the specified torque. Press-fit the inner race of the clutch case roller bearing to the large gear side.

Chapter 10 Reduction and Reversing Gear
3. Inspection and Servicing

SM/GM(F)(C)·HM(F)(C)

(3) Standard size of parts

mm (in.)

	A + A'	B	C + C'	D	E	F	Drive corn neutral center position
KM2P	123.40 ~ 123.75 (4.8583 ~ 4.8720)	122.20 ~ 123.10 (4.8110 ~ 4.8465)	129.80 ~ 130.15 (5.1102 ~ 5.1240)	128.07 ~ 129.53 (5.0421 ~ 5.0996)	20.50 ~ 21.10 (0.8071 ~ 0.8307)	53.59 ~ 54.41 (2.1098 ~ 2.1421)	48.3 (1.9016)
KM3P	132.40 ~ 132.75 (5.2126 ~ 5.2264)	131.20 ~ 132.10 (5.1654 ~ 5.2008)	141.20 ~ 141.55 (5.5591 ~ 5.5728)	139.56 ~ 141.00 (5.4945 ~ 5.5512)	23.50 ~ 24.10 (0.9252 ~ 0.9488)	57.83 ~ 58.65 (2.2768 ~ 2.3091)	47.3 (1.8622)

NOTE: Compare your measurements with the above standard size. If your measurements largely differ from the standard sizes, measurements may not be correct. Check and measure again.

(4) Adjusting shim set

	Part No.	Thickness.mm(in.)	No. of shims
Input shaft	177088-02350	0.5 (0.0197)	1
		0.4 (0.0157)	1
		0.3 (0.0118)	2
Output shaft	177088-02300	1.0 (0.0394)	1
		0.5 (0.0197)	1
		0.3 (0.0118)	2
		0.1 (0.0039)	3

4. Disassembly

4-1 Dismantling the clutch

(1) Remove the remote control cable.
(2) Remove the clutch assembly from the engine mounting flange.

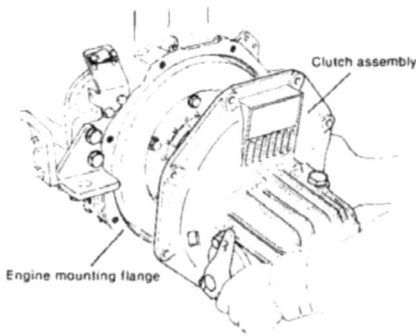

(3) Drain the lubricating oil.
Drain the lubricating oil by loosening the plug at the bottom of the clutch case.

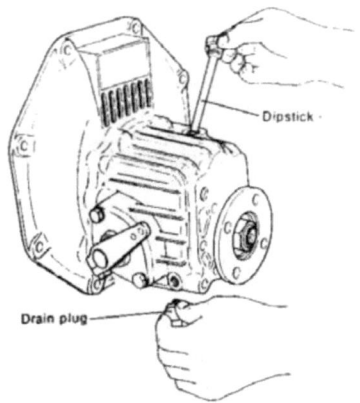

(4) Remove the end nut and output shaft coupling.

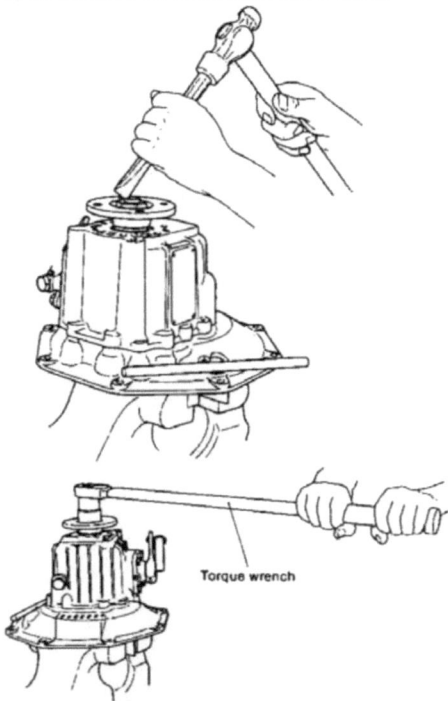

NOTE: *Take care as it has a left-handed thread.*

(5) Remove the oil dip stick and O-ring.
(6) Remove the fixing bolts on the side cover, and also remove the shift lever shaft, shift lever and shifter.

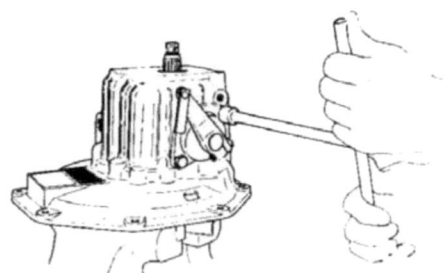

Chapter 10 Reduction and Reversing Gear
4. Disassembly

(7) Remove the bolts which secure the mounting flange to the case body, give light taps to the left and right with a plastic headed hammer while supporting the clutch case with your hand, then remove the mounting flange.

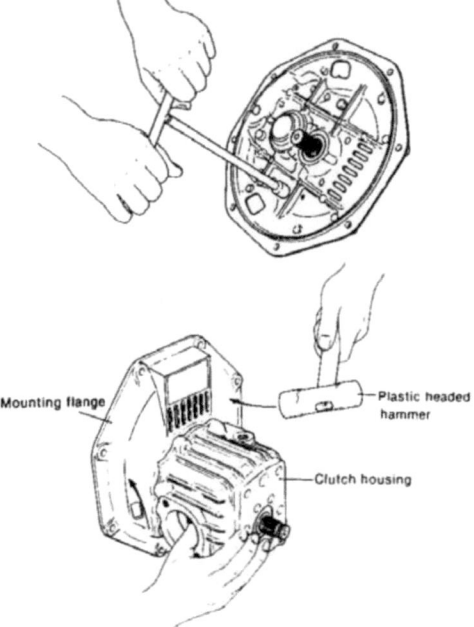

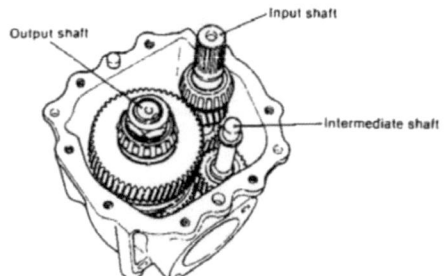

(8) Withdraw the output shaft assembly.

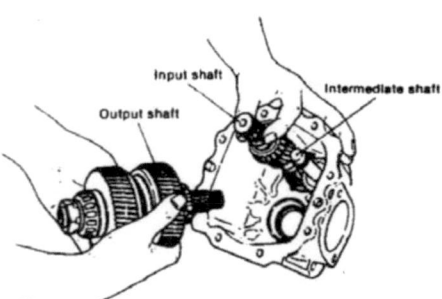

(9) Take out the intermediate shaft and input shaft. When taking out the intermediate shaft, place a bolt or spacer on the shaft hole of the case, and drive the shaft out by tapping it lightly.

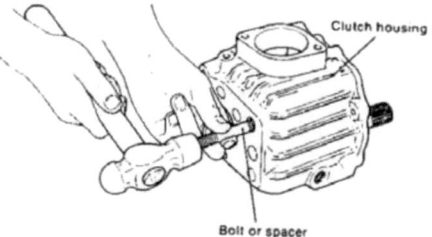

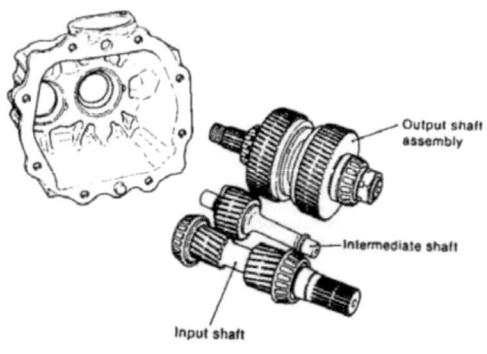

Chapter 10 Reduction and Reversing Gear
4. Disassembly

(10) Remove the oil seal of the output shaft from the case body.

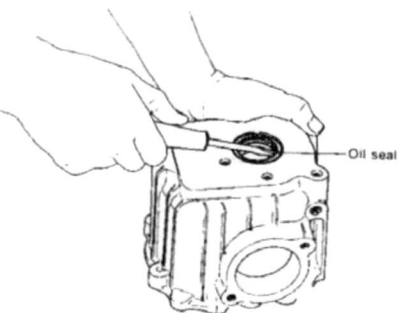

(11) Remove the outer bearing race from the case body by using the special tool.

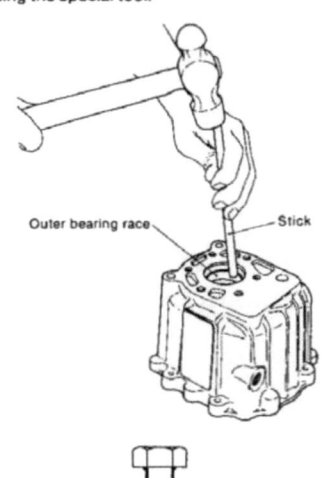

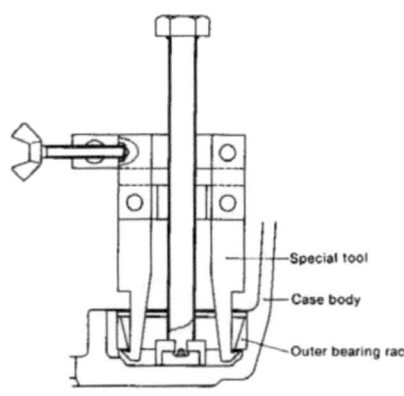

(12) Remove the oil seal of the input shaft from the mounting flange.
(13) Remove the outer bearing race from the mounting flange in the same way as with the case body.
(14) Remove each adjusting plate from the input or output shaft.

NOTE: *The same adjusting plates can be reused when the following parts are not replaced. When any part is replaced however, re-adjustment is necessary.*

4-2 Removal of the output shaft

(1) Take out the reverse large gear, thrust collar A, cup spring, spring retainer and inner bearing race.
The reverse large gear must be withdrawn using a pulley extracter, by fixing the nut at the forward end in a vice.

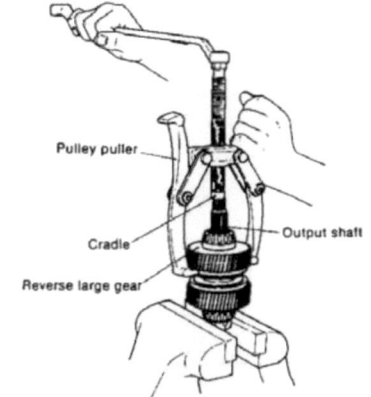

(2) Loosen the calking of the forward nut and remove the nut and spacer.
Remove the nut by using a torque wrench after setting the output shaft coupling and fixing the coupling bolt in a vice.

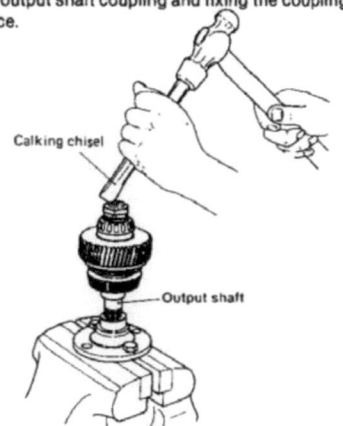

Chapter 10 Reduction and Reversing Gear
4. Disassembly

SM/GM(F)(C)-HM(F)(C)

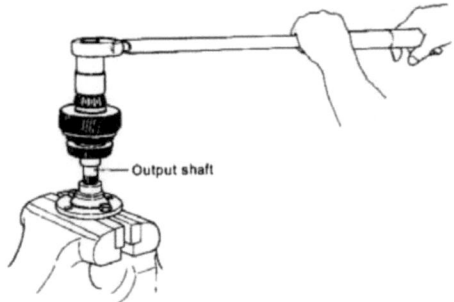

(3) Place the pulley extractor against the end surface of the forward large gear, and withdraw the forward large gear, thrust collar A, cup spring, spring retainer and inner bearing race.

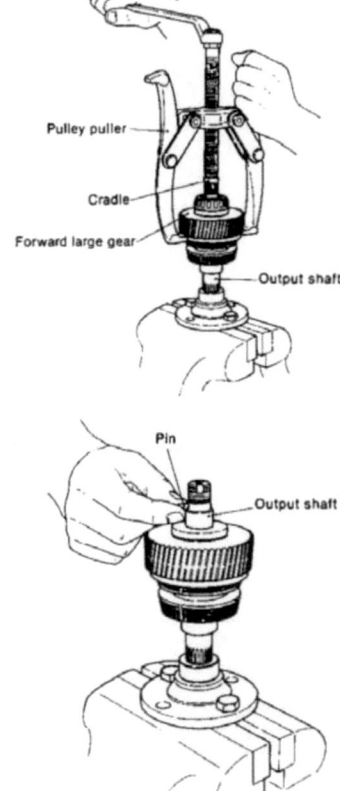

NOTE: Take care as the nut has left-handed thread.

(4) While gripping the drive cone, tap the end of the shaft with a plastic headed hammer, and withdraw the thrust collar B and inner needle bearing race. A pulley extractor may be used.

4-3 Removal of the intermediate shaft
(1) Remove the "O" ring.
(2) Remove the thrust washer.
(3) Remove the intermediate gear and needle bearing.

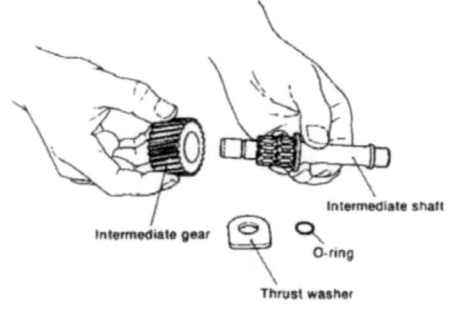

Chapter 10 Reduction and Reversing Gear
4. Disassembly

SM/GM(F)(C)-HM(F)(C)

4-4 Dismantling the shifting device
(1) Take out the shifter and shifter spring.

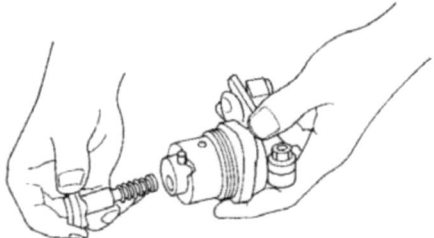

(4) Remove the shift lever to the anti-shift lever side.

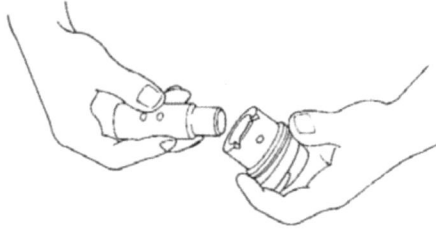

(2) Remove the stopper bolt of the shifter and shim.

(5) Remove the oil-seal and O-ring.

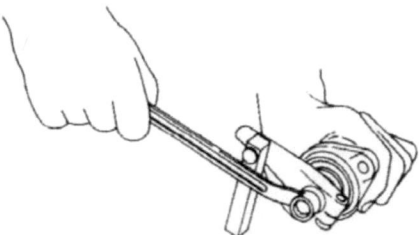

(3) Loosen the bolt of the shift lever and remove the shift lever from the shift lever shaft.

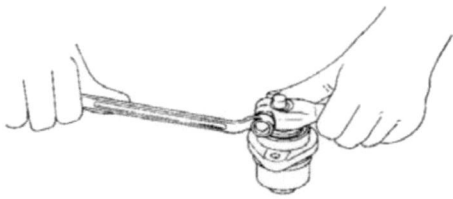

10-72

5. Reassembly

5-1 Reassembly of output shaft

(1) Fit the forward side thrust collar B onto the shaft.
(2) Drive in the forward end inner needle bearing race using a jig.

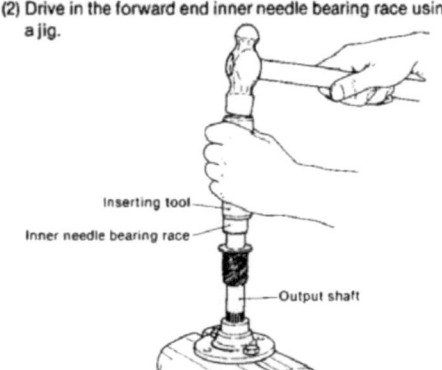

(3) Assemble the needle bearing and forward large gear.
NOTE: Check that the forward large gear rotates smoothly.
(4) Fit the cup spring, spring retainer, thrust collar A and pin, and drive in the inner bearing race using a jig.

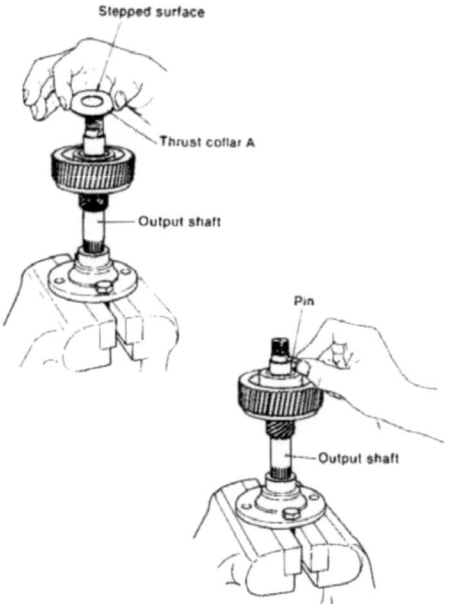

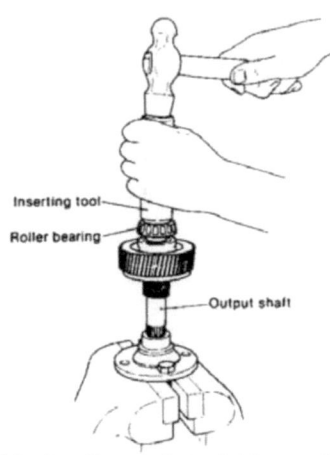

NOTE: 1) Drive in with a plastic headed hammer. Do not hit it hard.
2) When fitting the thrust collar A, note the fitting direction. Fit it keeping the stepped surface toward the roller bearing side.
3) Note that the pin cannot be fitted after the inner bearing race has been driven in.
4) Check that the forward large gear rotates smoothly.

(5) Assemble the collar and pin so that the pin is in the groove of the collar.
(6) Set and tighten the forward end nut. Insert the bolt into the coupling, and fix it in a vice, keeping the spline part upward.
Insert the shaft into the spline of the coupling, fit the spacer, and tighten the nut with a torque wrench.

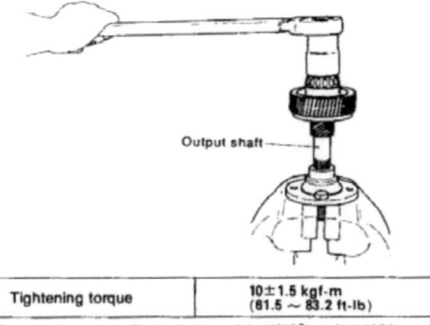

Tightening torque	10±1.5 kgf·m (61.5 ~ 83.2 ft-lb)

(The same torque applies to both models KM2P and KM3P)

NOTES: 1) Take care as it is a left-handed thread.
2) Use the reverse side nut used before dismantling as the forward end nut. This is so as not to match the calked portion to the same point.

Chapter 10 Reduction and Reversing Gear
5. Reassembly

SM/GM(F)(C)-HM(F)(C)

5-2 Reassembly of the clutch
(1) Fit the oil seal, bearing outer races and shim (output shaft side) in the clutch case.
(2) Insert the input shaft into the clutch case.
(3) Drive the intermediate shaft into the clutch case.

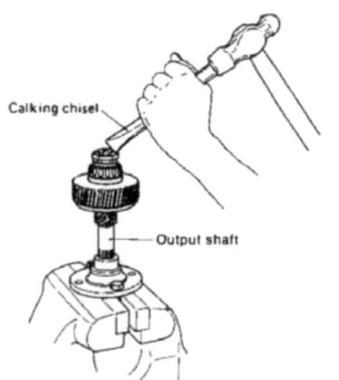

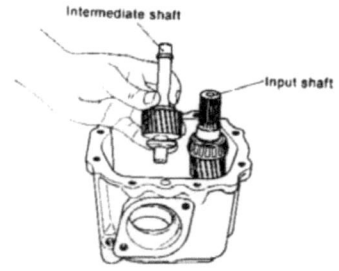

(7) Insert the drive cone while keeping the output shaft set for reverse.

NOTES: 1) If the output shaft is not fitted into the clutch case before driving-in the intermediate shaft, it cannot be assembled.
2) Note the assembly direction of the thrust washer.

(4) Insert the output shaft into the clutch case.

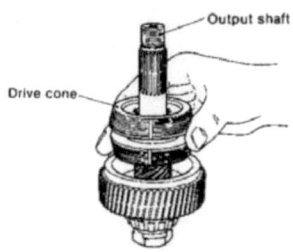

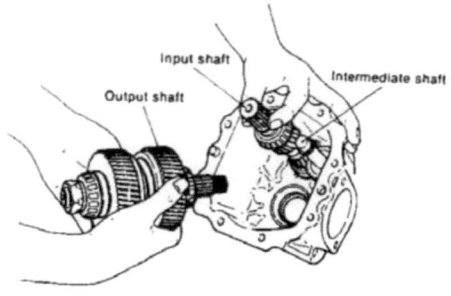

(8) Apply procedures 1 through 4 to the forward end.

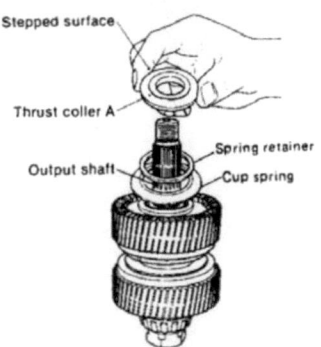

NOTE: 1) Fit thrust collar A so that the stepped surface faces the roller bearing side.
2) Check that the reverse large gear rotates smoothly.

Chapter 10 Reduction and Reversing Gear
5. Reassembly

SM/GM(F)(C)-HM(F)(C)

(5) Fit the adjusting plate to the mounting flange, and drive in the outer bearing race.
NOTE: *The outer bearing race can be easily driven in by heating the mounting flange to about 100°C, or by cooling the outer race with liquid hydrogen.*
(6) Apply non-drying liquid packing around the outer surface of the oil seal, and insert the oil seal into the mounting flange while keeping the spring part of the oil seal facing the inside of the case.
(7) Apply non-drying liquid packing to the matching surfaces of the mounting flange and the case body.

(8) Insert the input shaft and output shaft into the shaft holes of the mounting flange, assemble the mounting flange on the case body, and tighten the bolt.

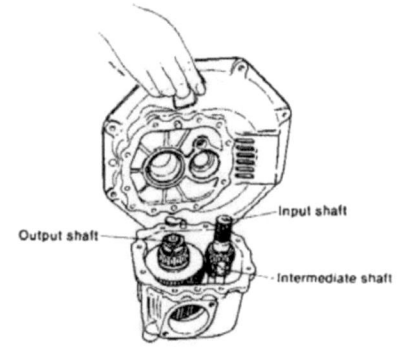

Output shaft — Input shaft — Intermediate shaft

NOTE: *Apply non-drying liquid packing to either the mounting flange or the case body.*

(9) Assemble the output shaft coupling on the output shaft, and fit the O-ring.
(10) Tighten the end nut by using a torque wrench, then calk it.

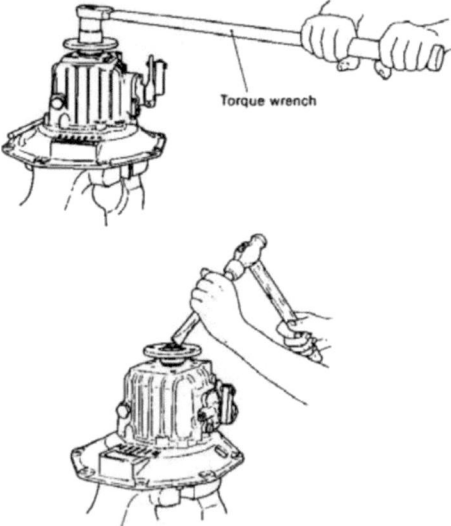

Torque wrench

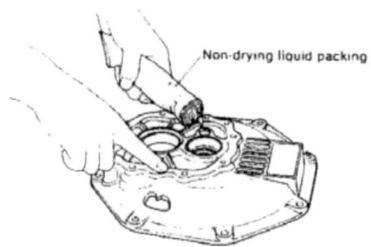

Non-drying liquid packing

NOTE: *Take care as it is a left-handed thread.*

Tightening torque	10±1.5 kgf-m (61.5 ~ 83.2 ft-lb)

(The same torque applies to both models KM2P and KM3P).

Chapter 10 Reduction and Reversing Gear
5. Reassembly

SM/GM(F)(C)·HM(F)(C)

5-3 Reassembly of the shifting device
(1) Fit the oil seal and O-ring to the side cover.

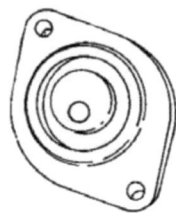
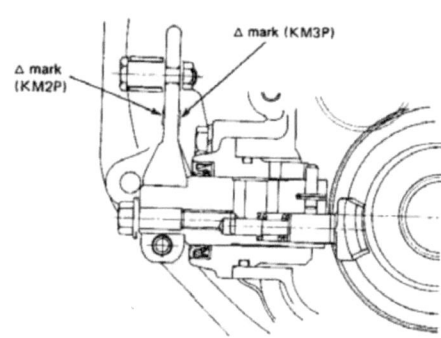

(2) Insert the shift lever shaft to the side cover.

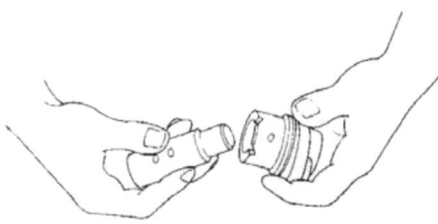

(4) Insert the shifter spring and shifter to the shift lever shaft.
(5) Fit the side cover assembly to the clutch case.

NOTE: 1) Check the direction of the shifter (Top and bottom side).
2) The shift lever may not turn smoothly if the clutch case is not filled with lubricating oil.

(6) Fit the shim and stopper bolt to the shift lever shaft.

NOTE: Apply non-drying liquid packing or seal-tape to the thread of the stopper bolt.

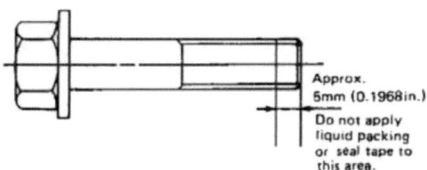

Approx. 5mm (0.1968in.)
Do not apply liquid packing or seal tape to this area.

(3) Fit the shift lever to the shift lever shaft.
NOTE: Check the direction of the shift lever △ mark.

(7) Fit the cable connector to the shift lever.

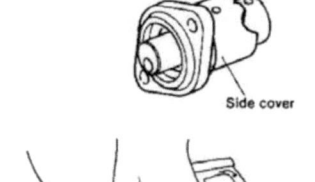
Side cover

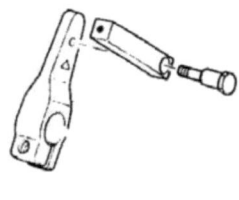

[D] V-drive Gear, Model KM3V

Applicable Engine Models: 1GM10V, 2GM20(F)V, 3GM30(F)V

1. Construction

1-1 Construction

Basically, the construction of the KM3V V-drive Gear is the combination of the KM3P mechanical cone type clutch, and the V-drive case. (However, there is no interchangeability between the KM3P clutch case and the KM3P.) The output shaft coupling of the KM3P is changed to the drive gear, and the power transmission is made through the drive gear connected with the conical gear inside the V-drive case. The conical gear is of mono-block construction with the V-drive output shaft.

The KM3V is equipped with the lube oil cooler which controls the lube oil temperature to a proper level in order to increase the durability of bearings, oil-seals, gears, and other relative parts.

The lube oil cooler is cooled by the cooling water coming from the engine's sea water pump. Cooling water flows as follows: Sea-water pump ⇒ Lube oil cooler ⇒ Engine side 3-way joint (for 1GM 10V), L-type joint (for 2GM20V, 3GM30V), or Heat exchanger (for 2GM20FV, 3GM30FV). In order to prevent cracking by freezing, the oil cooler is equipped with a drain cock which facilitates easy water draining.

Chapter 10 Reduction and Reversing Gear
1. Construction

1-1.1 KM3V Sectional View

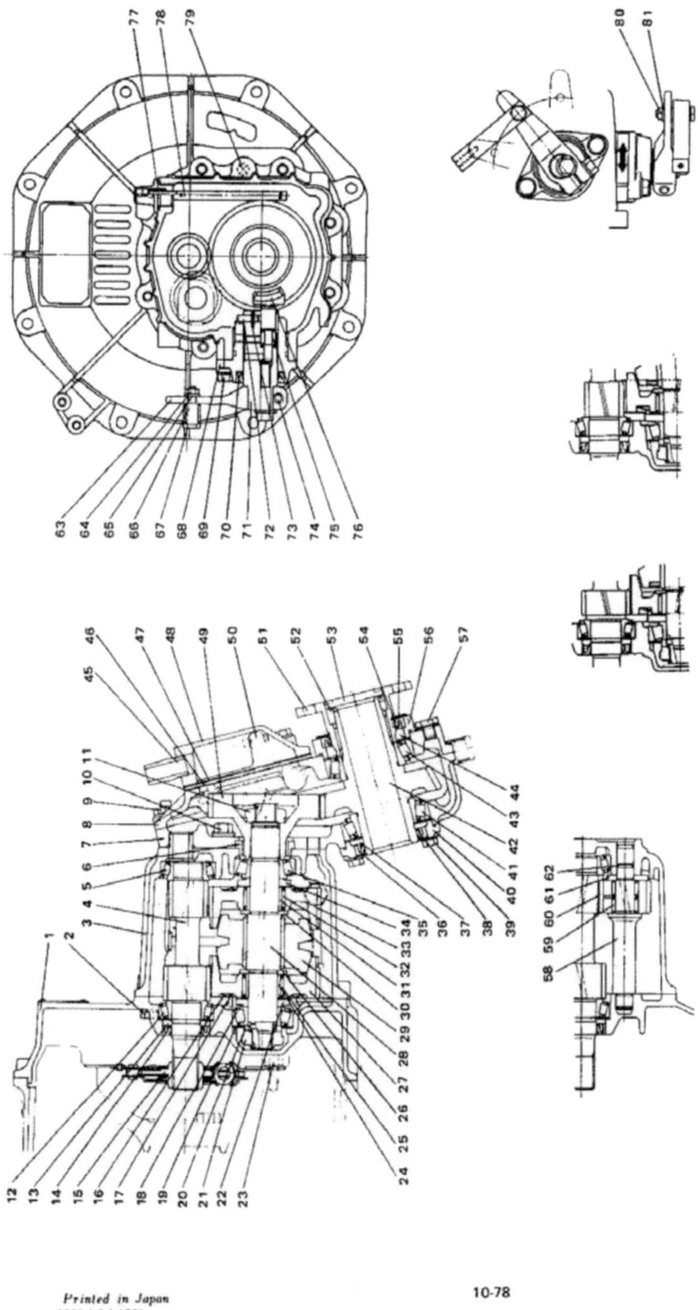

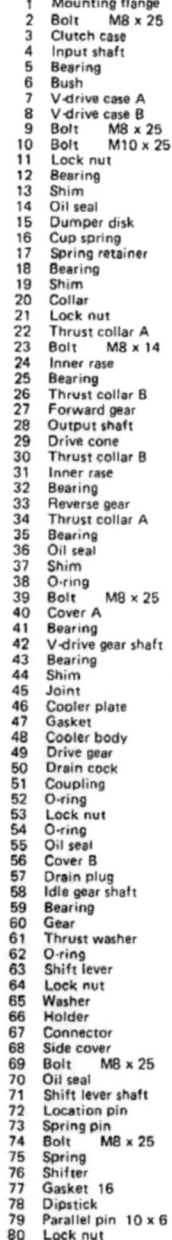

1. Mounting flange
2. Bolt M8 x 25
3. Clutch case
4. Input shaft
5. Bearing
6. Bush
7. V-drive case A
8. V-drive case B
9. Bolt M8 x 25
10. Bolt M10 x 25
11. Lock nut
12. Bearing
13. Shim
14. Oil seal
15. Dumper disk
16. Cup spring
17. Spring retainer
18. Bearing
19. Shim
20. Collar
21. Lock nut
22. Thrust collar A
23. Bolt M8 x 14
24. Inner rase
25. Bearing
26. Thrust collar B
27. Forward gear
28. Output shaft
29. Drive cone
30. Thrust collar B
31. Inner rase
32. Bearing
33. Reverse gear
34. Thrust collar A
35. Bearing
36. Oil seal
37. Shim
38. O-ring
39. Bolt M8 x 25
40. Cover A
41. Bearing
42. V-drive gear shaft
43. Bearing
44. Shim
45. Joint
46. Cooler plate
47. Gasket
48. Cooler body
49. Drive gear
50. Drain cock
51. Coupling
52. O-ring
53. Lock nut
54. O-ring
55. Oil seal
56. Cover B
57. Drain plug
58. Idle gear shaft
59. Bearing
60. Gear
61. Thrust washer
62. O-ring
63. Shift lever
64. Lock nut
65. Washer
66. Holder
67. Connector
68. Side cover
69. Bolt M8 x 25
70. Oil seal
71. Shift lever shaft
72. Location pin
73. Spring pin
74. Bolt M8 x 25
75. Spring
76. Shifter
77. Gasket 16
78. Dipstick
79. Parallel pin 10 x 6
80. Lock nut
81. Washer

Chapter 10 Reduction and Reversing Gear
1. Construction

1-1.2 Drawing

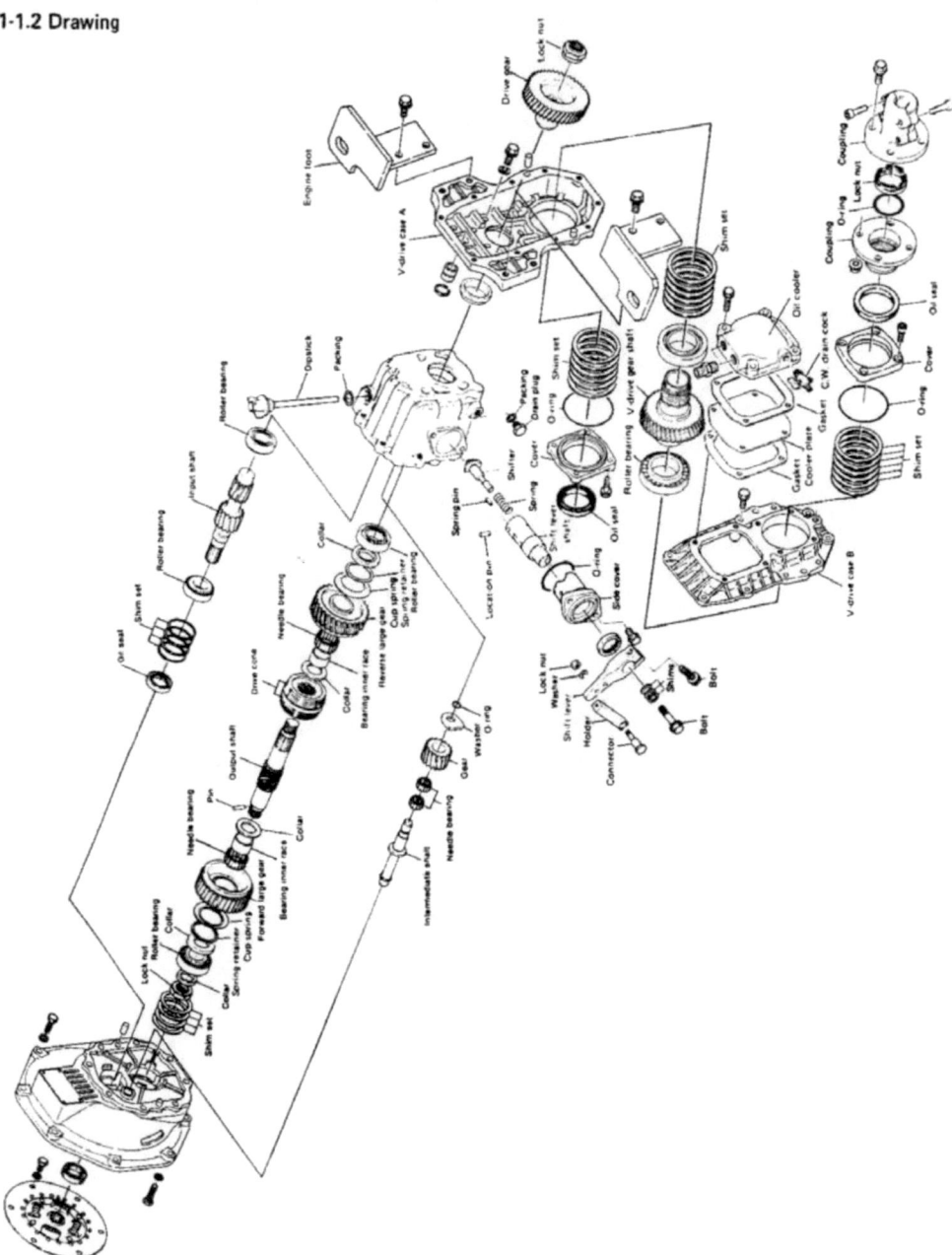

10-79

2. Specifications

Model			KM 3V		
For engine models			1GM10V, 2GM20(F)V, 3GM30(F)V		
Clutch			Constant mesh gear with servo cone clutch (wet type)		
Reduction ratio	Forward		2.36	2.61	3.20
	Reverse		3.16	3.16	3.16
Propeller shaft rpm (Forward) rpm			1441	1303	1063
Direction of rotation	Input shaft		Clockwise, viewed from stern		
	V-drive output shaft	Forward	Clockwise, viewed from stern		
		Reverse	Counter-clockwise, viewed from stern		
Remote control	Control head		Single lever control		
	Cable		Morse, 33-C (Cable travel 76.2mm or 3 in)		
	Clamp		YANMAR made, standard accessory		
	Cable connector		YANMAR made, standard accessory		
Output shaft coupling	Outer diameter		ø100mm (3.93")		
	Pitch circle diameter		ø78mm (3.07")		
	Connecting bolt holes		4 — ø10.5mm (4 — ø0.41")		
Position of shift lever			Right side, viewed from stern		
Lubricating oil			SAE #10W-30, CC class		
Lubricating oil capacity			0.8ℓ		
Dry weight			19.5 kg (43 lbs)		

3. Power transmission system

3-1 Arrangement of shaft and gears

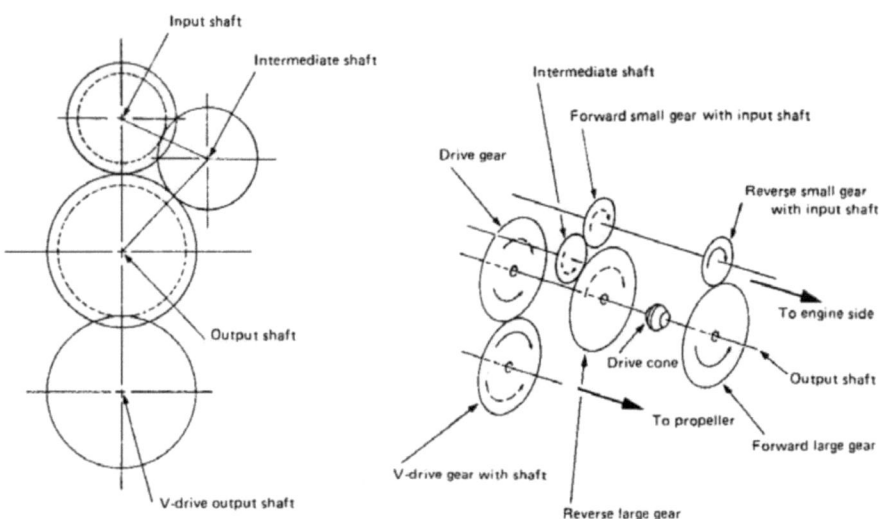

[Shaft arrangement viewed from the stern]

3-2 Reduction ratio

	Small gear	Large gear	Drive gear	V-drive gear	Reduction ratio
Forward	25	59	46	46	59/25 = 2.36
	23	60	46	46	60/23 = 2.61
	20	64	46	46	64/20 = 3.20
Reverse	19 / 26 Intermediate gear	60	46	46	60/19 = 3.16

4.Cooling system(Sea-water Cooling Engine)

4-1 Cooling water passage of engine model 1GM10V

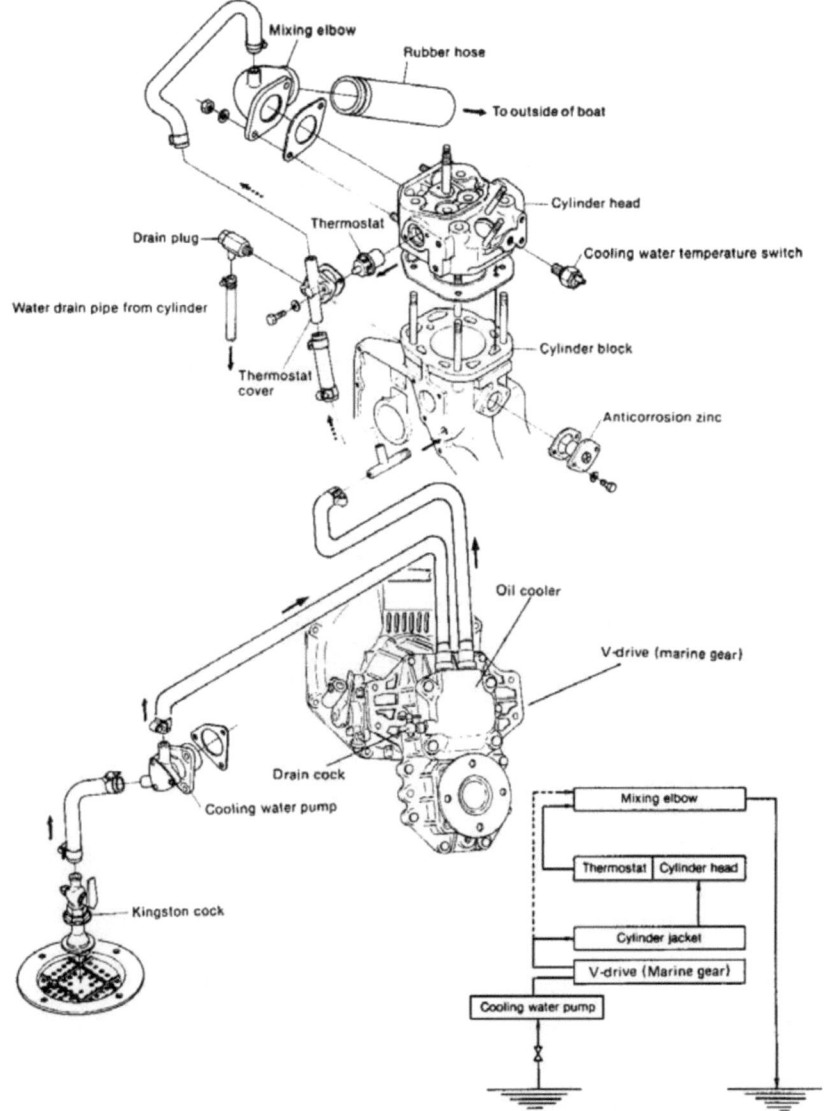

4-2 Cooling water passage of engine model 2GM20V and 3GM30V

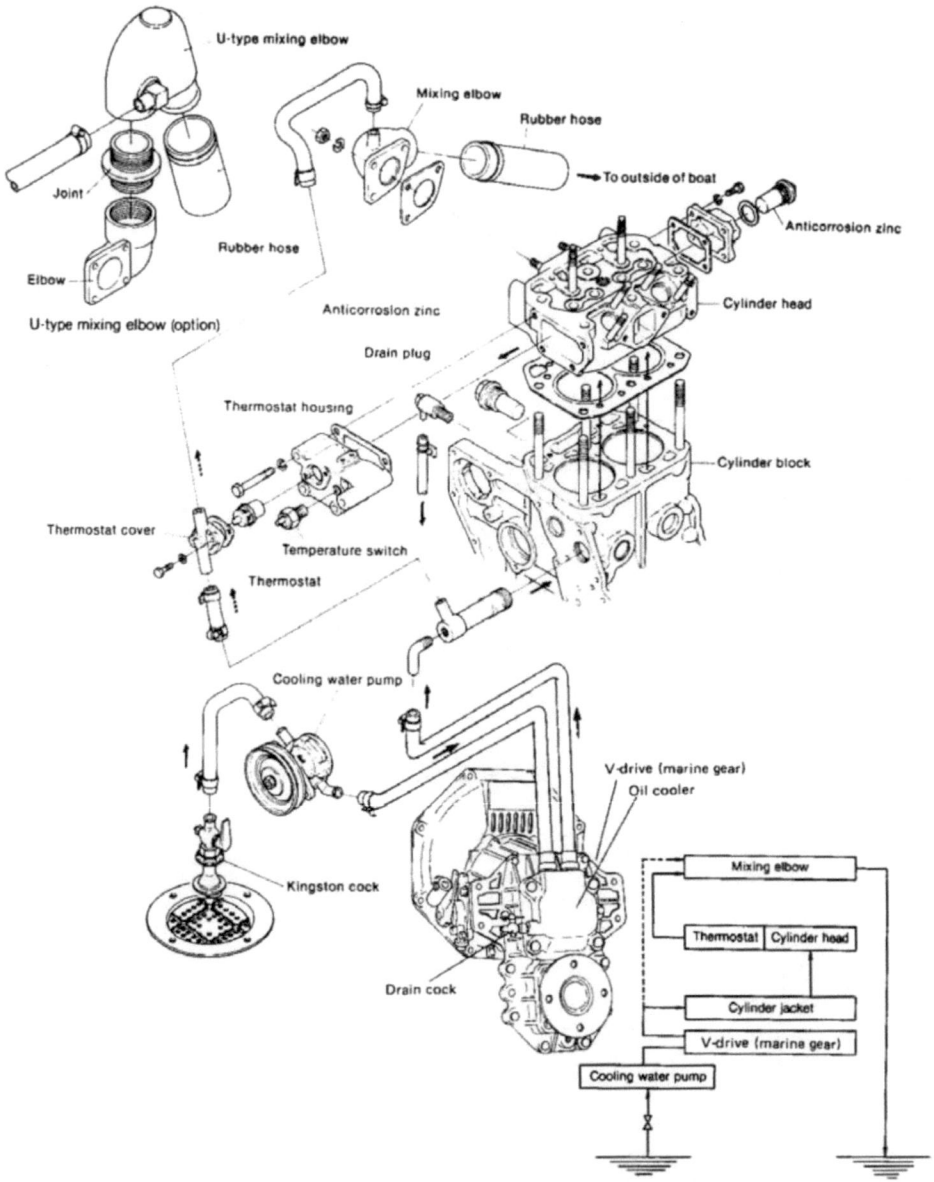

4-3 Cooling system (Fresh water cooling engine)
System Diagrams

Models 2GM20FV and 3GM30FV are constructed from different parts but use the same water flow.

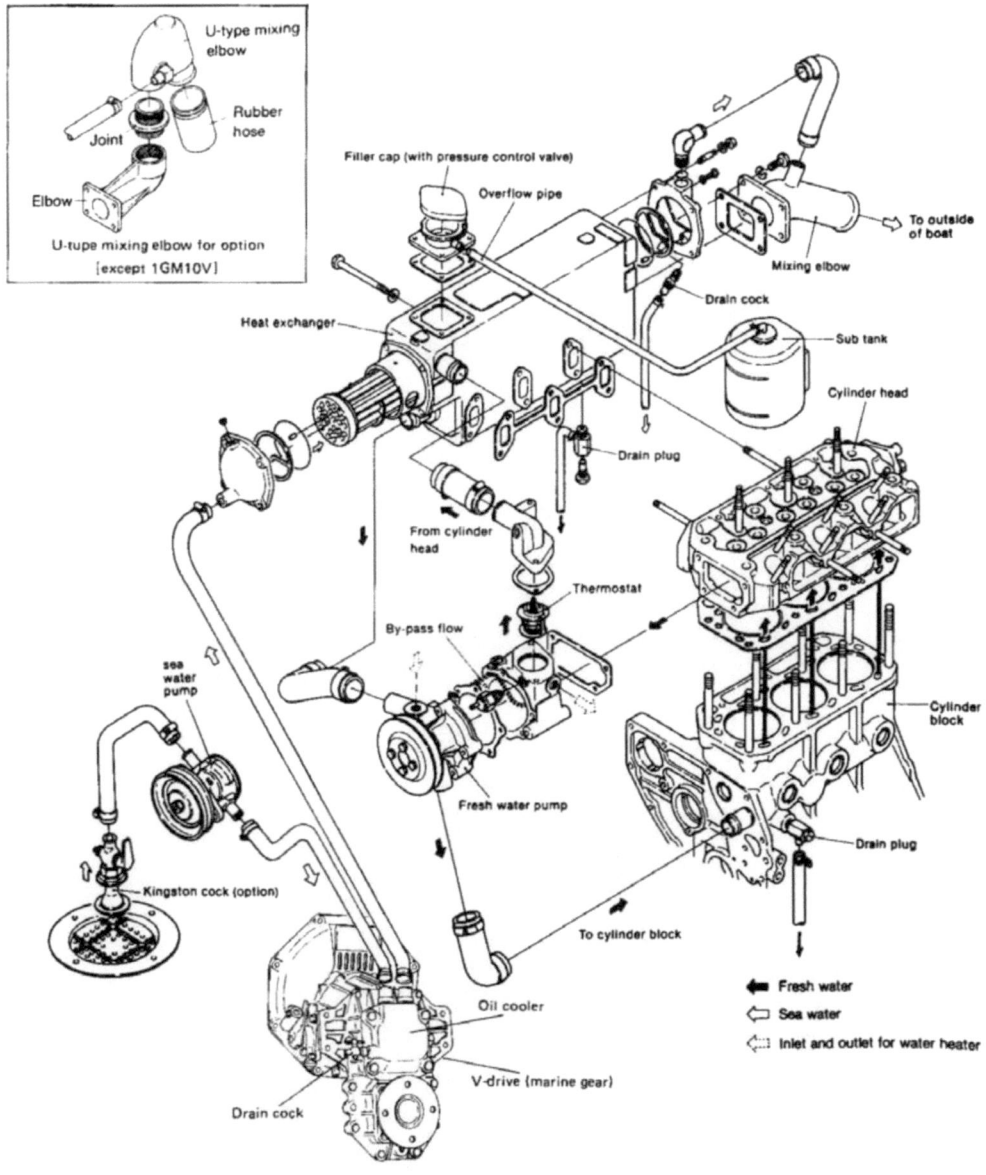

5. Piping Diagrams

5-1 1GM10V

Chapter 10 Reduction and Reversing Gear
5. Piping Diagrams

SM/GM(F)(C)-HM(F)(C)

5-2 2GM20V

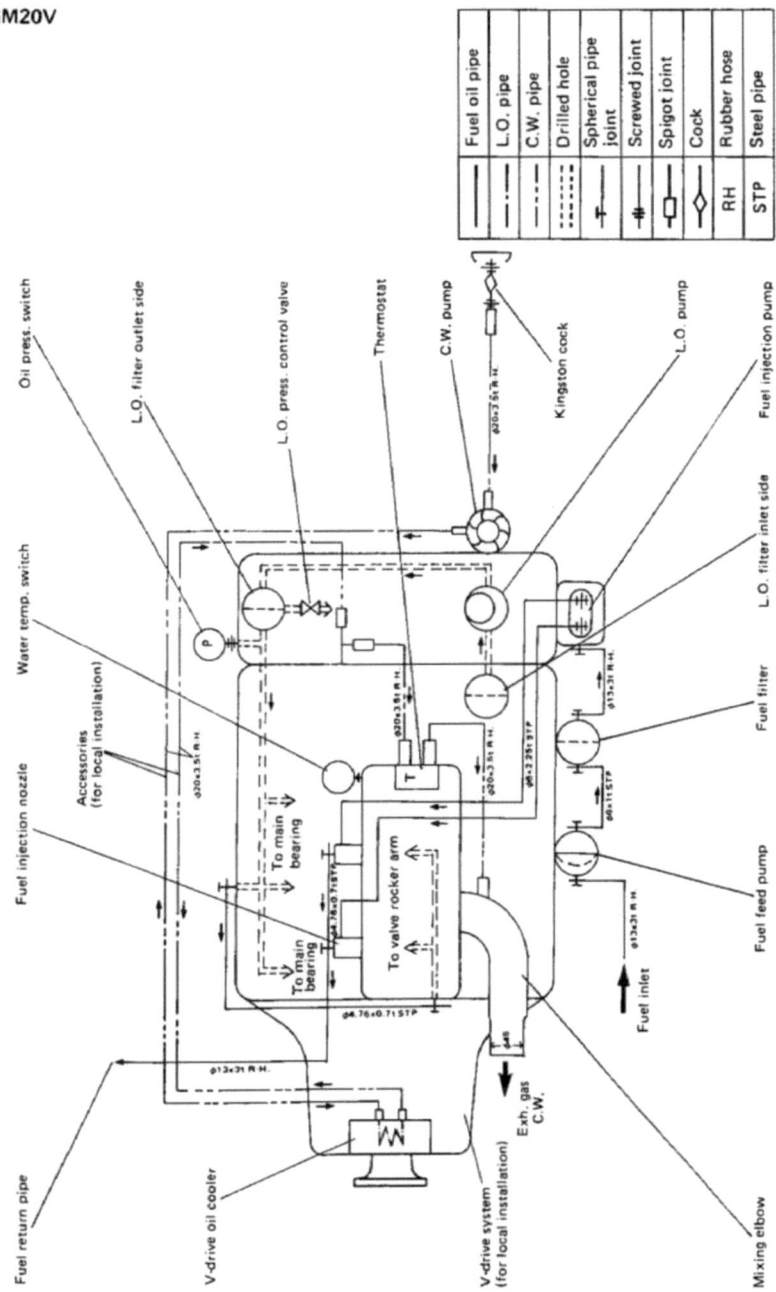

5-3 2GM20FV

Chapter 10 Reduction and Reversing Gear
5. Piping Diagrams

SM/GM(F)(C)-HM(F)(C)

5-4 3GM30V

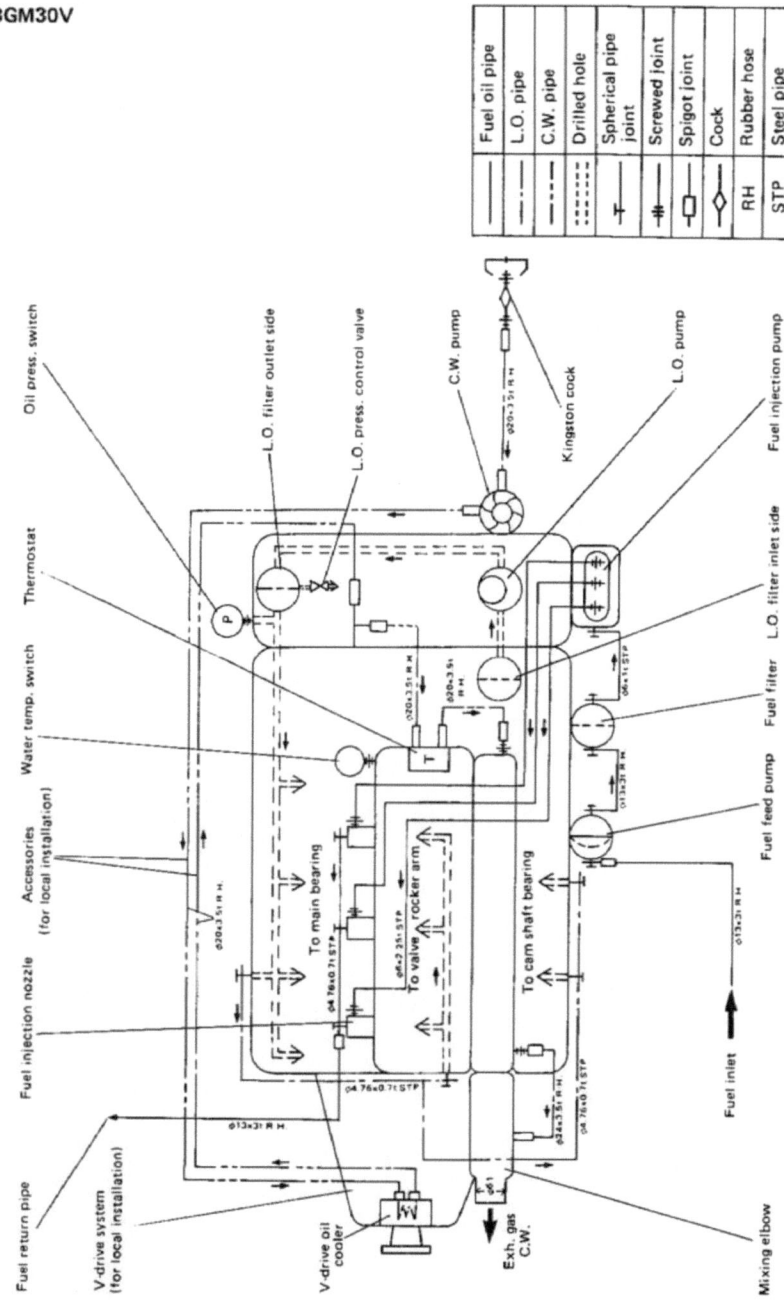

Chapter 10 Reduction and Reversing Gear
5. Piping Diagrams

SM/GM(F)(C)-HM(F)(C)

5-4 3GM30FV

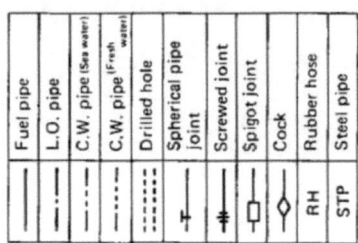

	Fuel pipe
	L.O. pipe
	C.W. pipe (Sea water)
	C.W. pipe (Fresh water)
	Drilled hole
	Spherical pipe joint
	Screwed joint
	Spigot joint
RH	Cock
RH	Rubber hose
STP	Steel pipe

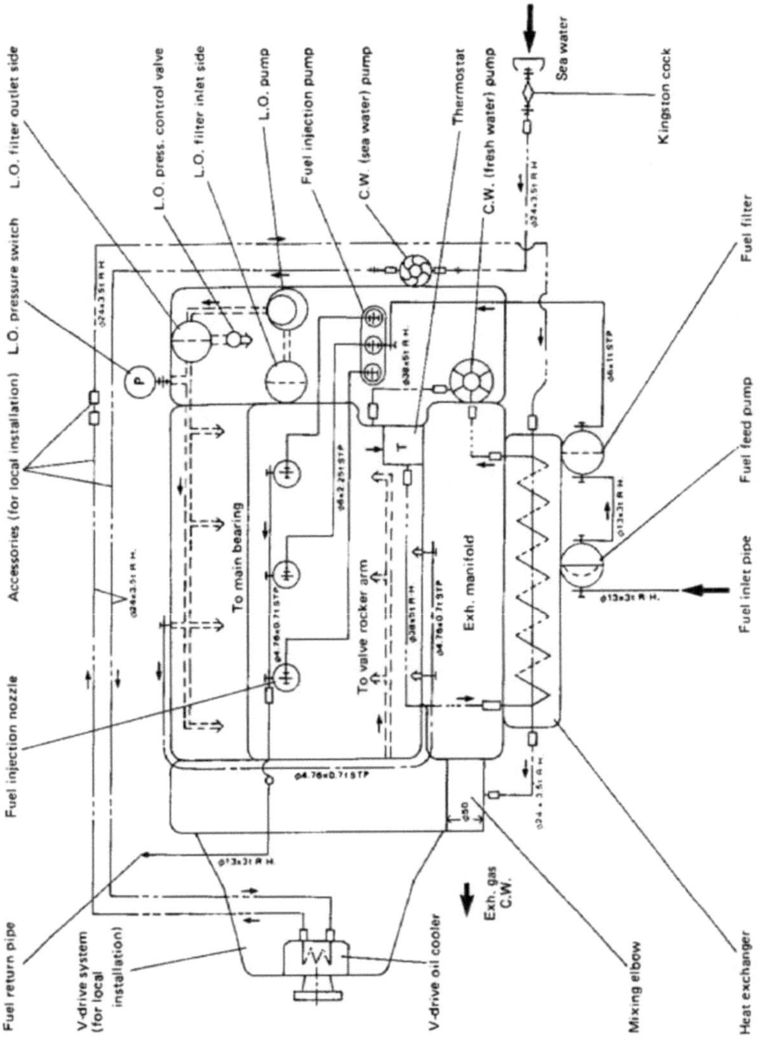

10-89

Printed in Japan
0000A0A1361

6. Inspection and Servicing

The KM3V V-Drive Gear is the combination of the V-drive system with the KM3P parallel drive gear. Accordingly, the explanation of this section is limited only to the V-drive system. Explanations for other parts excluding the V-drive system are identical to those described under the sections for model KM3P.

6-1 V-drive case
(1) Check the V-drive case with a test hammer for cracking. Perform a color check when required.
If the case is cracked, replace it.
(2) Check for staining on the inside surface of the bearing section.
Also, measure the inside diameter of the case.
Replace the case if it is worn beyond the wear limit.

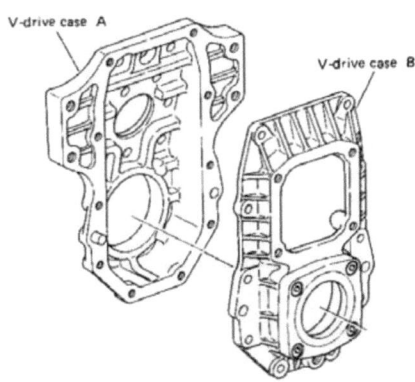

6-3 Gear
Check the surface, tooth face conditions and backlash of each gear. Replace any defective part.
(1) Tooth surface wear.
Check the tooth surface for pitting, abnormal wear, dents and cracks. Repair both the lightly damaged gears and replace heavily damaged gears.
(2) Tooth surface aontact.
Check the tooth surface contact. The amount of tooth surface contact between the tooth crest and tooth flank must be at least 70% of the tooth width.
(3) Backlash.
Measure the backlash of each gear, and replace the gear when it is worn beyond the wear limit.

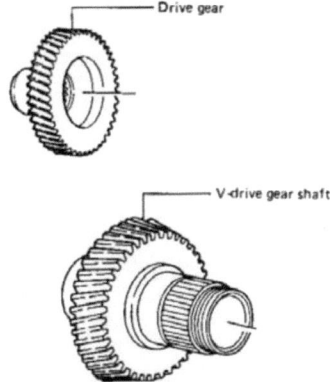

6-2 Bearing
(1) Rusting and damage.
If the bearing is rusted or the taper roller retainer is damaged, replace the bearing.
(2) Make sure that the bearings rotate smoothly.
If rotation is not smooth, if there is any binding, or if any abnormal sound is evident, replace the bearing.

Bearing of V-drive gear shaft

mm (in.)

Backlash	Maintenance standard	Wear limit
Drive gear and V-drive gear	0.08 ~ 0.16 (0.0031 ~ 0.0063)	0.3 (0.0118)

6-4 Oil seal and O-ring
Check the sealing surface of the oil seals and O-ring. If worn and/or scratched, replace.

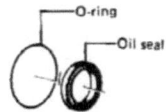

6-5 Packing

Check the oil cooler packings. If there are cracks, and/or scratches, repair.

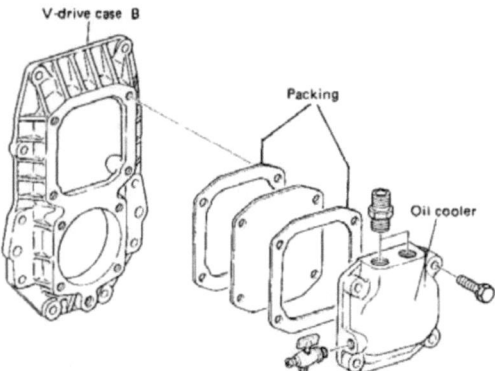

7. Shim adjustment for V-drive gear shaft, and backlash adjustment for V-drive gear shaft and drive gear

Check the thickness of shims for V-drive gear shaft. When the components are not replaced after dismantling, the same shims can be reused. When the clutch case, flange, V-drive case A and B or any of the following parts is replaced, the thickness of shim must be determined in the following manner.

For input shaft parts : input shaft, bearing
For output shaft parts : output shaft, thrust collar A, thrust collar B, gear, bearing
For V-drive gear shaft parts : V-drive gear shaft, cover A, cover B, bearing

7-1 Measuring method of shim thickness (T_1, T_2) of V-drive gear shaft

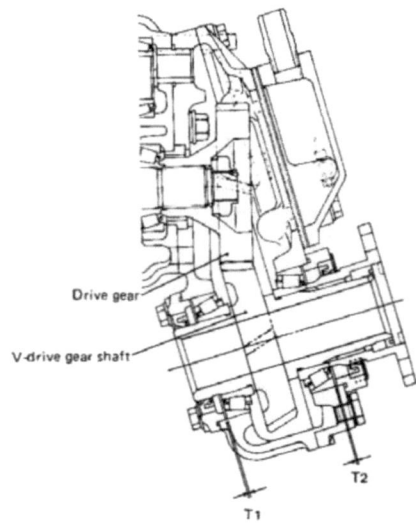

(1) Install the cover A to the V-drive case A, and measure the bearing insertion hole depth A.

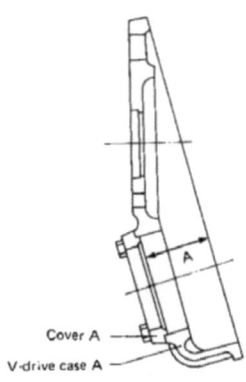

(2) Install the cover B to the V-drive case B, and measure the bearing insertion hole depth A'.

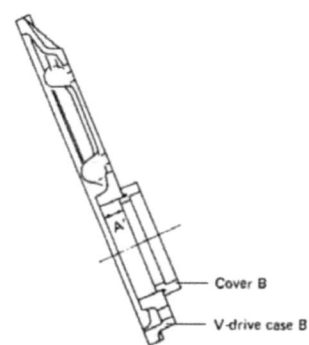

(3) Insert bearings into the V-drive gear shaft, and measure the B-length between the bearing outer races.

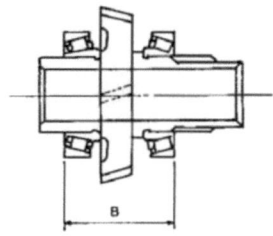

Chapter 10 Reduction and Reversing Gear
7. Shim Adjustment for V-drive Gear Shaft, etc.

SM/GM(F)(C)-HM(F)(C)

(4) Obtain the thickness T1, and T2 by the following formulas:

$$T = A + A' - B$$

Where: $T = T_1 + T_2$ Tolerance of T-dimension $\pm^{0.1}_{0}$

$$T_2 = T - T_1 (0.8 \text{ mm})$$

T1 standard dimension 0.8 mm

(5) Standard dimensions

	A	A'	B	T	T1
KM3V	49.8 ~ 50.2 (1.937 ~ 1.976)	20.8 ~ 21.2 (0.819 ~ 0.835)	68.7 ~ 69.3 (2.705 ~ 2.728)	1.3 ~ 2.7 (0.051 ~ 0.106)	0.8 (0.0315)

NOTE. If the measured values differ largely from the above values, your measurements may be incorrect. Measure again correctly.

(6) Adjusting shim set

Part No.	Thickness, mm(in.)	No. of shim
177070-02860	1.0 (0.0039)	1
	0.3 (0.0118)	3
	0.2 (0.0079)	2

7-2 Backlash adjustment for V-drive gear shaft and drive gear

(1) Insert the fuse wire into the interval between the Drive gear and the V-drive gear from the cooler installation window. Turn the gear, and pull out the fuse wire. Measure the pressed thickness of the fuse wire with a micrometer.

Backlash standard: 0.08 ~ 0.16 mm
(0.0032 ~ 0.0063 in.)

(2) When the backlash is larger than the standard value, move the T2 shim to the T1 side.
When the backlash is smaller than the standard value, move the T1 shim to the T2 side.

NOTE: By moving the T1, or T2 shim by 0.1mm, the backlash is varied by about 0.016 mm (0.0006 in.).

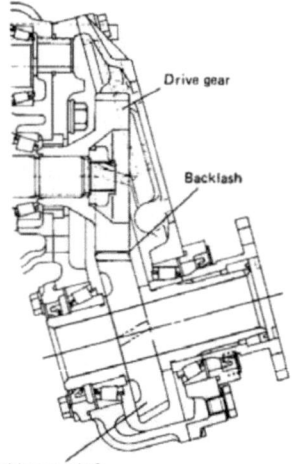

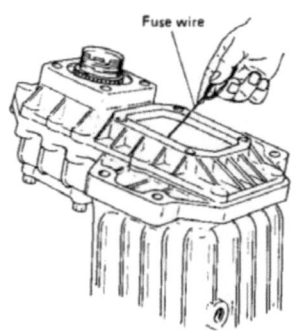

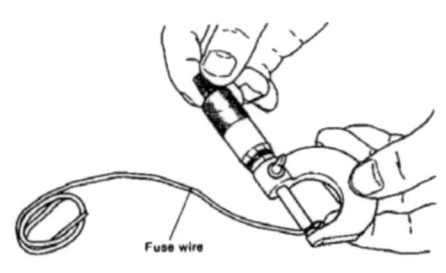

8. Disassembly

8-1 Dismantling the V-drive assembly
(1) Remove the remote control cable.
(2) Remove the cooling water hose of lube oil cooler.
(3) Remove the V-drive assembly from the engine mounting flange.

8-2 Drain the cooling water from the lube oil cooler.

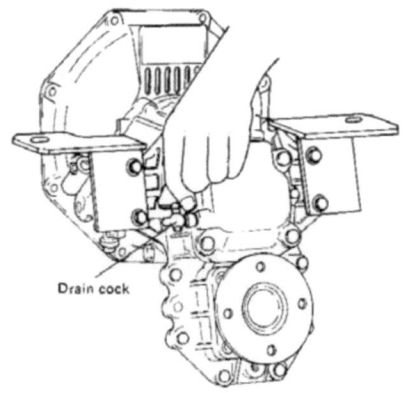

Drain cock

8-3 Loosen the plug at the bottom of the V-drive case, and drain the lube oil.

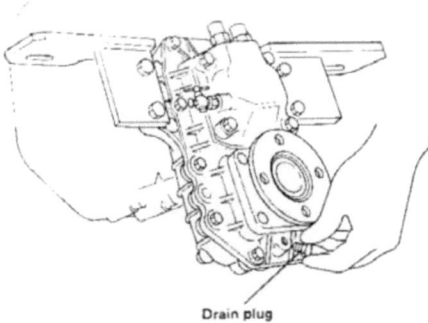

Drain plug

8-4 Remove the engine feet, lube oil dipstick, and O-ring.

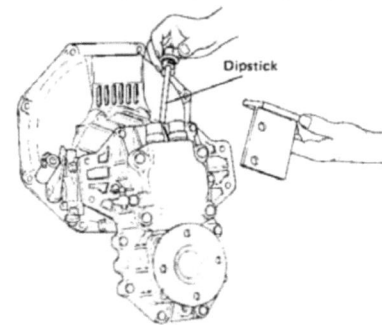

Dipstick

8-5 Remove the lube oil cooler fixing bolts, and then remove the lube oil cooler body, packings, and plate.

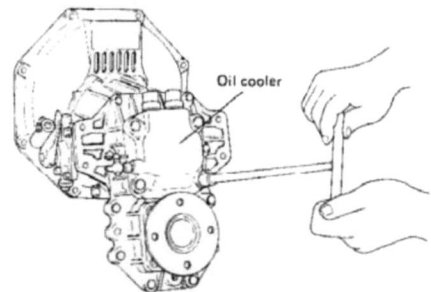

Oil cooler

8-6 Loosen the V-drive shaft end nut.

Chapter 10 Reduction and Reversing Gear
8. Disassembly

SM/GM(F)(C)-HM(F)(C)

8-7 Remove the end nut, output shaft coupling, and O-ring.

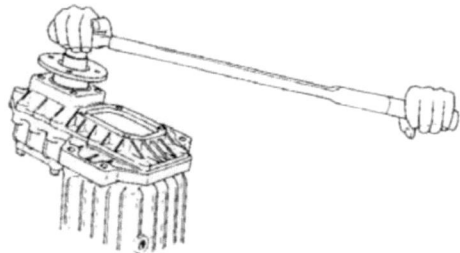

NOTE: 1) Use the special tool to remove the end nut from the V-drive shaft.
2) Special tool (option part) Part No. 177070-09010

```
Dimension: mm
Material: Carbon steel
Hardness: HB 203 ~ 258
         HS 31 ~ 38
(Hardening)
```

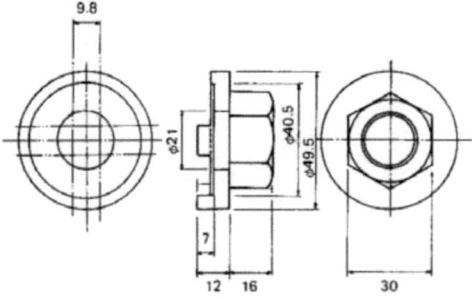

8-8 Remove the fixing bolts on the V-drive case B, and also remove the V-drive case B.

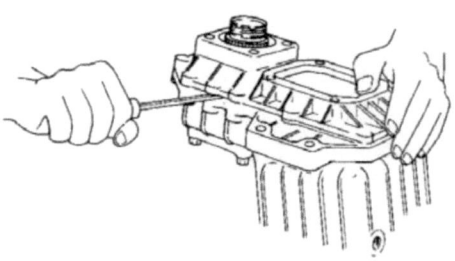

8-9 Remove the V-drive shaft.

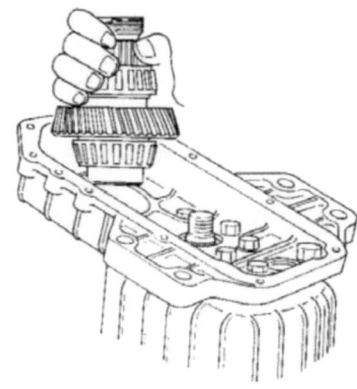

8-10 Remove the end nut, and drive gear from the output shaft.

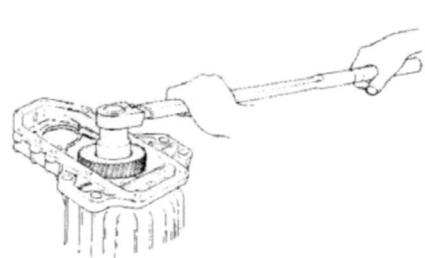

NOTE: Note that the end nut has a left-handed thread.

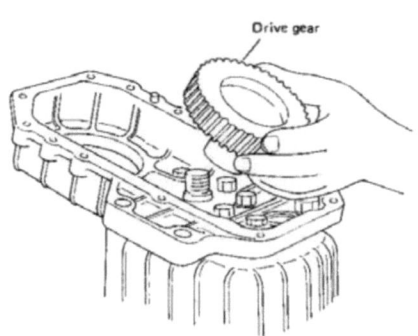

Drive gear

Chapter 10 Reduction and Reversing Gear
8. Disassembly

SM/GM(F)(C)-HM(F)(C)

8-11 Remove the V-drive case A fixing bolts, and remove the V-drive case A.

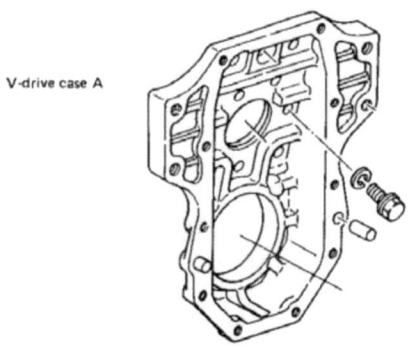

8-12 Disassembly of clutch assembly.
Follow to the same procedures described for model KM3P. (Refer to the sections under KM3P).

8-13 Remove the covers, O-rings, and oil-seals from the V-drive cases A and B.

8-14 Remove the bearing outer races from the V-drive cases A and B.

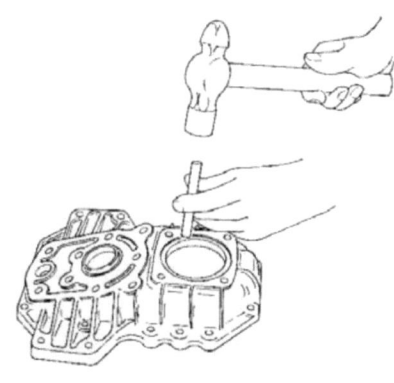

8-15 Remove the bearing inner races from the V-drive gear shaft.

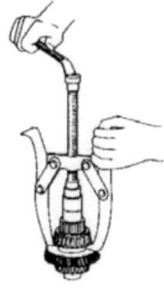

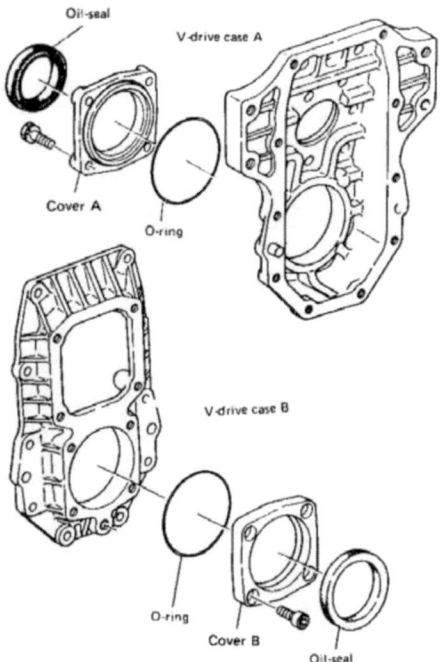

10-96

9. Reassembly

9-1 Reassembly of the clutch unit
Follow to the same procedures described for model KM3P. (Refer to sections under KM3P)

9-2 Fit the bearing inner races in the V-drive gear shaft.

9-3 Fit the bearing outer races in the V-drive cases A and B.

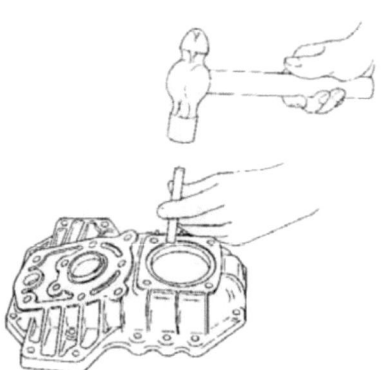

9-4 Insert the center bush and location pin to the clutch case end holes.

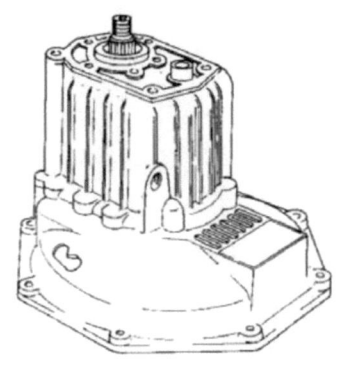

9-5 Fit the V-drive case A to the clutch case.

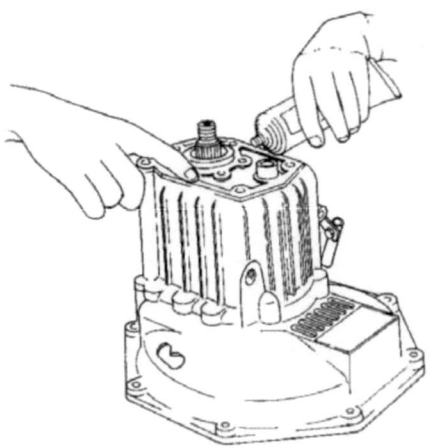

NOTE: Apply a non-drying type liquid packing to the matching surfaces of the case A and the clutch case.

Chapter 10 Reduction and Reversing Gear
9. Reassembly

9-6 Fit the covers, O-rings, oil seals and shims into the V-drive cases A and B.

9-9 Fit the V-drive case B to the V-drive case A.

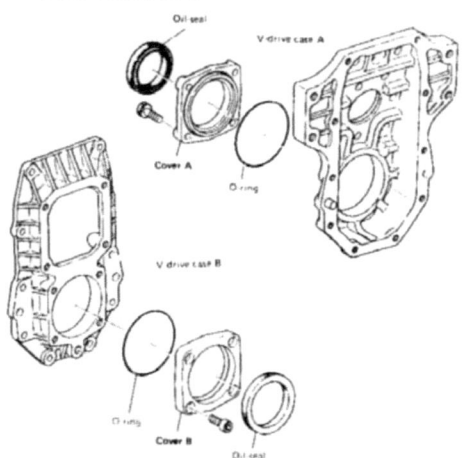

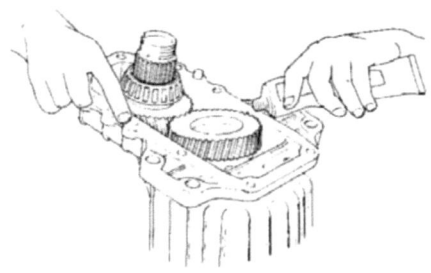

NOTES: 1) Apply a non-drying type liquid packing to the matching surfaces of the V-drive cases A and B.
2) Measure the backlash of the drive gear and V-drive gear. (Refer to the procedures under the foregoing 6-2.)

9-7 Insert the V-drive shaft into the V-drive case A, and insert the V-drive gear into the output shaft.

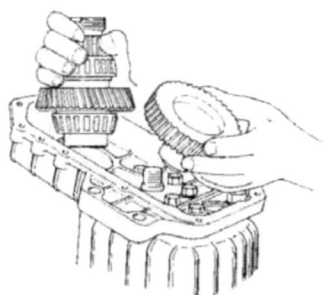

9-10 Insert the O-ring and output shaft coupling. Tighten the lock nut. Use the special tool and the torque wrench.

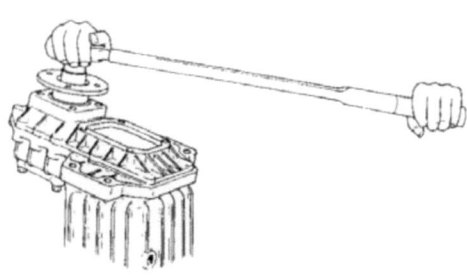

9-8 Tighten the end nut with a torque wrench

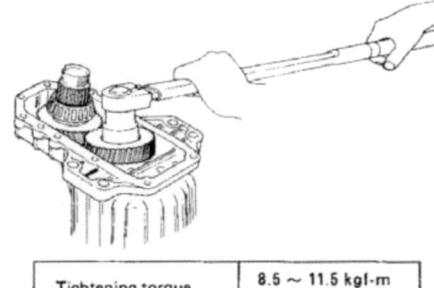

Tightning torque	18 ~ 22 kgf-m (130.2 ~ 159.1 ft-lb)

NOTE: Check that the output shaft coupling turns smoothly at the neutral position of the shift lever.

Tightening torque	8.5 ~ 11.5 kgf-m (61.5 ~ 83.2 ft-lb)

NOTE: Note that the end nut has a left-handed thread.

Chapter 10 Reduction and Reversing Gear
9. Reassembly

9-11 Caulk the lock nut

9-14 Fit the lube oil drain plug

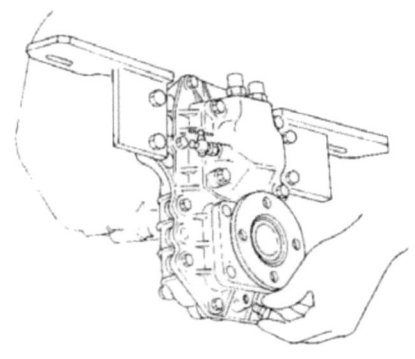

9-12 Fit the oil cooler

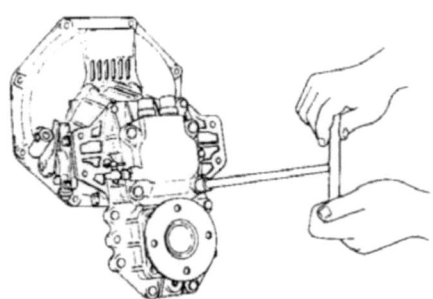

9-13 Fit the engine feet, lube oil dipstick and O-ring.

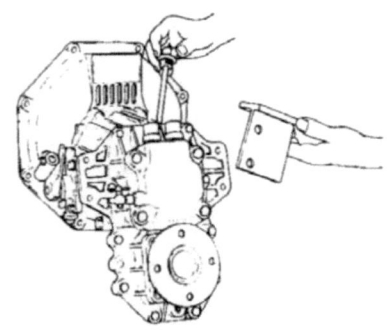

CHAPTER 11
REMOTE CONTROL SYSTEM

1. Construction 11-1
2. Clutch and Speed Regulator Remote Control 11-3
3. Engine Stop Remote Control 11-7

1. Construction

This engine is designed primarily for remote control operation. A remote control cable bracket can be installed by merely adding a remote control lever and link to the engine. Engine stop control and decompression remote control may also be installed, in addition to one-handle remote control, which permits engine speed adjustment and one-handle forward-astern switching.

For this engine, two-handle control cannot be used to replace one-handle control.

1-1 Model 1GM10, 2GM20(F) and 3GM30(F)

Model KM2-C reduction and reversing gear is used in model 1GM10 and 2GM20(F), 3GM30(F)(C) engines, therefore the forward and reverse lever is on the left when viewed from the stern. The construction for models 1GM10 and 2GM20(F) 3GM30(F) is the same except for the shape and mounting position of the bracket.

Chapter 11 Remote Control System
1. Construction

SM/GM(F)(C)·HM(F)(C)

1-2 Model 3HM35(F)

Model 3HM35(F) is built the same except for the shape and mounting position of the bracket.
The reduction and reversing gear for engine model 3HM35(F) is model KBW10E, therefore, the clutch lever is on the right when viewed from the stern.

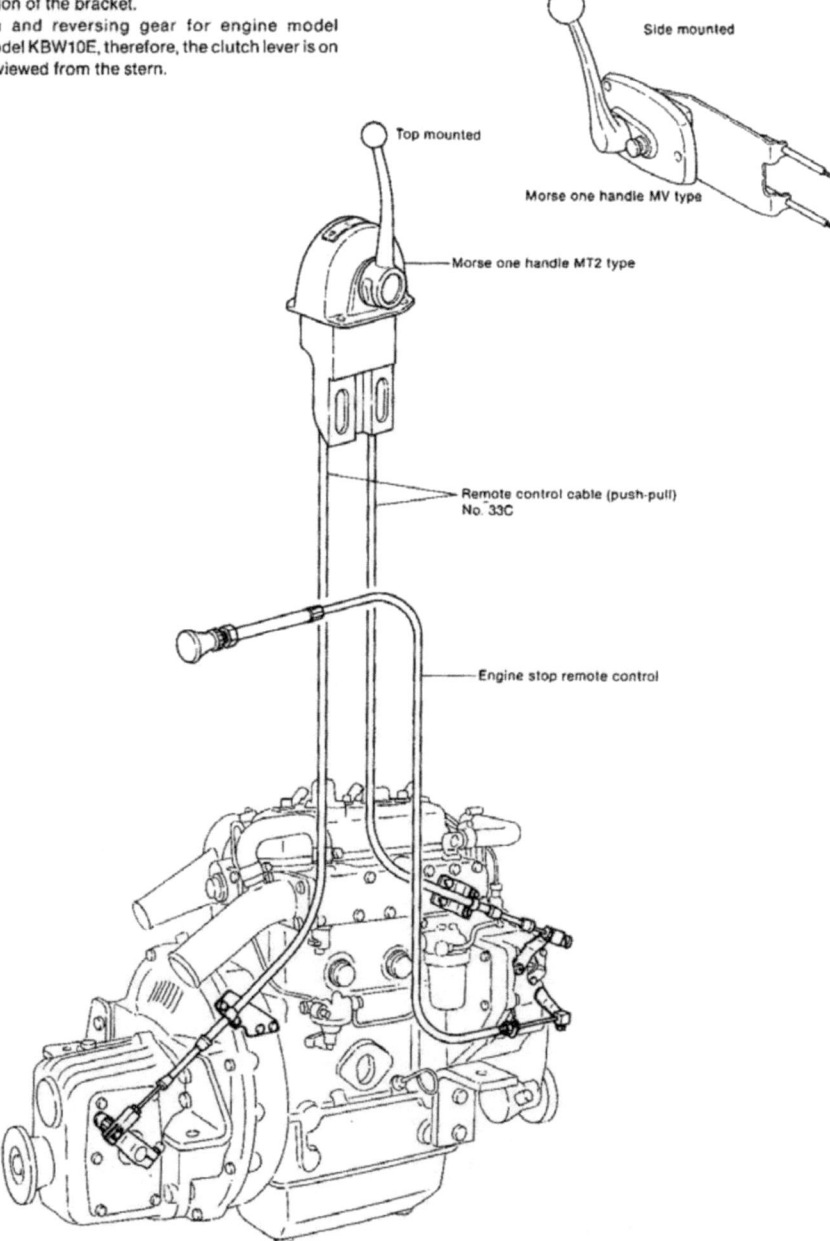

2. Clutch And Speed Regulator Remote Control

2-1 Construction
Both models of MT2 and MV morse one handle remote control can be used. They are optionally available.

2-1.1 MT2 type

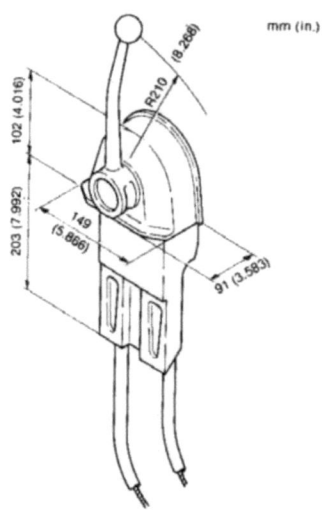

2-2.2 MV type
Newly expanded MV series controls include right and left hand models designed for easier installation and servicing. The MV control can be preassembled and installed without removing side panels.
Pull-out button disengages clutch for full throttle range in neutral for safe starting and warm-up.
MV controls have forward, neutral and reverse detents; built-in friction to prevent throttle creep.

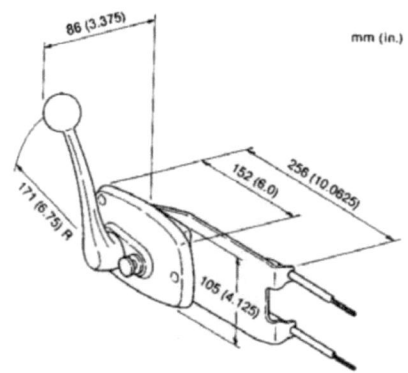

2-2 One-handle remote control composition

		1GM10, 2GM20(F), 3GM30(F)	3HM35(F)
Speed control	Remote control cable	33-C	
	Clamp	YANMAR made	
Clutch control	Remote contorl cable	33-C	
	Clamp	YANMAR made	
	Spring joint	YANMAR made	—
	Clevis	—	YANMAR made

(1) Control cable
Morse Type "33-C" push-pull control cables.

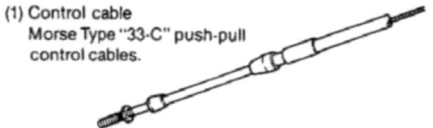

Use only Super-Responsive Morse Control Cables. They are designed specifically for use with Morse control heads. This engineered system of Morse cables, control head and engine connection kits ensures dependable, smooth operation with an absolute minimum of backlash. The thread size on cable ends is 10-32. Travel is up to 3". The core is a solid wire, with a 3/32" diameter.

(2) Clamp
YANMAR cable clamps are standard parts, and are fitted to the brackets on the engine and clutch.

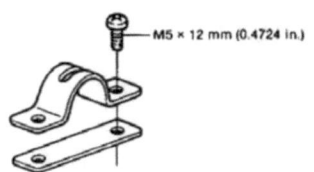

Chapter 11 Remote Control System
2. Clutch and Speed Regulator Remote Control

(3) Spring joint
The cone clutch is fitted to engine models 1GM10, 2GM20(F) and 3GM30(F). The spring joint is fitted to the clutch lever, and is also connected to the control cable.

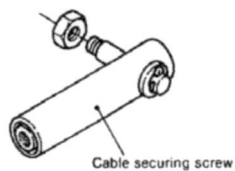

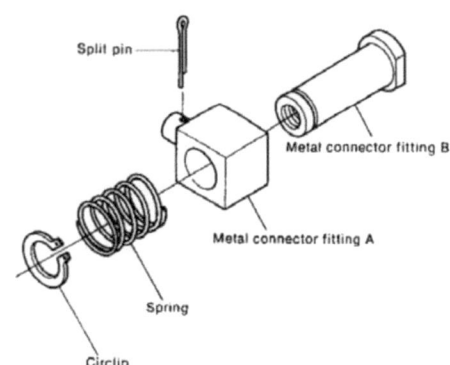

(4) Clevis
The YANMAR clevis is attached to the clutch lever on model 3HM35(F).
Cable securing screw.

NOTE: When the push-pull cable is fitted, it must be fitted at the spring side.

2-3.1 Movement of lever for model 1GM10(C)

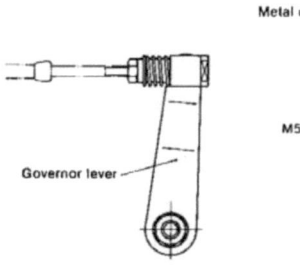

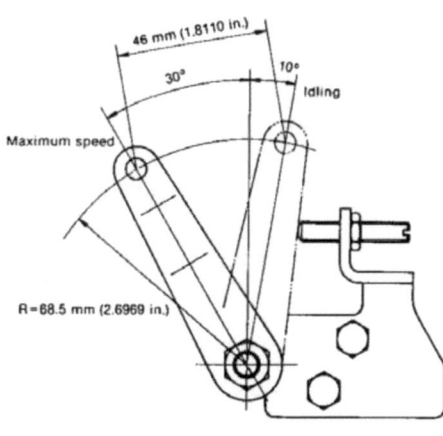

2-3 Engine side installation
The same governor lever is used in all 4 engine models, however, its operation angle is different depending on the model.
The connecting metal which fits with the damping spring is at the tip of the governor lever, and the cable has only to be screwed into this fitting.

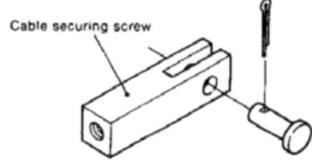
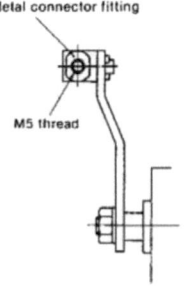
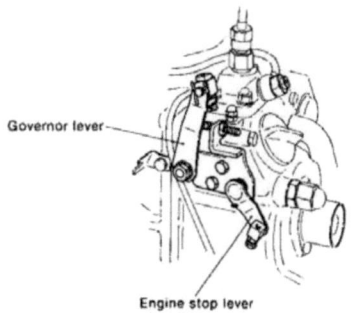

Chapter 11 Remote Control System
2. Clutch and Speed Regulator Remote Control

2-3.2 Movement of lever for models 2GM20(F)(C), 3GM30(F)(C), and 3HM35(F)(C)

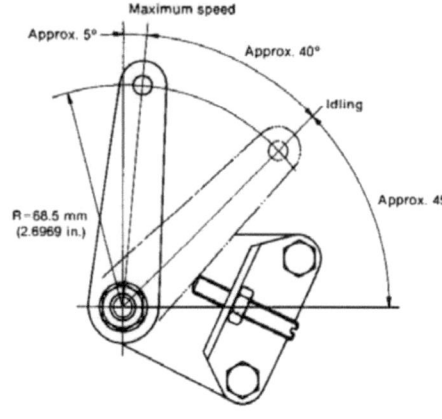

2-4.1 For models 1GM10, 2GM20(F) and 3GM30(F)

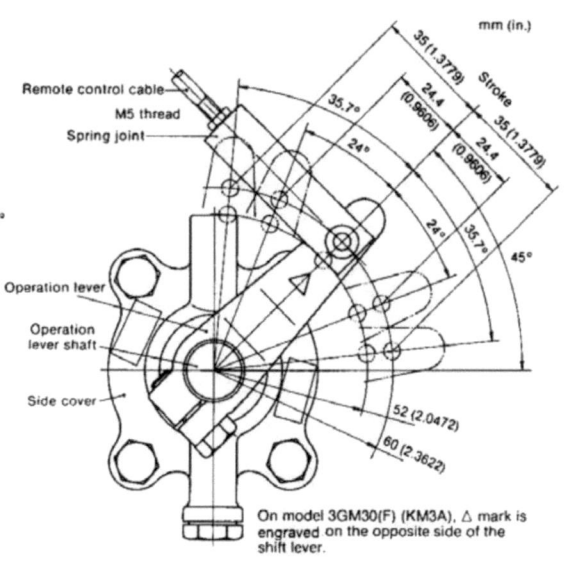

On model 3GM30(F) (KM3A), △ mark is engraved on the opposite side of the shift lever.

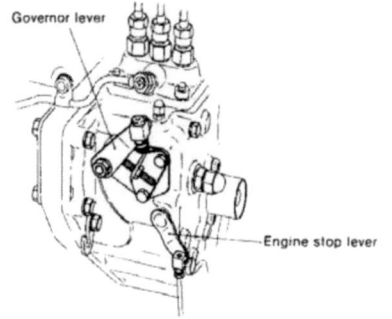

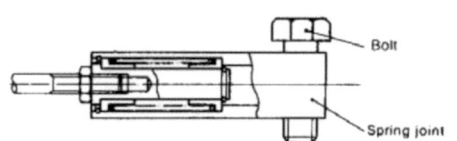

2-4 Setting the reduction and reversing gear side

Model KM2C reduction and reversing gear is used for engine models 1GM10 and 2GM20(F), and model KM3A for engine model 3GM30(F).
On these reduction and reversing gears, the spring joint is fitted to the control lever, and the remote control cable is connected to this joint.
Reduction and reversing gear model KBW10E is used on engine model 3HM35(F). On these reduction and reversing gears, the clevis is attached to the clutch operating lever, and the remote control cable is connected to the clevis.

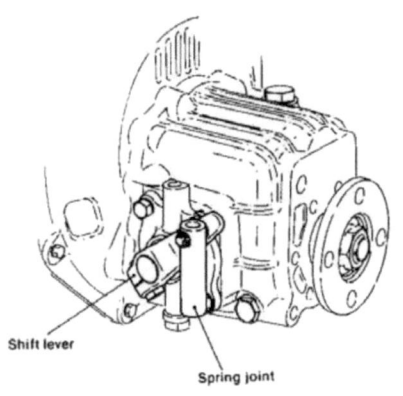

Chapter 11 Remote Control System
2. Clutch and Speed Regulator Remote Control

2-4.2 For model 3HM35(F)

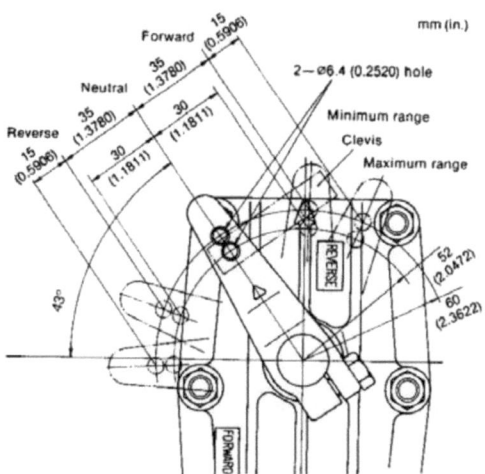

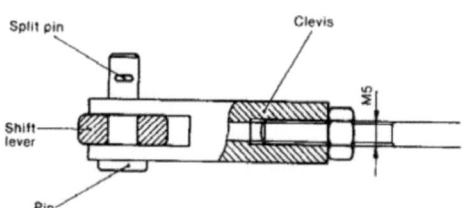

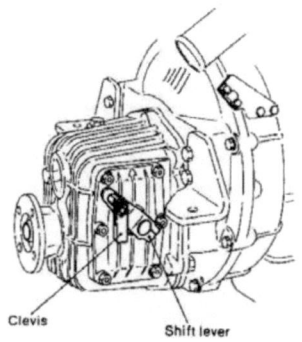

3. Engine Stop Remote Control

4-1 For model 1GM10(C)

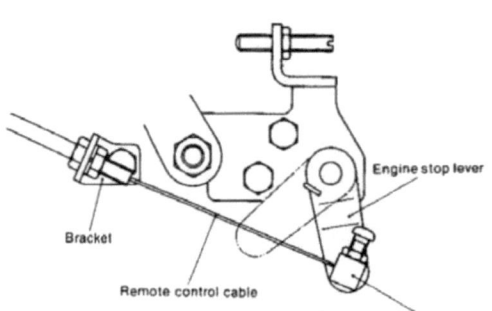

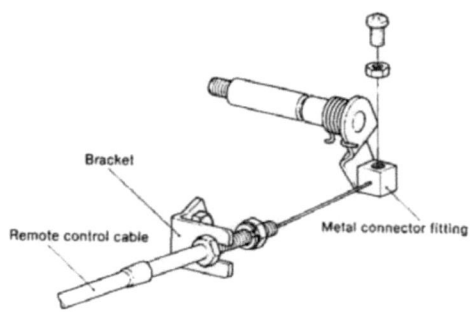

YANMAR made ø1.5mm (0.05906 in.) option
Cable dia. ø1.5 ~ ø2.5mm (0.05906 ~ 0.0984 in.)

The metal connector fitting has a hole of 2.5mm (0.0984 in.) dia. to accommodate the cable, and cable of 1.5 ~ 2.5mm (0.05906 ~ 0.0984 in.) dia. can be used in the connector.

4-2 For models 2GM20(F)(C), 3GM30(F)(C) and 3HM35(F)(C)

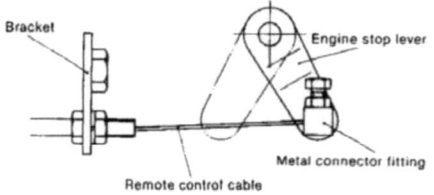

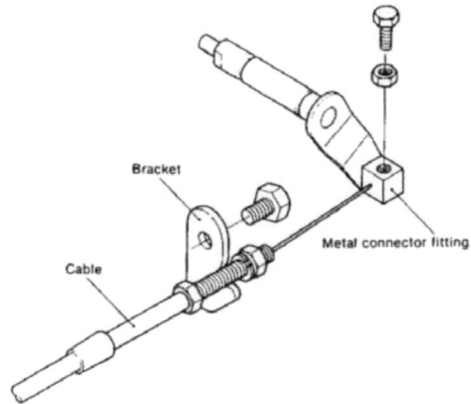

CHAPTER 12
ELECTRICAL SYSTEM

1. Electrical System 12-1
2. Battery 12-4
3. Starter Motor 12-7
4. Alternator Standard, 12V/55A 12-18
4A. Alternator Option, 12V/35A 12-28
5. Instrument Panel 12-37
6. Tachometer 12-43

1. Electrical System

1-1 System diagram of electric parts

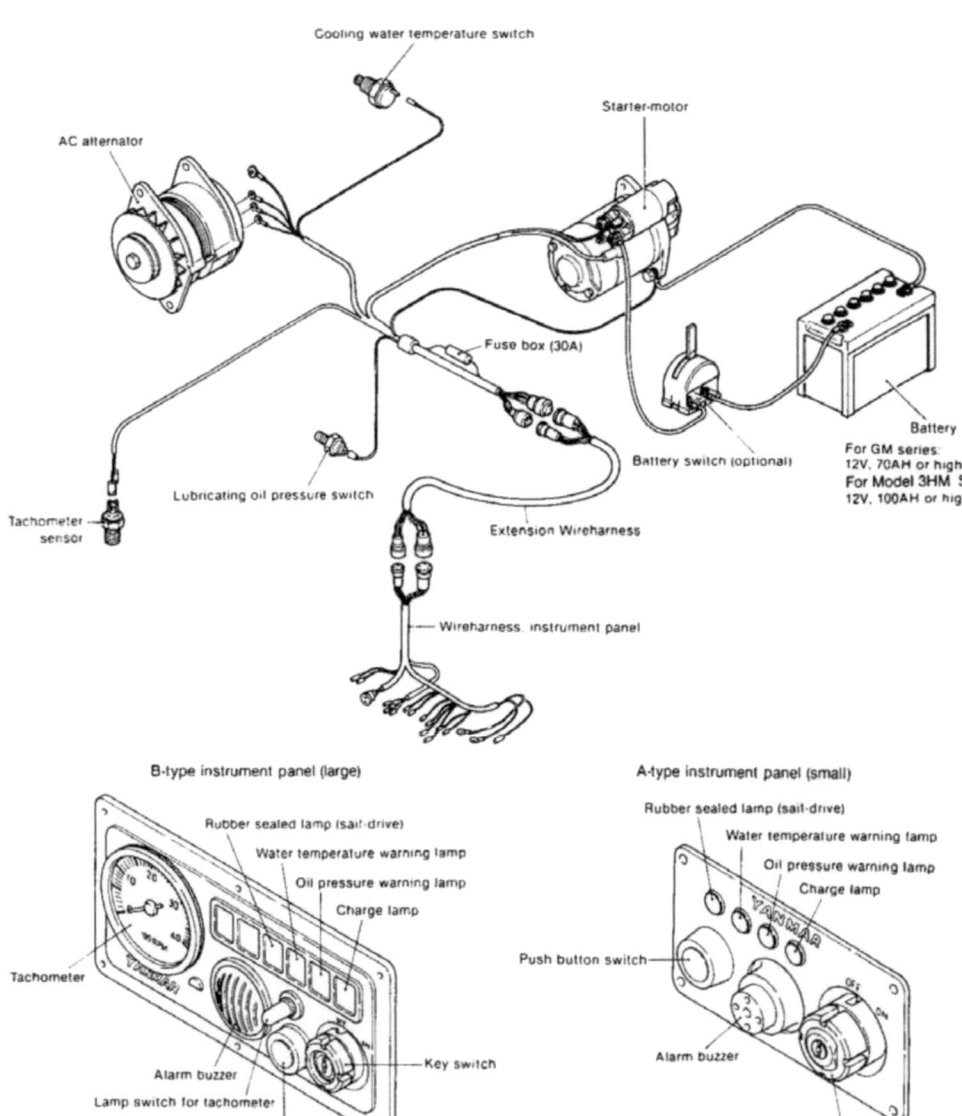

Chapter 12 Electrical System
1. Electrical System

SM/GM(F)(C)·HM(F)(C)

1-2. Wiring diagram
1-2.1 For the B-type (large) instrument board

Chapter 12 Electrical System
1. Electrical System

1-2.2 For the A-type (small) instrument board

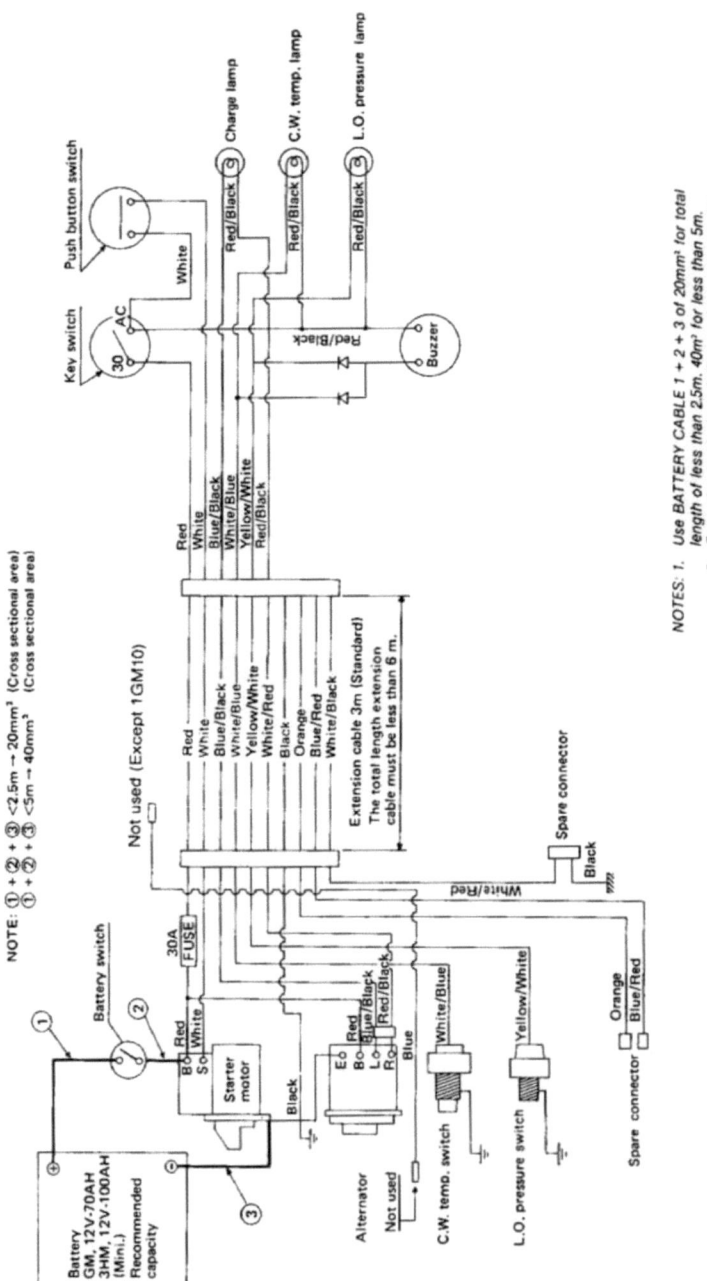

2. Battery

2-1 Construction

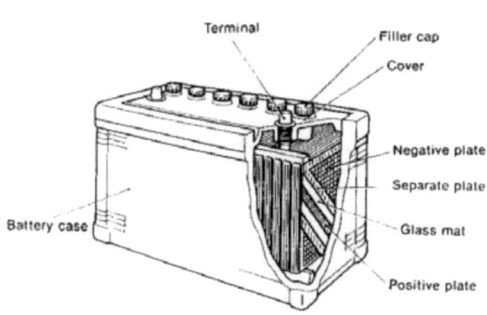

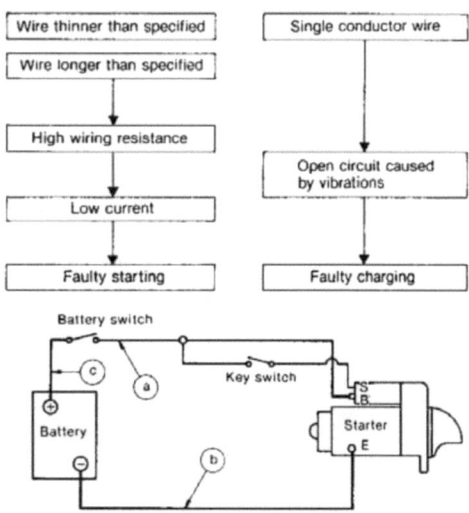

The battery utilizes chemical action to convert chemical energy to electrical energy. This engine uses a lead acid battery which stores a fixed amount of power that can be used when required. After use, the battery can be recharged and used again.
As shown in the figure, a nonconductive container is filled with dilute sulfuric acid electrolyte. Lead dioxide positive plates and lead dioxide negative plates separated by glass mats are stacked alternately in the electrolyte. The positive and negative plates are connected to their respective terminals.
Power is removed from the battery by connecting the load across these two terminals.
When the battery is discharging, an electric current flows from the positive plates to the negative plates. When the battery is being charged, electric current is passed through the battery in the opposite direction by an external power source.

2-2 Battery capacity and battery cables

2-2.1 Battery capacity
Since the battery has a minimum capacity of 12V, 70AH, it can be used for 100 ~ 150AH.

	1GM10(C) 2GM20(F)(C) 3GM30(F)(C)	3HM35(F)(C)
Minimum battery capacity	12V 70AH	12V 100AH
Fully charged specific gravity	1.26	1.26

2-2.2 Battery cable
Wiring must be performed with the specified electric wire. Thick, short wiring should be used to connect the battery to the starter, (soft automotive low-voltage wire [AV wire]). Using wire other than that specified may cause the following troubles:

The overall lengths of the wiring between the battery (+) terminal and the starter (B) terminal, and between the battery (−) terminal and the starter (E) terminal should be based on the following table.

Voltage system	Allowable wiring voltage drop	Conductor cross-section area	a+b+c allowable length
12V	0.2V or less/100A	20mm² (0.0311 in.²)	Up to 2.5m (98.43 in.)
		40mm² (0.062 in.²)	Up to 5m (196.87 in.)

NOTE: Excessive resistance in the key switch circuit (between battery and start (S) terminals) can cause improper pinion engagement. To prevent this, follow the wiring diagram exactly.

2-3 Inspection
The quality of the battery governs the starting performance of the engine. Therefore the battery must be routinely inspected to assure that it functions perfectly at all times.

2-3.1 Visual inspection
(1) Inspect the case for cracks, damage and electrolyte leakage.
(2) Inspect the battery holder for tightness, corrosion, and damage.
(3) Inspect the terminals for rusting and corrosion, and check the cables for damage.
(4) Inspect the caps for cracking, electrolyte leakage and clogged vent holes.
Correct any abnormal conditions found. Clean off rusted terminals with a wire brush before reconnecting the battery cable.

2-3.2 Checking the electrolyte
(1) Electrolyte level

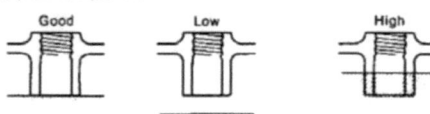

Good　　　　Low　　　　High

Check the electrolyte level every 7 to 10 days. The electrolyte must always be 10 ~ 20mm over the tops of the plates.

NOTES: 1) The "LEVEL" line on a transparent plastic battery case indicates the height of the electrolyte.
2) Always use distilled water to bring up the electrolyte level.
3) When the electrolyte has leaked out, add dilute sulfuric acid with the same specific gravity as the electrolyte.

(2) Measuring the specific gravity of the electrolyte
1) Draw some of the electrolyte up into a hydrometer.

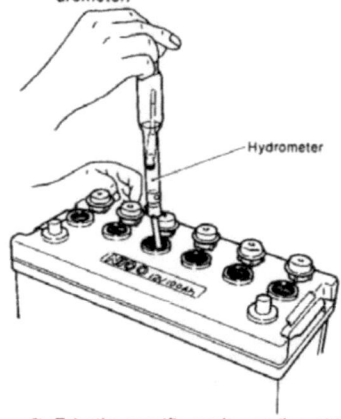

Hydrometer

2) Take the specific gravity reading at the top of the scale of the hydrometer.

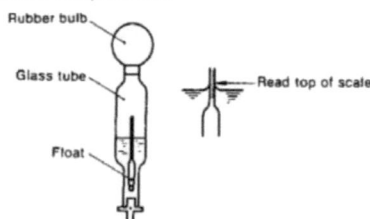

Rubber bulb
Glass tube
Read top of scale
Float

3) The battery is fully charged if the specific gravity is 1.260 at an electrolyte temperature of 20°C. The battery is discharged if the specific gravity is 1.200 (50%). If the specific gravity is below 1.200, recharge the battery.
4) If the difference in the specific gravity among the cells of the battery is ±0.01, the battery is OK.
5) Measure the temperature of the electrolyte.
Since the specific gravity changes with the temperature, 20°C is used as the reference temperature.
Reading the specific gravity at 20°C
$S_{20} = St + 0.0007(t - 20)$
S_{20}: Specific gravity at the standard temperature of 20°C
St: Specific gravity of the electrolyte at t°C
0.0007: Specific gravity change per 1°C
t: Temperature of electrolyte

2-3.3 Voltage test
Using a battery tester, the amount of discharge can be determined by measuring the voltage drop which occurs while the battery is being discharged with a large current.

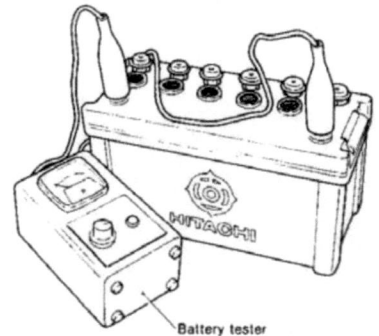

Battery tester

(1) Connect the tester to the battery.
12V battery tester
Adjust the current (A).
(2) Connect the (+) lead of the tester to the (+) battery terminal, and the (−) tester lead to the (−) battery terminal.
(3) Push the TEST button, wait 5 seconds, and then read the meter.
• Repeat the test twice to make sure that the meter indication remains the same.

2-3.4 Washing the battery
(1) Wash the outside of the battery with a brush while running cold or warm water over the battery. (Make sure that no water gets into the battery.)
(2) When the terminals or other metal parts are corroded due to exposure to electrolyte leakage, wash off all the acid.
(3) Check the vent holes of the caps and clean if clogged.
(4) After washing the battery, dry it with compressed air, connect the battery cable, and coat the terminals with grease. Since the grease acts as an insulator, do not coat the terminals before connecting the cables.

2-4 Charging

2-4.1 Charging methods

There are two methods of charging a battery: normal and rapid.

Rapid charging should only be used in emergencies.

- Normal charging...Should be conducted at a current of 1/10 or less of the indicated battery capacity (10A or less for a 100AH battery).
- Rapid charging...Rapid charging is done over a short period of time at a current of 1/5 ~ 1/2 the indicated battery capacity (20A ~ 50A for a 100AH battery). However, since rapid charging causes the electrolyte temperature to rise too high, special care must be exercised.

2-4.2 Charging procedure

(1) Check the specific gravity and adjust the electrolyte level.
(2) Disconnect the battery cables.
(3) Connect the red clip of the charger to the (+) battery terminal and connect the black clip to the (−) terminal.

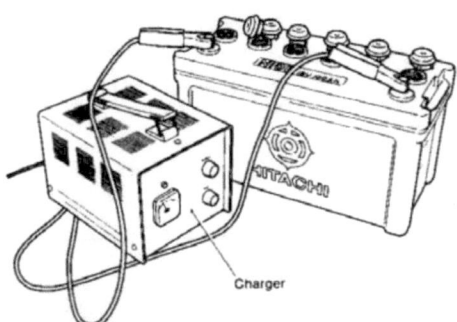
Charger

(4) Set the current to 1/10 ~ 1/5 of the capacity indicated on the outside of the battery.
(5) Periodically measure the specific gravity during charging to make sure that the specific gravity remains at a high fixed value. Also check whether gas is being generated.

2-4.3 Charging precautions

(1) Remove the battery caps to vent the gas during charging.
(2) While charging, ventilate the room and prohibit smoking, welding, etc.
(3) The electrolyte temperature should not exceed 45°C during charging.
(4) Since an alternator is used on this engine, when charging with a charger, always disconnect the battery (+) cable to prevent destruction of the diodes.
(Before disconnecting the (+) battery cable, disconnect the (−) battery cable [ground side].)

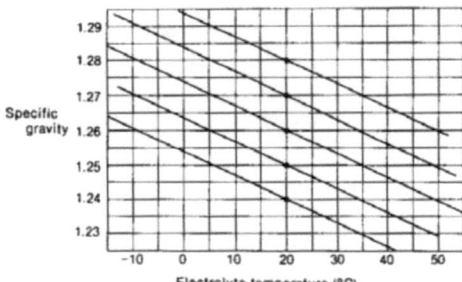
Electrolyte temperature and specific gravity

2-5 Battery storage precautions

The life of a battery depends considerably on how it is handled. Generally speaking, however, after about two years its performance will deteriorate, starting will become difficult, and the battery will not fully recover its original charge even after recharging. Then it must be replaced.

(1) Since the battery will self-discharge about 0.5%/day even when not in use, it must be charged 1 or 2 times a month when it is being stored.

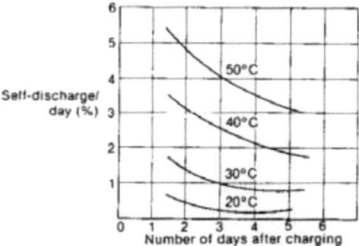

(2) If charging by the engine alternator is insufficient because of frequent starts and stops, the battery will rapidly lose power.
Charge the battery as soon as possible after it is used under these conditions.
(3) An easy-to-use battery charger that permits home charging is available from Yanmar. Take proper care of the battery by using the charger as a set with a hydrometer.
When the specific gravity has dropped to about 1.16 and the engine will not start, charge the battery up to a specific gravity of 1.26 (24 hours).
(4) Before putting the battery in storage for long periods, charge it for about 8 hours to prevent rapid aging.

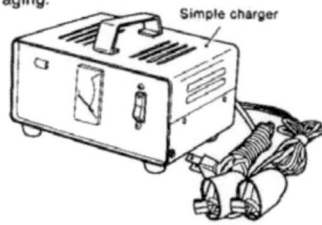
Simple charger

3. Starter Motor

The starter motor is installed on the flywheel housing. When the starting button is pushed, the starter motor pinion flies out and engages the ring gear of the flywheel. Then the main contact is closed, current flows, and the engine is started.
After the engine starts, the pinion automatically returns to its initial position when the starting button is released. Once the engine starts, the starting button should be released immediately. Otherwise, the starter motor may be damaged or burned out.

3-1 Specifications and Performance.

Engine Model		1GM10(C) 2GM20(F)(C) 3GM30(F)(C)	3HM35(F)(C)
Model		S114-303	S12-77A
Rating (sec)		30	30
Output (kW)		1.0	1.8
Direction of rotation (viewed from pinion side)		Clockwise	Clockwise
Weight kg (lb)		4.4 (9.7)	9.3 (20.5)
Clutch system		Overrunning	Overrunning
Engagement system		Magnetic shift	Magnetic shift
No. of pinion teeth		9	15
Pinion flyout voltage (V)		8 or less	8 or less
No-load	Terminal voltage (V)	12	12
	Current (A)	60 or less	90 or less
	Speed (rpm)	7000 or greater	4000 or greater
Loaded characteristics	Terminal voltage (V)	6.3	8.5
	Current (A)	460 or less	420
	Torque kgf-m(ft-lb)	0.9 (6.51) or greater	1.35 (9.76) or greater

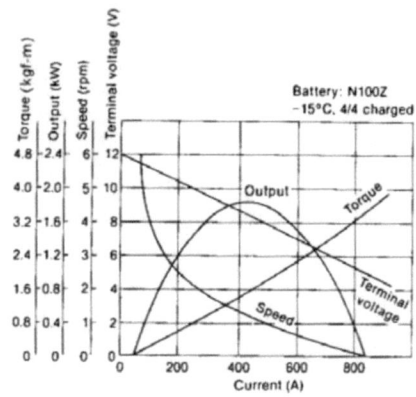

S12-77A Performance curves

3-2 Construction

The starter motor described in this section is a conventional pre-engaged 4-brush 4-pole starter motor with a screw roller drive clutch.
The starter motor is composed of three major parts, as follows:
(1) Magnetic switch
Moves plunger to engage and disengage pinion, and through the engagement lever, opens and closes main contact (moving contact) to stop the starter motor.
(2) Motor
A continuous current series motor which generates rotational drive power.
(3) Pinion
Transfers driving power from motor to ring gear. An overspeed clutch is employed to prevent damage if the engine should run too fast.

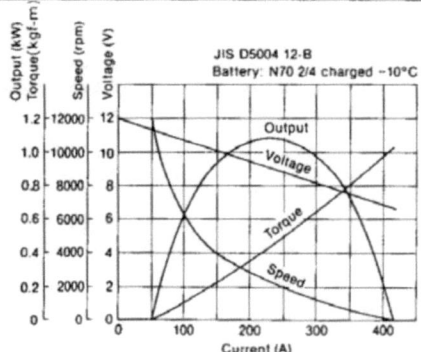

S114-303 Performance curves

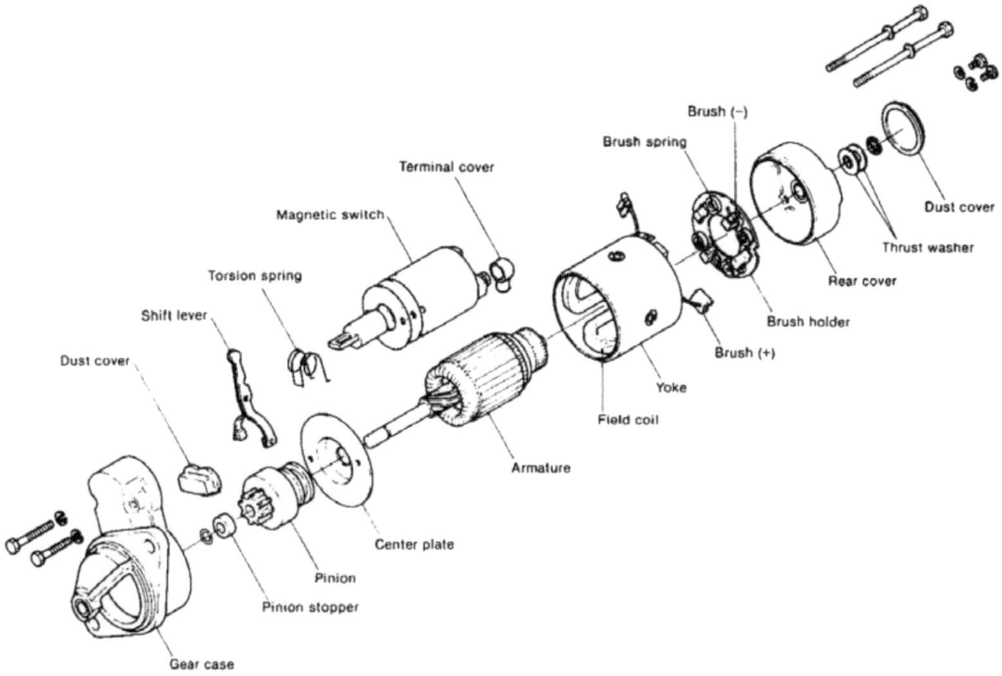

Chapter 12 Electrical System
3. Starter Motor

SM/GM(F)(C)·HM(F)(C)

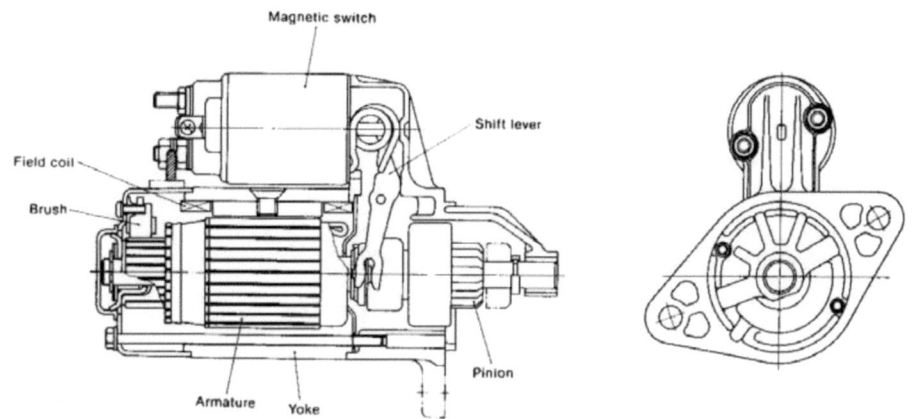

12-8

Printed in Japan
0000A0A1361

Chapter 12 Electrical System
3. Starter Motor

To prevent the motor receiving a shock which will occur as the engine starts and over-runs, this starter motor is installed with an over-running clutch.

Over-running clutch

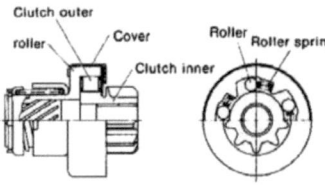

3-3 Operation

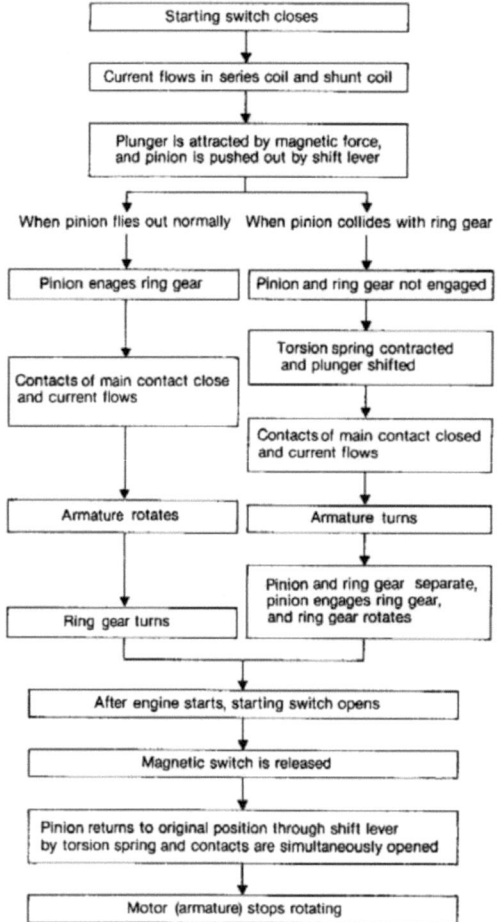

3-4 Adjustment and performance test

3-4.1 L-size measurement (gap between pinion and pinion stopper)

When the pinion is at the projected position, measure between pinion and pinion stopper. This check should be made with the pinion pressed back lightly to take up any play in the engagement linkage.

mm (in.)

Starter motor	l dimension
S114-303	0.3 ~ 2.5 (0.0118 ~ 0.0984)
S12-77A	0.2 ~ 1.5 (0.0079 ~ 0.0591)

Pressing the pinion

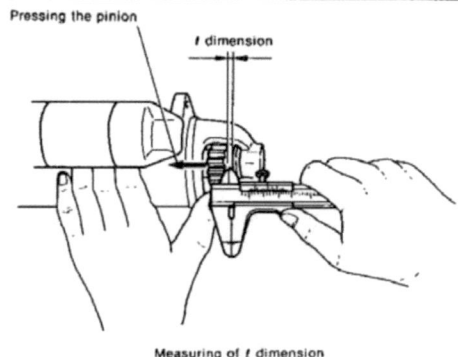

Measuring of l dimension

3-4.2 Pinion movement

After complete assembly of the starter motor, connect up the motor as in Fig.

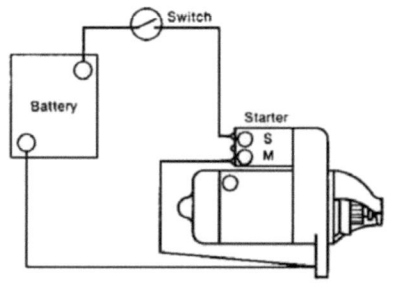

3-4.3 Plunger movement
Adjustment made by adjusting stroke of magnetic plunger to the prescribed value.
(1) Shim adjusting type (S114-303)
Adjust the l-dimension installing shim (Adjusting plate) at the magnetic switch attach section.
There are two kind of shim [Thickness 0.5mm (0.0197in.), 0.8mm (0.0315in.)]

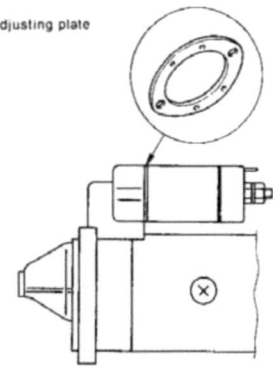

Adjusting plate

3-4.4 Pinion lock torque measurement

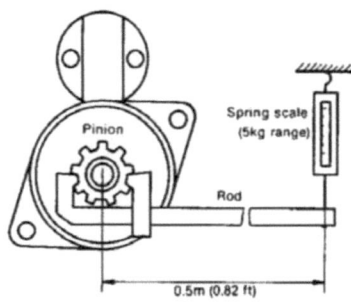

3-4.5 Mesh clearance
Mesh clearance is the distance between the flywheel ring gear and starter motor pinion in the rest position. This clearance should be between 3mm (0.1181in.) to 5mm (0.1969in.).

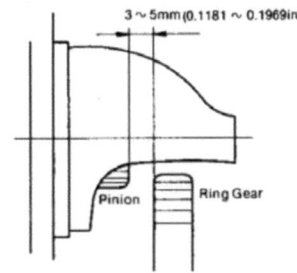

(2) Adjusting screw type (S12-77A)
Adjust the l-dimension by adjusting screw and nut.

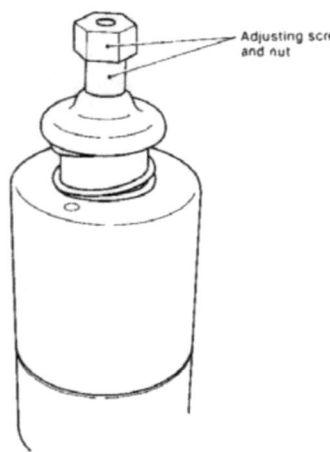

Adjusting screw and nut

3-5 Disassembly

3-5.1 Magnetic switch
(1) Disconnect magentic switch wiring.
(2) Remove through bolt mounting magnetic switch.
(3) Remove magnetic switch.

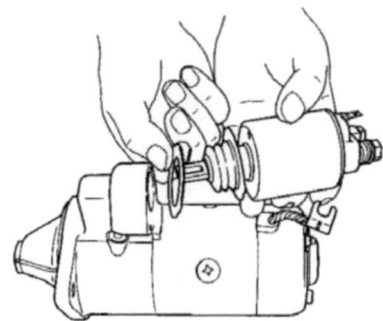

Chapter 12 Electrical System
3. Starter Motor

SM/GM(F)(C)-HM(F)(C)

3-5.2 Rear cover
(1) Remove dust cover.

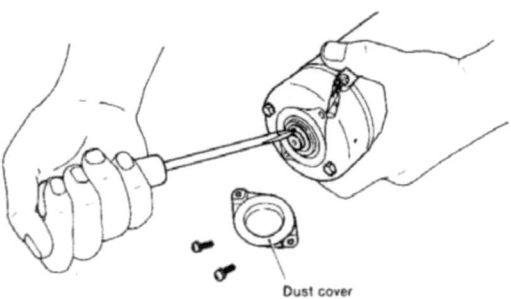
Dust cover

(2) Remove E-ring, and remove thrust washer (be careful not to lose the washer and shim).
(3) Remove the two through bolts holding the rear cover and the two screws holding the brush holder.
(4) Remove rear cover.

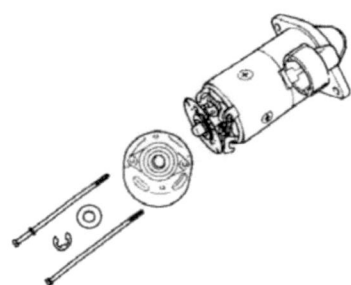

3-5.3 Brush holder
(1) Float (−)brush from the commutator.
(2) Remove (+)brush from the brush holder.
(3) Remove brush holder.

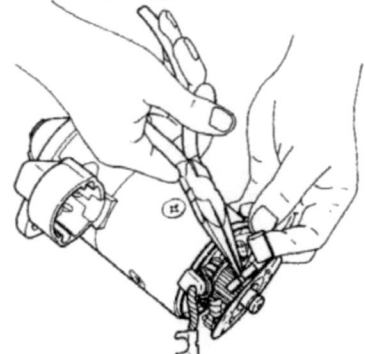

3-5.4 Yoke
(1) Remove yoke. Pull it out slowly so that it does not strike against other parts.

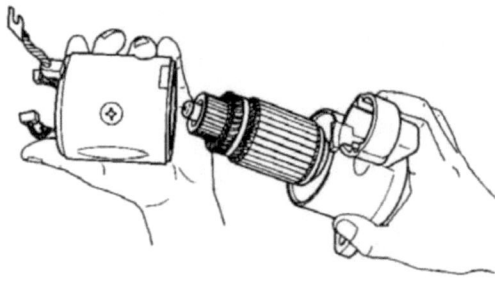

3-5.5 Armature
(1) Slide pinion stopper to pinion side.

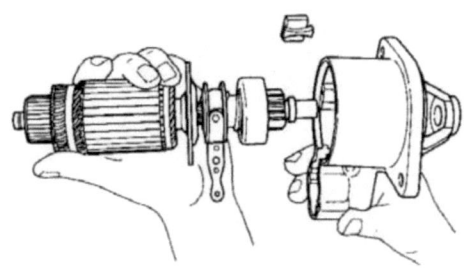

(2) Remove the pinion stopper clip.

3-5.6 Pinion
(1) Slide the pinion stopper to the pinion side.
(2) Remove the pinion stopper clip.
(3) Remove the pinion from the armature.

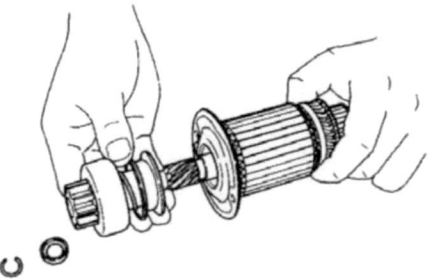

3-6 Inspection
3-6.1 Armature
(1) Commutator

Inspect the surface of the commutator. If corroded or pitted, sand with #500 ~ #600 sandpaper. If the commutator is severely pitted, grind it to within a surface roughness of at least 0.4 by turning it on a lathe. Replace the commutator if damage is irreparable.

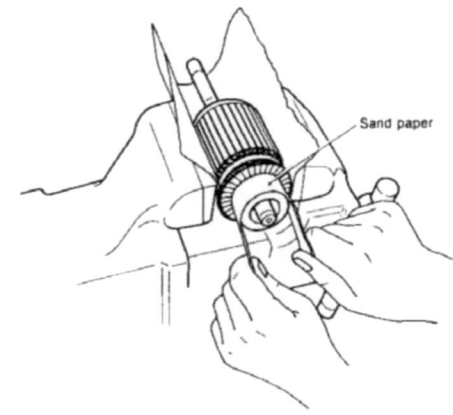

mm (in.)

	S114-303		S12-77A	
	Maintenance standard	Wear limit	Maintenance standard	Wear limit
Commutator outside diameter	ø33 (1.299)	ø32 (1.260)	ø43 (1.693)	ø40 (1.575)
Commutator run-out	Within 0.03 (0.0012)	0.2 (0.0079)	Within 0.03 (0.0012)	0.2 (0.0079)
Difference between maximum diameter and minimum diameter	Repair limit 0.4 (0.0157)	Repair accuracy 0.05 (0.002)	Repair limit 0.4 (0.0157)	Repair accuracy 0.05 (0.002)

(2) Mica undercut

Check the mica undercut, correct with a hacksaw blade when the undercut is too shallow.

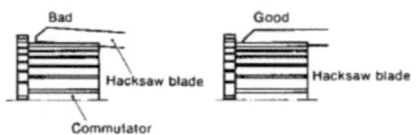

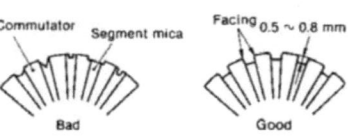

mm (in.)

	Maintenance standard	Repair limit
Mica undercut	0.2 (0.0079)	0.5 ~ 0.8 (0.0197 ~ 0.0315)

(3) Armature coil ground test

Using a tester, check for continuity between the commutator and the shaft (or armature core). Continuity indicates that these points are grounded and that the armature must be replaced.

1) Short test...existence of broken or disconnected coil.
2) Insulation test...between commutator and armature core or distortion shaft.

Checking commutator for insulation defects.

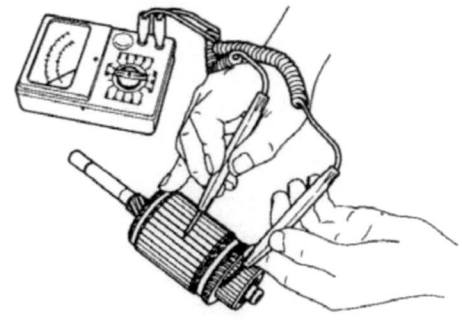

Checking armature windings for insulation faults.

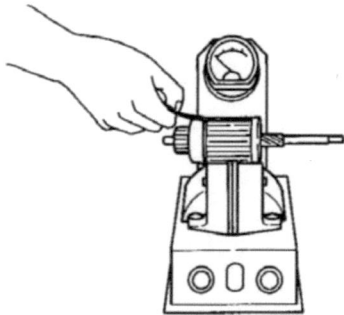

(4) Armature shaft outside diameter
Measure the outside diameter of the armature shaft at four locations: front, center, end, and pinion. Replace the armature if the shaft is excessively worn.
Check the bend of the shaft; replace the armature if the bend exceeds 0.08mm (0.0031in.)

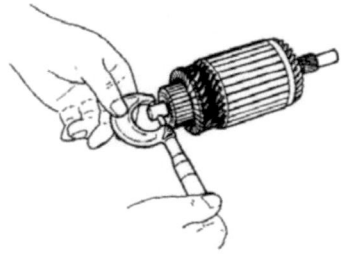

3-6.2 Field coil
(1) Open test
Check for continuity between the terminals connecting the field coil brushes. Continuity indicates that the coil is open and that the coil must be replaced.

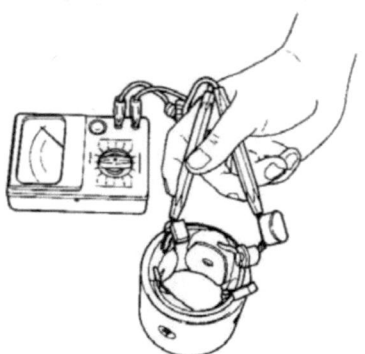

(2) Short test
Check for continuity between the yoke and any field coil terminal. Continuity indicates that the coil is shorted and that it must be replaced.
(3) Cleaning the inside of the yoke
If any carbon powder or rust has collected on the inside of the yoke, blow the yoke out with dry compressed air.
*Do not remove the field coil from the yoke.

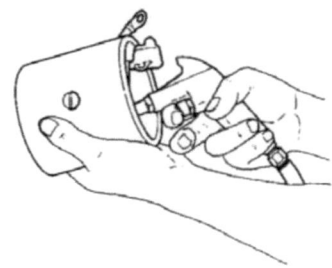

3-6.3 Brush
The brushes are quickly worn down by the motor. When the brushes are defective, the output of the motor will drop.

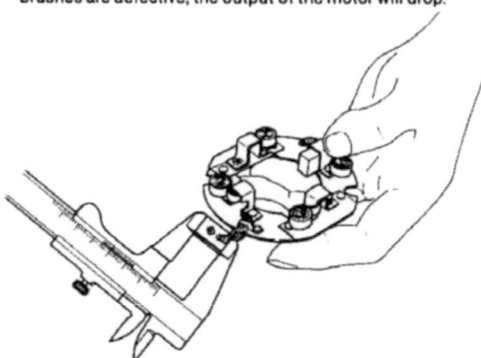

(1) Brush dimensions
Replace brushes which have been worn beyond the specified wear limit.

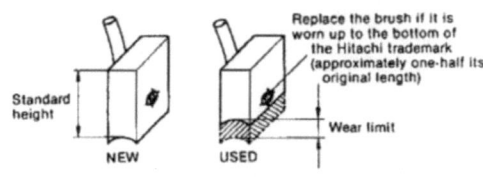

mm (in.)

	S114-303	S12-77A
Brush standard height	16 (0.6299)	22 (0.8661)
Wear limit	4 (0.1575)	8 (0.3150)

Chapter 12 Electrical System
3. Starter Motor

SM/GM(F)(C)·HM(F)(C)

(2) Brush appearance and movement in brush holder
If the outside of the brush is damaged, replace it. If the movement of the brushes in the brush holder is hampered because the holder is rusted, repair or replace the holder.

(3) Brush spring
Since the brush spring pushes the brush against the commutator while the motor is running, a weak or defective spring will cause excessive brush wear, resulting in sparking between the brush and the commutator during operation. Measure the spring force with a spring balance; replace the spring when the difference between the standard value and the measured value exceeds ±0.2kg.

3-6.4 Magnetic switch
(1) Shunt coil continuity test
Check for continuity between the S terminal and the magnetic switch body (metal part). Continuity indicates that the coil is open and that the switch must be replaced.

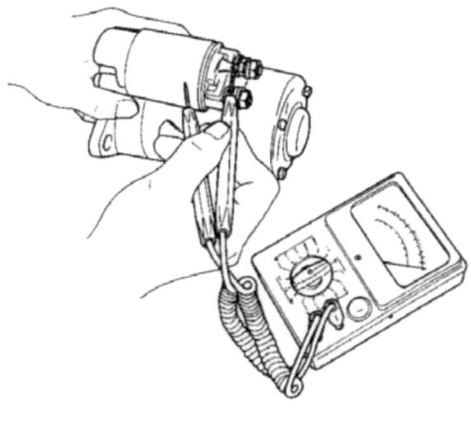

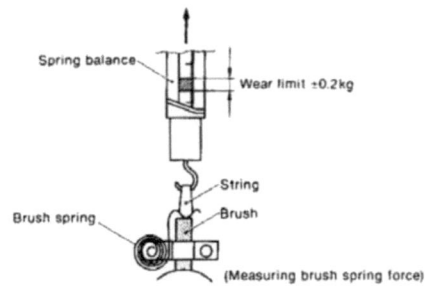

(Measuring brush spring force)

	S114-303	S12-77A
Standard spring load	1.6kg (3.527 lb)	0.85kg (1.8737 lb)

	S114-303	S12-77A
Coil resistance (at 20°C)	0.694Ω	0.590Ω

(2) Series coil continuity test
Check for continuity between the S terminal and M terminal. Continuity indicates that the coil is open and that it must be replaced.

(4) Brush holder ground test
Check for continuity between the insulated brush holder and the base of the brush holder assembly. Continuity indicates that these two points are grounded and that the holder must be replaced.

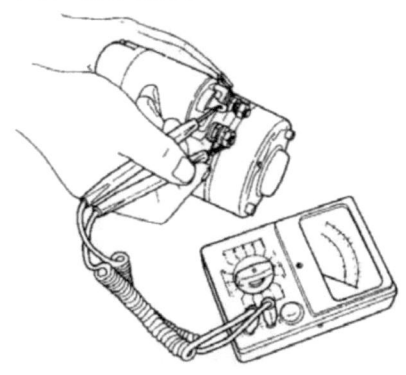

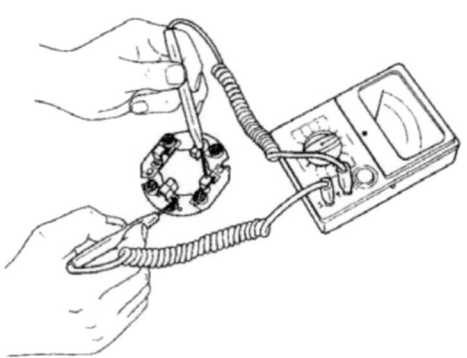

	S114-303	S12-77A
Resistance value (at 20°C)	0.324Ω	0.267Ω

Chapter 12 Electrical System
3. Starter Motor

SM/GM(F)(C)·HM(F)(C)

(3) Contactor contact test
Push the plunger with your finger and check for continuity between the M terminal and B terminal. Continuity indicates that the contact is faulty and that the contactor must be replaced.

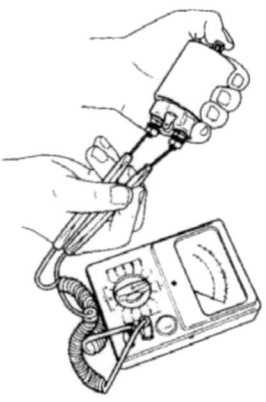

(2) Mounting the magentic switch
Attach the shift lever to the pinion; assemble the gear case as shown below.
Do not forget to install the dust cover before assembling the gear case.
After reassembly, check by conducting no-load operation.

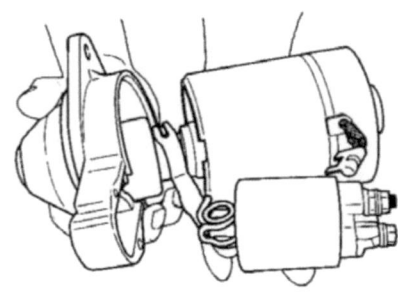

(3) Lubrication
Lubricate each bearing and spline (points indicated in the figure below) with high quality "Hitachi Electrical Equipment Grease A".
The following lubricants may be used in place of Hitachi Electrical Equipment Grease A.

3-6.5 Pinion
(1) Inspect the pinion teeth and replace the pinion if the teeth are excessively worn or damaged.
(2) Check if the pinion slides smoothly; replace the pinion if faulty.
(3) Inspect the springs and replace if faulty.
(4) Replace the clutch if it slips or seizes.

Magnetic switch plunger	Shell	Aeroshell No. 7
Bearing and spline	Shell	Albania Grease No. 2

3-7 Reassembly precautions
Reassemble the starter motor in the reverse order of disassembly, paying particular attention to the following:
(1) Torsion spring and shift lever
Hook the torsion spring into the hole in the magnetic switch and insert the shift lever into the notch in the plunger of the magnetic switch through the torsion spring.

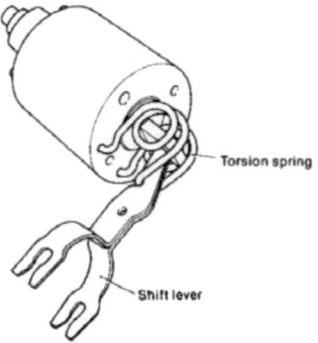
Torsion spring
Shift lever

3-8 Testing

3-8.1 No load test

Test procedure

(1) Connect the positive side of the ammeter (A) to the positive terminal of the battery, and connect the negative side of the ammeter to the B terminal of the starter.

(2) Connect the negative terminal of the battery to the body of the starter.
(3) Connect the positive side of the voltmeter (V) to the B terminal of the starter, and connect the negative side of the voltmeter to the body of the starter.
(4) Attach the tachometer.
(5) Connect the B terminal of the starter to the S terminal of the magnetic switch.

- The magnetic switch should begin operating, and the speed, current, and voltage should be the prescribed values.
- A fully charged battery must be used.
- Since a large current flows when the starter is operated, close the protection circuit switch before initial operation, then open the switch and measure the current after the starter reaches a constant speed.

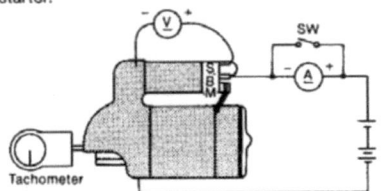

Tachometer

3-9 Maintenance standard

				S114-303	S12-77A
Brush	Standard spring load		kg (lb)	1.6 (3.527)	0.85 (1.8737)
	Standard height		mm (in.)	16 (0.6299)	22 (0.8661)
	Wear limit		mm (in.)	12 (0.472)	8 (0.3150)
Magnetic switch	Series coil resistance		Ω	0.324	0.267
	Shunt coil resistance		Ω	0.694	0.590
Commutator	Outside diameter	Maintenance standard	mm (in.)	⌀33 (1.299)	⌀43 (1.193)
		Wear limit	mm (in.)	⌀32 (1.260)	⌀40 (1.575)
	Difference between maximum diameter and minimum diameter	Repair limit	mm (in.)	0.4 (0.0157)	
		Repair accuracy	mm (in.)	0.05 (0.002)	
	Mica undercut	Maintenance standard	mm (in.)	0.2 (0.0079)	
		Repair limit	mm (in.)	0.5 ~ 0.8 (0.0197 ~ 0.0315)	
Standard dimension	Rear side bearing	Shaft diameter	mm (in.)	12.450 ~ 12.468 (0.4902 ~ 0.4909)	14.950 ~ 14.968 (0.5886 ~ 0.5893)
		Bearing inside diameter	mm (in.)	12.500 ~ 12.527 (0.4921 ~ 0.4932)	15.000 ~ 15.018 (0.5906 ~ 0.5913)
	Intermediate bearing	Shaft diameter	mm (in.)	—	20.250 ~ 20.268 (0.7972 ~ 0.7980)
		Bearing inside diameter	mm (in.)	—	20.500 ~ 20.518 (0.8071 ~ 0.8080)
	Pinion sliding section	Shaft diameter	mm (in.)	12.450 ~ 12.468 (0.4902 ~ 0.4909)	13.950 ~ 13.968 (0.5492 ~ 0.5499)
		Pinion inside diameter	mm (in.)	12.530 ~ 12.550 (0.4933 ~ 0.4941)	14.030 ~ 14.050 (0.5524 ~ 0.5531)
	Pinion side bearing	Shaft diameter	mm (in.)	12.450 ~ 12.468 (0.4902 ~ 0.4909)	13.950 ~ 13.968 (0.5492 ~ 0.5499)
		Bearing inside diameter	mm (in.)	12.500 ~ 12.527 (0.4921 ~ 0.4932)	14.000 ~ 14.018 (0.5512 ~ 0.5519)

Chapter 12 Electrical System
3. Starter Motor

3-10 Various problems and their remedies

(1) Pinion fails to advance when the starting switch is closed

Problem	Cause	Corrective action
Wiring	Open or loose battery or switch terminal	Repair or retighten
Starting switch	Threaded part connected to pinion section of armature shaft is damaged, and the pinion does not move	Repair contacts, or replace switch
Starter motor	Threaded part connected to pinion section of armature shaft is damaged, and the pinion does not move	Replace
Magnetic switch	Plunger of magnetic switch malfunctioning or coil shorted	Repair or replace

(2) Pinion is engaged and motor rotates, but rotation is not transmitted to the engine

Problem	Cause	Corrective action
Starting motor	Overrunning clutch faulty	Replace

(3) Motor rotates at full power before pinion engages ring gear

Problem	Cause	Corrective action
Starter motor	Torsion spring permanently strained	Replace

(4) Pinion engages ring gear, but starter motor fails to rotate

Problem	Cause	Corrective action
Wiring	Wires connecting battery and magnetic switch open or wire connecting ground, magnetic switch and motor terminals loose	Repair, retighten, or replace wire
Starter motor	Pinion and ring gear engagement faulty Motor mounting faulty Brush worn or contacting brush spring faulty Commutator dirty Armature, field coil faulty Field coil and brush connection loose	Replace Remount Replace Repair Repair or replace Retighten
Magnetic switch	Contactor contact faulty Contactor contacts pitted	Replace Replace

(5) Motor fails to stop when starting switch is opened after engine starts

Problem	Cause	Corrective action
Starting switch	Switch faulty	Replace
Magnetic switch	Switch faulty	Replace

4. Alternator Standard, 12V/55A

The alternator serves to keep the battery constantly charged. It is installed on the cylinder block by a bracket, and is driven from the V-pulley at the end of the crankshaft by a V-belt.
The type of alternator used in this engine is ideal for high speed engines with a wide range of engine speeds. It contains diodes that convert AC to DC, and an IC regulator that keeps the generated voltage constant even when the engine speed changes.

4-1 Features

The alternator contains a regulator using an IC, and has the following features.
(1) The IC regulator is self-contained, and has no moving parts (mechanical contact point). It therefore has superior features such as freedom from vibration, no fluctuation of voltage during use, and no need for readjustment.
Also, it is of the over-heating compensation type and can automatically adjust the voltage to the most suitable level depending on the operating temperature.
(2) The regulator is integrated within the alternator to simplify external wiring.
(3) The alternator is designed for compactness, lightness of weight, and high output.
(4) A newly developed U-shaped diode is used to provide increased reliability and easier checking and maintenance.
(5) As the alternator is to be installed on board, the following measures are taken to provide salt-proofing.
 1) The front and rear covers are salt-proofed.
 2) Salt-proof paint is applied to the diode.
 3) The terminal, where the inboard harness is connected to the alternator, is nickel plated.

4-2 Specifications

Model of alternator	LR155-20 (HITACHI)
Model of IC regulator	TRIZ-63 (HITACHI)
Battery voltage	12V
Nominal output	12V/55A
Earth polarity	Negative earth (⊖)
Direction of rotation (viewed from pulley end)	Clockwise
Weight	4.3kg (9.5lb.)
Rated speed	5000 rpm
Operating speed	1000 ~ 9000
Speed for 13.5V	1000 or less
Output current at 20°C	over 53A/5000 rpm
Regulated voltage	14.5 ±0.3V (Standard temperature voltage gradient, −0.01/°C)

4-3 Characteristics

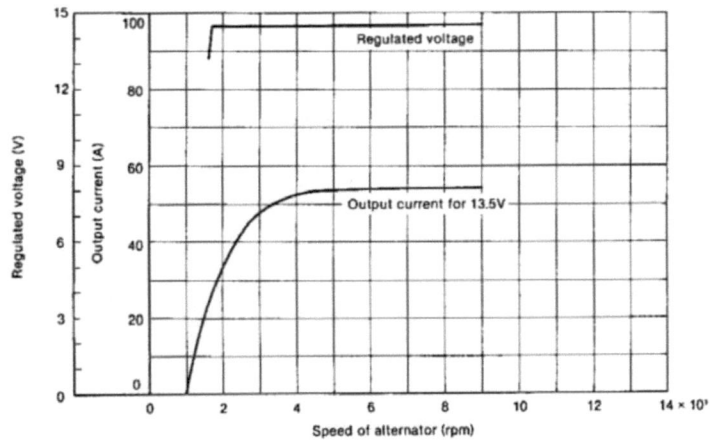

Speed of alternator (rpm)

Chapter 12 Electrical System
4. Alternator Standard, 12V/55A

SM/GM(F)(C)-HM(F)(C)

4-4 Construction

This is a standard rotating field type three-phase alternator. It consists of six major parts: the pulley, fan, front cover, rotor, stator and rear cover. The IC regulator is an integral part of the alternator.

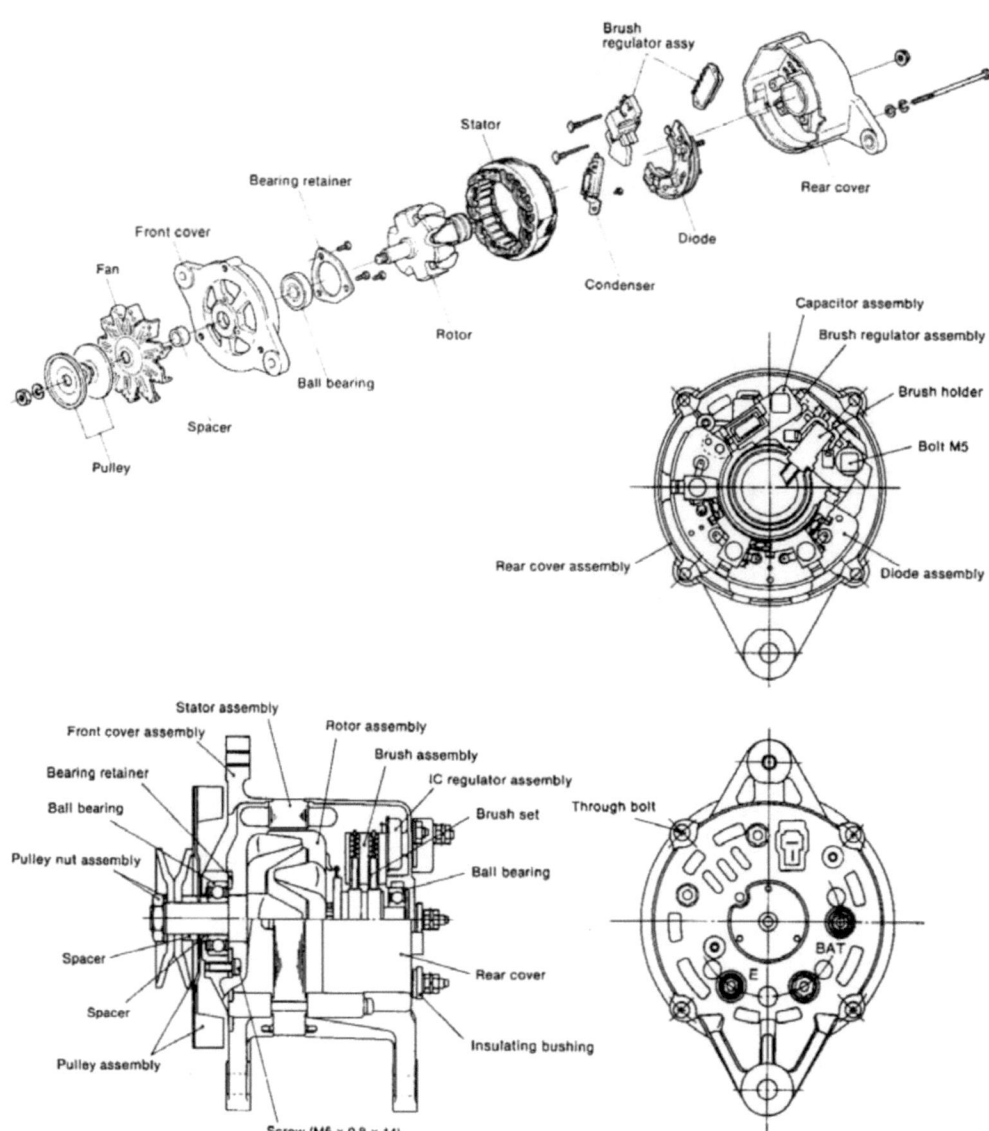

4-5 Alternator functioning

(1) IC regulator
The IC regulator is the transistor (Tr_2) which is series-connected with the rotor. The IC regulator controls the output voltage of the generator by breaking or conducting the rotor coil (exciting) current.
When the output voltage of the generator is within the standard value, the transistor (Tr_2) turns on. When the voltage exceeds the standard value, the Zener diode goes on and the transistor (Tr_2) turns off.
With the repeated turning on and off of the transistor, the output voltage is kept at the standard value. (Refer to the circuit diagram below.)

(2) Charge lamp
When the transistor (Tr_2) is on, the charge lamp key switch is turned to ON, and current flows to R_1, R_4 and to Tr_1 to light the lamp. When the engine starts to run and output voltage is generated in the stator coil, the current stops flowing to this circuit, turning off the charge lamp.

(3) Circuit diagram

4-6 Handling precautions

(1) Be careful of the battery's polarity (+, − terminals), and do not connect the wrong terminals to the wrong cables or the battery will be short-circuited by the generator diode.
In this case too much current will flow, the IC regulator and diodes burn out, and the wire harness will burn.
(2) Make sure of the correct connection of each terminal.
(3) When quick-charging, etc., disconnect either the battery terminal on the AC generator or the terminal on the battery.
(4) Do not short-circuit the terminals.
(5) Do not conduct any tests using high tension insulation resistance. (The diodes and IC regulator will burn out.)

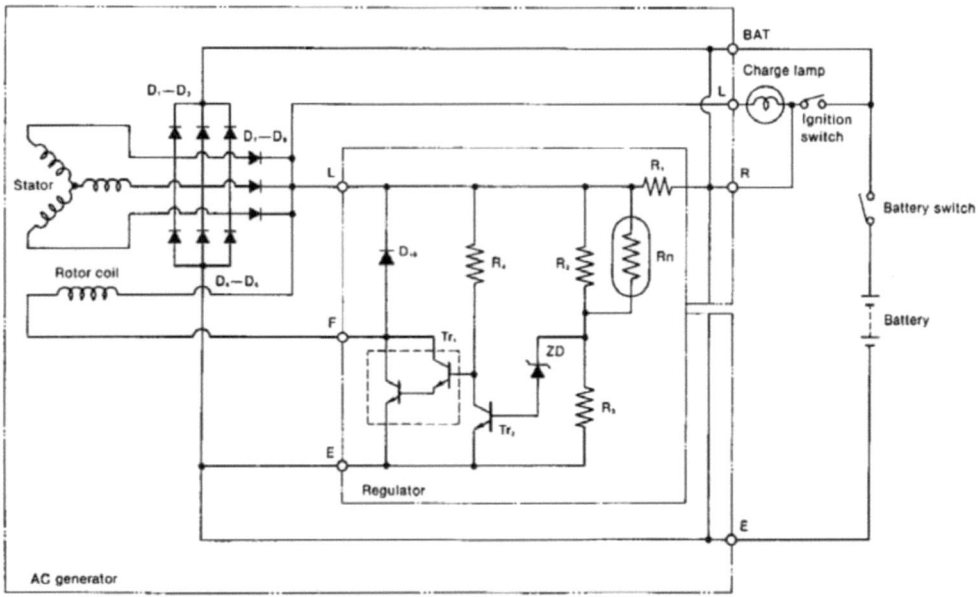

BAT:	Generator output terminal	$D_1 - D_6$:	Output commutation diode
D_{10}:	IC protecting diode	$R_1 - R_5$:	Resistor
L:	Charge lamp terminal	$D_7 - D_9$:	Charging lamp switching diode
ZD:	Zener diode	F:	To supply current to rotor coil
E:	Earth	Rn:	Thermistor
Tr_1, Tr_2:	Transistor		(Temperature gradient resistance)

4-7 Disassembling the alternator

(1) Remove the through-bolt, and separate the front assembly from the rear assembly.

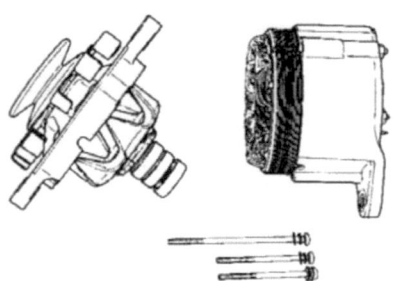

(2) Remove the pulley nut, and pull out the rotor from the front cover.

(3) Remove the ø5mm screw from the front cover, and then remove the ball bearing.

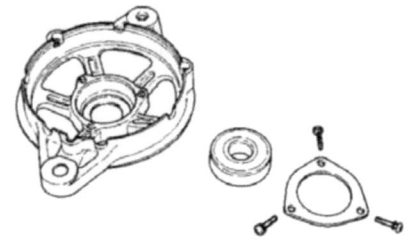

(4) Remove the nut, the brush-holder, and diode fixing nut at the BAT, and the terminal screws of the rear cover. Separate the rear cover from the stator (with the diode and brush holder).

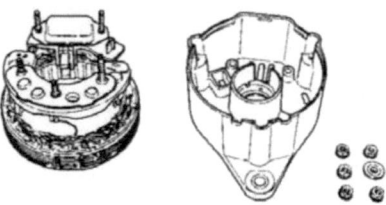

(5) Disconnect the soldered joint of the stator lead wire, and remove the diode and brush regulator assemblies from the stator at the same time.

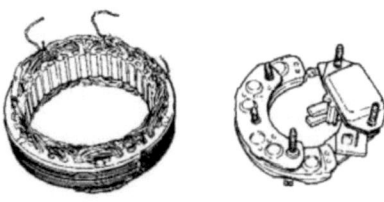

(6) Separating the regulator
1) To separate the regulator, remove the ø3.0mm rivet which keeps the diode assembly and the brushless regulator in place, and the soldered joint of the L-terminal.

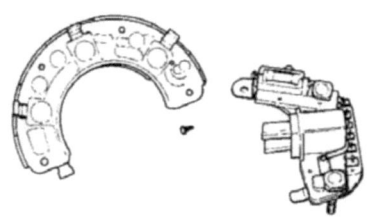

Chapter 12 Electrical System
4. Alternator Standard, 12V/55A

2) To replace the IC regulator, disconnect the soldered joint of the IC regulator and pull out the two bolts. Do not remote these two bolts except when replacing the IC regulator.

After repeating the above test, if any diode is found to be defective, replace the diode assembly. Since there is no terminal on the auxiliary diode, check the continuity between both ends of the diode.

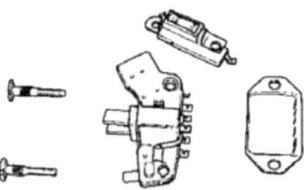

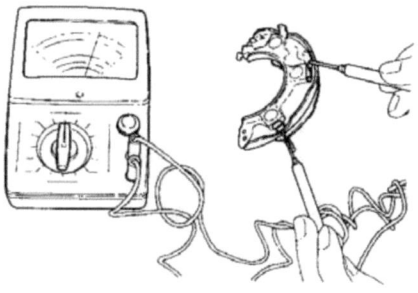

4-8 Inspection and adjustment

(1) Diode

Between terminals		BAT (+ side diode)	
U.V.W.	Tester wire	+ side	− side
	+ side		No continuity
	− side	Continuity	

Between terminals		E (− side diode)	
U.V.W.	Tester wire	+ side	− side
	+ side		Continuity
	− side	No continuity	

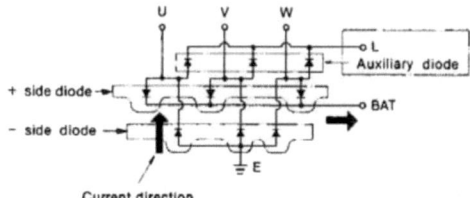

+ side diode
− side diode
U
V
W
L
Auxiliary diode
BAT
E
Current direction

U.V.W.: terminal from the stator coil

Current flows only in one direction in the diode as shown in Fig. 181. Accordingly, when there is continuity between each terminal (e.g. BAT and U), the diode is in normal condition (photo). When there is no continuity, the diode is defective.
When the tester is connected in the reverse of above, there should be no continuity. If there is, the diode is defective.

CAUTION: Do not use high tensile insulation resistance such as meggers, etc. for testing. Otherwise, the diode may burn out.

(2) Rotor
Inspect the slip ring surface, rotor coil continuity and insulation.

1) Inspecting the slip ring surface
Check if the surface of the slip ring is sufficiently smooth. If the surface is rough, grind the surface with No. 500—600 sand paper. If it is contaminated with oil, etc., wipe the surface clean with alcohol.

	Standard	Wear limit
Slip ring outer dia.	ø31.6mm (1.2441in.)	ø30.6mm (1.2049in.)

2) Rotor coil continuity test
Check the continuity in the slip ring with the tester. If there is no continuity, there is a wire break. Replace the rotor coil.

Resistance value	Approx. 3.34Ω at 20°C

Chapter 12 Electrical System
4. Alternator Standard, 12V/55A

3) Rotor coil insulation test
Check the continuity between the slip ring and the rotor core, or the shaft. If there is continuity, insulation inside the rotor is defective, causing a short-circuit with the earth circuit. Replace the rotor coil.

4) Check the rear side ball bearing. If the rotation of the bearing is heavy, or produces abnormal sounds, replace the ball bearing.

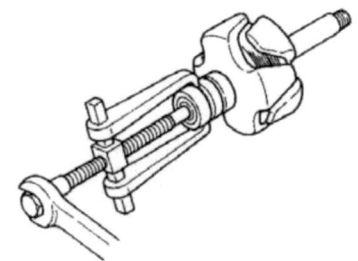

(3) Stator
1) Stator coil continuity test
Check the continuity between each terminal of the stator coil. If there is no continuity, there is a wire break in the stator coil. Replace the stator coil.

Resistance value	Approx. 0.077Ω at 20°C 1-phase resistance

2) Stator coil insulation test
Check the continuity between the terminals and the stator core. If there is continuity, insulation of the stator coil is defective. This will cause a short-circuit with the earth core. Replace the stator coil.

(4) Brush
The brush is hard and wears slowly, but when it is worn beyond the allowable limit, replace it. When replacing the brush also check the strength of the brush spring. To check, push the spring down to 2mm from the end surface of the brush holder, and read the gauge.

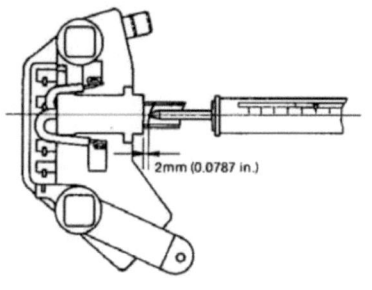
2mm (0.0787 in.)

Brush spring strength	255–345g (0.56 ~ 0.76lb.)

(5) Brush wear
Check the brush length.
The brush wears very little, but replace the brush if worn over the wear limit line printed on the brush.

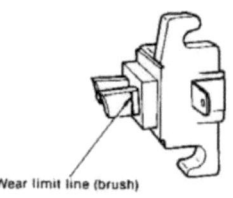
Wear limit line (brush)

mm (in.)

	Maintenance standard	Wear limit
Brush length	16 (0.6299)	9 (0.3543)

Chapter 12 Electrical System
4. Alternator Standard, 12V/55A

(6) IC regulator

Connect the variable resistance, two 12V batteries, resistor, and voltmeter as shown in the diagram.

1) Use the following measuring devices.

Resistor (R_1)	100Ω, 2W, 1pc.
Variable resistor (Rv)	0—300Ω, 12W, 1pc.
Battery (BAT_1, BAT_2)	12V, 2pcs.
DC voltmeter	0—30V, 0.5 class 1pc. (to measure at 3 points)

2) Check the regulator in the following sequence, according to the diagram.

   a) Check V_3 (BAT_1 + BAT_2 voltage). If the voltage is 20—26V, both BAT_1 and BAT_2 are normal.

   b) While measuring V_2 (F-E terminal voltage), move Rv gradually from the 0-position. Check if there is a point where the V_2 voltage rises sharply from below 2.0V to over 2.0V. If there is no such point, the regulator is defective. Replace the regulator. If there is a sharp voltage rise when testing, return the Rv to the 0-position, and connect the voltmeter to the V_1 position.

   c) While measuring V_1 (voltage between L·E terminals), move Rv gradually from the 0-position. There should be a point where the voltage of V_1 rises sharply by 2—6V. Measure the voltage of V_1 just before this sharp voltage rise. This is the regulating voltage of the regulator. If this voltage of V_1 is within the standard limit, the regulator is normal. If the voltage deviates from the limit, the regulator is defective. Replace the regulator.

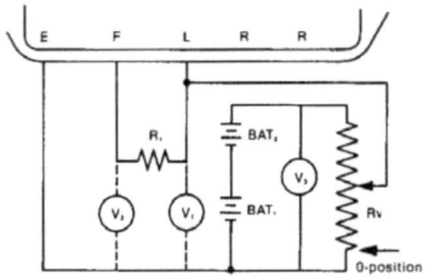

4-9 Reassembling the alternator

Reassembly is done in the reverse order of disassembly. For reassembly, be careful of the following points. (Refer to 4—7 disassembling alternator).

(1) Assembling the brush regulator

1) Solder the brushes.

   Position the brush as shown in the drawing and solder it. Be careful not to let the solder drip into the pig tail (lead wire).

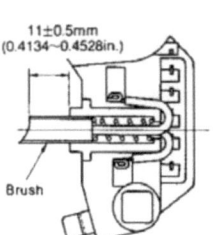

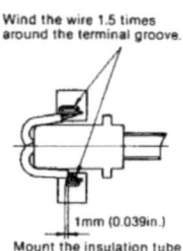

11±0.5mm (0.4134~0.4528in.)

Brush

Wind the wire 1.5 times around the terminal groove.

1mm (0.039in.)

Mount the insulation tube on the terminal surface.

NOTES: 1. Use non-acid type paste.
2. The soldering iron temperature is 300 ~ 350°C.

2) Mount the IC regulator on the brush holder as illustrated, and press in the M5 bolt. Do not forget to assemble the bushing and the connecting plate at the same time.
(If the bushing is left out, the output terminal will be earthed and the battery short-circuited).

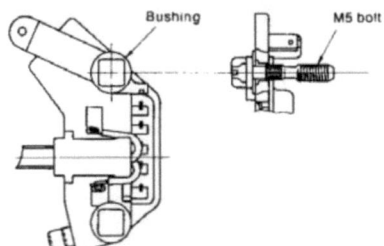

Bushing   M5 bolt

NOTES: 1. Insertion pressure is 100kg (220.5 lbs.)
2. Insert vertically.

(2) Connecting the brush regulator assembly and diode

1) Check the rivets

   Place the rivets as shown in the figure, and then calk them using the calking tool.

Calking torque	500kg (1102 lbs.)

2) Connect the brush to the diode.

   Insert the brush side terminal into the diode terminal, calk it, and then solder into place.

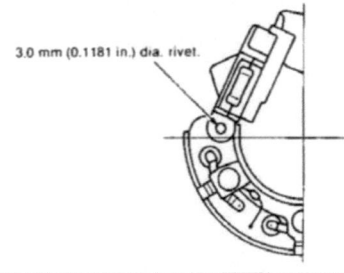

3.0 mm (0.1181 in.) dia. rivet.

Rivetting pressure	500kg (1102 lbs.)

Chapter 12 Electrical System
4. Alternator Standard, 12V/55A

(3) Assembling the rear cover

Insert pins from the outside of the rear cover. Install the brush on the brush holder, then attach the rear cover. After assembly, pull out the pins.

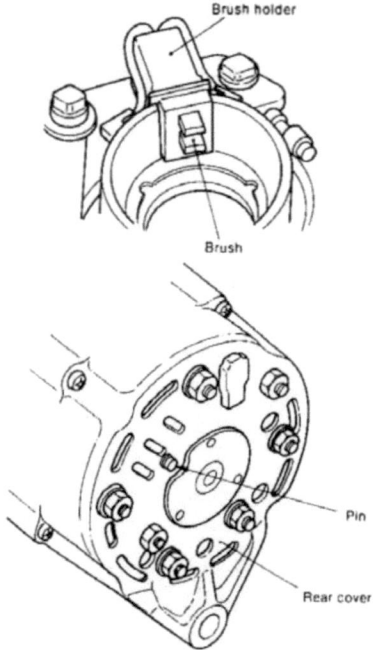

(1) Measuring devices

DC voltmeter	0—15V or 0—30V, 0.5 Class, 1pc.
DC ammeter	0—100A, 1.0 Class, 1pc.
Variable resistor	0—0.25Ω, 1kW, 1pc.
Lamp	12V, 3W
100Ω resistor	3W
0.25Ω resistor	25W

(2) Measuring the regulating voltage
1) When measuring devices are connected in the performance test circuit as shown above, the charge lamp lights.
2) Close SW_2 while keeping SW_1 open and run the AC generator. When the revolutions of the generator are gradually raised, the charge lamp goes off.
3) Raise the revolutions of the AC generator, and read the voltmeter gauge when the revolutions reach about 5,000 rpms.

NOTES: 1. Make sure that the ammeter indication at this time is less than 5A. If the indication is over 5A, connect the 0.25Ω resistor. The voltmeter indication at this time must be within the prescribed regulating voltage value.
2. Raise the AC generator revolutions high to make sure the regulating voltage does not fluctuate along with changes in the revolution speed.

(3) Precautions for measuring the regulating voltage
1) When measuring the voltage, measure the voltage between the AC generator BAT terminal, or Battery + terminal, and AC generator E-terminal.
2) Use a fully charged battery.
3) Measure the voltage quickly.
4) Keep SW_1 open for measurement.

(4) Tightening torques

Positions	Tightening torque kgf-cm(ft-lb)
Brush holder fixing	32—40 (2.31 ~ 2.89)
Diode fixing	32—40 (2.31 ~ 2.89)
Bearing retainer fixing	32—40 (2.31 ~ 2.89)
Pulley nut tightening	400—600 (28.93 ~ 43.40)
Through-bolt tightening	32—40 (2.31 ~ 2.89)

4-10 Performance test

Conduct a performance test on the reassembled AC generator as follows. The following is the circuit for the performance test.

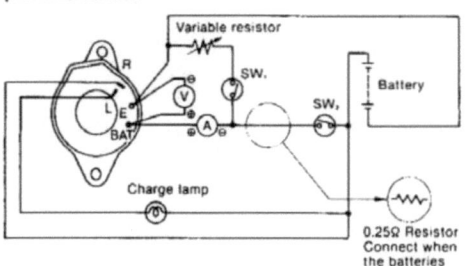

Chapter 12 Electrical System
4. Alternator Standard, 12V/55A

SM/GM(F)(C)-HM(F)(C)

4-11 Troubleshooting

(1) Charging failure

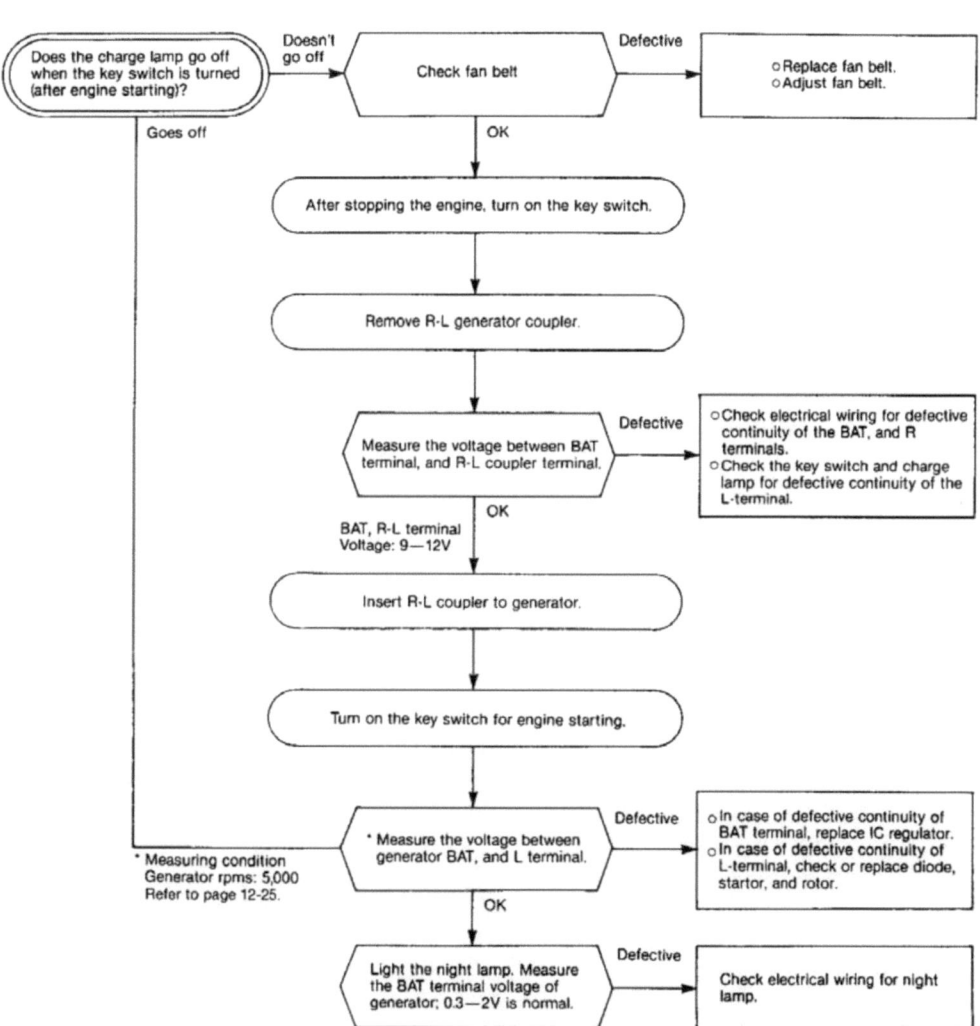

(2) Overcharging

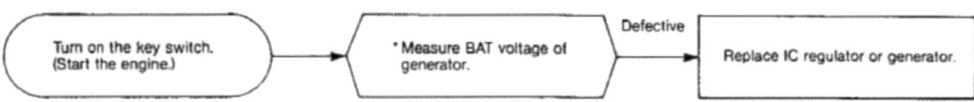

(3) Charge lamp failure

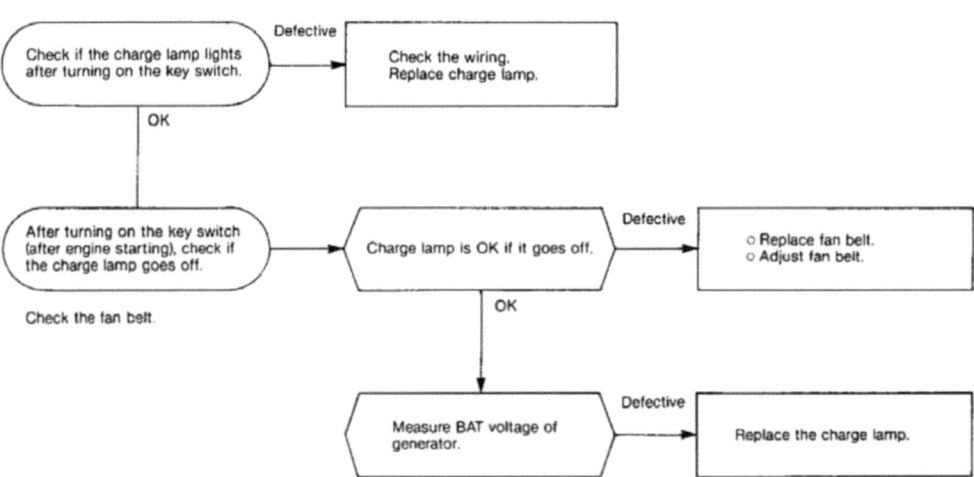

4A. Alternator, Option, 12V/35A [Except 1GM10(C)]

The alternator serves to keep the battery constantly charged. It is installed on the cylinder block by a bracket, and is driven from the V-pulley at the end of the crankshaft by a V-belt.

The type of alternator used in this engine is ideal for high speed engines having a wide range of engine speeds. It contains diodes that convert AC to DC, and an IC regulator that keep the generated voltage constant even when the engine speed changes.

4A-1. Features

The alternator contains a regulator using an IC, and has the following features.

(1) The IC regulator, which is self-contained, has no moving part (mechanical contact point), therefore it has superior features such as, freedom from vibration, no fluctuation of voltage during use, and no need for readjustment.
Also, it is of the over-heating compensating type and can automatically adjust the voltage to the most suitable level depending on the operating temperature.
(2) The regulator is integrated within the alternator to simplify external wiring.
(3) It is an alternator designed for compactness, light weight, and high output.
(4) A newly developed U-shaped diode is used to provide increased reliability and easier checking and maintenance.
(5) As the alternator is to be installed on board, the following countermeasures are taken to provide salt-proofing.
 1) The front and rear covers are salt-proofed.
 2) Salt-proof paint is applied to the diode.
 3) The terminal, where the harness inboard is connected to the alternator, is nickel plated.

4A-3. Characteristics

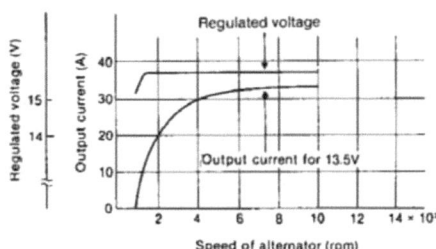

Speed of alternator (rpm)

4A-2. Specifications

Model of alternator	LR135-105 (HITACHI)
Model of IC regulator	TR1Z-63 (HITACHI)
Battery voltage	12V
Nominal output	12V, 35A
Earth polarity	Negative earth
Direction of rotation (viewed from pulley end)	Clockwise
Weight	3.5 kg (7.7 lb)
Rated speed	5000 rpm
Operating speed	900 ~ 8000 rpm
Speed for 13.5V	900 rpm or less
Output current (when heated)	5000 rpm 32±2A
Regulated voltage	14.5±0.3V (at 20°C, Full battery)
Standard temperature/ voltage gradient	–0.01V/°C

Chapter 12 Electrical System
4A. Alternator Option, 12V/35A

4A-4. Construction

This is a standard rotating field type three-phase alternator. It consists of six major parts: the pulley, fan, front cover, rotor, stator and rear cover. The IC regulator is an integral part of the alternator.

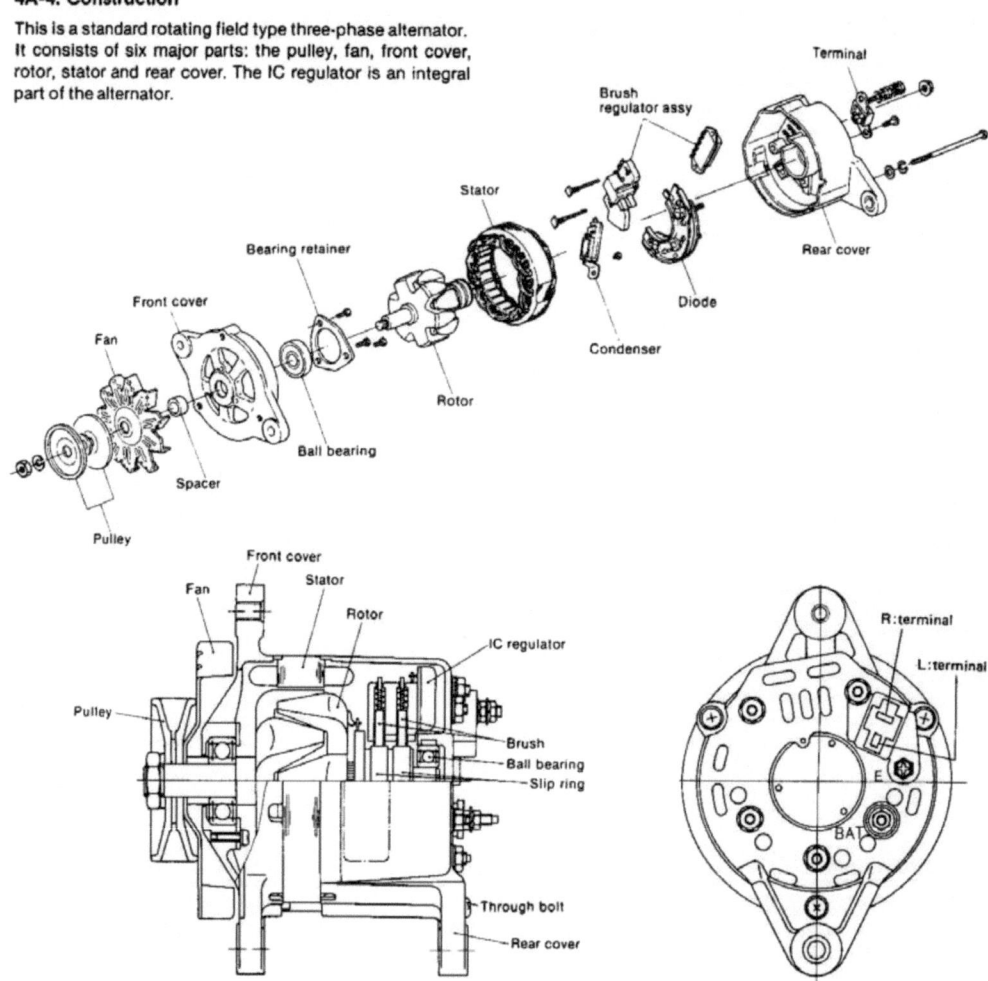

4A-5. Wiring

(1) Wiring diagram

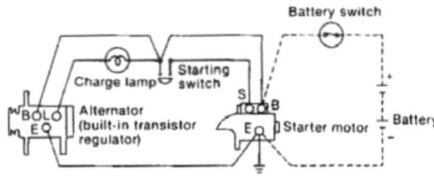

(2) Terminal connections

The alternator has the following terminals. Connect these terminals as indicated below.

Symbol	Terminal name	Connection to external wiring
B	Battery terminal	To battery (+) side
E	Ground terminal	To battery (−) side
L	Lamp (charge) terminal	To charge lamp terminal

4A-6. Circuit diagram
4A-6.1 Circuit diagram

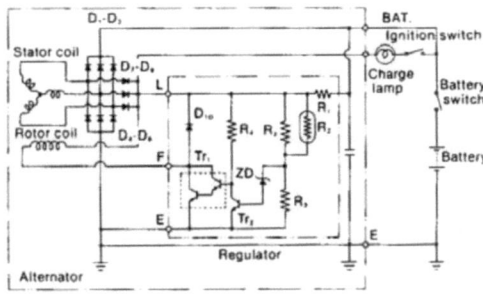

BAT: Battery output terminal
L: Charge lamp terminal
E: Earth
$D_1 \sim D_6$: Diodes for rectifying the output current
$D_7 \sim D_9$: Diodes for switching the charge lamp
D_{10}: Diode for protecting the IC
ZD: Zener diode
Tr_1, Tr_2: Transistors
$R_1 \sim R_5$: Resistors
F: Rotor current
Rn: Thermistor (resistors with current/temperature gradient)

4A-6.2 Principle of IC regulator function

The IC regulator controls the output voltage of the alternator by switching the rotor current (exciting current) on or off by means of the transistor Tr_1 which is connected in series with the rotor coil.
When the output voltage of the alternator is within the regulated values, transistor Tr_1 is "ON" but when the voltage is outside the regulated value, the Zener diode ZD comes "ON", and regulates the output voltage rise by turning transistor Tr_1 "OFF".
The output voltage is kept within the regulated values by repeating the "ON"–"OFF" operation.

4A-7. Alternator handling precautions

(1) Pay attention to the polarity of the battery; be careful not to connect it in reverse polarity. If the battery is connected in reverse polarity, the battery will be shorted by the diode of the alternator, an overcurrent will result, the diodes and transistor regulator will be destroyed, and the wiring harness will be burned.
(2) Connect the terminals correctly.
(3) When charging the battery from outside, such as during rapid charging, disconnect the alternator B terminal or the battery terminals.
(4) Do not short the terminals.
(5) Never test the alternator with a high voltage meter.

4A-8. Alternator disassembly

Disassemble the alternator as follows.
The major points of disassembly are the removal of the cover, the separation of the front and rear sides, and detailed disassembly.
(1) Remove the cover attached to the rear cover, remove the through bolts, and disassemble into front and rear sides.

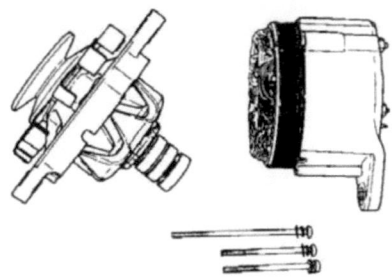

Chapter 12 Electrical System
4A. Alternator Option, 12V/35A

SM/GM(F)(C)·HM(F)(C)

(2) When disassembling the front side pulley and fan, front cover and rotor, clamp the rotor in a vice within copper plates and loosen the pulley nut, as shown in the figure.

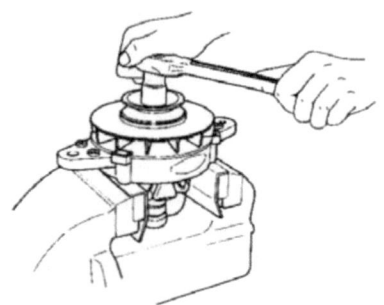

(3) When the fan and pulley have been removed, the rotor can be pulled from the front cover by hand.

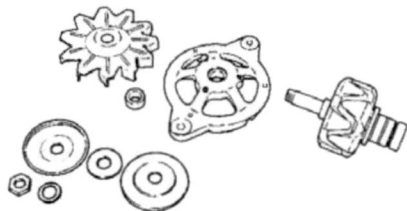

(4) Next, remove the bearing attached to the front cover. Loosen the bearing protector mounting bolts (M4) and pull the bearing by applying pressure to the bearing from the front cover.

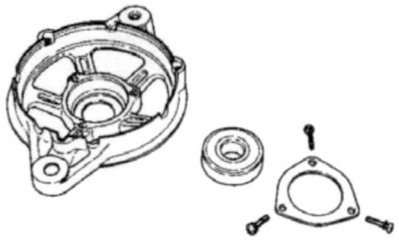

(5) Remove the nut at the threaded part of the BAT terminal on the rear cover, the fixing nut of the diode, and the bolt of E terminal.
After removing the L terminal assembly, separate the alternator into rear cover and stator (with attached diode and brush holder).

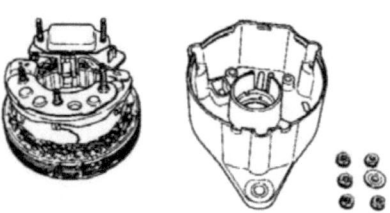

(6) Unsolder the lead wire connection and remove the diode assembly together with the regulator assembly.

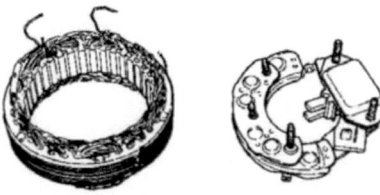

(7) Separate the diode assembly and the brush regulator assembly by removing the 3mm dia rivet which connects these two parts and then unsolder the L terminal connection.

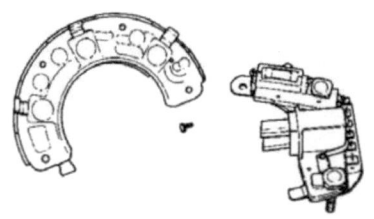

Chapter 12 Electrical System
4A. Alternator Option, 12V/35A

SM/GM(F)(C)·HM(F)(C)

(8) When replacing the IC regulator, it can be removed by unsoldering the regulator's terminals and removing two bolts. Never remove these two bolts except when the regulator is replaced.

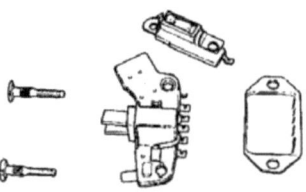

(9) When (1)—(8) above are completed, the alternator is completely disassembled.

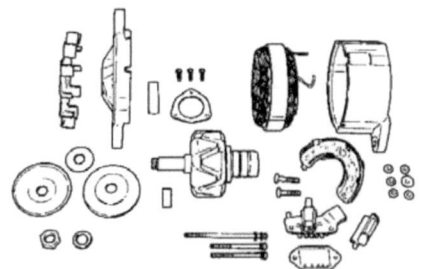

Electric current flows only in one direction in the diode as shown on the previous page. By testing the continuity between terminals (e.g. BAT and U) with the continuity tester, (as shown in the picture), the diode is determined as usable when the continuity is "Yes", but is faulty when it "No".
Connect the tester in the reverse way, and then the diode is usable when continuity is "No", but faulty when "Yes". If a faulty diode is found in this test, replace it with a complete new diode assembly.
As the auxiliary diode does not have a terminal, check the continuity between its ends.

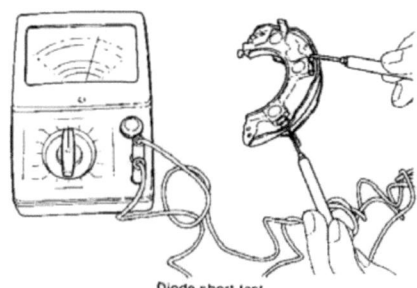

Diode short test

CAUTION: If a high voltage meter is used, a high voltage will be applied to the diode and the diode will be destroyed. Therefore, never test the diodes with a high voltage meter, etc.

4A-9. Inspection and adjustment
4A-9.1 Diodes

Between terminal		BAT (+ side diode)	
	Tester pin	(+)side	(−)side
U.V.W	(+)side	—	Continuity No
	(−)side	Continuity Yes	—

Between terminal		E (− side diode)	
	Tester pin	(+)side	(−)side
U.V.W	(+)side	—	Continuity Yes
	(−)side	Continuity No	—

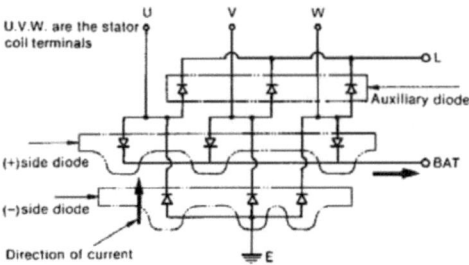

4A-9.2 Rotor

(1) Slip ring wear
Because the slip rings wear very little, the diameter of the rings must be measured with a micrometer. Replace the rings (rotor assembly) when wear exceeds the maintenance standard by 1mm. (0.0393in.)

mm (in.)

	Maintenance standard	Wear limit
Slip ring outside diameter	⌀31.6 (1.2441)	⌀30.6 (1.2047)

(2) Slip ring roughness
The slip ring should be smooth with no surface oil, etc. If the surface of the rings is rough, polish with #500 ~ #600 sandpaper, and if the surface is soiled, clean with a cloth dipped in alcohol.

(3) Rotor coil short test
Check the continuity between the rotor coil and slip ring with a tester. The resistance should be near the prescribed value.
If the resistance is extremely low, there is a layer short at the rotor coil; if the resistance is infinite, the coil is open. In either case, replace the rotor.

Chapter 12 Electrical System
4A. Alternator Option, 12V/35A

Resistance value	Approx. 3.1Ω (at 20°C)	LR135-105

(4) Rotor coil ground test
Check the rotor coil for grounding with a tester, or by checking the continuity between one slip ring and the rotor core or shaft.
Usable if the continuity is "No".
If "Yes", replace it as the rotor coil is grounded.

4A-9.3 Stator coil

(1) Stator coil short test
Check the continuity between the terminals of the stator coil. Measure the resistance between the output terminals with a tester. The resistance should be near the prescribed value.
If the stator coil is open, indicated by infinite resistance, it must be replaced.

Resistance value	Approx. 0.16Ω (at 20°C) 1-phase resistance	LR135-105

(2) Stator coil ground test
Check the continuity between one of the stator coil leads and the stator core.
The stator coil is good if the resistance is infinite. If the stator core is grounded, indicated by continuity, it must be replaced.

4A-9.4 Brush

(1) Brush wear
Check the brush length.
The brush wears very little, but replace the brush if worn over the wear limit line printed on the brush.

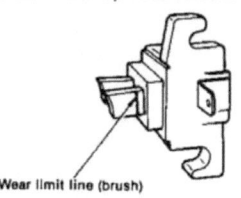

Wear limit line (brush)

mm (in.)

	Maintenance standard	Wear limit
Brush length	16 (0.6299)	9 (0.3543)

(2) Brush spring pressure measurement.
Measure the pressure with the brush protruding 2mm from the brush holder, as shown in the figure. The spring is normal if the measured value is over 150 gr.
Confirm that the brush moves smoothly in the holder.

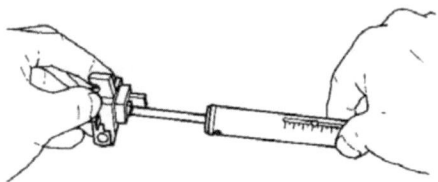

Chapter 12 Electrical System
4A. Alternator Option, 12V/35A

SM/GM(F)(C)·HM(F)(C)

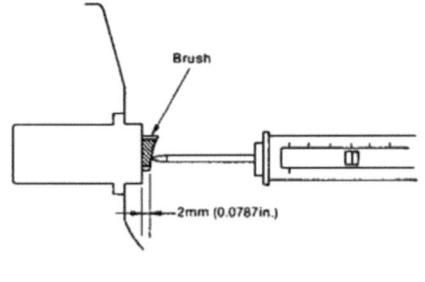

Brush spring strength	300 ±45g (0.562 ~ 0.761 lb) (New brush)

4A-9.5 Checking IC regulator unit

Connect the wiring as shown in the diagram below using a variable register, two 12V batteries, register and ammeter.

(1) Prepare the following measuring devices
 1) Resistor (R_1) 100Ω 2W − 1
 2) Variable resistor (Rv) 0-300Ω 12W − 1
 3) Battery (BAT_1, BAT_2) 12V − 2
 4) DC voltmeter 0 ~ 30V 0.5 class − 1
  (to measure at 3 points)

(2) Check the regulator in the following sequence.
 1) Check V_3 (total voltage of BAT_1 plus BAT_2).
  When the value is between 20V and 26V, BAT_1 and BAT_2 are normal.
 2) When measuring V_2 (Voltage between F − E terminals), shift the variable resistor gradually from the "0" position. Check if the V_2 voltage changes sharply from below 2.0V to over 2.0V.
  If there is no sharp voltage change, the regulator is faulty and must be replaced.
  When there is sharp voltage change, stop the variable registor at that point.
 3) Measure V_1 (voltage between L − E terminals).
  The V_1 voltage is the regulated voltage of the regulator
  ...Confirm that the value is within the standard range.

Adjusted voltage	14.3±0.3V (at 20°C, with 2 batteries)

4A-10 Reassembly precautions

After inspection and servicing, reassemble the parts in the reverse order of disassembly, paying careful attention to the following items:

(1) Brush regulator assembly
 1) Soldering the brush
  Solder the brush after setting it as shown in the figure. Take care that solder does not flow into the pig-tail (lead wire).

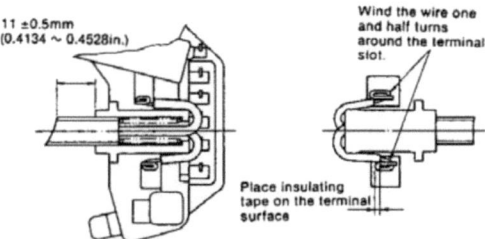

NOTES: 1) Use non-acid flux for soldering.
2) The temperature of the soldering bit is to be 300 to 350°C.

 2) Assembly of IC regulator
  Place the IC regulator on the brush holder as shown in the figure, and insert the M5 bolt.
  After inserting the bolt, solder the brush holder to the IC regulator.

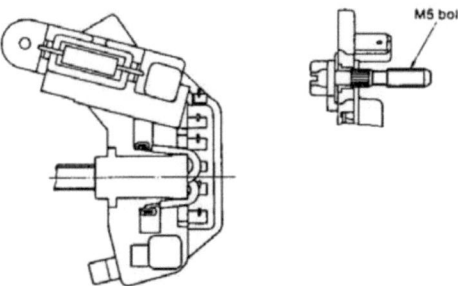

NOTES: 1) Insertion pressure is 100 kg (220.5 lbs)
2) Insert vertically.

(2) Connecting the brush regulator assembly to the diode.
 1) Fixing with rivet
  Insert a 3mm dia. rivet as shown in the figure, and fix it by using the appropriate tool

12-34

Printed in Japan
0000A0A1361

Chapter 12 Electrical System
4A. Alternator Option, 12V/35A

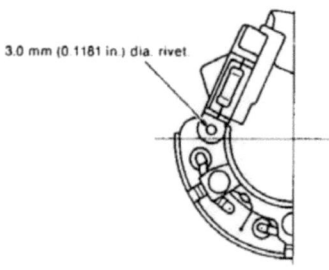

3.0 mm (0.1181 in.) dia. rivet

Rivetting pressure	500 kg (1102 lbs)

(3) Assembling rear cover
Assemble the rear cover after inserting the pin from outside and fitting the brush into the brush holder.

(4) Tightening torque of each part

kgf-cm(ft-lb)

Fixing flange holder	32 ~ 40 (2.31 ~ 2.89)
Fixing diode	32 ~ 40 (2.31 ~ 2.89)
Fixing bearing retainer	16 ~ 20 (1.16 ~ 1.45)
Tightening pulley nut	350 ~ 400 (25.32 ~ 28.93)
Tightening through bolt	32 ~ 40 (2.31 ~ 2.89)

4A-11 Alternator performance test
4A-11.1 Test equipment

Test equipment	Quantity	Specifications
Battery	1	12V
DC voltmeter	1	0 ~ 30V Range 0.5
DC ammeter	1	0 ~ 50A Range 1.0
Variable resistor	1	0 ~ 0.25Ω capacity: 1 kW
Switch	2	Switch capacity: 40A
Tachometer	1	
0.25Ω resistor	1	25W

4A-11.2 Performance test circuit
When the circuit is connected the charge lamp will light.

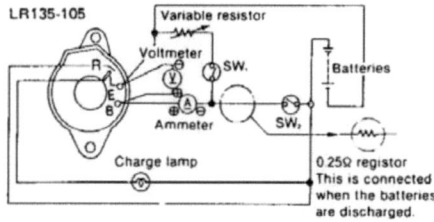

LR135-105
Variable resistor
Voltmeter
SW₁
Batteries
Ammeter
SW₂
Charge lamp
0.25Ω resistor
This is connected when the batteries are discharged.

4A-11.3 Performance test
(1) Speed measurement at 13.5V.
 1) Run the alternator up to a speed of approx. 1500 rpm with SW₁ and SW₂ open.
    Then reduce speed gradually and measure the rpm when the voltage reaches 13.5V.
 2) This value is called the "rpm at 13V" and is acceptable if 1000 rpm or below.
    (The alternator speed at which the lamp goes on or off is 1500 rpm, or 1000 rpm or below, respectively, and there are different conditions for each of the two cases.)
(2) Voltage measurement. Acceptable within the range of 14.3 ±1.3V and when the generator rpm is 5000, SW₁ is open and SW₂ is closed, the temperature is 20°C and using two batteries.
 (Confirm that the ammeter is 5A or below. If over 5A, connect the 0.25Ω resistor.)
(3) Measurement of output current
 1) In the circuit shown in figure, set the variable resistor at the minimum value, close SW₁ and SW₂, and run the alternator.
 2) While keeping the voltage at 13.5V by adjusting the variable resistor, increase the alternator speed, and measure the current at 2500 rpm and 5000 rpm.

Acceptable current values	32A at 5000 rpm	LR135-105

(4) Remarks on performance test
a) For the test leads, use cable with a cross-sectional area of 8mm² or more and with a length not exceeding 2.5m between the alternator B terminal and the positive terminal of the battery, and between the S terminal and the negative terminal of the battery.
b) Switches with low contact resistance are to be used in the circuit.

4A-12. Standards of adjustment

	LR135-105
Standard height of brush	16mm (0.6299in.)
Limit of reduced height	9mm (0.3543in.)
Strength of brush spring	255 ~ 345g (0.56 ~ 0.76 lb)
Standard dimension of shaft at front end	15mm (0.5906in.)
Part No. of ball bearing	6302 BM
Standard dimension of shaft at rear end	12mm (0.4724in.)
Part No. of ball bearing	6201 SD
Resistance of rotor coil (at 20°C)	3.1Ω
Resistance of stator coil single phase (at 20°C)	1.6Ω
Standard O.D. of slip ring	31.6mm (1.244in.)
Limit of reduced size (diameter)	1mm (0.0394in.)
Limit of swing correction	0.3mm (0.0118in.)
Accuracy of swing correction	0.05mm (0.0070in.)

4A-13. Alternator troubleshooting and repair

(1) Failure to charge

Problem	Cause	Corrective action
Wiring, current	Open, shorted, or disconnected	Repair or replace
Alternator	Open, grounded, or shorted coil Terminal insulator missing Diode faulty	Replace Repair Replace
Transistor regulator	Transistor regulator faulty	Replace regulator

(2) Battery charge insufficient and discharge occurs easily

Problem	Cause	Corrective action
Wiring	Wiring shorted or loose, wiring thickness or length unsuitable	Repair or replace Replace
Generator	Rotor coil layer short Stator coil layer short; One phase of stator coil open Slip ring dirty V-belt loose Brush contact faulty Diode faulty	Replace Replace Clean or polish Retighten Repair Replace

(3) Battery overcharged

Problem	Cause	Corrective action
Battery	Electrolyte low or unsuitable	Add distilled water Adjust specific weight Replace
Transistor regulator	Regulator transistor shorted	Replace regulator

(4) Current charge unstable

Problem	Cause	Corrective action
Wiring	Wiring shorted at a break in the covering due to hull vibration or intermittent contact at break	Repair or replace
Alternator	Layer short Balance spring damaged Slip ring dirty Coil open	Replace Replace Replace Repair or replace

Chapter 12 Electrical System
5. Instrument Panel

SM/GM(F)(C)-HM(F)(C)

5. Instrument Panel

5-1 B-type (large) instrument panel

5-2 A-type (small) instrument panel

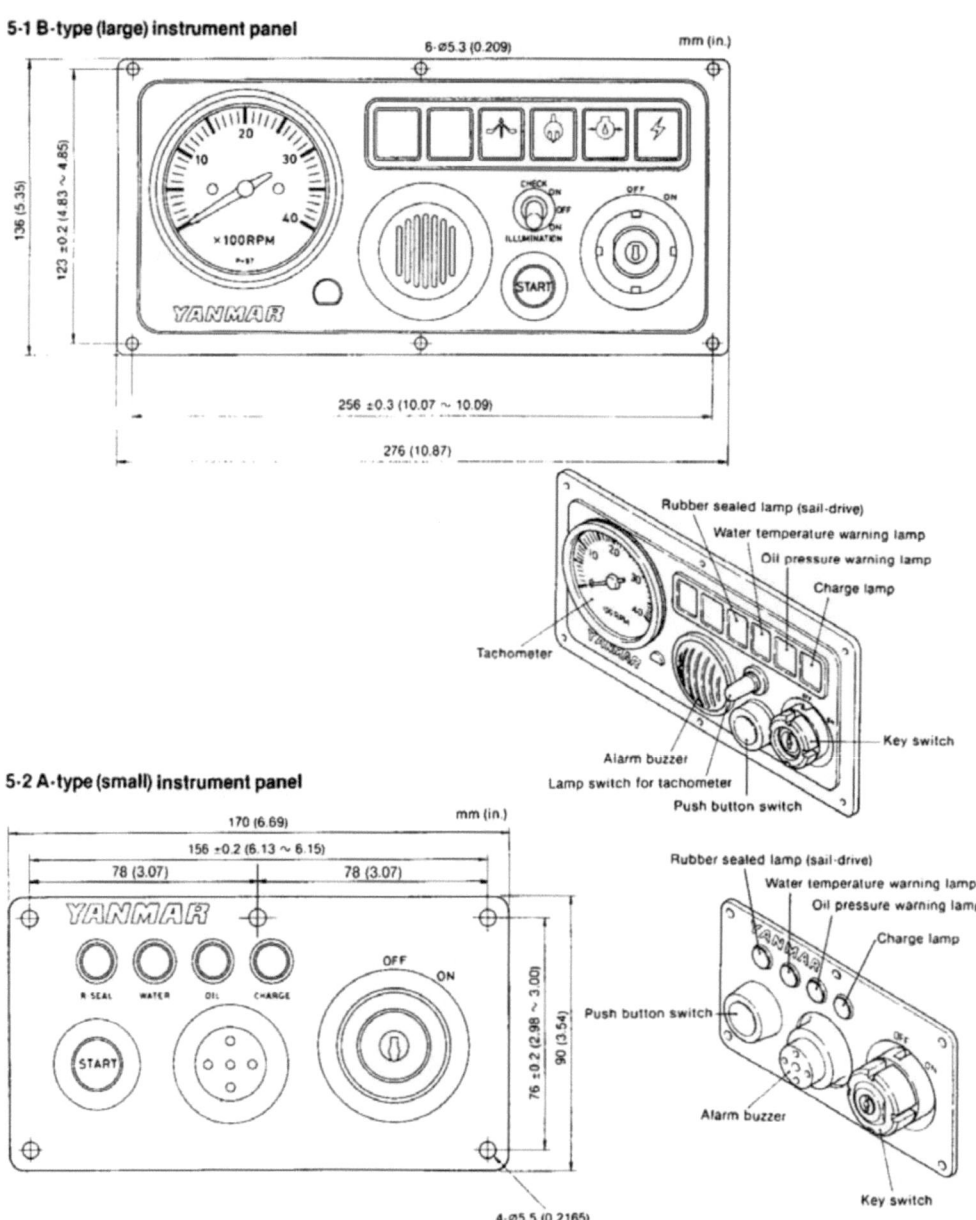

12-37

5-3 Key switch

(1) Construction and dimensions of key switch.

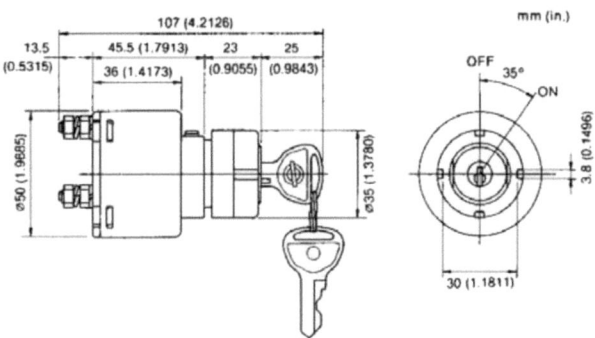

(2) Specifications of key switch

	All models
Rated voltage	DC 12V
Rated current	25A
Range of operating voltage	DC 10 ~ 30V
Part No.	124070-91250

5-4 Push button switch

(1) Construction and dimensions of key switch.

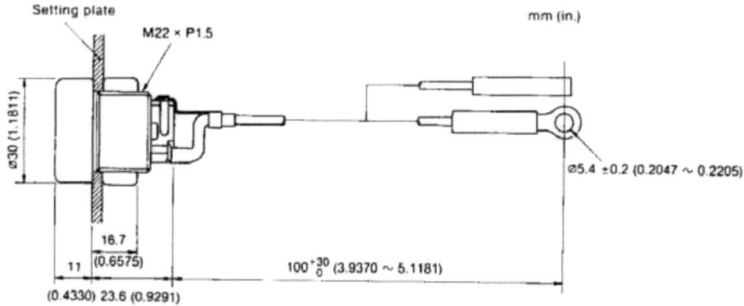

(2) Specifications of push button switch

	All models
Rated voltage	DC 12V
Rated load	20A (within 30 seconds)
Part No.	124070-91300

Chapter 12 Electrical System
5. Instrument Panel

SM/GM(F)(C)·HM(F)(C)

5-5 Warning devices
5-5.1 Oil pressure alarm

If the engine oil pressure is below $0.2 \pm 0.1 \text{kgf/cm}^2$ (1.422 ~ 4.266 lb/in²), with the main switch in the ON position, the contacts of the oil pressure switch are closed by a spring, and the lamp is illuminated through the lamp → oil pressure switch → ground circuit system. If the oil pressure is normal, the switch contacts are opened by the lubricating oil pressure and the lamp remains off.

Oil pressure unit specifications

	All models
Part No.	124060-39451
Rated voltage	12V
Operating pressure	$0.2 \pm 0.1 \text{kgf/cm}^2$ (1.422 ~ 4.266 lb/in.²)
Lamp capacity	5W

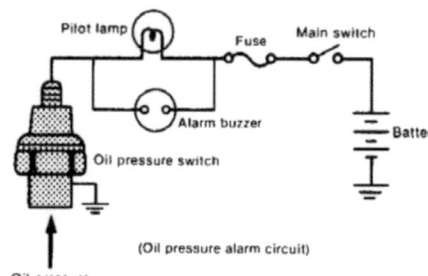

(Oil pressure alarm circuit)

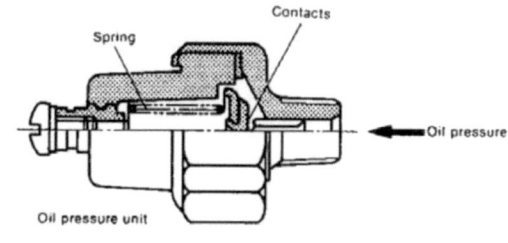

Inspection

Problem	Inspection item	Inspection method	Corrective action
Lamp not illuminated when main switch set to ON	1. Oil pressure lamp blown out	(1) Visual inspection (2) Lamp not illuminated even when main switch set to ON position and terminals of oil pressure switch grounded	Replace lamp
	2. Operation of oil pressure switch	Lamp illuminated when checked as described in (2) above	Replace oil pressure switch
Lamp not extinguished while engine running	1. Oil level low	Stop engine and check oil level with dipstick	Add oil
	2. Oil pressure low	Measure oil pressure	Repair bearing wear and adjust regulator valve
	3. Oil pressure faulty	Switch faulty if abnormal at (1) and (2) above	Replace oil pressure switch
	4. Wiring between lamp and oil pressure switch faulty	Cut the wiring between the lamp and switch and wire with separate wire	Repair wiring harness

5-5.2 Cooling water temperature alarm

A water temperature lamp and water temperature gauge, backed up by an alarm in the instrument panel, are used to monitor the temperature of the engine cooling water. A high thermal expansion material is set on the end of the water temperature unit. When the cooling water temperature reaches a specified high temperature, the contacts are closed, and an alarm lamp and buzzer are activated at the instrument panel.

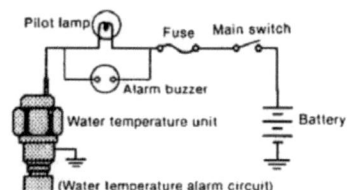

(Water temperature alarm circuit)

12-39

Chapter 12 Electrical System
5. Instrument Panel

Water temperature switch

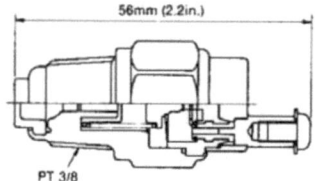

Direct Sea Water Cooling Type

Operating temperature		Current capacity	Response time	Indication color	Parts code
ON	OFF				
65±2°C (154~148°F)	58°C (140°F) or more	DC12V 1A	Within 60 sec.	White	128275-91340

Fresh Water Cooling Type

Operating temperature		Current capacity	Response time	Indication color	Parts code
ON	OFF				
95°C(202~193°F)	88°C(187°F) or higher	DC 12V 1A	Within 60 sec.	Green	127610-91350

Pilot lamp: 12V, 3.4W

The parts of the alarm circuit which must be checked are the open pilot bulb, fuse, and wiring. To check, disconnect the wiring at the water temperature unit side and ground the cord—the pilot lamp is normal if the pilot lamp illuminates. Moreover, be sure the check the color of the code after replacing.

5-6 Alarm buzzer

The alarm buzzer sounds when the engine oil pressure, cooling water temperature, or charging becomes abnormal. The trouble source is indicated by illumination of the appropriate alarm lamp simultaneously with the sounding of the buzzer.

5-6.1 Buzzer for B-type instrument boad

(1) Construction

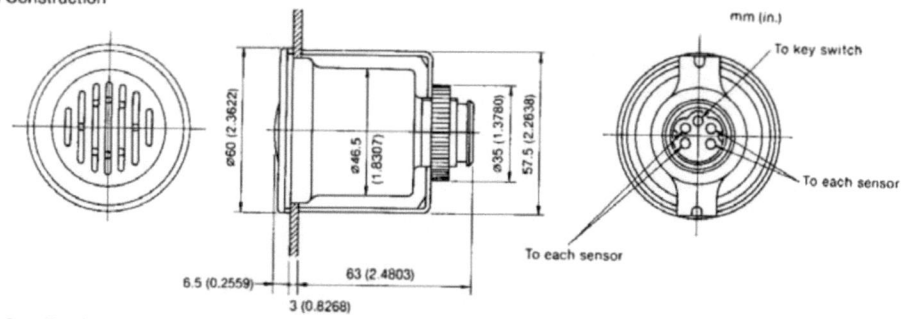

(2) Specifications

Model	WI1-05
Voltage	12V
Current consumption	100mA or below [at 12V, 15~30°C (59~86°F)]
Range of operating voltage	10~15V
Sound output	75dB (A) [at 1m, 12V, 15~30°C (59~86°F)]
Frequency	3 ±0.5kHz [at 12V, 15~30°C (59~86°F)]
Weight	0.2kg (0.44 lb)
Part No.	104271-91351

(3) Wiring diagram

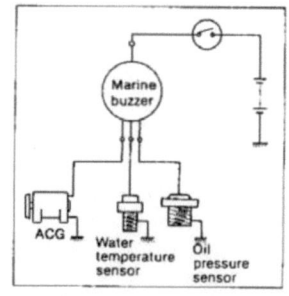

5-6.2 Buzzer for A-type instrument panel

(1) Construction

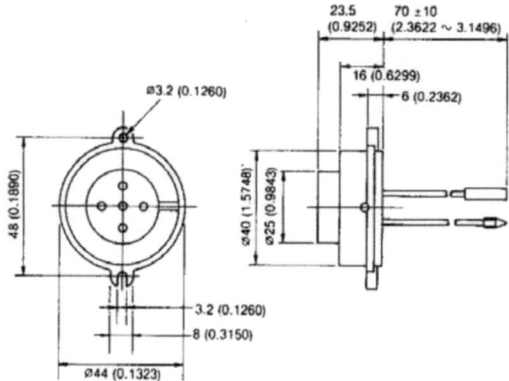

(2) Specifications

Operating voltage	DC 10 ~ 15V
Rated voltage	DC 12V
Current	50 mA or below
Lead wire	49N (5kgf) or more, 15 seconds
Voltage for starting action	1V or more
Basic frequency of sound	$3.0^{+110}_{-0.5}$ kHz
Sound output	$\theta = 0 \sim 45°$ 70dB or below
Current consumption	50 mA or below
Part No.	128270-91350

(3) Wiring diagram

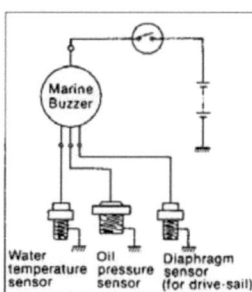

5-6.3

Normal operation is as follows:

	Alarm buzzer	Charge lamp	Oil pressure lamp	Water temperature lamp
Main switch ON, engine stopped	Alarm	Illuminated	Illuminated	Extinguished
Main switch ON, engine running	No alarm	Extinguished	Extinguished	Extinguished
Key switch OFF, engine stopped	No alarm	Extinguished	Extinguished	Extinguished

*The condition of the lamp can be checked by using the check switch.

5-7

Fault	Diagnosis		Remedy
Warning lamp does not light.	Check if there is a loose or open-circuit connection at the coupling connector between the instrument panel and the relay harness.	Yes	Make good the connection.
	↓ No		
	Take out the lamp from P box case and check if it is unserviceable.	Yes	Replace the lamp. (G-1 amp 12V 3.4W)
	↓ No		
	It must be an open-circuit connection in the harness.		Replace the harness.
Buzzer does not sound.	Check if there is a loose or open-circuit connection at the coupling connector between the instrument panel and the relay harness.	Yes	Make good the connection.
	↓ No		
	Check if the buzzer is serviceable. (Fig.) ↓ Yes		Replace the buzzer.
	It must be an open-circuit connection in the harness. DC 12V		Replace the harness.
Other switches and items do not operate.	Check if there is a loose or open-circuit connection at the coupling connector between the instrument panel and the relay harness.		Make good the connection.
	↓ No		
	Check the continuity of the individual switch when the switch is closed by the tester.		Replace the defective item.
	↓ OK		
	It must be an open-circuit connection in the harness.		Replace the harness.

6. Tachometer

6-1 Construction of tachometer

The tachometer indicates the number of revolutions per minute by means of an electrical input signal which is generated as a pulse signal from the magnetic pickup sender (MPU sender).
The function of the sender is to convert the rotary motion into an electrical signal by means of counting by the number of teeth of the ring gear fitted to the flywheel housing.

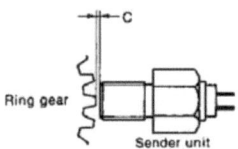
Ring gear
Sender unit

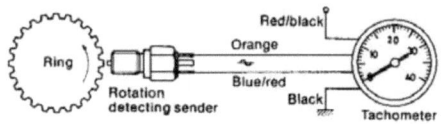

6-2 Specifications and dimensions of tachometer

(1) Specifications

		1GM10(C) 2GM20(F)(C) 3GM30(F)(C)	3HM35(F)(C)
Rated voltage		DC 12V	
Range of operating voltage		10 ~ 15V	
Illumination		3.4W/12V	
Ring gear	No. of teeth	97	114
	Module	2.54	2.54
Part No. of tachometer		128170-91100	128670-91100
Part No. of sender unit		128170-91160	128170-91160

(2) Sensitivity limit of sender unit

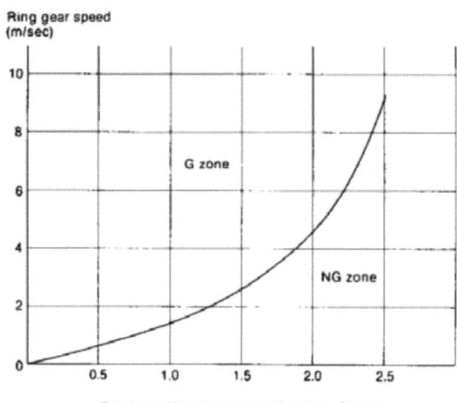

Sender unit and ring gear clearance C (mm)

(3) Dimensions of sender unit

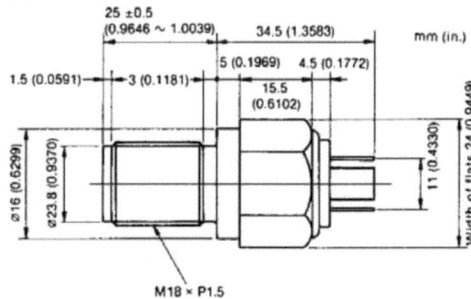

(4) Dimensions and shape of tachometer

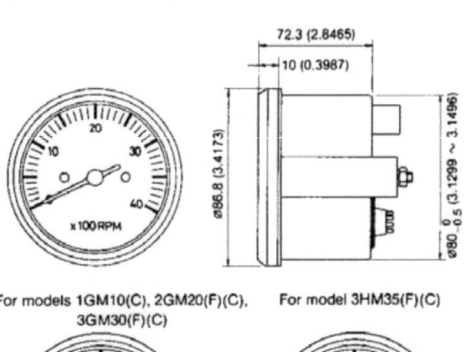

For models 1GM10(C), 2GM20(F)(C), 3GM30(F)(C)

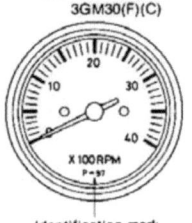

Identification mark

For model 3HM35(F)(C)

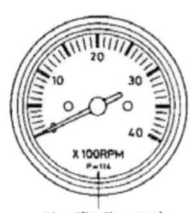

Identification mark

6-3 Measurement of sensor unit characteristics

(1) Measurement of output voltage

Output voltage	1.0V or higher

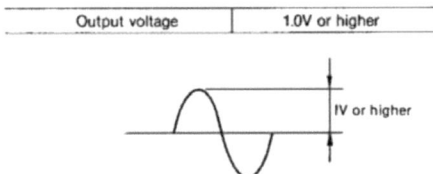

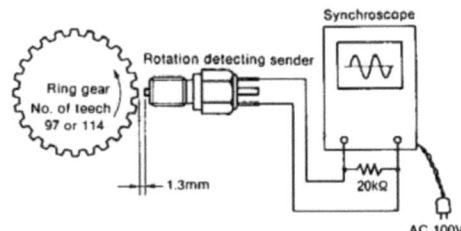

Measuring conditions

Number of teeth of ring gear:	97, 114
Gap between the ring gear and sender:	1.3mm (0.0511in.)
Resistance:	20kΩ
Speed of ring gear:	500 rpm (approx. 800Hz)
Measuring temperature:	20°C
Measuring instrument:	Synchroscope

*Check the output wave pattern and number of pulses when carrying out the output voltage measurement.

(2) Measurement of internal resistance

Measuring conditions

Measuring temperature:	20°C
Measuring instrument:	Digital tester

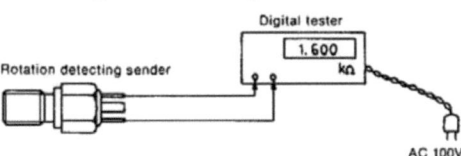

6-4

Fault	Diagnosis		Remedy
Does not function well. 1) Pointer does not move. 2) Functions intermittently.	Check if there is an open-circuit cable connection at the rear of the meter, a loose or disconnected terminal, or bad continuity due to corrosion.	Yes	Make good the connection.
	↓		
	Disconnect at the instrument terminals, and measure the voltage between the cable terminals. (To be 10 ~ 16V) ↓ Satisfactory	No	If the input voltage is abnormal, check the cause. (e.g. shot-circuit, disconnection, or blown fuse, etc.)

Chapter 12 Electrical System
6. Tachometer

Check if the sender is loosely fitted. ↓ No —Tachometer sender	Yes	Fix the sender securely.
Measure the internal resistance of the sender. (To be 1.6 ±0.1kΩ at 20°C) ↓	No	Replace the sender.
Measure the output voltage of the sender. (To be 1V or higher at 20°C)	No	Replace the sender.

CHAPTER 13
OPERATING INSTRUCTIONS

1. Fuel Oil and Lubricating Oil 13-1
2. Engine Operating Instructions 13-8
3. Troubleshooting and Repair 13-13

1. Fuel Oil and Lubricating Oil

Selection of and proper attention to fuel and lubricating oils has a substantial effect on engine performance, and these are vital factors governing engine life.

The use of low quality fuel and lubricating oils will lead to various engine troubles. Yanmar diesel engines will display satisfactory performance and ample reliability if the fuel and lubricating oil recommended by Yanmar are used correctly. For the engine to have long-term high performance, sufficient knowledge of the properties of the fuel and lubricating oils and their selection, management and usage are necessary.

1-1 Fuel

1-1.1 Properties of fuel

Numerous kinds of fuels are used with diesel engines, and the properties and composition of each differ somewhat according to the manufacturer.

Moreover, the various national standards are introduced here for reference purposes.

1-1.2 Recommended fuels

Manufacturer	Brand name
Caltex	Caltex Diesel Oil
Shell	Shell Diesoline or local equivalent
Mobil	Mobil Diesel Oil
Esso	Esso Diesel Oil
British Petroleum	BP Diesel Oil

1-1.3 Fuel selection precautions

Pay careful attention to the following when selecting the fuel.

(1) Must have a suitable specific gravity

Fuel having a specific gravity of 0.88 ~ 0.94 at 15°C is suitable as diesel engine fuel. Specific gravity has no relation to spontaneous combustibility, but does give an idea of viscosity and combustibility or mixing of impurities.

Generally, the higher the specific gravity, the higher the viscosity and the poorer the combustibility.

(2) Must have a suitable viscosity

When the viscosity is too high, the fuel flow will be poor, operation of the pump and nozzle will be inferior, atomization will be faulty and fuel combustion will be incomplete.

If the viscosity is too low, the plunger, nozzle, etc. will wear rapidly because of insufficient lubrication. Generally, however, the higher the viscosity, the lower the quality of the fuel.

(3) Cetane value must be high.

The most important indicator of fuel's combustibility is its cetane value (also represented by cetane index or diesel index), The cetane value is particularly important for fuels used in high-speed engines. The relationship among the cetane value, startability and firing delay is shown in the figure below. Firing delay becomes smaller and starting characteristics better as the cetane value becomes higher.

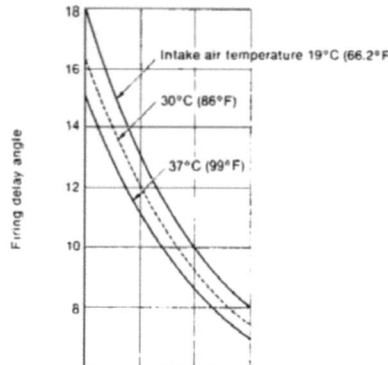

Relationship between cetane value and firing delay

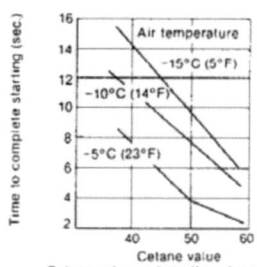

Cetane value and starting characteristic

The use of a fuel with an unsuitable cetane value will cause the following troubles:
1) Difficult starting.
2) Poor operation.
3) High combustion pressure and diesel knock.
4) Lower output and engine damage because of overheating caused by knocking.
5) Sticking of nozzles and exhaust valves.
6) Severe smoking, carbon build-up inside the engine, and oil contamination.
7) Deterioration of the oil and excessive wear in the piston rings, ring grooves, and cylinder liner.

(4) The level of impurities must be low
1) Sulfur

With proper combustion sulfur in the fuel turns to nitrous acid gas (SO_2) and sulfuric anhydride (SO_3). When combustion is imperfect, it becomes sulfuric acid containing water that corrodes and wears the cylinder liners, pistons, exhaust valve and exhaust pipe.

Chapter 13 Operating Instructions
1. Fuel Oil and Lubricating Oil

SM/GM(F)(C)·HM(F)(C)

Properties and compositions of fuel of various national standards

Properties and components		National standard	Japan JIS-K-2204-1965		U.S.A. ASTM-D975-74	U.K. BS-2689-70	
			Class No.1 light oil	Class No.2 light oil	No. 2D Diesel oil	Class A1	Class A2
Specific gravity	15/4°C		—	—	—	—	—
Kinetic viscosity	30°C cst (86°F cst)		2.7 or more	2.5 or more	(~ 5.2)	(~ 7.5)	(~ 7.5)
	37.8°C (100°F) cst		(2.3 or more)	(2.2 or more)	2.0~4.3	1.6 ~ 6.0	1.6 ~ 6.0
Reaction			Neutral	Neutral	—	—	—
Flash point	°C (°F)		50 or more (122 or more)	50 or more (122 or more)	51.7 or more (125 or more)	55 or more (131 or more)	55 or more (131 or more)
Flow point	°C (°F)		−5 or less (23 or less)	−10 or less (14 or less)	−12 or less (10.4 or less)	—	—
Residual carbon	Weight %		(10% residual oil) 0.15 or less	(10% residual oil) 0.15 or less	0.35 or less	0.2 or less	0.2 or less
Moisture	Volume %		—	—	—	0.05 or less	0.05 or less
Ash	Weight %		—	—	0.01 or less	0.01 or less	0.01 or less
Sulfur	Weight %		1.2 or less	1.2 or less	0.5 or less	0.5 or less	1.0 or less
Cetane valve			50 or more	45 or more	40 or more	50 or more	45 or more
Sludge or sedimentation	%		—	—	0.05 or less	0.01 or less	0.01 or less
Distillation properties, temperatures at 90% distillation	°C (°F)		350 or below (662 or below)	350 or belw (662 or below)	282.21 ~ 338 (540 ~ 640)	357 or below (675 or below)	357 or below (675 or below)

Chapter 13 Operating Instructions
1. Fuel Oil and Lubricating Oil

2) Water content
   A high water content causes sludge, resulting in lower output, imperfect combustion and trouble in the fuel injection system.
3) Carbon content
   If the carbon content is high, carbon will remain inside the combustion chamber, causing accelerated cylinder liner and piston wear and corrosion of the pistons and exhaust valves.
4) Residual carbon (coke content)
   Coke becomes a carbide that sticks to the end of the nozzle, causing faulty injection. In addition, unburned carbon will build up on the pistons and liners, causing piston ring wear and sticking.

1-1.4 Simple methods of identifying fuel properties
(1) Fuel that is extremely odorous and smoky contains a large amount of volatile components and impurities.
(2) Fuel that emits little smoke when used in a lamp is of good quality.
(3) Fuel that emits a crackling sound when soaked into paper and ignited contains a high water content.
(4) If a transparent film of diesel oil is squeezed between two pieces of glass, the water content and impurities can be determined.

(5) If cracked by mixing with an equal amount of sulfuric acid in a glass tube, numerous black particles and impurities will appear. These are mainly carbon and resin.
(6) Discoloration of litmus test paper indicates the presence of acids.

1-1.5 Troubles caused by bad fuel
(1) Clogging of exhaust valve
   In addition to faulty compression, incomplete combustion, and high fuel consumption, a clogged exhaust valve will cause fuel to be mixed in the exhaust, leading to corrosion of the exhaust valve seat.
(2) Clogging of piston ring grooves
   Clogged piston ring grooves will cause accelerated cylinder liner and piston wear due to sticking rings, fuel gas blowback, faulty lubrication, incomplete combustion, high fuel consumption, contaminated lubricating oil, and combustion gas blowback.
(3) Clogged or corroded injection valve hole
   This will cause incomplete combustion and piston and liner wear, fuel injection mechanism wear, corrosion, and groove wear and corrosion.
(4) Sediment inside crankcase
   Since sediment in the crankcase is often mistakenly judged as coming from the lubricating oil, care must be taken in determining its true origin.

1-1.6 Relationship between fuel properties and engine performance

Fuel property	Starting characteristic	Lubrication characteristic	Smoke generation	Exhaust odor	Output	Fuel consumption	Clogging of combustion chamber
Firing Cetane value	Directly related— Starting characteristic improves as cetane value increases	Directly related— Lubrication improves as cetane value rises	Closely related— Smoke increases as cetane value decreases	Directly related— Decreased by increasing cetane value	Irrelevant	Related	Related— Decreased by reducing cetane value
Volatility 90% end point	No clear relationship	Related— Becomes poor when volatility is poor	Directly related— Increases as volatility decreases	No direct relationship	Irrelevant	Irrelevant	Related— Increases as volatility decreases
Viscosity	No clear relationship	Some relationship— Becomes poor when viscosity increases	Related— Increases as viscosity increases	No independent relationship	Irrelevant	Irrelevant	Related— Increases with viscosity
Specific gravity	Irrelevant	Irrelevant	Related— Increases as specific gravity increases	No independent relationship	Directly related— Associated with calorific value	Related— Associated with calorific value	Related— Depends on properties of engine
10% residual carbon	Irrelevant	Irrelevant	Related— Improves as residual carbon decreases	No independent relationship	Irrelevant	Irrelevant	Related— Decreases as residual carbon decreases
Sulfur				No independent relationship			
Flash point				No independent relationship			

Chapter 13 Operating Instructions
1. Fuel Oil and Lubricating Oil

1-1.7 Fuel handling precautions
(1) Fill the fuel tank after work to prevent condensation of water in the tank.
(2) Always use a tank inlet strainer. Water mixed in the fuel can be removed by removing the strainer quickly.
(3) Remove the plug at the bottom of the fuel tank and drain out the water and sediment after every 100 hours of operation, and when servicing the pump and nozzle.
(4) Do not use fuel in the bottom of the fuel tank because it contains large amounts of dirt and water.

1-2 Lubricating oil
Selection of the lubricating oil is extremely important with a diesel engine. The use of unsuitable lubricating oil will cause sticking of the piston rings, accelerated wear and seizing of the piston and cylinder liner, rapid wear of the bearings and other moving parts, and reduced engine durability. Since this engine is a high-speed engine, always follow the lubricating oil replacement interval.

1-2.1 Action of the lubricating oil
(1) Lubricating action: Builds a film of oil on each moving part reduces wear and its accompanying damage.
(2) Cooling action: Removes heat generated at moving parts by carrying it away with the lubricating oil flow.
(3) Sealing action: Maintains the air tightness of the pistons and cylinders by the oil film on the piston rings.
(4) Cleaning action: Carries away carbon produced at the cylinders as well as dust that has entered from the outside.
(5) Rustproofing action: Prevents corrosion by coating metal surfaces with a thin film of oil.

Various additives are added to the lubricating oil to ensure that adequate performance is assured under the high-speed, high-load and other severe operating conditions met by modern diesel engines. While these additives differ with each manufacturer, commonly used additives include:
1) Flow point reduction additive
2) Viscosity index improvement additive
3) Oxidation prevention additive
4) Cleaning dispersent
5) Lubrication additive
6) Anticorrosion additive
7) Bubble elimination additive
8) Alkali neutralizer

1-2.2 Required lubricating oil conditions
(1) Must be of suitable viscosity
If the viscosity is too low, the oil film will be too thin and the lubricating action insufficient. If the viscosity is too high, the friction resistance will be increased and starting will become especially difficult.
(2) Viscosity change with temperature must be small. While the lube oil temperature goes from low at starting to high during operation, the viscosity index should be high at all temperatures.
(3) Must have good lubricating capability
That is, it must coat metal surfaces as a thin film. In other words, the lubricating oil must coat the metal surfaces so that metal-to-metal contact caused by breaking of the oil film at the top dead center and bottom dead center piston position does not occur, and that the oil film is not broken by collision, even at the bearings.
(4) Mixability with water must be low
Since water can mix with the oil because of the presence of cooling water in the engine, emulsification of water and oil, which causes the oil to lose its lubricating properties, must be prevented.
(5) Must be neutral and difficult to oxidize
Since acids and alkalis corrode metal, the lubricating oil must be neutral. Moreover, since even a neutral oil will be oxidized easily by contact with the combustion gas, the oil must be stable with few oxidizing elements.
(6) Must withstand high temperature and must evaporate or combust with difficulty
Oil must have a high flash point. If it is evaporated by heat or is not burned completely, carbon will be produced. This carbon is toxic.
(7) Must not contain any water or dirt and must have a low sulfur and coke content

1-2.3 Classification by viscosity

SAE No.	−17.8°C (6°F)		98.9°C (210°F)		Applicable temperature range (outside temperature)
	Saybolt universal viscosity (sec)	Dynamic viscosity (cst)	Saybolt universal viscosity (sec)	Dynamic viscosity (cst)	
5W	Under 4,000	Under 869	—	—	20°C or less
10W	6,000 ~ 12,000	1,303 ~ 2,606	—	—	(68°F or less)
20W	12,000 ~ 48,000	2,606 ~ 10,423	—	—	
20	—	—	45 ~ 58	5.73 ~ 9.62	20°C ~ 35°C
30	—	—	58 ~ 70	9.62 ~ 12.93	(68°F ~ 95°F)
40	—	—	70 ~ 85	12.93 ~ 16.77	35°C or greater
50	—	—	85 ~ 110	16.77 ~ 22.68	(95°F or greater)

Chapter 13 Operating Instructions
1. Fuel Oil and Lubricating Oil

SM/GM(F)(C)·HM(F)(C)

Since only 98.9°C viscosity is stipulated for S.A.E. No. 20 ~ 50 oil in the table, and only −17.8°C viscosity is stipulated for S.A.E. No. 5W ~ 20W oil, they are not guaranteed at other temperatures. On the other hand, S.A.E. No.10W viscosity is stipulated. Oil having viscosity equal to that of S.A.E. No.30 even at 98.9°C is called S.A.E. 10W—30, or multigrade oil. Multigrade oil comprises S.A.E. No.5W—20, 10W—30, and 20W—40. In arctic regions, oil from S.A.E. No.20W to 10W—30 can be used.

1-2.4 SAE service classification and API service classification

SAE new classification (1970)	API service classification (1960)
CA	DG
CB·CC	DM
CD	DS

(1) DG grade: Used when deposits and engine wear must be controlled when the engine is normally operated at a light load using low sulfur fuel.
(2) DM grade: Used when the generation of deposits and wear caused by sulfur in the fuel is possible under severe conditions.
(3) DS grade: Used under extremely severe operating conditions or when excessive wear or deposits are caused by the fuel.

Classification	Engine service (API)
CA	Light duty diesel engine service: Mild, moderate operation diesel engine service with high-performance fuel, and mild gasoline engine service. The oil designed for this service was mainly used in the 1940s and 50s. This oil is for high performance fuel use and has bearing corrosion and high temperature deposit prevention characteristics.
CB	Moderate duty diesel engine service: Mild, moderate operation diesel engine service using low performance fuel requiring bearing corrosion and high temperature deposit prevention characteristics. Includes mild gasoline engine service. Oil designed for this service was introduced in 1949. The oil is used with high sulfur fuels and has bearing corrosion and high temperature deposit prevention characteristics.
CC	Moderate duty diesel engine service and gasoline engine service: Applicable to low supercharged diesel engines for moderate to severe duty. The oil designed for this service was introduced in 1961 and is widely used in trucks and agricultural equipment, construction machinery, farm tractors, etc. The oil features high deposit prevention characteristics in low supercharged diesel engines, and rust, corrosion and low temperature sludge prevention characteristics in gasoline engines.
CD	Severe duty diesel engine service: Applicable to high-speed, high-output high supercharged diesel engines which are subjected to considerable wear and deposits. This oil was introduced in 1955, and is used as a wide property-range fuel in high supercharged engines. It also has bearing corrosion and high temperature deposit prevention characteristics.

1-2.5 Lubricatlong oil
SAE new classification CB grade or CC grade fuel having suitable viscosity for the atmospheric temperature must be used in this engine.

1-2.6 Recommended lubricating oils

Supplier	Brand Name	SAE No.			
		Below 10°C (Below 50°F)	10~20°C (50~68°F)	20~35°C (68~95°F)	Over 35°C (Over 95°F)
SHELL	Shell Rotella Oil	10W, 20/20W	20/20W	30 40	50
	Shell Talona Oil	10W	20	30 40	50
	Shell Rimula Oil	20/20W	20/20W	30 40	—
CALTEX	RPM Delo Marine Oil	10W	20	30 40	50
	RPM Delo Multi-Service Oil	20/20W, 10W	20	30	50
MOBIL	Delvac Special	10W	20	30	—
	Delvac 20W—40	20W—40	20W—40	—	—
	Delvac 1100 Series	10W, 20/20W	20/20W	30 40	50
	Delvac 1200 Series	10W, 20/20W	20/20W	30 40	50
ESSO	Estor HD	10W	20	30 40	—
	Esso Lube HD	—	20	30 40	50
	Standard Diesel Oil	10W	20	30 40	50
B.P. (British Petroleum)	B.P. Energol ICMB B.P. Energol DS-3	20W	20W	40	50

1-2.7 Engine oil replacement and handling

(1) Necessity of replacement

Since the engine oil is exposed to high temperatures during use and is mixed with air at high temperatures, it will oxidize and its properties will gradually change. In addition, its lubricating capabilities will be lost through contamination and dilution by water, impurities, and the fuel. Emulsification and sludge are produced by heat and mixing when the lubricating oil contains water and impurities, causing its viscosity to increase. Moreover, if the carbon in the cylinders enters the crankcase, the oil will turn pure black and the change in its properties can be seen at a glance. The continued use of deteriorated oil will not only cause wear and corrosion of moving parts, but will ultimately cause the bearings and cylinders to seize. Therefore, deteriorated oil must be replaced.

(2) Replacement period

Although the engine oil change interval differs with the engine operating conditions and the quality of the lubricating oil and fuel used, the oil change interval should be as follows when CB grade oil is used in a new engine:

1st time .... After approximately 20 hours of use
2nd time .... After approximately 30 hours of use
From 3rd time ... After every 100 hours of use

Drain the old oil completely and replace it with new oil while the engine is still warm.

CAUTION: Never mix different brands of lubrication oil.

1-2.8 Adding oil

The crankcase and clutch case are not connected. For the crankcase, add one of the lubricating oils described in chapter 1.2.6. For the clutch case, add the lubrication oil described below. Be sure not to mix up the oils.

Supplier	1GM10	2GM20 (F)	3GM30 (F)	3HM35(F)
SHELL				SHELL DEXRON
CALTEX	Same lube oil as for crankcase			TEXAMATIC FLUID(DEXRON)
MOBIL				MOBILE ATF 220
ESSO				ESSO ATF
B.P.				B.P. AUTRAN DX

Chapter 13 Operating Instructions
1. Fuel Oil and Lubricating Oil

(1) Remove the clutch case clutch and head cover filler plug (engine), and fill with specified lubricating oil up to the top marks on the respective dipsticks. (Oil levels must not drop below the lower marks on the dipsticks.)

1-2.9 Oil capacity

Lubricating oil capacity at an engine mounting angle (rake) of 8° is given below.

	Crankcase	Clutch case (except sail-drive)
1GM10(C)	1.3∫	0.25∫
2GM20(F)(C)	2.0∫	
3GM30(F)(C)	2.6∫	0.3∫
3HM35(F)(C)	5.4∫	0.7∫

- Check the crankcase oil level by completely inserting the dipstick. Check the clutch case oil level without screwing in the cap.
  The oil levels must be between the upper and lower limit marks on both dipsticks.

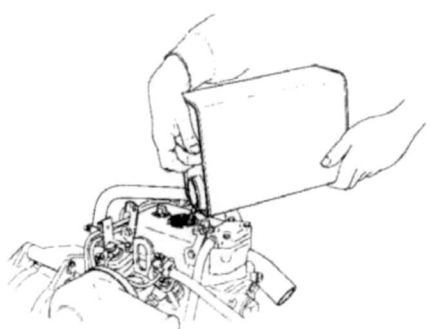

Engine

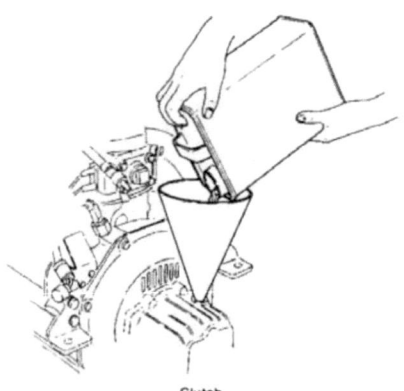

Clutch

(2) Since it takes sometime for the oil to flow completely into the clutch case and oil pan, wait for 2 ~ 3 minutes after filling before checking the oil levels. Moreover, check the oil while the boat is afloat.

2. Engine Operating Instructions

2-1 Preparations before starting

2-1.1 Fueling up
(1) Check the fuel level in the fuel tank and add fuel if necessary.
(2) Remove water and dirt collected in the bottom of the tank using the fuel tank drain cock.
(3) Add clean fuel to the tank.
Since dirt and water sink to the bottom of the fuel drum, do not turn the drum upside down and do not pump the fuel from the bottom of the drum.

2-1.2 Adding lubricating oil
(1) Check the oil level with the dipstick, and add oil, if necessary, to bring the level up to the full mark of the dipstick.
The level must be neither too low nor too high.
(2) The crankcase and clutch case require different oil. Check both and add oil separately, being careful not to mix the oils.
(3) Since the crankcase oil flows into the crankcase through the camshaft and valve chambers, wait 2 ~ 3 minutes before checking its level.

2-1.3 Lubricating each part
(1) Lubricate each pin of the remote control lever.

2-1.4 Checking fuel priming and injection
(1) Operate the priming lever of the fuel pump.
(2) Set the regulator handle to the full speed position and check for injection sound by turning the engine over several times.
(3) If there is no fuel injection sound, bleed the air from the fuel system.

2-1.5 Bleeding the fuel system
Since the presence of air in the fuel system anywhere between the fuel tank and the injection valve will cause faulty fuel injection, always bleed the air from the system when the fuel system is disassembled and reassembled.

Bleeding the fuel system
(1) Open the fuel tank cock.
(2) Bleed the air from the fuel filter.
Loosen the air bleeding plug at the top of the fuel filter body and operate the manual handle of the fuel pump until no more bubbles appear in the fuel flowing from the filter.
Then install and tighten the air bleeding plug.

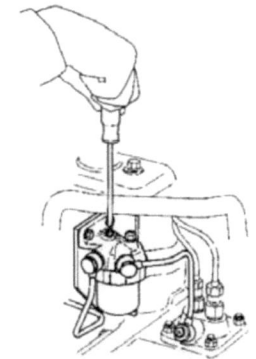

(3) Bleed the air from the fuel return pipe.
Loosen the connnector bolt of the fuel return pipe installed on the fuel injection valve, and bleed the air by operating the manual handle of the fuel pump.
Bleed the air in the No.1 cylinder (timing gear case side) and No.2 cylinder (clutch side), in that order.

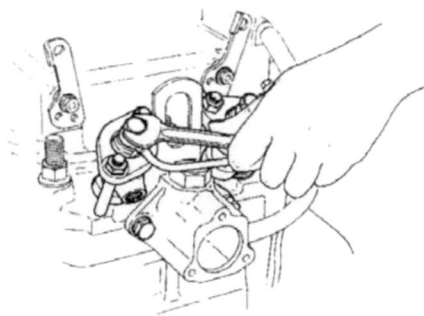

(4) Bleed the air from the fuel injection pipe.

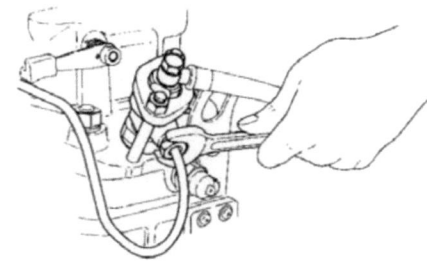

Chapter 13 Operating Instructions
2. Engine Operating Instructions

Loosen the nipple on the fuel injection valve side, set the regulator handle to the operating position and the decompression lever to the decompression position, and crank the engine. When no more bubbles appear in the fuel flowing from the end of the injection pipe, retighten the nipple.

(5) Check injection.
After bleeding the air, set the regulator handle to the operating position, set the decompression lever to the decompression position, and crank the engine. When fuel is being injected from the injection valve, an injection sound will be heard and you can feel resistance if you place your hand on the fuel injection pipe. This check must not be performed more than two or three times since overchecking will flood the combustion chamber with fuel, and faulty combustion will occur at starting.

2-1.6 Checking for abnormal sounds by cranking
(1) Set the regulator handle to the STOP position, release the compression of the engine by setting the decompression lever, and crank the engine about 10 times to check for abnormal sounds.
(2) Crank the engine with the starting handle
(Always turn the engine in the proper direction of rotation.)

2-1.7 Checking the cooling system
(1) Open the Kingston cock.
(2) Check for bending and cross-sectional deformation of the cooling water inlet pipe.
(3) Set all water drain cocks to the CLOSED position.

2-1.8 Checking the remote control system
(1) Check that the remote control handle operates correctly.
(2) Check that the engine stop remote control operates smoothly.

2-1.9 Checking the electrical system
(1) Check the battery electrolyte level and add distilled water if low.
(2) Check that the wiring is connected correctly.
(Especially for polarity.)
(3) Turn the battery switch on, set the main switch to the ON position, and check if the oil pressure lamp and charge lamp are illuminated and if the alarm buzzer sounds when the engine is stopped.
(The charge lamp should be on while the engine is stopped and should be off while the engine is running.)

2-1.10 Checking appearance and exterior
(1) Check for loose or missing bolts and nuts.
(2) Check for loose or disconnected piping and hoses.
(3) Check that there are no tools or other articles near rotating parts or on the engine.

2-2 Starting and warm-up
2-2.1 Starting
(1) Starting procedure
1) Set the clutch handle to the "NEUTRAL" position.
2) Set the governor lever to the "MEDIUM SPEED".
3) Keep the decompression lever in the "OPERATION" position.
4) Set the main switch to the ON position.
The alarm buzzer will sound.
5) Push the starting button to start the engine.
Release the start button after the engine has started.
6) When the engine has started, the alarm lamps and buzzer will go off.
If the lamps or buzzer stay on, immediately stop the engine and check for trouble.

A type

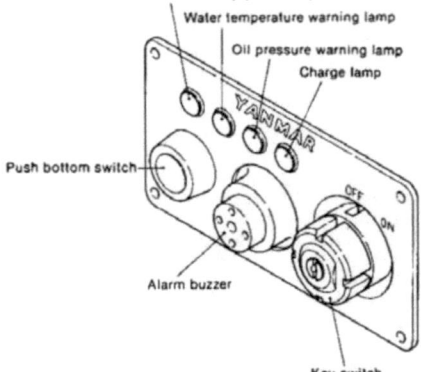

B type

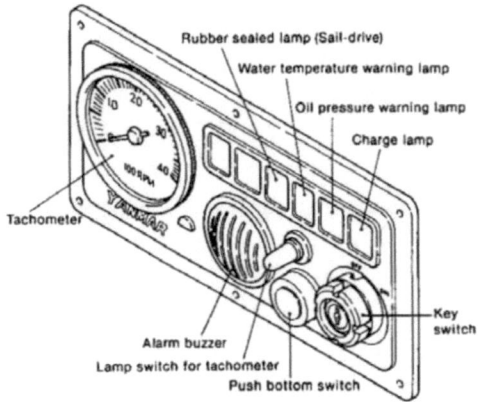

Chapter 13 Operating Instructions
2. Engine Operating Instructions

SM/GM(F)(C)·HM(F)(C)

(2) Starting precautions
1) Don't continue to push the starting button over 15 seconds.
 If the engine doesn't start, wait 30 seconds or more.
2) When restarting the engine, always confirm the flywheel is stopped.
 If you re-start the engine while the flywheel is rotating, the pinion gear of the starter motor and the ring gear of the flywheel will be damaged.
3) When starting is difficult in cold weather lift the decompression lever to decompress the engine, and turn the starting motor. Once the engine has reached a certain speed, return the decompression lever to the "OPERATION" position. In this way, starting is made easier while current consumption is reduced.

2-2.2 Starting with one-handle remote control (option)

(1) Starting procedure
1) Pull the neutral knob and set the control lever to the "MEDIUM SPEED" position.

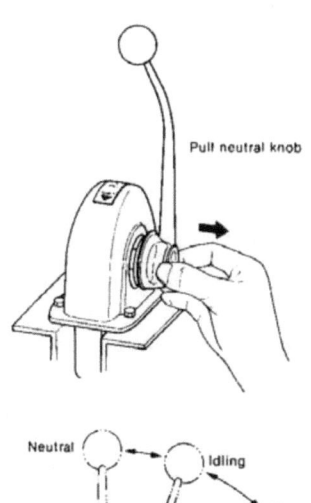
Pull neutral knob

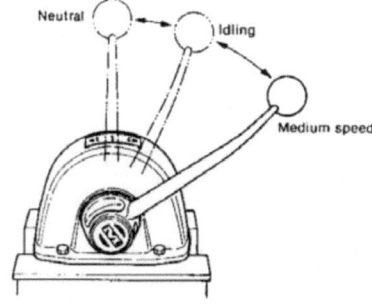
Neutral — Idling
Medium speed

2) Set the main switch to the "ON" position, and push the starting button to start the engine.

(2) Starting in cold weather
1) Pull the neutral knob, and set the control lever to the HIGH SPEED position.

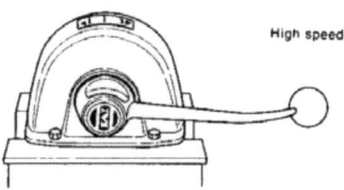
High speed

2) Set the decompression handle to the DECOMPRESSION position.
3) Set the main switch to the ON position and start the engine by pushing the starting button, at the same time putting the decompression lever to the COMPRESSION position. After the engine has started, return the control lever to the MEDIUM SPEED position.
*When the control lever is set in the HIGH SPEED position, injection timing is automatically delayed to facilitate starting.

CAUTION: *When the engine is started with the control lever in the HIGH SPEED position, the starting button must be released immediately and the control lever must be returned to the idling position after the engine has started.
If the starting button is not released, the starter motor will overrun, causing it to be damaged or burnt out.*

2-2.3 After starting

(1) Warm-up operation
The engine must not be suddenly operated at full load immediately after starting. Warm up the engine for about 5 minutes after starting by running the engine at about half speed, and begin full load operation only after the temperature of each part has risen to a uniform value. Neglecting to warm up the engine will result in:
1) Seizing of the piston and liner due to sudden heat expansion of the piston.
2) Burning of piston rings and seizing of bearings/bushings because of insufficient lubrication.
3) Faulty intake and exhaust valve seat contact and shortening of the life of each part due to sudden heating.

Warm-up time (no-load operation)
1,000 ~ 1,200 rpm 3 minutes
1,600 ~ 1,800 rpm 2 minutes

CAUTION: *Do not run the engine at full speed for 50 hours after installation to assure proper break-in.*

Chapter 13 Operating Instructions
2. Engine Operating Instructions

(2) Checking after starting
Check the following with the clutch in the NEUTRAL position:
1) Meters and lamps on the instrument panel
   - Check that all alarm lamps are off (1,000 rpm or higher).
   - Alarm buzzer must be off.
2) Cooling water discharge
   (Check that the cooling water temperature reaches 45 ~ 55°C before beginning operation.)
3) Check for abnormal sounds and heating.
4) Check for oil and water leakage from piping.
5) Check the state of lubrication of the valve arms.

2-3 Operation
If warm-up operation is normal, engage the clutch and begin normal operation. Check the following during operation and stop the engine and take suitable corrective action if there are any abnormalities.

2-3.1 Checks during operation
(1) Oil pressure
Check that the lubricating oil pressure and operating oil pressure lamps are off.
Lubrication oil pressure during operation: 2.5 ~ 3.5 kgf/cm^2

(2) Cooling water
Periodically check whether water is being discharged from the cooling water outlet pipe.
If the cooling water is being discharged intermittently or if only a small amount of water is being discharged during high speed operation, immediately stop the engine and check if air is being sucked into the cooling system, the impeller of the water pump is abnormal, or the water pipes and Kingston cock are clogged.
Cooling water temperature during operation: 45 ~ 55°C.
Check that the water temperature alarm lamp is off.

(3) Fuel
Check the fuel level in the fuel tank and add fuel before the tank becomes too low. If the fuel level is low, air will enter the fuel injection system and the engine will stop.

(4) Charging
Check that the charge lamp is off.
If the charge lamp is still on even when the engine is run at 1,000 rpm or above, the charging system is faulty and the battery is not being charged.

(5) Temperature of each part
At full power operation, the surface temperature of each engine part is about 50 ~ 60°C and hot to the touch. If engine temperature is too high, the oil will be used up, the propeller shaft will not be centered, or other troubles may occur.

(6) Leakage and abnormalities
Check for water leakage, oil leakage, gas leakage, loose bolts, abnormal sounds, abnormal heating, and vibration.

(7) Exhaust color
Black exhaust smoke indicates that the engine is being overloaded and that the lives of the intake and exhaust valves, piston rings, cylinder liners, and injection nozzle will be shortened. Do not run the engine for long periods when exhaust is this colour.

(8) Abnormal sounds, abnormal heating
When abnormal sounds or abnormal heating occur during operation, immediately stop the engine and check for trouble.

2-3.2 Operating precautions
(1) Always set the battery switch and main switch to the ON position during operation.
Since the diodes of the alternator will be damaged, don't set the switches OFF position.
(2) Do not touch the starting button during operation. Operation of the starter motor pinion will damage the gears.
(3) Since the boat will resonate and vibrate at a certain speed, depending on the structure of the hull, do not operate it at that speed.
(4) Always set the clutch in the neutral position and wait for the propeller to stop rotating before raising the propeller shaft (if hoisting type stern gears are installed).
(5) Do not suddenly apply a full load to the engine or operate it at full load for long periods.

2-4 Stopping

2-4.1 Stopping procedure
(1) Before stopping, put the clutch in NEUTRAL and run the engine at approximately 1,000 rpm for about 5 minutes.
(2) Before stopping, temporarily raise the speed to the rated speed to blow out residue in the cylinders. Then stop the engine by pulling the engine stop lever to cut the fuel.

2-4.2 Stopping precautions
(1) Do not stop the engine with the decompression lever.
If the engine is stopped with the decompression lever, fuel will remain in the combustion chamber and abnormal combustion will occur when the engine is started again, perhaps damaging the engine.
(2) If the engine is stopped immediately after full-load operation, the temperature of each part will rise suddenly, leading to trouble.

2-4.3 Inspection and procedures after stopping
(1) Always close the Kingston cock after the engine is stopped.
Water may enter because of a faulty water pump, etc.
(2) In cold weather, the cooling water should always be drained after engine use to prevent freezing. There are water drain cocks on the cylinders and the exhaust manifold. (Drain the water after the engine has cooled.)
(3) Check for oil leakage and water leakage, and repair as required.
(4) Check for loose bolts and nuts, and repair as required.

2-5 Storage when moored for an extended period

(1) Securely close engine room windows and doors so that rain and snow cannot enter.
Also plug the exhaust outlet since water that enters the cylinder from the exhaust pipe will be compressed when the engine is started, causing serious trouble.
(2) The boat may also sink because of water leakage at the stern tube stuffing box packing. This can be prevented by tightening the packing.
(3) Change the lubricating oil before cranking the engine.
(4) Wipe off each part and coat with oil to prevent rusting of the engine exterior.
(5) Coat the regulator handle stand and each link with a thin film of lube oil or grease.
(6) Run the engine once a week to lubricate each part. This will prevent rusting of the bearings, pistons, and cylinder liners.

2-6 Emergency stop

(1) Loosen the fuel valve high-pressure pipe to release the fuel.
(2) Pull the decompression lever (decompression mechanism) so that compression is not applied to the combustion chamber.
(3) Block the air intake port so that air does not enter the combustion chamber.

3. Troubleshooting and Repair

If trouble occurs in the engine, the engine must be immediately stopped or run at low speed until the cause of the trouble is located.

If even extremely small troubles are not detected and corrected early, they can lead to serious trouble and even disaster. Detecting and correcting troubles quickly is extremely important.

3-1 Troubles and corrective action at starting

Trouble	Cause	Corrective action
Flywheel fails to rotate correctly	(1) Battery not charged (2) Starter motor faulty (3) Moving parts seized (4) Lubricating oil viscosity too high	1) Recharge battery 2) Disassemble and repair starter motor 3) Inspect and repair 4) Replace with lubricating oil of suitable viscosity
Starter motor rotates, but engine fails to start	(1) Fuel not injected, or injection faulty	1) Prime and bleed air from fuel lines 2) Inject fuel through injection valve and replace needle if required 3) Clean fuel filter 4) Check operation of fuel pump, plunger, plunger spring, and delivery valve, and replace if required 5) The remote control system or governor is faulty, so check if fuel is cut off, and adjust if required
	(2) Fuel injection timing incorrect	1) Correct the fuel injection timing 2) Check if alignment mark of timing gear is aligned
	(3) Compression pressure low	1) Lap valves when air tightness of intake and exhaust valve is poor 2) Replace cylinder head packing if gas is leaking 3) Clean or replace piston rings when sticking occurs 4) Readjust timing when intake and exhaust valve closing is very slow.
	(4) Drop in compression ratio	1) Replace piston pin bearing and crank pin bearing if worn 2) Replace piston rings if worn

3-2 Troubles and corrective action during operation

Trouble	Cause	Corrective action
Engine stops suddenly	(1) Fuel injection cut off due to trouble in the governor or governor system (2) Air in fuel tank (3) Air in fuel system (4) Piston, bearing, or other moving parts seized	1) Inspect, and repair or replace 2) Add fuel 3) Bleed air 4) Inspect and repair or replace the parts
Speed decreases unexpectedly	(1) Governor maladjusted (2) Overload (3) Piston seized (4) Bearing seized (5) Fuel filter clogged (6) Fuel injection pump or injection valve sticking Dirt in fuel pump delivery valve (7) Air in fuel system (8) Water in fuel	1) Adjust 2) Lighten the load (check propeller system and power take-off system) 3) Stop the engine, and repair or replace 4) Stop the engine, and repair or replace 5) Clean the fuel filter 6) Stop the engine, and repair or replace 7) Prime and bleed air 8) Drain the fuel tank and fuel filter Add fuel if insufficient
Exhaust color is bad	(1) Load unsuitable (2) Fuel injection timing off (3) Fuel unsuitable. (4) Injection valve faulty (5) Intake and exhaust valve adjustment faulty (6) Intake and exhaust valves leaking. (7) Output of cylinders uneven (8) Injection pressure too low (9) Precombustion chamber melted	1) Adjust the load (check propeller system and power take-off system) 2) Adjust injection timing 3) Change the fuel type 4) Test injection and replace valve if required 5) Adjust valve head clearance 6) Lap or grind valves 7) Check the fuel injection pump and injection valve and replace if necessary 8) Set injection pressure with shims 9) Replace the precombustion chamber...Perform item (1) above
Full load operation impossible	(1) Fuel filter clogged (2) Fuel pump plunger worn	1) Check and replace filter element 2) Replace plunger and barrel as a set
Output of cylinders uneven	(1) Air in fuel pump or fuel line (2) Water in fuel (3) Fuel injection volume uneven (4) Fuel injection timing uneven (5) Intake and exhaust valves sticking (6) Injection valve faulty	1) Prime and bleed air from the fuel pump and fuel lines 2) Drain the fuel tank and fuel filter and add fuel 3) Check and adjust injection volume 4) Check and adjust injection timing 5) Disassemble and clean 6) If nozzle is clogged, clean; replace nozzle if necessary If the needle is sticking, inspect and replace

Chapter 13 Operating Instructions
3. Troubleshooting and Repair

Trouble	Cause	Corrective action
Engine knocks	(1) Bearing clearance too large (2) Connecting rod bolt loose (3) Flywheel bolt, coupling bolt loose (4) Injection timing faulty (5) Too much fuel injected because of faulty fuel pump or injection nozzle	1) Inspect, and repair or replace parts 2) Check and retighten 3) Check and retighten or replace bolt as required 4) Check and adjust 5) Check fuel injection pump and injection nozzle and replace if required
Engine oil pressure low	(1) Lubricating oil leakage (2) Bearing, crankpin bearing clearance too large (3) Oil filter clogged (4) Oil regulator valve loose. (5) Oil temperature high; cooling water flow insufficient (6) Lubricating oil viscosity low (7) Excessive gas leaking into crankcase	1) Check engine interior and exterior piping, replenish oil 2) Check clearance, and replace bearing if necessary 3) Check and replace filter element 4) Check and readjust oil pressure 5) Check oil pump, and replace if necessary 6) Replace with oil having a high viscosity index 7) Check pistons, piston ring, and cylinder liners and replace if necessary
Lubricating oil temperature too high	(1) Cooling water flow insufficient (2) Excessive gas leaking in to crankcase (3) Overload	1) Check water pump 2) Check piston rings and cylinder liners 3) Lighten the load
Cooling water temperature high	(1) Air sucked in with cooling water (2) Cooling water flow insufficient (3) Cooling system dirty (4) Thermostat faulty	1) Check water pump inlet side pipe connections 2) Check water pump 3) Flush cooling system with cleaner 4) Replace thermostat
Propeller shaft rotates even when clutch is in neutral position	(1) Neutral position adjustment faulty (2) Friction plate seized (3) Steel plate warped	1) Reset neutral position adjusting bolt 2) Check and repair 3) Repair or replace
Ahead, neutral, astern switching faulty	(1) Clutch face seized (2) Moving parts, lever system malfunctioning (3) Remote control system malfunctioning	1) Replace 2) Readjust 3) Repair or replace
Abnormal heating	(1) Clutch slipping because of overload operation (2) Bearing damaged (3) Excessive oil (4) Oil deteriorated	1) Reduce load 2) Replace 3) Check oil level and adjust to prescribed level 4) Replace oil
Abnormal sound	(1) Gear noise caused by torsional vibration (2) Gear backlash excessive	1) Avoid high speeds 2) Replace

CHAPTER 14
DISASSEMBLY AND REASSEMBLY
(Direct Sea-Water Cooling Engine)

1. Disassembly and Reassembly Precautions 14-1
2. Disassembly and Reassembly Tools 14-2
3. Others 14-13
4. Disassembly 14-14
5. Reassembly 14-28

DISASSEMBLY AND REASSEMBLY

This chapter covers the most efficient method of disassembling and reassembling the engine. Some parts may not have to be removed, depending on the maintenance and inspection objective. In this case, removal is unnecessary and disassembling in accordance with this section is not required.
However, if you follow the disassembly and reassembly procedures, adjustment methods, and precautions described in this chapter, you should be able to prevent subsequent troubles and a loss in engine performance after reassembly. The engine must be test-run to confirm that the engine is functioning properly and delivering full performance.
Since this chapter does not cover detailed disassembly and reassembly procedures for each part, refer to pertinent chapters for details.

1. Disassembly and Reassembly Precautions

(1) Record the parts that require replacement, and replace them with new parts during reassembly.
Be careful not to reassemble with the old parts.
(2) Do not forget adhesives and packing agents for sealing during reassembly.
Packing of the specified quality and packing agents matched to the packing material must be used.
(3) Arrange the disassembled parts into groups, such as individual cylinders, intake and exhaust, etc.
Cylinder No. is indicated No. 1, No. 2 and/or No. 3 cylinder from Flywheel side.
(4) The prescribed tightening torque must be observed when tightening bolts and nuts. Moreover, since the strength of the bolts and nuts depends on their material, be sure to use the correct bolts and nuts at their proper places.
Special bolts, nuts. . . . Head cover, rod bolts, flywheel, etc.
Strong bolts. . . . . Bolts marked (7) (JIS.7T)
Common bolts, nuts . . Unmarked (JIS.4T)
In addition, check the disassembly and reassembly precautions for each engine model.

2. Disassembly and Reassembly Tools

The following tools are necessary when disassembling and reassembling the engine. These tools must be used according to disassembly process and location.

2-1. General handtools

Name of tool	Illustration	Remarks
Wrench		YANMAR standard Code no.: 28110-100130 Size: 10 × 13
Wrench		YANMAR standard Code no.: 28110-120140 Size: 12 × 14
Wrench		YANMAR standard Code no.: 28110-170190 Size: 17 × 19
Wrench		YANMAR standard Code no.: 28110-220240 Size: 22 × 24
Screwdriver		YANMAR standard Code no.: 104200-92350

Chapter 14 Disassembly and Reassembly
2. Disassembly and Reassembly Tools

SM/GM(F)(C)·HM(F)(C)

Name of tool	Illustration	Remarks
Steel hammer		Local supply
Copper hammer		Local supply
Mallet		Local supply
Nippers		Local supply
Pliers		Local supply
Offset wrench		Local supply 1 set
Box spanner		Local supply 1 set

Chapter 14 Disassembly and Reassembly
2. Disassembly and Reassembly Tools

SM/GM(F)(C)·HM(F)(C)

Name of tool	Illustration	Remarks
Scraper		Local supply
Lead rod		Local supply
File		Local supply 1 set
Rod spanner for hexagon socket head screws		Local supply Size: 4mm (0.1575in.) 5mm (0.1969in.)

Chapter 14 Disassembly and Reassembly
2. Disassembly and Reassembly Tools

SM/GM(F)(C)-HM(F)(C)

2-2 Special handtools

Name of tool	Shape and size	Application
Main bearing replacer	Lock nut, Plate B, Spacer cylinder, Insertion guiding and extracting bolt, Plate A, Insertion guide and extraction seat, Nut | Model	Assembly code no.
---	---
1GM10(C), 2GM20(F)(C), 3GM30(F)(C)	124085-92400
3HM35(F)(C)	128670-92400		Removal — Insertion extraction bolt, Plate A, Spacer, Insertion guide, Crank bearing Installation — Plate A, Spacer, Insertion guide, Plate B, Crank bearing
Lubricating oil No. 2 filter case remover	(wrench shape)	(filter removal illustration)

14-5

Chapter 14 Disassembly and Reassembly
2. Disassembly and Reassembly Tools

SM/GM(F)(C)·HM(F)(C)

Name of tool	Shape and size	Application
Piston pin insertion/ extraction tool	mm (in.) **Table 1:** | Model | d | D | *l* | Code No. | |---|---|---|---|---| | 1GM10 (C) 2GM20 (F)(C) 3GM30 (F)(C) | $\phi 9.5 \pm 0.1$ (0.3700 −0.3780) | $\phi 20^{-0.15}_{-0.25}$ (0.7776 ~7815) | 97 (3.8189) | 124085 -92260 | | 3HM35 (F)(C) | $\phi 12$ (0.4724) | $\phi 22$ (0.8661) | 80 (3.1496) | 128670 -92260 |	Piston pin extractor Extraction of piston pin Insertion of piston pin
Connecting rod small end bushing insertion/ extraction tool	mm (in.) | Model | d | D | *l* | Code No. | |---|---|---|---|---| | 1GM10 (C) 2GM20 (F)(C) 3GM30 (F)(C) | $\phi 20^{-0.3}_{-1.0}$ (0.7480 −0.7756) | $\phi 22^{-0.3}_{-0.0}$ (0.8268 ~8543) | 20 (0.7874) | 124085 -92270 | | 3HM35 (F)(C) | $\phi 23^{-0.3}_{-0.6}$ (0.8819 −0.8937) | $\phi 25^{-0.3}_{-0.6}$ (0.9606 ~9724) | 3.0 (1.1811) | 128670 -92270 |	Extraction
Intake and exhaust valve insertion/ extraction tool	65 (2.5591) 55 (2.1654) 25 (0.9843) $\phi 7^{-0.3}_{-0.5}$ (0.2599 ~ 0.2638) $\phi 12^{-0.3}_{-0.5}$ (0.4526 ~ 0.4606) $\phi 15$ (0.5906) All models Code no.: 124085-92250	

Chapter 14 Disassembly and Reassembly
2. Disassembly and Reassembly Tools

SM/GM(F)(C)·HM(F)(C)

Name of tool	Shape and size	Application
Piston ring compressor	Model / Code No. 3GM35(F)(C) / 102700-92140 1GM10(C), 2GM20(F)(C), 3GM30(F)(C) / 101200-92140	Piston insertion guide
Valve lapping handle	All models Code no.; 28210-000031	Lapping tool
Valve lapping powder	All models Code no.; 28210-000070	
Feeler gauge 0.2mm (0.0079)	All models Code no.; 28312-200750	
Fuel injection valve replacer	mm (in.) M14 × 1.5 (0.55) Ø13 (0.51) All models Code no.; 101104-92180	

14-7

Chapter 14 Disassembly and Reassembly
2. Disassembly and Reassembly Tools

SM/GM(F)(C)·HM(F)(C)

Name of tool	Shape and size	Application
Pulley puller	Rocal supply	Removing the coupling
Tool for turning crankshaft gear nut Tightening the crankshaft gear nut	mm (in.) 140 (5.5118), ⌀49 (1.9291), 36 (1.4173), ⌀38 (1.4961), 12.7 (0.5000) Width across flats of hexagonal hole All models Code no.: 124085-92700	
Driving tool for bearing inner race (for models 1GM10, 2GM20(F), 3GM30(F)	mm (in.) 88 (3.4646), ⌀30 (1.181), ⌀26 (1.0236) Code no.: 177088-09150	Tool Bearing inner race Output shaft The bearing inner race of the drive output shaft.
Operation lever locating jig [for models 1GM10 2GM20(F) and 3GM30(F)]	mm (in.) Approx. 118 (4.6457), 74 (2.9134) t = 10.0 (0.3937) Code no.: 177088-09160	Adjusting the operation lever

14-8

Chapter 14 Disassembly and Reassembly
2. Disassembly and Reassembly Tools

SM/GM(F)(C)·HM(F)(C)

Name of tool	Shape and size	Application
Extractor tool for the bearing outer race (for models 1GM10, 2GM20(F), 3GM30(F)	Code no.; 177088-09160	Extracting the bearing outer race from the housing. (Tool, Case body, Bearing outer race)
Output shaft nut wrench for 3HM35(F)	mm (in.) A: ø55 (2.1654), B: 230 (9.0551), C: 45 (1.7717) Code No. 177099-09010	Output shaft nut wrench Output shaft coupling lock
Output shaft coupling lock for 3HM35(F)	mm (in.) A: 290 (11.4173), B: ø100 (3.9370) Code No. 177099-09020	For removing and tightening the output shaft nut.
Puller cradle for 3HM35(F)	mm (in.) ø29 (1.1417), 17 (0.6693) Code no.; 177095-09170	Cradle, Pulley puller For removing the output shaft when using a pulley puller.

14-9

Chapter 14 Disassembly and Reassembly
2. Disassembly and Reassembly Tools

SM/GM(F)(C)·HM(F)(C)

Name of tool	Shape and size	Application
Pulling support for 3HM35(F)	⌀100 (3.9370), 33 (1.2992), 12 (0.4724) Code no.: 177099-09030	For removing the needle bearing inner race, thrust collar and thrust bearing of the output shaft (forward gear side).
Plate for spring retainer for 3HM35(F)	(3.1496) 80, 80 (3.1496) Code no.: 177095-09070	For removing and installing the plate spring, retainer and circlip of the large gears (forward and reverse).
Assembly spacer for 3HM35(F)	(3.1496) ⌀80, 20 (0.7874) Code no.: 177090-09010	For determining the thickness of the adjusting plate.

14-10

Chapter 14 Disassembly and Reassembly
2. Disassembly and Reassembly Tools

SM/GM(F)(C)·HM(F)(C)

Name of tool	Shape and size	Application
Inserting tool for 3HM35(F)	mm (in.) (1.7717) Ø45 250 (9.8425) Code no.; 177095-09020	Inserting tool — Driving plate — Pressure plate — Forward large gear For installing the spacer and needle bearing inner race of the output shaft. (reverse small gear side).
Inserting tool for 3HM35(F)	mm (in.) (1.4961) Ø38 100 (3.9370) Ø45 (1.7717) Code no.; 177099-09040	Inserting tool — Reverse large gear — Shift ring — Forward large gear For installing the thrust bearing and thrust collar (reverse large gear side).

2-3 Measuring instruments

Nomenclature		Accuracy and range
Vernier calipers		1/20 mm, 0 ~ 150 mm.
Micrometer		1/100 mm, 0 ~ 25 mm, 25 ~ 50 mm, 50 ~ 75 mm, 75 ~ 100 mm,
Cylinder gauge		1/100 mm, 18 ~ 35 mm, 35 ~ 60 mm, 50 ~ 100 mm,
Thickness gauge		0.05 ~ 2mm (0.0020 ~ 0.0787 in.)
Torque wrench		0 ~ 13 kgf-m (0 ~ 94 ft-lb)
Nozzle tester		0 ~ 500 kgf-cm^2 (0 ~ 7111.7 lb/in.^{2})

3. Others

Supplementary packing agent

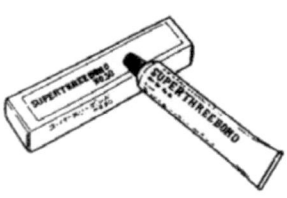

Type	Use
"Three Bond 3B8-005"	White. Since "Three Bond 3B8-005" is a nonorganic solvent, it does not penetrate asbestos sheets made principally or completely of asbestos. Always use it with grey asbestos sheet packing for complete oil tightness. When "Three Bond 3B8-005" is difficult to obtain, use silicone nonsolvent type "Three Bond No. 50."
"Three Bond No. 50"	Grey. Silicone nonsolvent type liquid packing. Semidry type packing agent coated on mating faces to prevent oil and gas leakage. Does not penetrate asbestos sheet and assures complete oil tightness.
"Three Bond No. 1"	Reddish brown. Paste type wet viscous liquid packing. Ideal for mating faces which are removed but reinstalled. Particularly used to prevent water leakage and to prevent seizing of bolts and nuts.

The surface to be coated must be thoroughly cleaned with thinner or benzene and completely dry. Moreover, coating must be thin and uniform.

Products of Three Bond Co., Ltd.

Paint

Color spray

Metallic Ecole Silver is used entirely on this engine.

Wipe off the surface to be painted with thinner or benzene, shake the spray can well, push the button at the top of the can and spray the paint onto the surface from a distance of 30 ~ 40 cm.

Paint

Type
White paint
(Mixed oil paint)

Usage point
Cylinder liner
insertion hole

Use
Paint parts that contact the cylinder body when inserting the cylinder liner to prevent rusting and water leakage.

Yanmar cleaner (Ref.)

Cooling passage cleaner is made by adding one part "Unicon 146" to about 16 parts water (specific gravity ratio). To use, drain the water from the cooling system, fill the system with cleaner, allowing it to stand overnight (10 ~ 15 hours). Then drain out the cleaner, fill the system with water, and operate the engine for at least one hour.

NEJI LOCK SUPER   203M: a locking agent for screws (Ref.)

For coating on screws and bolts to prevent loosening, rusting, and leaking. To use, wipe off all oil and water on the threads of studs, coat the threads with screw lock, tighten the stud bolt, and allow to stand until the screw lock hardens. Use screw lock on the oil intake pipe threads, oil pressure switch threads, fuel injection timing shim faces, and front axle bracket mounting bolts.

4. Disassembly

4-1. General Precautions

Maintenance and inspection should be done as effectively as possible, avoiding unnecessary disassembling except for general overhauls.
At the time of disassembly, record the presence of parts which require repair or replacement, and make arrangements beforehand for procurement of such parts so that problems will not occur during the reassembling operation.

4-2. Dismantling engine model 1GM10(C)

4-2.1 Open the cooling water drain cock and drain the cooling water

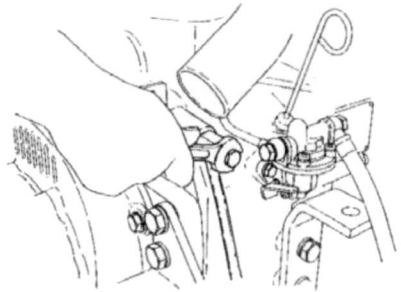

4-2.2 Drain the lubricating oil

(1) Engine side
   Insert a suction tube into the dipstick hole and pump out the oil with a waste oil pump (option).
   Alternatively remove the plug of oil pan and oil intake pipe, and drain the oil.

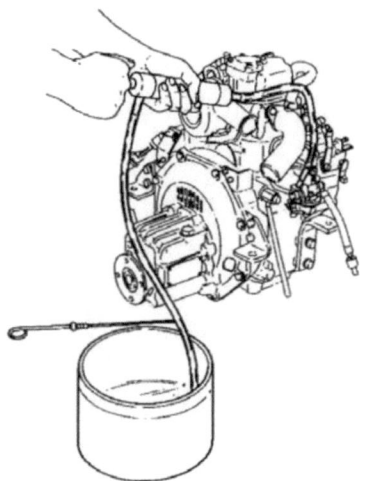

(2) Clutch side
   Pump out the oil from the filler/dipstick hole using a waste oil pump or remove the drain plug at the bottom stern side of the clutch case and drain the oil.

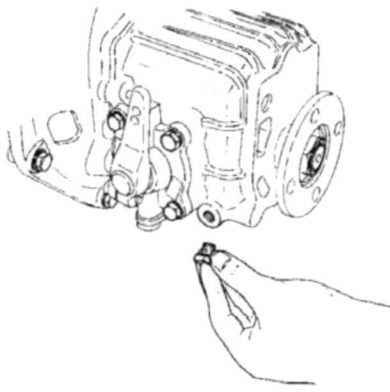

4-2.3 Disconnect the remote control cables

(1) Clutch remote control cable and bracket
(2) Speed remote control cable and bracket
(3) Engine stop remote control cable and bracket
(4) Decompression remote control cable

4-2.4 Disconnect the electrical wiring

(1) Alternator wiring
(2) Starter motor wiring
(3) Water temperature switch wiring
(4) Oil pressure switch wiring
(5) Tachometer sender wiring

4-2.5 Disconnect the cooling water inlet pipe
NOTE: Always close the Kingston cock.

Chapter 14 Disassembly and Reassembly
4. Disassembly

4-2.6 Remove the air intake silencer

Remove the intake silencer clip and the filter element. Then remove the set screw and the cover.

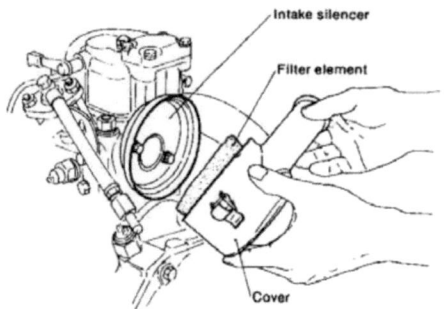

4-2.7 Disconnect the fuel piping

(1) Fuel tank to feed pump
(2) Feed pump to fuel filter
(3) Fuel filter to fuel injection pump
(4) Fuel high pressure pipe
(5) Fuel return pipe

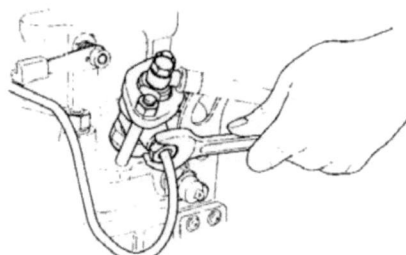

4-2.8 Remove the starter motor

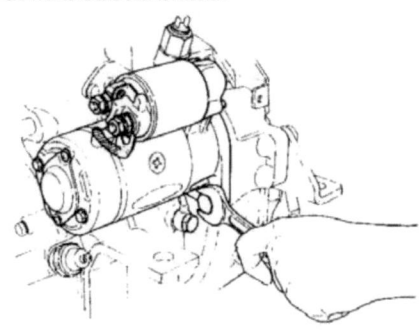

4-2.9 Remove the alternator

(1) Loosen the adjusting bolt and remove the V-belt
(2) Remove the alternator and bracket

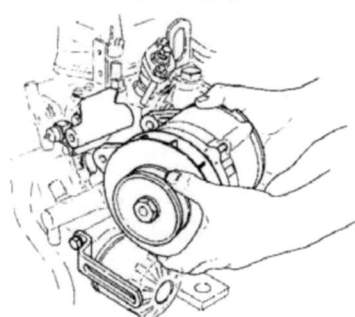

4-2.10 Remove the water pump

(1) Disconnect the hose between the water pump and cooling water cylinder inlet joint.

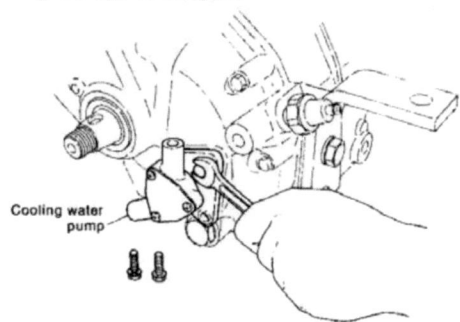

(2) Loosen the water pump mounting bolts and remove the water pump.

4-2.11 Remove the oil filter and bracket.

(1) Remove the oil filter using the remover.
(2) Loosen the joint bolts and remove the oil pipes.
(3) Remove the oil filter bracket.

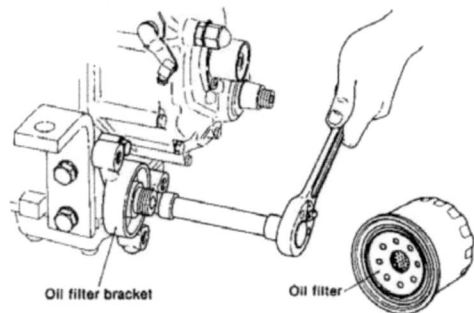

4-2.12 Remove the rocker arm chamber

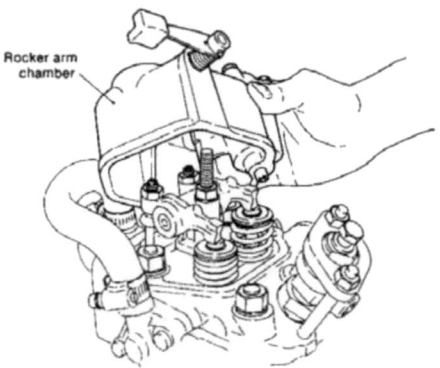

4-2.13 Remove the mixing elbow

(1) Disconnect the cooling water bypass hose
(2) Remove the mixing elbow

4-2.14 Remove the rocker arms

(1) Remove the rocker arm ass'y

(2) Pull the push rods
(3) Remove the cotter pins of the intake and exhaust valve springs.
NOTE: Arrange parts by intake and exhaust.

4-2.15 Remove the cylinder head

(1) Disconnect the lubricating oil pipe located at the cylinder block and the cylinder head.
(2) Remove the cylinder head nuts in the prescribed order, and remove the cylinder head.

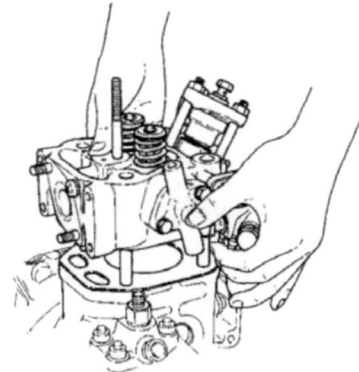

(3) Remove the gasket packing
NOTE: Clearly identify the front and back of the gasket packing.

4-2.16 Remove the crankshaft pulley

Remove the crankshaft pulley end nut and remove the V-pulley and key.

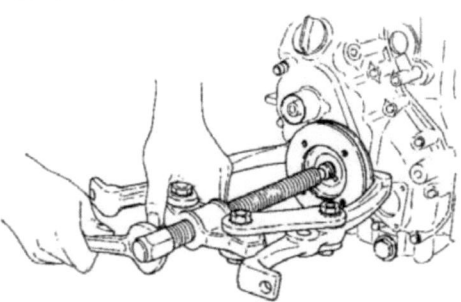

4-2.17 Remove the injection pump

(1) Remove the fixing nut of the fuel injection pump
(2) Open the oil supply hole, move the governor lever 2, and take out the fuel injection pump by matching the control rack with the cut-off part of the gear case.

Chapter 14 Disassembly and Reassembly
4. Disassembly

SM/GM(F)(C)·HM(F)(C)

(3) Remove the injection timing adjustment shims

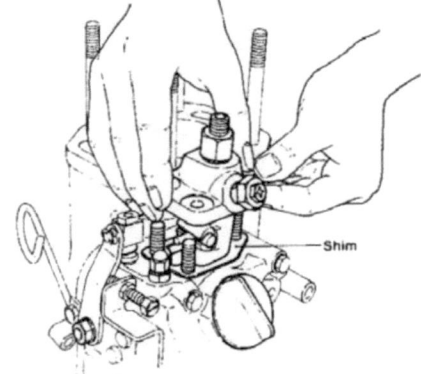

CAUTION: Note the number and total thickness of the timing adjustment shims.

4-2.18 Remove the timing gear case

(1) Remove the starting shaft cover, loosen the bolt with the hexagonal socket head, and withdraw the pin for handle fitting.
(2) Remove the gear case

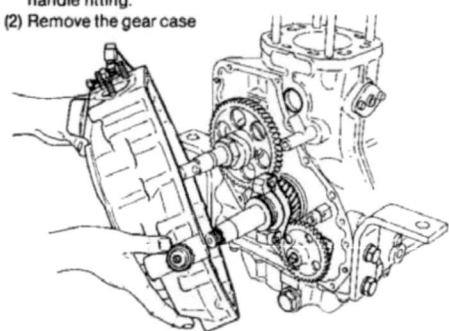

(3) Remove the thrust collar, thrust needle bearing, and governor sleeve.

4-2.19 Remove the clutch assembly

Loosen the mounting flange bolts and remove the clutch assembly.

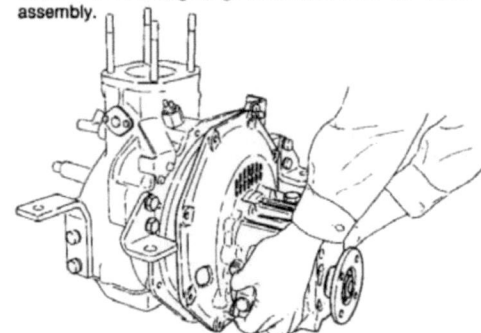

4-2.20 Remove the flywheel

(1) Remove the clutch disk

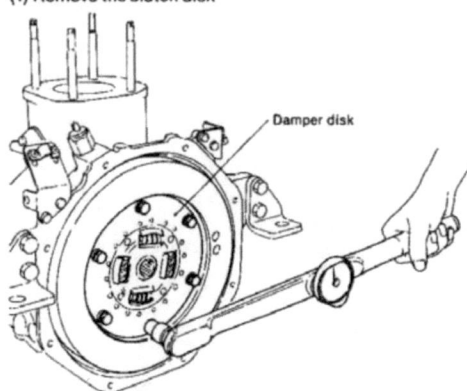

(2) Remove the flywheel
Screw-in the two bolts for securing the clutch disc (slightly to the left and right sides of the flywheel) and remove it by pulling on the bolts.

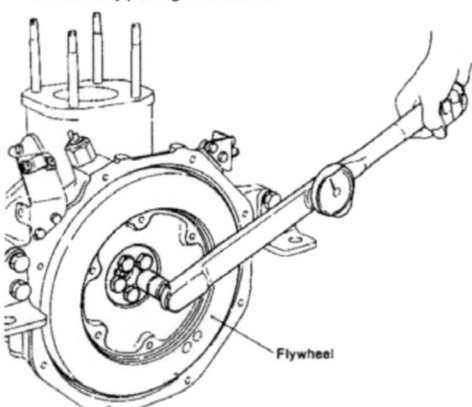

4-2.21 Remove the flywheel housing

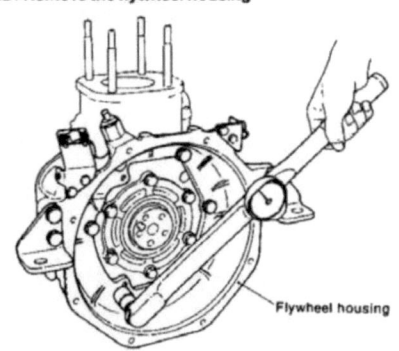

14-17

Chapter 14 Disassembly and Reassembly
4. Disassembly

SM/GM(F)(C)-HM(F)(C)

4-2.22 Remove the feed pump

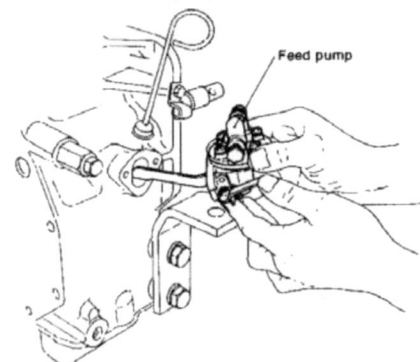

4-2.23 Remove the lubricating oil dipstick
4-2.24 Remove the governor weight assembly

Remove the crankshaft end nut and remove the governor weight assembly.

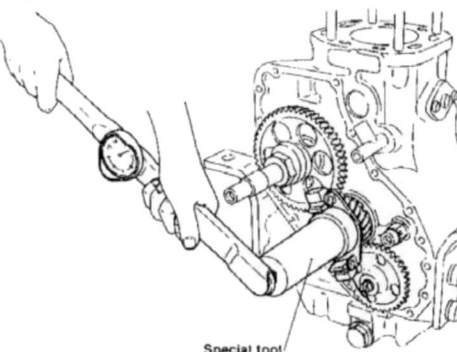

4-2.25 Remove the camshaft gear

(1) Remove the camshaft end nut and remove the fuel cam
(2) Remove the camshaft gear

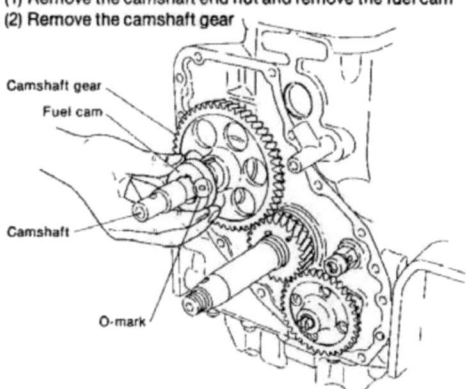

4-2.26 Remove the crankshaft gear and the lubricating oil pump

(1) Remove the crankshaft gear
(2) Remove the lubricating oil pump and gear assembly

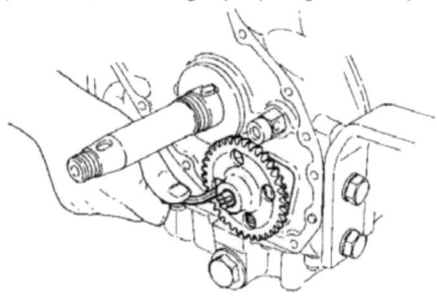

(3) Remove the thrust metal and the thrust washer from the crankshaft.

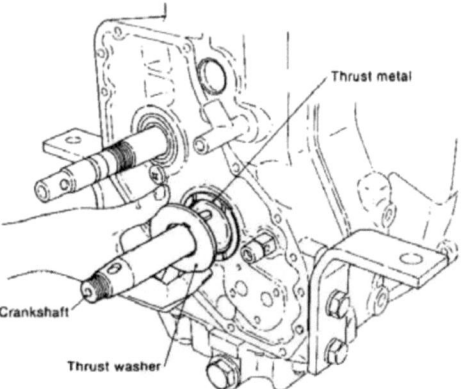

(4) Remove the lubricating oil pressure control valve.

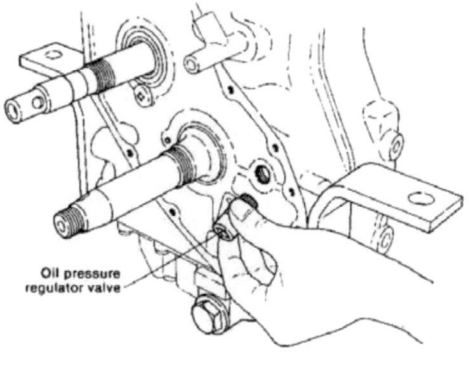

14-18

Printed in Japan
0000A0A1361

Chapter 14 Disassembly and Reassembly
4. Disassembly

SM/GM(F)(C)·HM(F)(C)

4-2.27 Turn the engine onto its side
(1) Remove the engine feet of the camshaft side
(2) Turn the cylinder block over so that the camshaft side is on the bottom.

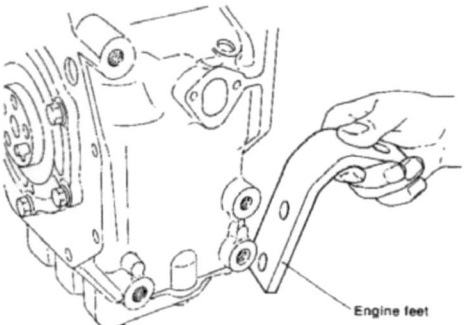
Engine feet

4-2.28 Remove the oil pan and the oil intake pipe

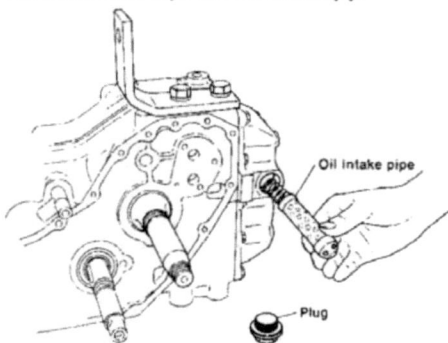
Oil intake pipe
Plug

4-2.29 Remove the piston connecting rod assembly
(1) Set the piston to bottom dead center and remove the connecting rod bolts.

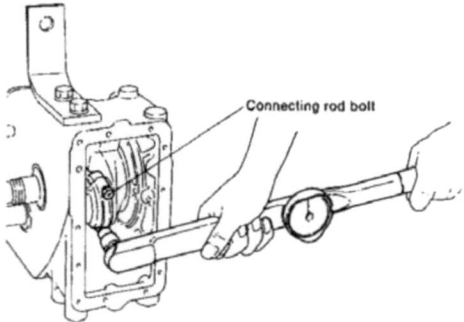
Connecting rod bolt

(2) Set the piston to top dead center, turning the crankshaft so that the connecting rod does not separate from the crank pin. Pull out the piston connecting rod assembly by pushing the large end of the rod with a pusher.

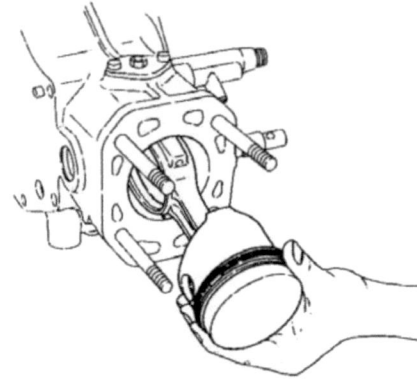

4-2.30 Remove the main bearing housing
Remove the main bearing housing bolt and remove the main bearing housing.

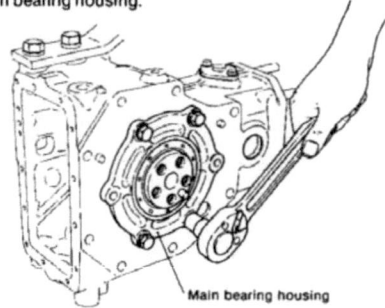
Main bearing housing

4-2.31 Pull the crankshaft
(1) Pull the crankshaft
(2) Remove the thrust metal

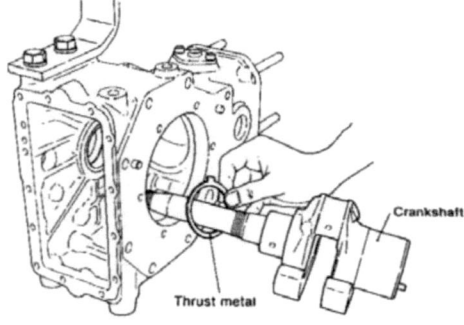
Crankshaft
Thrust metal

4-2.32 Remove the camshaft

(1) Remove the camshaft bearing set screw
(2) Check that all the tappets are separated from the cam, and pull the camshaft out.

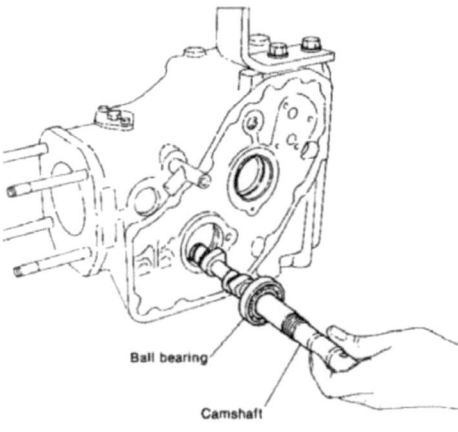

Ball bearing
Camshaft

4-2.33 Remove the tappets

NOTE: Arrange the removed tappets by intake and exhaust.

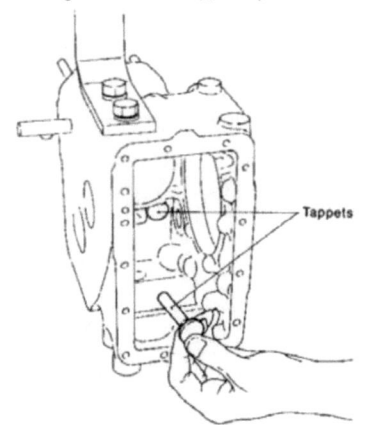

Tappets

4.3 Dismantling engine models 2GM20(C) and 3GM30(C)

For the model 3HM35 engine, refer to the model 3GM30(C) instructions as the procedure is almost the same for both engine models.

4-3.1 Open the cooling water drain cocks and drain the cooling water

(1) Cylinder body water drain cock

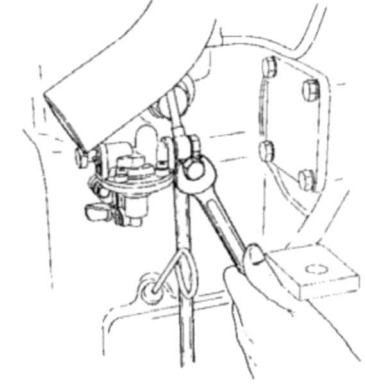

(2) Exhaust pipe water drain cock [only for model 3GM30(C)]

4-3.2 Drain the lubricating oil

(1) Engine side
Insert a suction tube into the dipstick hole and pump out the oil with a waste oil pump (option).

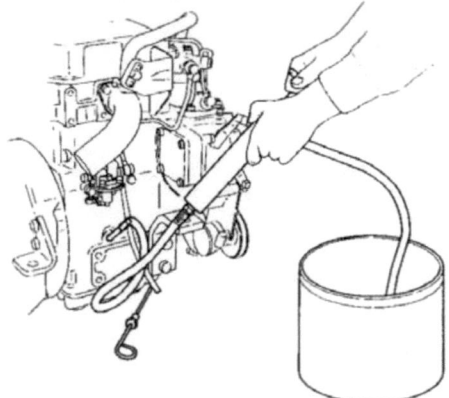

Chapter 14 Disassembly and Reassembly
4. Disassembly

SM/GM(F)(C)·HM(F)(C)

(2) Clutch side
Pump out the oil from the filler/dipstick hole using a waste oil pump or remove the drain plug at the bottom stern side of the clutch case and drain the oil.

(1) Alternator wiring
(2) Starter motor wiring
(4) Water temperature switch wiring
(4) Oil pressure switch wiring
(5) Tachometer sender wiring

4-3.5 Disconnect the cooling water inlet pipe and bilge pipe

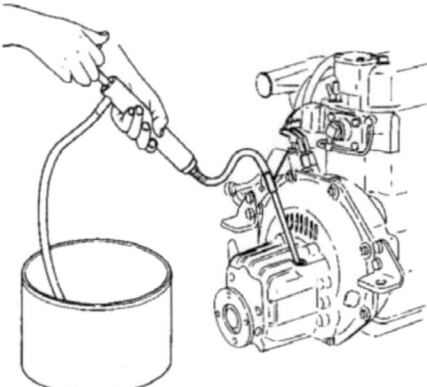

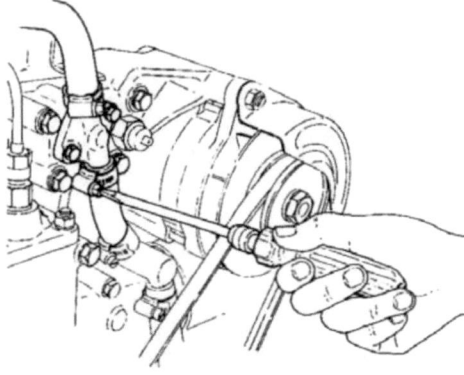

4-3.3 Disconnect the remote control cables

NOTE: Always close the Kingston cock

4-3.6 Remove the air intake silencer

Remove the intake silencer clip and the filter element. Then remove the set screw and the cover.

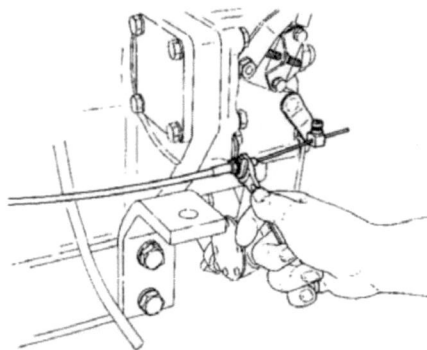

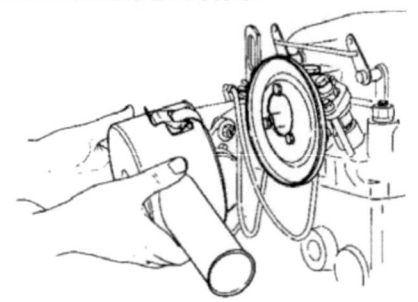

(1) Clutch remote control cable and bracket
(2) Speed remote control cable and bracket
(3) Engine stop remote control cable and bracket
(4) Decompression remote control cable

4-3.4 Disconnect the electrical wiring

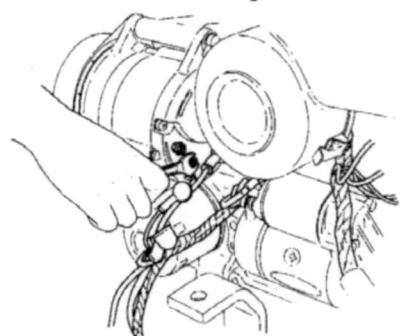

4-3.7 Disconnect the fuel piping

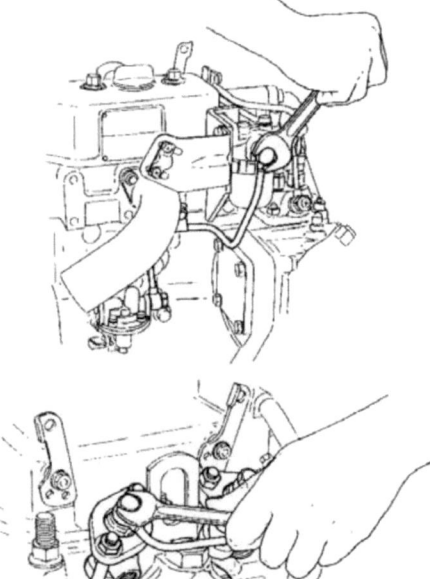

(1) Fuel tank to feed pump
(2) Feed pump to fuel filter
(3) Fuel filter to fuel injection pump
(4) Fuel high pressure pipe
(5) Fuel return pipe

4-3.8 Remove the starter motor

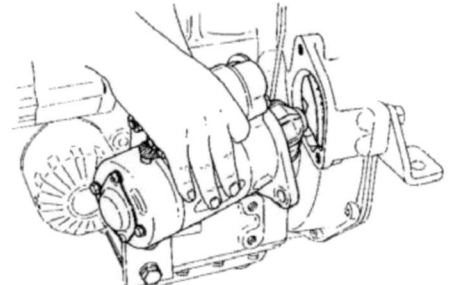

4-3.9 Remove the alternator

(1) Loosen the adjusting bolt and remove the V-belt
(2) Remove the alternator and bracket

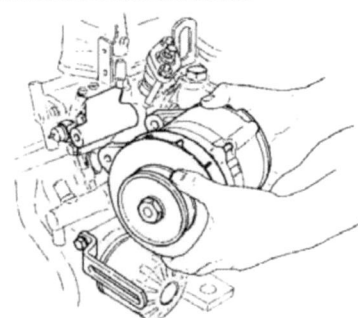

4-3.10 Remove the oil filter

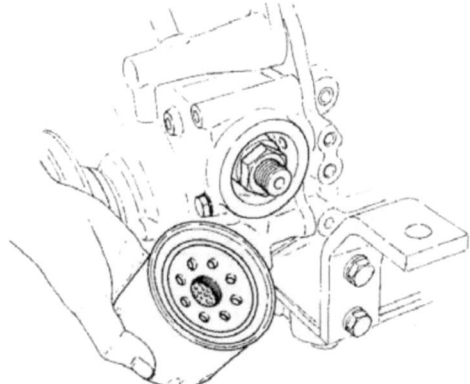

4-3.11 Remove the water pump

(1) Disconnect the hose between the water pump and cooling water cylinder inlet joint.

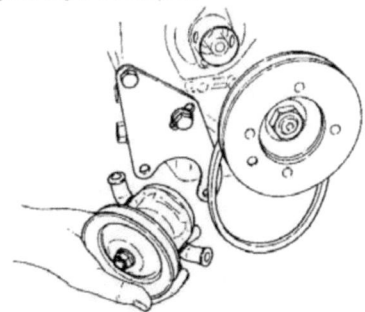

(2) Loosen the water pump mounting bolts, remove the V-belt by sliding it toward the crankshaft side, and remove the water pump.

Chapter 14 Disassembly and Reassembly
4. Disassembly

4-3.12 Remove the rocker arm chamber

(1) Remove the breather pipe at the side of the intake pipe [intake manifold for model 3GM30(C)].
(2) Remove the rocker arm chamber

4-3.13 Remove the exhaust manifold
[only for model 3GM30(C)] and the mixing elbow

(1) Disconnect the cooling water bypass hose at the thermostat cover side.
(2) Remove the mixing elbow [2GM20(C)].
(3) Remove the exhaust manifold together with the fuel filter and mixing elbow [3GM30(C)].

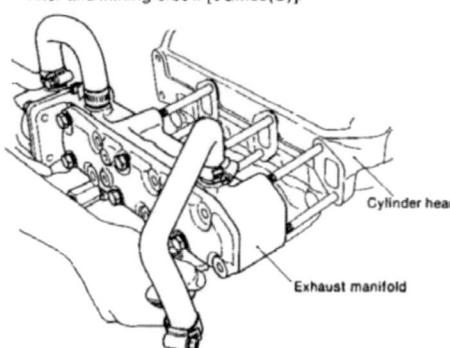

4-3.14 Remove the rocker arms

(1) Remove the mounting nut and remove the rocker arm shaft assembly.

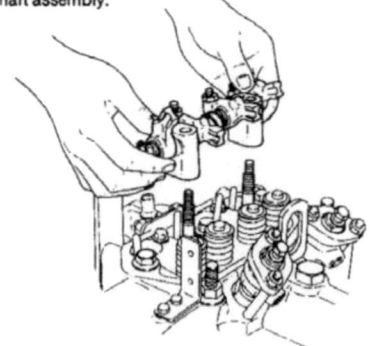

(2) Pull the push rods.

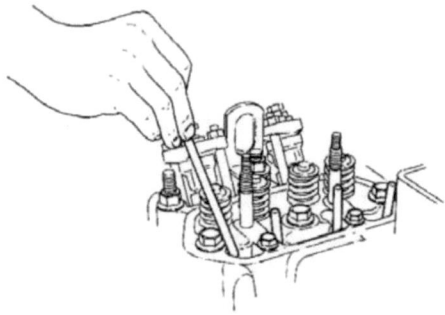

(3) Remove the cotter pins of the intake and exhaust valve springs.
NOTE: Arrange the parts by cylinder no., intake and exhaust.

4-3.15 Remove the cylinder head

(1) Disconnect the lubricating oil pipe.
(2) Remove the cylinder head nuts in the prescribed order, and remove the cylinder head.
(3) Remove the gasket packing

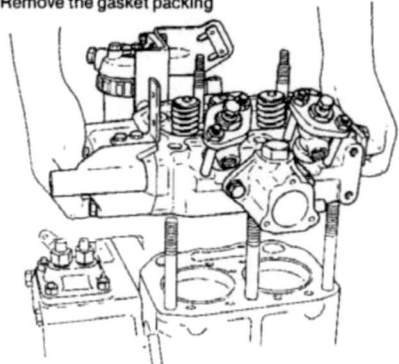

NOTE: Clearly identify the front and back of the gasket packing.

4-3.16 Remove the crankshaft pulley

Remove the crankshaft pulley end nut and remove the V-pulley and key.

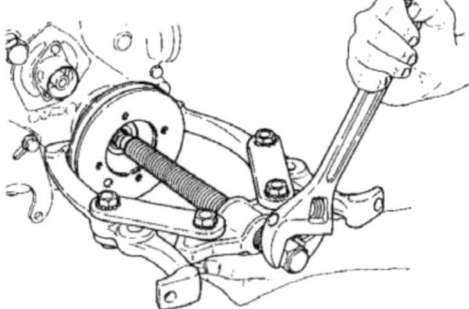

4-3.17 Remove the injection pump

(1) Remove the injection pump nut.
(2) Remove the gear case side cover, move the governor lever 2, take out the fuel injection pump by matching the control rack with the cut-off part of the gear case.

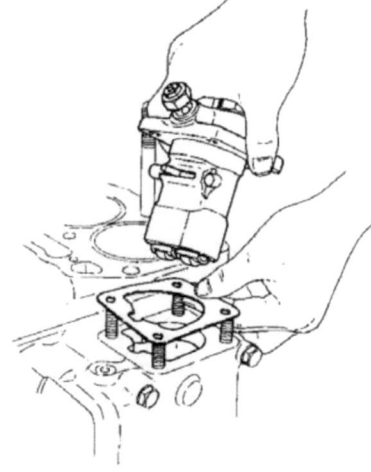

(3) Remove the injection timing adjustment shims.
CAUTION: Note the number and total thickness to the timing adjustment shims.

4-3.18 Remove the timing gear case

(1) Remove the starting shaft cover, loosen the bolt with the hexagonal socket head, and withdraw the pin for handle fitting.

(2) Remove the gear case

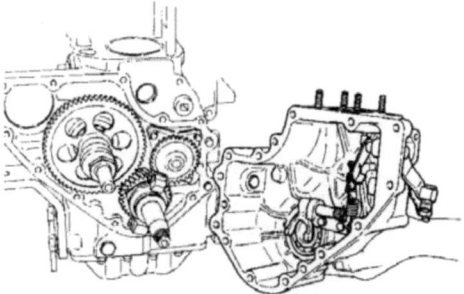

(3) Remove the thrust collar, thrust needle bearing, and governor sleeve.

4-3.19 Remove the clutch assembly

Loosen the mounting flange bolts and remove the clutch assembly.

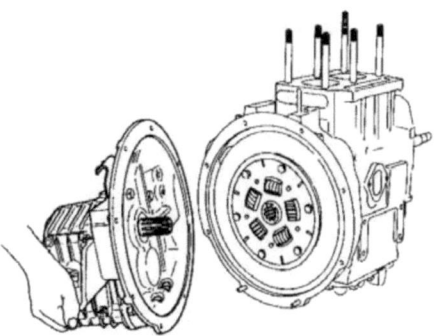

Chapter 14 Disassembly and Reassembly
4. Disassembly

4-3.20 Remove the flywheel
(1) Remove the damper disk

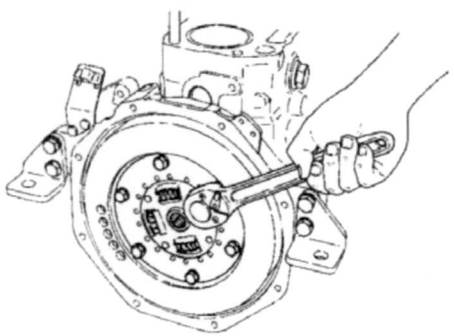

(2) Remove the flywheel
Screw-in the two bolts to secure the clutch disk (slightly to the left and right sides of the flywheel) and remove it by pulling on the bolts.

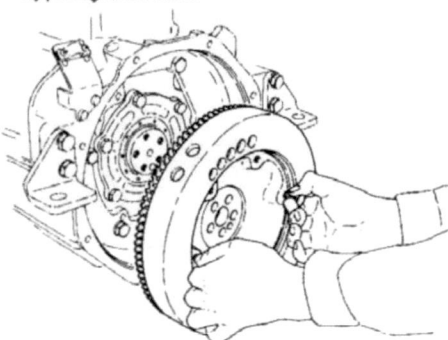

4-3.21 Remove the flywheel housing

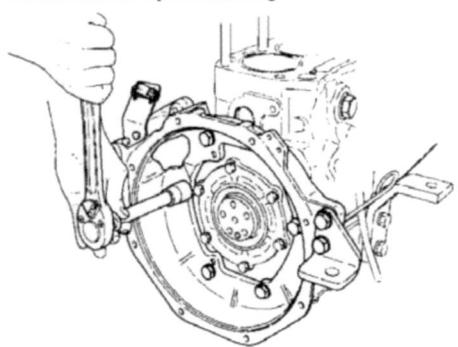

4-3.22 Remove the lubricating oil dipstick
4-3.23 Remove the feed pump

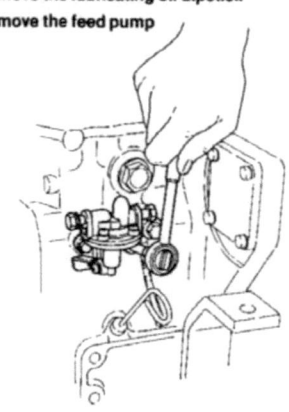

4-3.24 Remove the fuel cam
Remove the camshaft end nut and remove the fuel cam

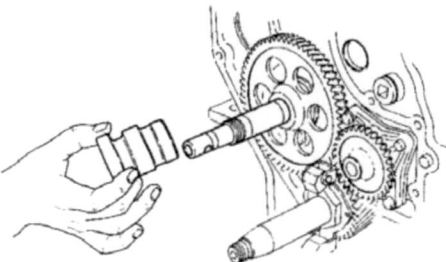

4-3.25 Remove the governor weight assembly

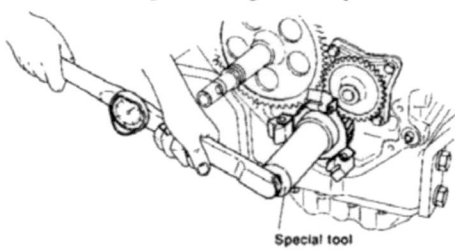

Special tool

Remove the crankshaft end nut and remove the governor weight assembly.

Chapter 14 Disassembly and Reassembly
4. Disassembly

SM/GM(F)(C)·HM(F)(C)

4-3.26 Remove the lubricating oil pump and driving gear assembly

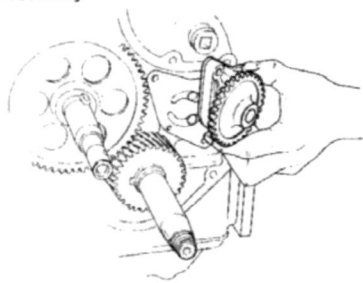

4-3.27 Remove the camshaft gear and the crankshaft gear

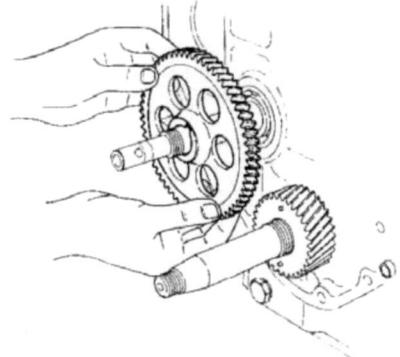

4-3.28 Turn the engine onto its side

(1) Remove the engine feet of the crankshaft side
(2) Turn the cylinder block over so that the crankshaft side is on the bottom.

4-3.29 Remove the oil pan and the oil intake pipe

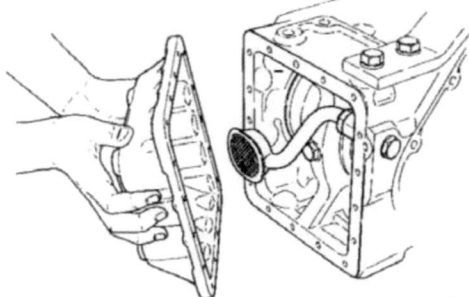

4-3.30 Remove the piston connecting rod assembly

(1) Set the piston to bottom dead center and remove the connecting rod bolts.

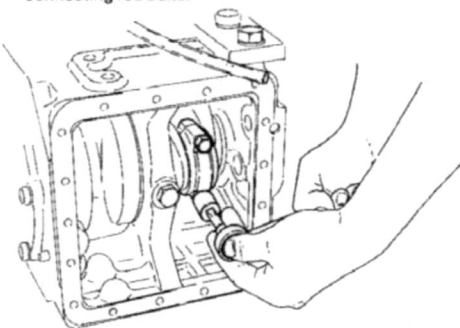

(2) Set the piston to top dead center, turning the crankshaft so that the connecting rod does not separate from the crank pin. Pull out the piston connecting rod assembly by pushing the large end of the rod with a pusher.

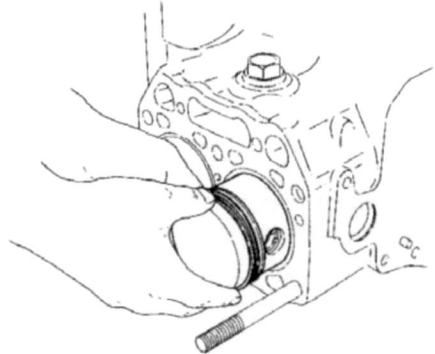

4-3.31 Remove the main bearing housing

Remove the main bearing housing bolt and remove the main bearing housing.

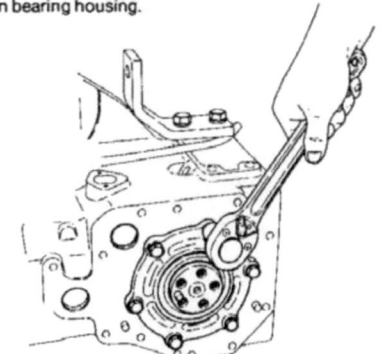

14-26

Printed in Japan
0000A0A1361

Chapter 14 Disassembly and Reassembly
4. Disassembly

4-3.32 Remove the mounting bolt of the intermediate main bearing

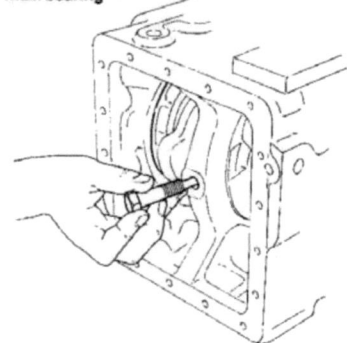

Two intermediate main bearings, viz. No.1 and No.2, for engine model 3GM30(C).

4-3.33 Pull the crankshaft

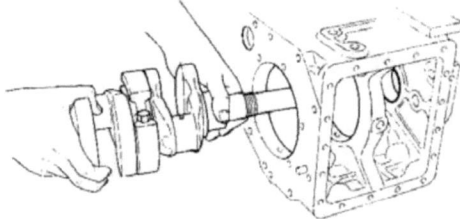

4-3.34 Remove the camshaft

(1) Remove the camshaft bearing set screw.

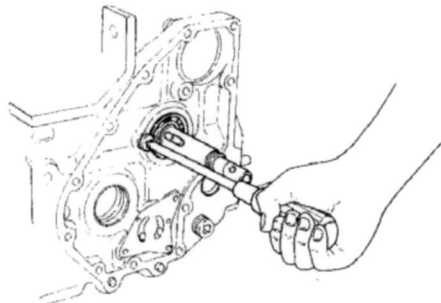

(2) Place the cylinder block upside down or raise the cylinder block by inserting a plate beneath it in order to prevent contact between the tappet and the cam.

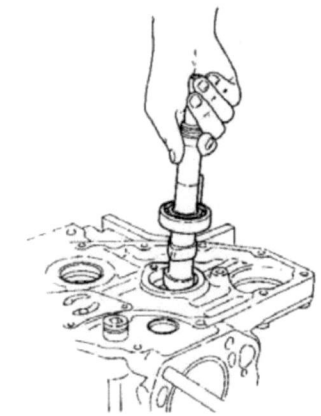

(3) Check that all the tappets are separated from the cam, and pull the camshaft out.

4-3.35 Remove the tappets

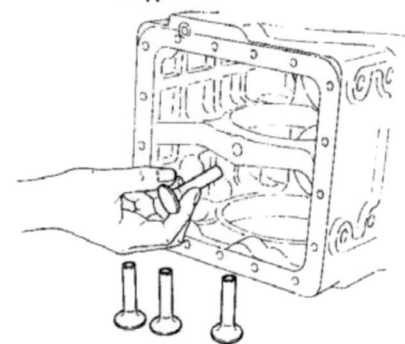

NOTE: *Arrange the removed tappets by cylinder no. and intake and exhaust groups.*

4-3.36 Remove the liners

Set the engine upright and pull the liners with a liner puller.

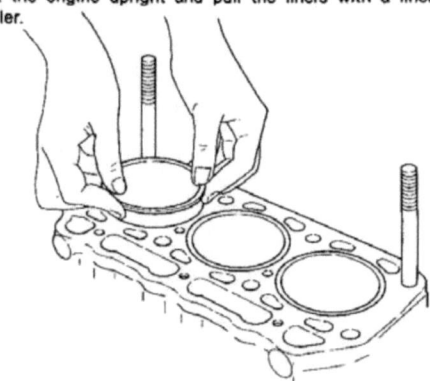

14-27

5. Reassembly

5.1 General Precautions

Warped washers and packings must necessarily be replaced with new ones.

In assembling, sealing must be applied to all designated parts. Omission may cause serious trouble during a trial running of the engine after completion of reassembly. Adjustments should be performed in accordance with the instructions given.

After completion of engine reassembly, recheck any deficiencies which might have appeared during maintenance and inspection, conduct a trial running of the engine and then submit it to the user.

5.2 Reassembly of engine model 1GM10(C)

5-2.1 Insert the tappets

(1) Turn the cylinder block over or turn it upside down.
(2) Coat the tappets with oil and insert into the tappet holes.

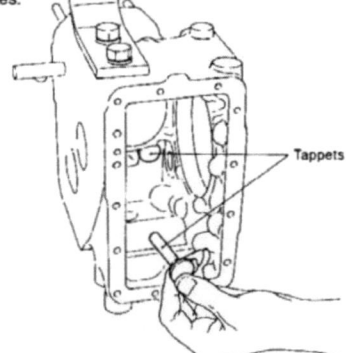

Tappets

NOTE: Assemble the tappets at their original positions, paying careful attention to intake and exhaust.

5-2.2 Insert the camshaft

(1) Coat the camshaft bearing section with oil and insert the camshaft into the cylinder block by tapping the shaft end with a plastic hammer.

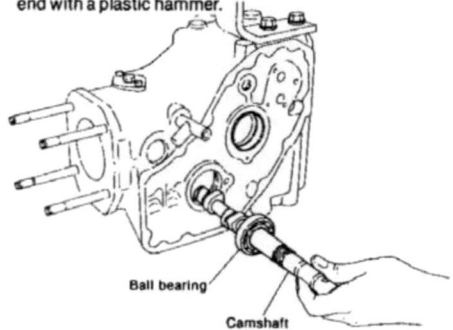

Ball bearing
Camshaft

NOTE: Be careful not to damage the groove in the end of the shaft.

(2) After inserting the camshaft, check that it rotates smoothly before tightening the camshaft bearing set screw.

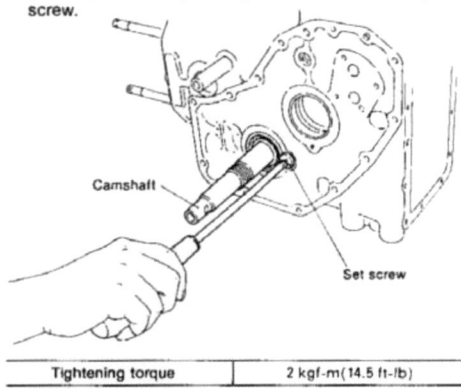

Camshaft
Set screw

| Tightening torque | 2 kgf-m (14.5 ft-lb) |

5-2.3 Install the crankshaft

(1) Coat the cam gear side thrust metal with oil and install.

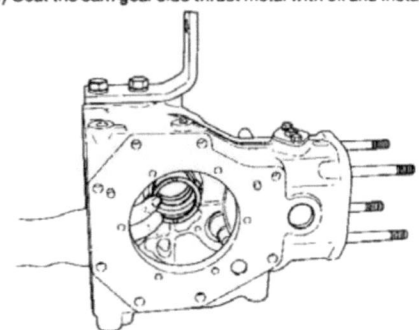

CAUTION: Install so that the thrust metal oil groove is at the crankcase side, being careful not to damage the tab.

Chapter 14 Disassembly and Reassembly
5. Reassembly

(2) Insert the crankshaft

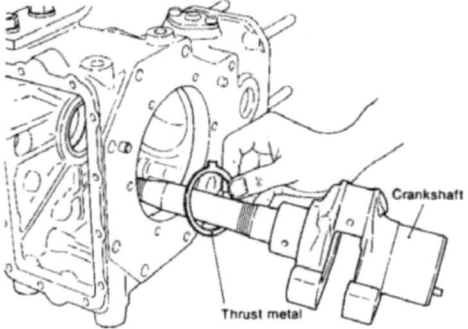
Crankshaft
Thrust metal

5-2.4 Install the main bearing housing

(1) Coat the oil seal section with oil
(2) Insert the main bearing housing and tighten.

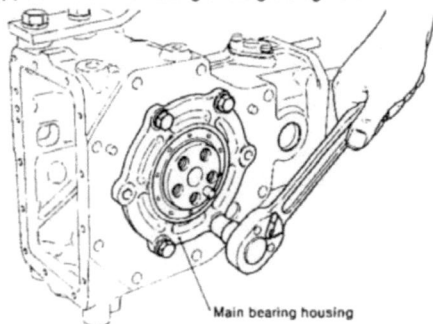
Main bearing housing

Tightening torque	2.5 kgf-m (18 ft-lb)

(3) Check that the crankshaft rotates smoothly
(4) Measure the crankshaft side gap, and adjust it to the prescribed value by the thickness of the packing.

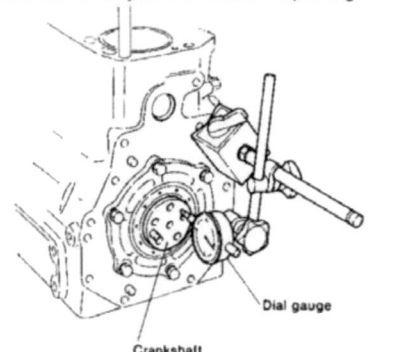
Dial gauge
Crankshaft

Crankshaft side gap	0.06 ~ 0.19mm (0.0024 ~ 0.0075in.)

5-2.5 Assemble the piston and connecting rod assembly

(1) Coat the crankpin section with oil and position so that the crank is at the top.
(2) Coat the piston and crankpin bearing with oil.
(3) Position the piston rings so that the gaps are 120° apart, being sure that there is no gap at the side pressure section.

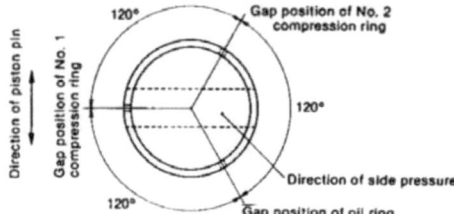

(4) Insert the piston connecting rod assembly so that the side of the connecting rod big end with the identification number is on the camshaft side.
Install the piston rings with a piston ring inserter.

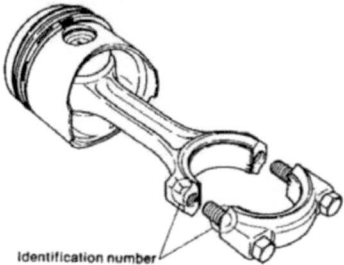
Identification number

(5) After the connecting rod large end contacts the crankpin, push the piston crown down slowly to turn the crankshaft to bottom dead center.

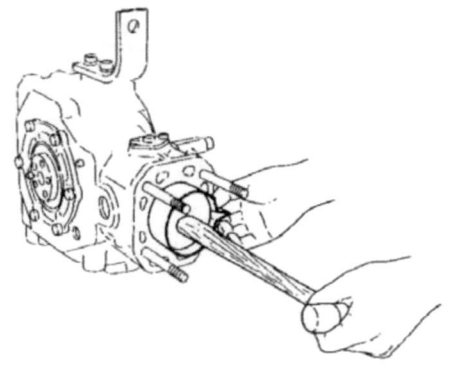

Chapter 14 Disassembly and Reassembly
5. Reassembly

SM/GM(F)(C)·HM(F)(C)

(6) Align the connecting rod cap and connecting rod large end matching mark and tighten the connecting rod bolts.

CAUTION: 1. Be careful to tighten the connecting rod bolts evenly.
2. Coat the bolt threads and washer face with oil.

| Tightening torque | 0.9 kgf-m(6.5 ft-lb) |

5-2.8 Install the mounting flange
(1) Set the engine upright.
(2) Align the positioning pins and tighten the flange.

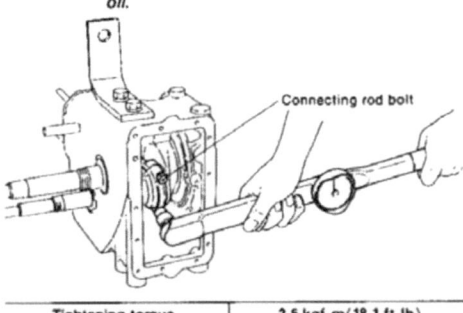

| Tightening torque | 2.5 kgf-m(18.1 ft-lb) |

(7) Measure the side clearance.

| Side clearance | 0.2 ~ 0.4mm (0.0079 ~ 0.0157in.) |

(8) Check that the crankshaft rotates smoothly.

5-2.6 Intall the lubricating oil intake pipe to the oil pan

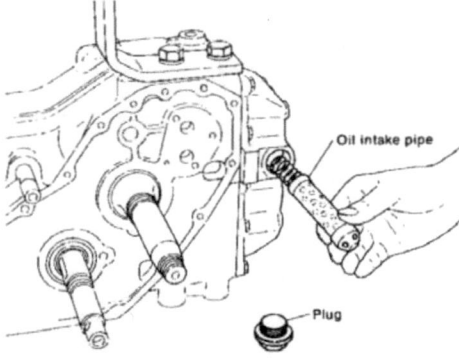

5-2.7 Install the engine bottom cover (oil pan)
(1) Change the packing.
(2) Install the bottom cover.

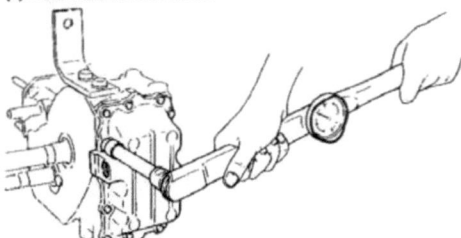

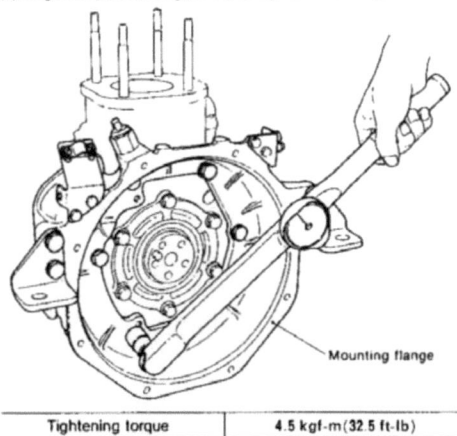

| Tightening torque | 4.5 kgf-m(32.5 ft-lb) |

5-2.9 Install the flywheel
(1) Align the reference pins
(2) Install the flywheel

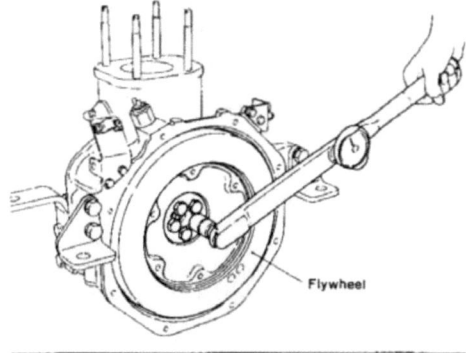

| Tightening torque | 6.5 ~ 7.0 kgf-m (47 ~ 50.6 ft-lb) |

NOTE: After tightening, check the end run-out

5-2.10 Install the clutch assembly
(1) Install the clutch disc on the flywheel

| Tightening torque | 2.5 kgf-m(18 ft-lb) |

(2) Align the disk and input shaft spline, and install the clutch assembly on the mounting flange.

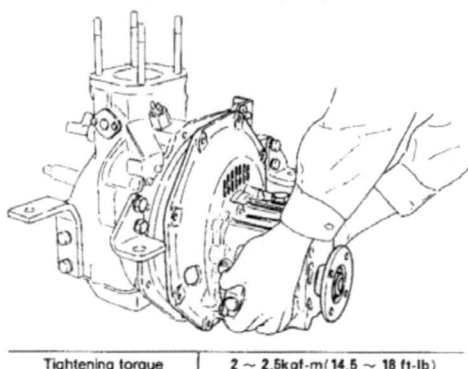

| Tightening torque | 2 ~ 2.5kgf-m(14.5 ~ 18 ft-lb) |

5-2.11 Install the engine feet and set the engine in position

(1) Dipstick
(2) Fuel feed pump

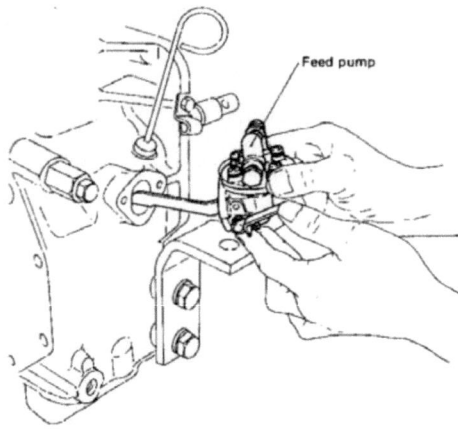

Feed pump

5-2.12 Assemble the thrust metal and thrust washer

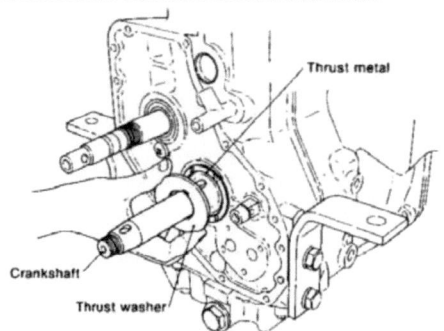

Thrust metal

Crankshaft

Thrust washer

5-2.13 Install the lubricating oil pump and gear assembly

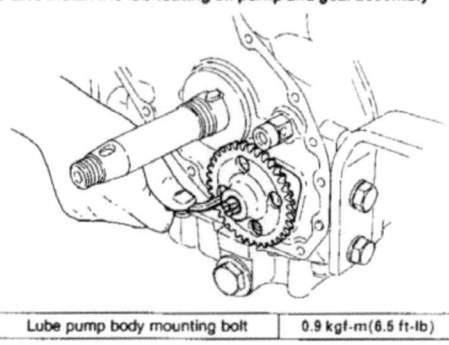

| Lube pump body mounting bolt | 0.9 kgf-m(6.5 ft-lb) |

5-2.14 Assemble the crankshaft gears

(1) Coat the crankshaft section and the inside of the gear with oil.
(2) Insert the crankshaft gear

5-2.15 Assemble the camshaft gear and fuel cam

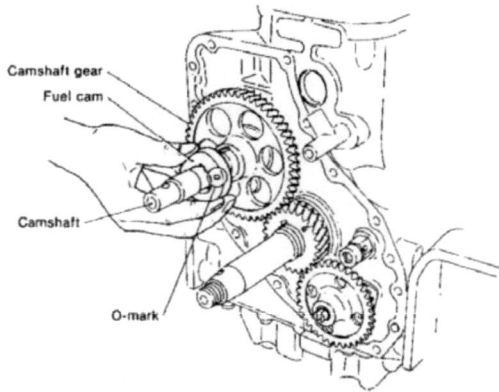

Camshaft gear
Fuel cam
Camshaft
O-mark

(1) Coat the shaft hole of the camshaft gear with oil, and align the matching marks of the camshaft gear and the crankshaft gear and insert the camshaft gear.
(2) Coat the fuel cam with oil and insert the cam by aligning the "0" mark opposite the camshaft gear.
(3) Tighten the camshaft end nut

| Tightening torque | 7 ~ 8kgf-m(50.6 ~ 57.9 ft-lb) |

(4) Check the backlash

mm (in.)

	Maintenance standard	Wear limit
Crankshaft gear and camshaft gear backlash		
Crankshaft gear and lubrication oil pump driven gear backlash	0.05 ~ 0.13 (0.0020 ~ 0.0051)	0.3 (0.0118)
Camshaft gear and fuel feed pump driven gear backlash		

Chapter 14 Disassembly and Reassembly
5. Reassembly

5-2.16 Install the governor weight assembly and tighten the crankshaft end nut

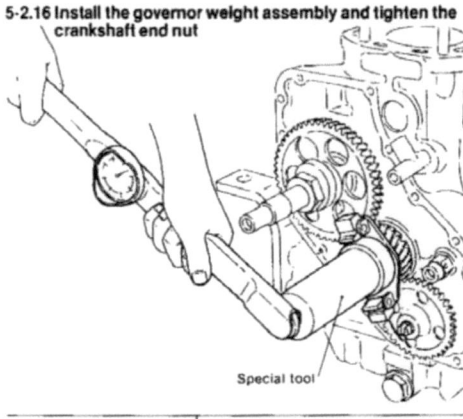

Special tool

Tightening torque	8 ~ 10 kgf-m (57.9 ~ 72.3 ft-lb)

5-2.17 Install the governor sleeve

Install the governor sleeve, thrust needle bearing and thrust collar.

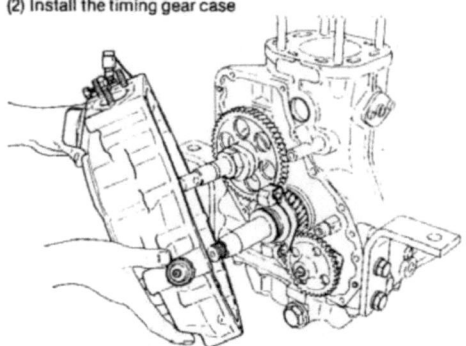

Governor sleeve

Needle bearing
Thrust cover

5-2.18 Install the timing gear case

(1) Coat both sides of the new packing with "Three Bond 3B8-005" and install.
(2) Install the timing gear case

Tightening torque	0.9 kgf-m (6.5 ft-lb)

(3) Insert the pin for fitting the handle into the camshaft and fix it by means of the bolt with the hexagonal socket head, then fit the starting shaft cover.

5-2.19 Install the oil filter and bracket

(1) Install the oil filter bracket on the gear side of the cylinder block.
(2) Install the oil pipes.
(3) Install the oil filter.

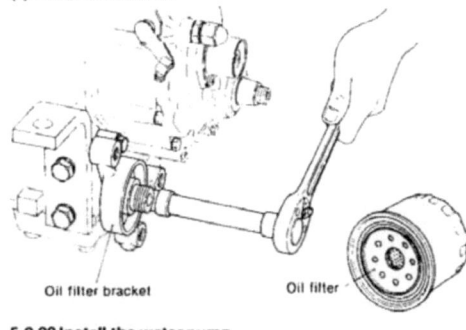

Oil filter bracket — Oil filter

5-2.20 Install the water pump

(1) Install the water pump

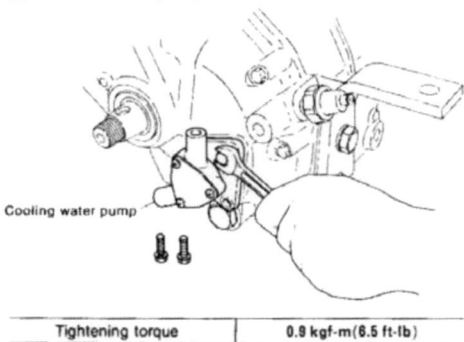

Cooling water pump

Tightening torque	0.9 kgf-m (6.5 ft-lb)

(2) Install the water pipe (pump to cylinder inlet joint)

5-2.21 Install the crankshaft V-pulley

(1) Install the crankshaft key
(2) Coat the crankshaft V-pulley and the inside of the oil seal with oil.
(3) Insert and tighten the V-pulley, making sure that the lip of the oil seal is not distorted.

Tightening torque	10 kgf-m (72.3 ft-lb)

5-2.22 Install the fuel injection pump

(1) Remove grease from both sides of the fuel injection timing adjustment shims with thinner, and coat the shims with "Screw Lock Super 203M."

Chapter 14 Disassembly and Reassembly
5. Reassembly

(2) Insert the pump by looking through the oil filler and align the governor No.2 lever and rack connecting part.

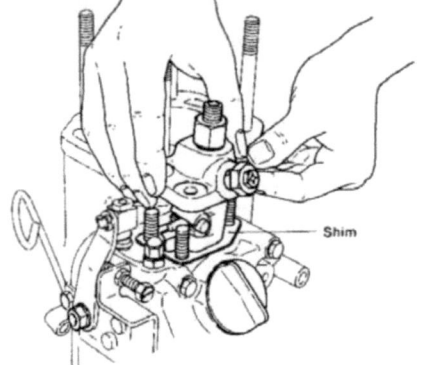

(3) Tighten the fuel pump

Tightening torque	2.5 kgf-m (18 ft-lb)

5-2.23 Install the cylinder head

(1) Install the gasket packing
CAUTION: Take particular note of the fitting surfaces. Fit the side with the recessed part of the cooling water passage to the cylinder block side.
(2) Insert the cylinder head, being careful not to damage the threads of the tightening bolts, and tighten the nuts in the tightening sequence.

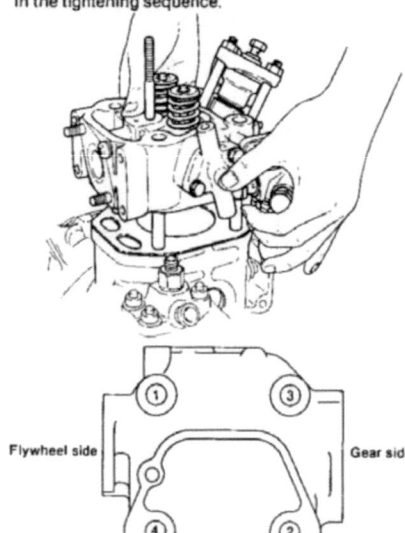

Tightening torque	7.5 kgf-m (54.2 ft-lb)

(3) Install the water pipe (from the thermostat cover to the cylinder inlet joint)

5-2.24 Install the rocker arms

(1) Install the push rods on the tappets
(2) Coat the inside of valve spring retainer with oil.
(3) Install the rocker arm shaft assembly and tighten the nut.

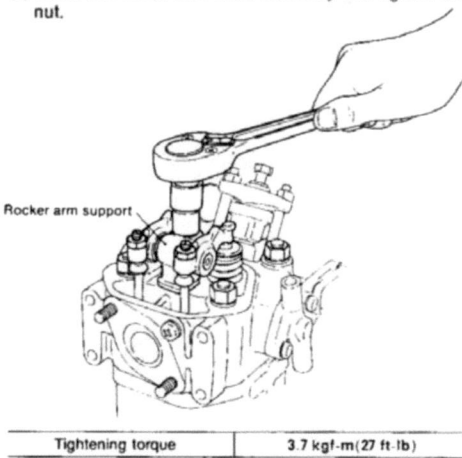

Tightening torque	3.7 kgf-m (27 ft-lb)

CAUTION: 1. Loosen the valve head clearance adjusting screw in advance.
2. Check that the arm moves smoothly.
(4) Adjust the intake and exhaust valve head clearance and lock with the nut.

Intake and exhaust valve head clearance (engine cold):	0.2mm (0.008in.)

5-2.25 Install the rocker arm cover

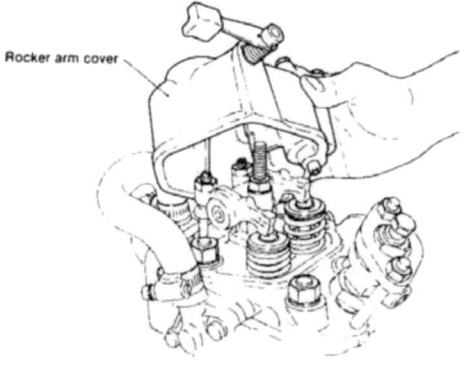

5-2.26 Install the mixing elbow

(1) Install the mixing elbow

(2) Install the cooling water bypass hose
(from the mixing elbow to the thermostat cover)

5-2.27 Install the fuel pipe

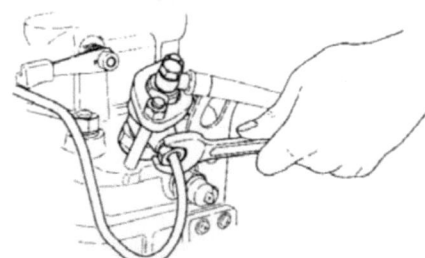

(1) Install the feed pump to fuel filter pipe
(2) Install the fuel filter to fuel injection pump pipe
(3) Install the fuel high pressure pipe
(4) Install the fuel return pipe

5-2.28 Install the starter motor

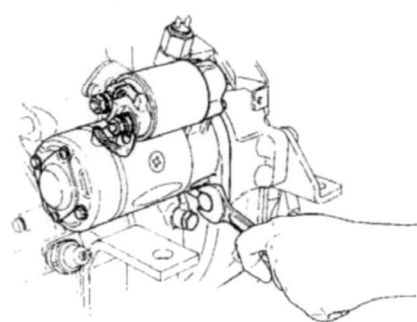

5-2.29 Install the intake silencer

(1) Install the intake silencer cover to the intake port.
(2) Install the intake silencer and tighten it with the clip.

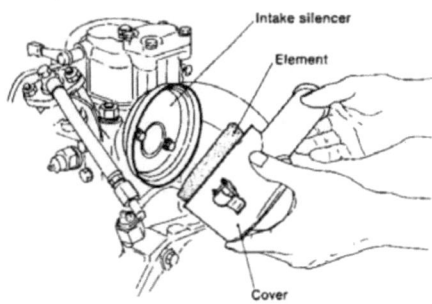

5-2.30 Install the alternator

(1) Install the alternator to the bracket.

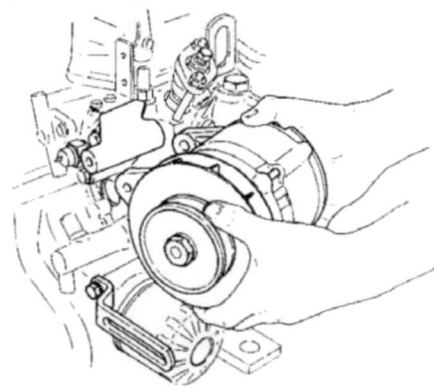

(2) Install the V-belt and tighten the adjusting bolt while adjusting the V-belt tension.

5-2.31 Connect the electrical wiring

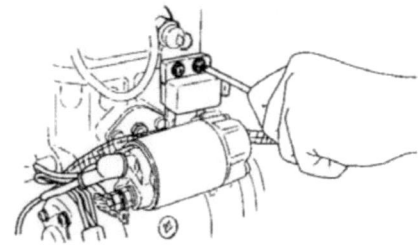

5-2.32 Install the remote control cables
5-2.33 Connect the interior piping

5.3 Reassembly of engine models 2GM20(C) and 3GM30(C)

Refer to the model 3GM30(C) instructions, as the models 3HM35(C) and 3GM30(C) are almost the same.

5-3.1 Assemble the cylinder liners

(1) Remove any rust from the cylinder block where it contacts the cylinder liners.
(2) Coat the outside periphery of the liners with waterproofing paint.
(3) Insert the liners into the cylinder block, making sure to check that the cylinder liner protrusion is correct.

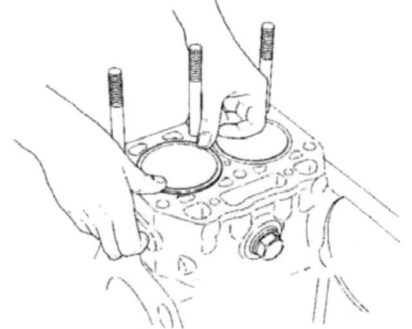

5-3.2 Insert the tappets

(1) Turn the cylinder block over or turn it upside down.
(2) Coat the tappets with oil and insert into the tappet holes.

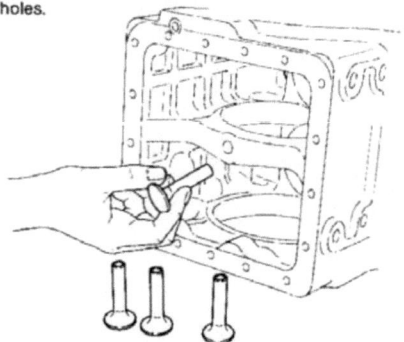

NOTE: Assemble the tappets in their original positions, paying careful attention to the cylinder numbers and intake and exhaust groupings.

5-3.3 Insert the camshaft

(1) Coat the camshaft bearing section with oil and insert the camshaft into the cylinder block by tapping the shaft end with a plastic hammer.

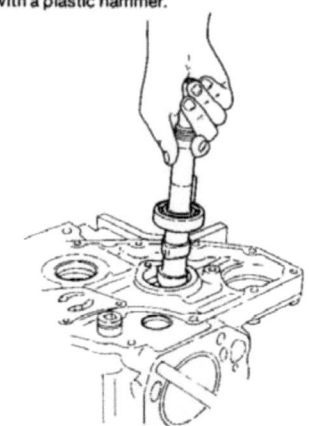

NOTE: Be careful not to damage the groove in the end of the shaft.

(2) After inserting the camshaft, check that it rotates smoothly before tightening the camshaft bearing set screw.

Tightening torque	2 kgf-m (14.5 ft-lb)

5-3.4 Install the crankshaft

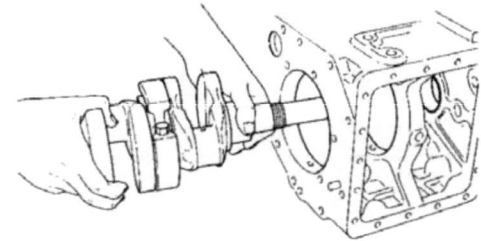

Chapter 14 Disassembly and Reassembly
5. Reassembly

5-3.5 Tighten the set bolt of the intermediate main bearing

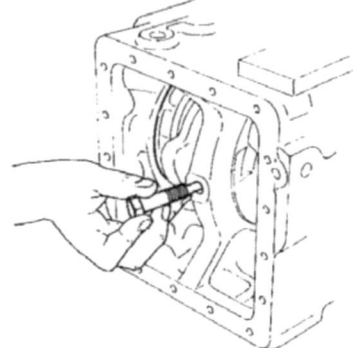

The two intermediate main bearings, viz No.1 and No.2, for model 3GM30(C).

	2GM20(C), 3GM30(C)	3HM35(C)
Tighten torque	4.5 ~ 5.0 (32.5 ~ 36.2)	7.0 ~ 7.5 (50.6 ~ 54.2)

kgf-m (ft-lb)

5-3.6 Install the main bearing housing
(1) Coat the oil seal section with oil
(2) Insert the main bearing housing and tighten

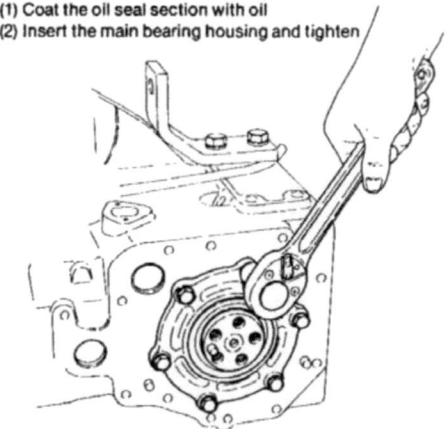

Tightening torque	2.5 kgf-m (18 ft-lb)

(3) Check that the crankshaft rotates smoothly

(4) Measure the crankshaft side gap, and adjust it to the prescribed value by the thickness of the packing.

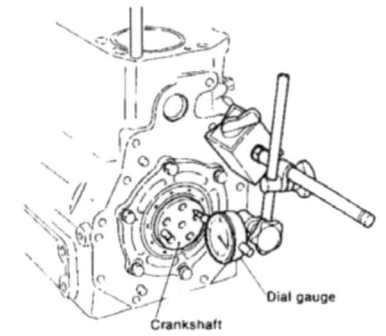

Crankshaft side gap	0.09 ~ 0.18mm (0.035 ~ 0.0071in.)

5-3.7 Assemble the piston and connecting rod assembly
(1) Coat the crankpin section with oil and position so that the insertion side crank is at the top.
(2) Coat the piston and crankpin bearing with oil.
(3) Position the piston rings so that the gaps are 120° apart, being sure that there is no gap at the side pressure section.

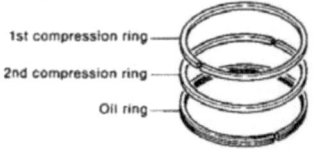

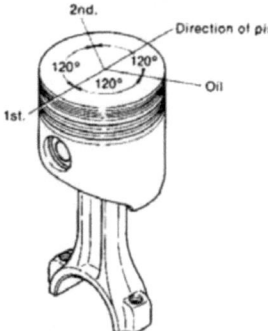

Chapter 14 Disassembly and Reassembly
5. Reassembly

(4) Insert the piston connecting rod assembly so that the side of the connecting rod big end with the identification number is on the exhaust side.
Install the piston rings with a piston ring inserter.

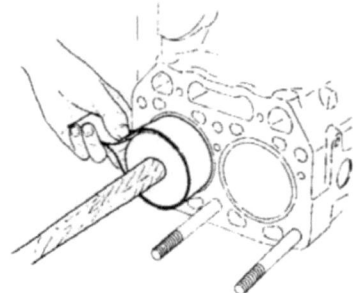

(5) After the connecting rod large end contacts with the crankpin, push the piston crown down slowly to turn the crankshaft to bottom dead center.
(6) Align the connecting rod cap and the connecting rod large end matching mark and tighten the connecting rod bolts.

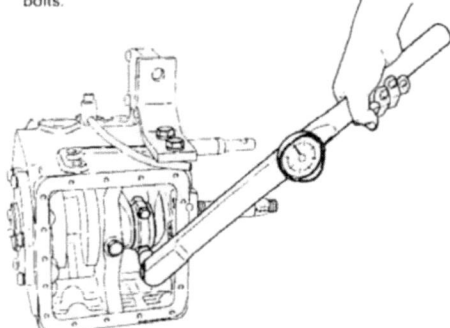

CAUTION: 1. Be careful to tighten the connecting rod bolts evenly.
2. Coat the bolt threads and washer face with oil.

kgf-m (ft-lb)

	2GM20(C), 3GM30(C)	3HM35(F)
Tightening torque	2.5(18.1)	4.5(32.5)

(7) Measure the side clearance

Side clearance	0.2 ~ 0.4mm (0.0079 ~ 0.0157in.)

(8) Check that the crankshaft rotates smoothly

5-3.8 Install the lubricating oil intake pipe
Coat the threads with "Screw Lock Super 203M", screw the pipe in and lock with the nut.

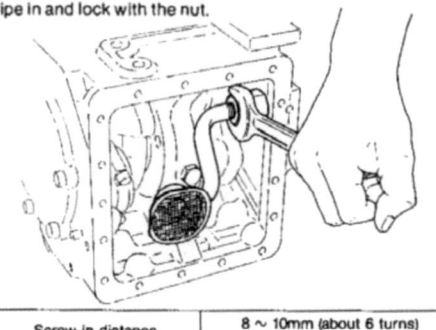

Screw-in distance	8 ~ 10mm (about 6 turns) (0.3149 ~ 0.3937in.)

5-3.9 Install the engine bottom cover (oil pan)
(1) Change the packing
(2) Install the bottom cover

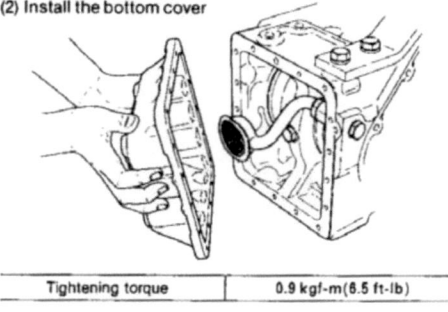

Tightening torque	0.9 kgf-m (6.5 ft-lb)

5-3.10 Install the mounting flange
(1) Set the engine upright
(2) Align the positioning pins and tighten the flange

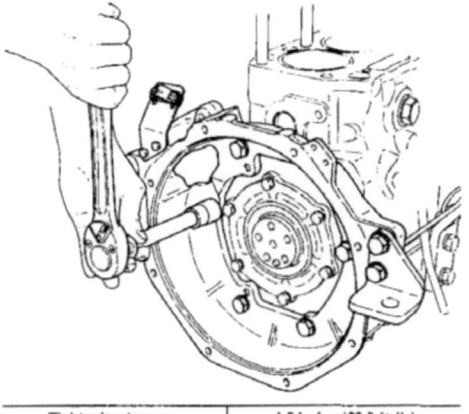

Tightening torque	4.5 kgf-m (32.5 ft-lb)

Chapter 14 Disassembly and Reassembly
5. Reassembly

5-3.11 Install the flywheel
(1) Align the reference pins
(2) Install the flywheel

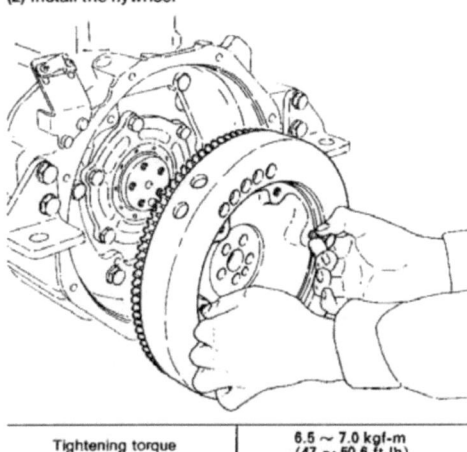

Tightening torque	6.5 ~ 7.0 kgf-m (47 ~ 50.6 ft-lb)

NOTE: After tightening, check the end run-out

5-3.12 Install the clutch assembly
(1) Install the clutch disc on the flywheel

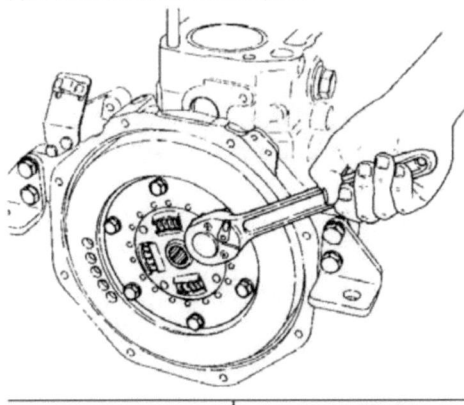

Tightening torque	2.5 kgf-m (18 ft-lb)

(2) Align the disc and input shaft spline, and install the clutch assembly on the mounting flange.

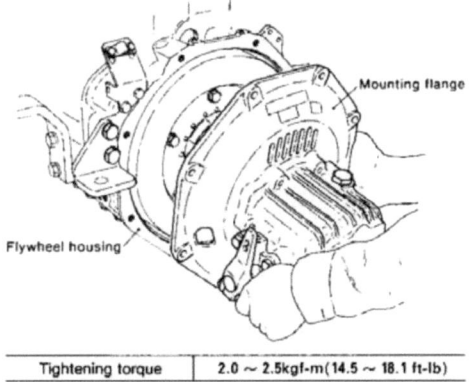

Tightening torque	2.0 ~ 2.5kgf-m (14.5 ~ 18.1 ft-lb)

5-3.13 Install the engine feet and set the engine in position
(1) Dipstick flange and dipstick
(2) Fuel pump

5-3.14 Install the lubricating oil pump
Install the lubricating oil pump and driving gear assembly.

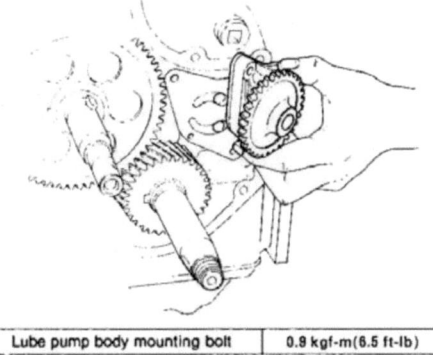

Lube pump body mounting bolt	0.9 kgf-m (6.5 ft-lb)

5-3.15 Assemble the camshaft gear and fuel cam
(1) Coat the shaft hole of the camshaft gear with oil and insert the gear.

Chapter 14 Disassembly and Reassembly
5. Reassembly

(2) Coat the fuel cam with oil and insert the cam by aligning the "0" mark opposite the camshaft gear.

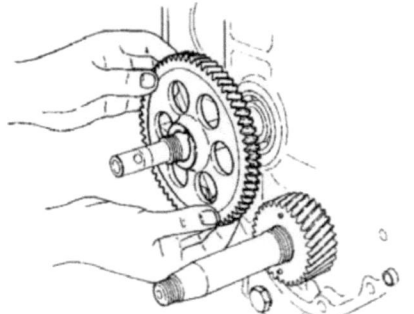

(3) Tighten the camshaft end nut.

Tightening torque	7~8kgf-m(50.6~57.9ft-lb)

5-3.16 Assemble the crankshaft gears

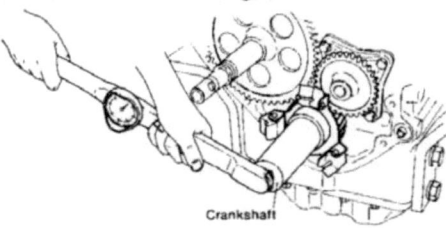

Crankshaft

(1) Coat the crankshaft section and the inside of the gear with oil.
(2) Align the matching marks of the camshaft gear and the crankshaft gear and insert the crankshaft gear.
(3) After inserting the crankshaft gear, check the backlash.

Backlash	0.05~0.13mm (0.0020~0.0051in.)

(4) Install the governor weight assembly and tighten the crankshaft end nut.

Tightening torque	8~10kgf-m(57.9~72.3ft-lb)

5-3.17 Install the governor sleeve

Install the governor sleeve, thrust needle bearing and thrust collar.

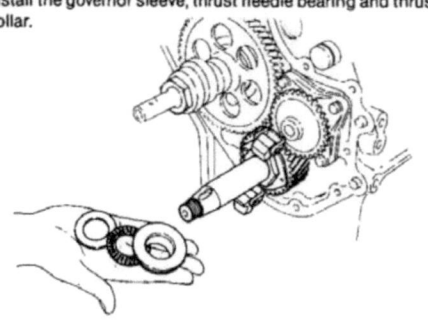

5-3.18 Install the timing gear case

(1) Coat both sides of the new packing with "Three Bond 3B8-005" and install.
(2) Install the timing gear case

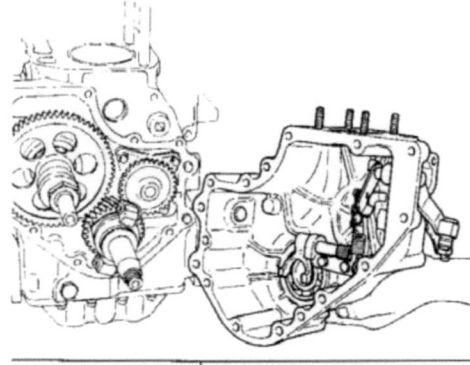

Tightening torque	2.5kgf-m(18ft-lb)

(3) Insert the pin for fitting the handle into the camshaft and fix it by means of the bolt with the hexagonal socket head, then fit the starting shaft cover.

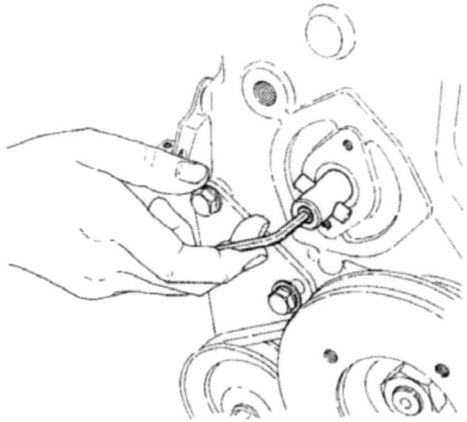

5-3.19 Install the crankshaft V-pulley

(1) Install the crankshaft key
(2) Coat the crankshaft V-pulley and the inside of the oil seal with oil.
(3) Insert and tighten the V-pulley, making sure that the lip of the oil seal is not distorted.

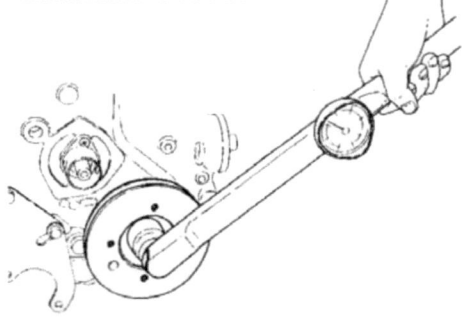

Tightening torque	10kgf-m (72.3ft-lb)

5-3.20 Install the water pump

(1) Install the V-belt to the crankshaft V-pulley and install the water pump.

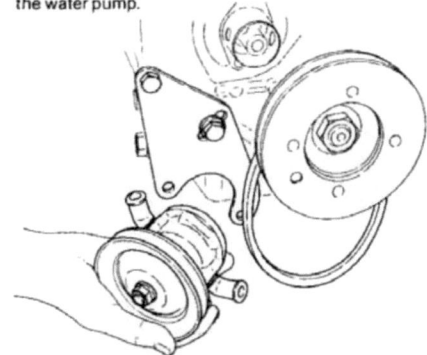

(2) Tighten while adjusting the V-belt tension

Tightening torque	2.5kgf-m (18ft-lb)

(3) Install the water pipe (pump to cylinder inlet joint)

5-3.21 Install the fuel injection pump

(1) Remove grease from both sides of the fuel injection timing adjustment shims with thinner, and coat the shims with "Screw Lock Super 203M."

(2) Insert the pump by looking through the gear case side cover, and align the governor No.2 lever and rack connecting part.

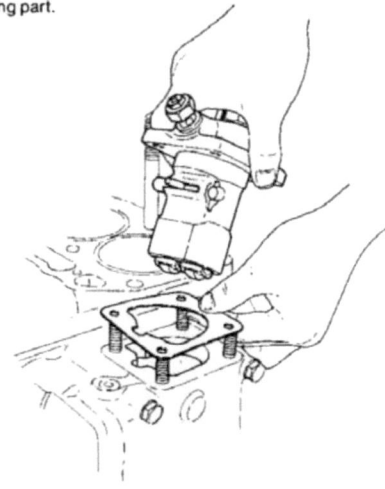

(3) Tighten the fuel pump

Tightening torque	2.5kgf-m (18ft-lb)

(4) Install the gear case side cover

5-3.22 Install the cylinder head

(1) Install the gasket packing

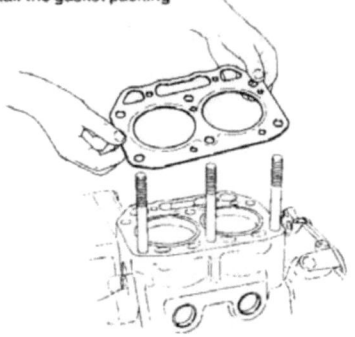

CAUTION: Take particular notice of the surfaces to be fitted.
Fit it keeping the TOP mark to the cylinder head side.

Chapter 14 Disassembly and Reassembly
5. Reassembly

(2) Insert the cylinder head, being careful not to damage the threads of the tightening bolts, and tighten the nuts in the tightening sequence.

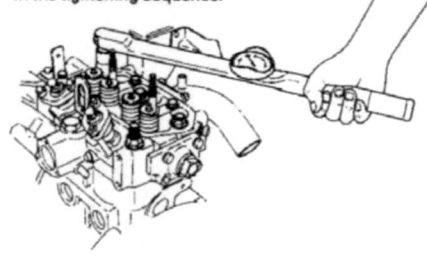

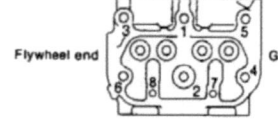

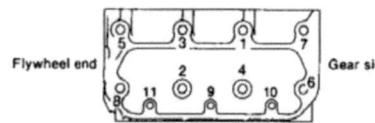

		kgf-m(ft-lb)	
		2GM20(C),3GM30(C)	3HM35(C)
Tightening torque	Main(M12)	12.0(86.8)	13(94.0)
	Sub(M8)	3.0(21.7)	3(21.7)

(3) Install the water pipe
(from the thermostat cover to the cylinder inlet joint)

5-3.23 Install the rocker arms

(1) Install the push rods on the tappets

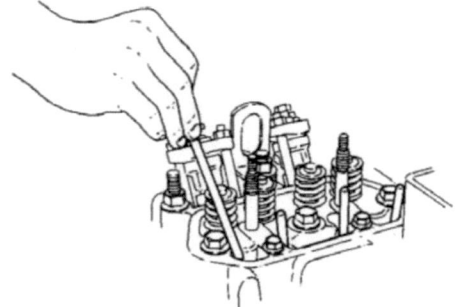

(2) Coat the inside of valve spring retainer with oil
(3) Install the rocker arm shaft assembly and tighten the nut.

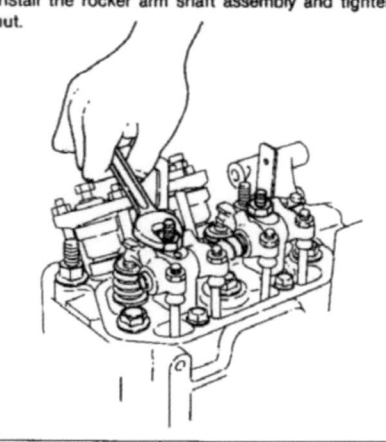

Tightening torque	3.7kgf-m(27ft-lb)

CAUTION: 1. Loosen the valve head clearance adjusting screw in advance.
2. Check that the arm moves smoothly.

(4) Adjust the intake and exhaust valve head clearance and lock with the nut.

Intake and exhaust valve head clearance (engine cold)	0.2mm (0.008in.)

5-3.24 Install the rocker arm cover

(1) Install the rocker arm cover

(2) Install the breather pipe to the air intake pipe [intake manifold ... 3GM30(C).]

5-3.25 Install the exhaust manifold (only for 3-cylinder engine) and the mixing elbow

(1) Install the exhaust manifold with mixing elbow [3GM 30(C)].
(2) Install the mixing elbow. [2GM20(C)].
(3) Install the cooling water bypass hose to the thermostat cover.

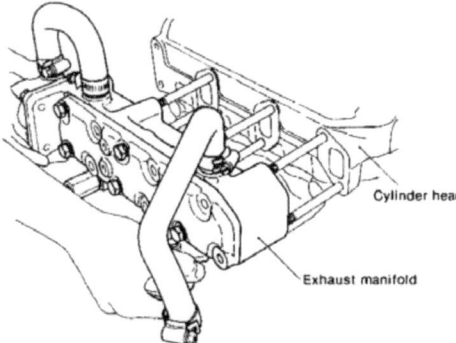

Cylinder head
Exhaust manifold

5-3.26 Install the fuel pipe

(1) Install the feed pump to fuel filter pipe
(2) Install the fuel filter to fuel injection pump pipe
(3) Install the fuel high pressure pipe
(4) Install the fuel return pipe

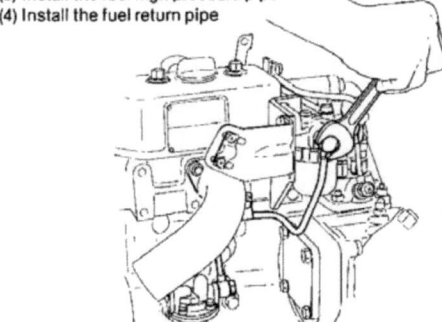

5-3.27 Install the starter motor

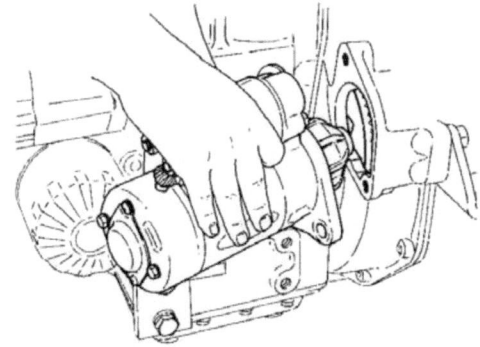

5-3.28 Install the oil filter

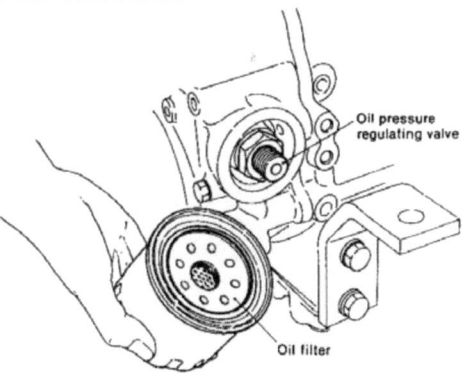

Oil pressure regulating valve
Oil filter

5-3.29 Install the intake silencer

(1) Install the intake silencer cover to the air intake pipe. [intake manifold . . . 3GM30(C)].
(2) Install the intake silencer and tighten it with the clip

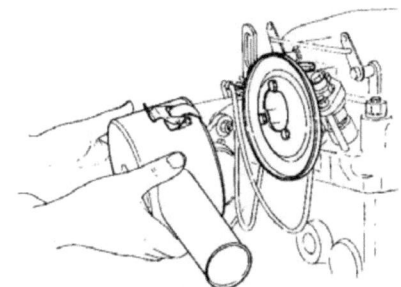

5-3.30 Install the alternator

(1) Install the alternator to the bracket

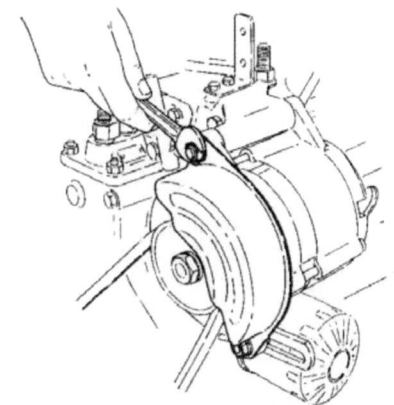

(2) Install the V-belt and tighten the adjusting bolt while adjusting the V-belt tension.

5-3.31 Connect the electrical wiring

5-3.32 Install the remote control cables

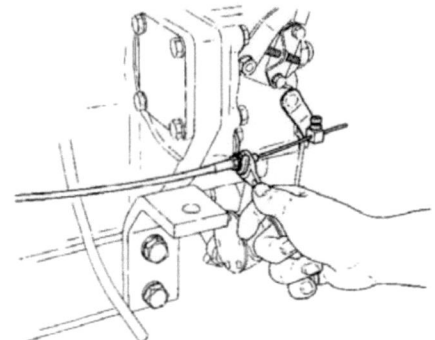

5-3.33 Connect the interior piping

CHAPTER 15
DISASSEMBLY AND REASSEMBLY
(Fresh Water Cooling Engine)

1. Disassembly of Fresh Water-Cooled Engine..............15-1
2. Reassembly of Fresh Water-Cooled Engine..............15-11
3. Tightening Torque......................................15-21
4. Packing Supplement and Adhesive Application Point 15-24

1. Disassembly of Fresh Water-Cooled Engine

In general, the disassembly sequence for a fresh water-cooled engine is the same as that for a sea water-cooled engine, except that the sequence for parts related to the cooling water system are slightly different. Refer to the disassembly section of the sea water-cooled engine manual for the latter steps.

1-1. Draining the cooling water

(1) Drain the sea water from the heat exchanger. The sea water drain cock is installed on the side cover of the heat exchanger at the rear.

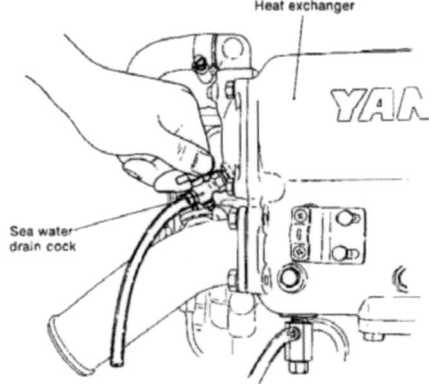

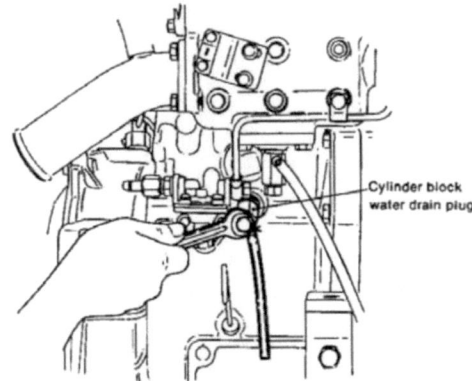

Cylinder block water drain plug

(2) Drain the fresh water from the heat exchanger. Loosen the fresh water drain plug installed at the bottom of the heat exchanger.

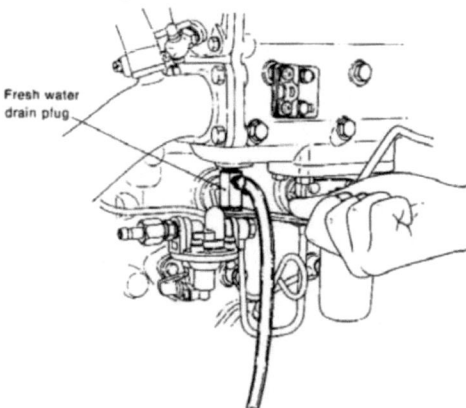

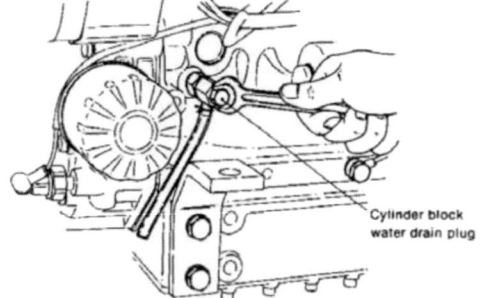

Cylinder block water drain plug

(4) Draining sea water from the CSW pump
Loosen the CSW pump cover fixing screws, and drain the sea water from the CSW pump and CSW hose.

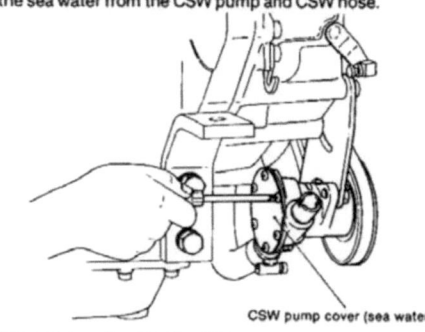

CSW pump cover (sea water)

(3) Drain the fresh water from the cylinder block.
Loosen the cylinder block water drain plug and drain the fresh water. The water drain plug is installed on the block wall surface at the exhaust side in a model 2GM20F engine, and on the block wall surface at the intake side in models 3GM30F and 3HM35F engines.

NOTE: CSW = Cooling Sea Water

1-2. Drain the lubricating oil

(1) Engine side
Insert a suction tube into the dipstick hole and pump out the oil with a waste oil pump (option).

1-3. Disconnect the remote control cables

(1) Clutch remote control cable
(2) Speed remote control cable
(3) Engine stop remote control cable
(4) Decompression remote control cable

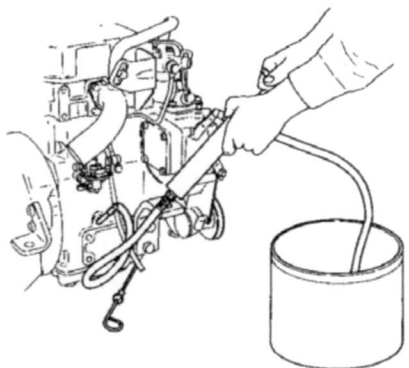

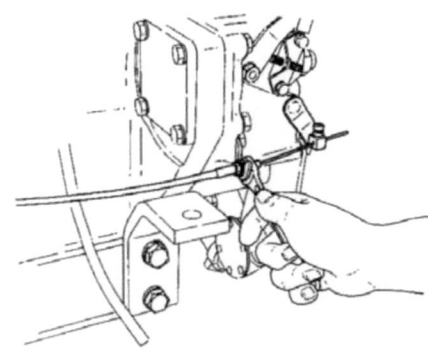

(2) Clutch side
Pump out the oil from the filler/dipstick hole using a waste oil pump or remove the drain plug at the bottom stern side of the clutch case and drain the oil.

1-4. Disconnect the electrical wiring

(1) Alternator wiring
(2) Starter motor wiring
(3) Water temperature switch wiring
(4) Oil pressure switch wiring
(5) Tachometer sender wiring

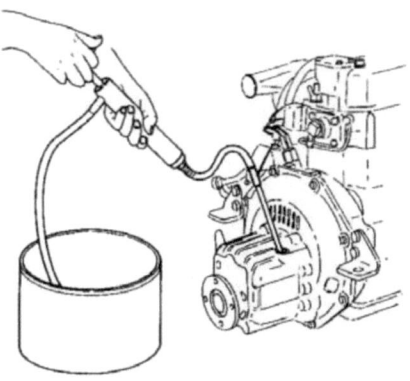

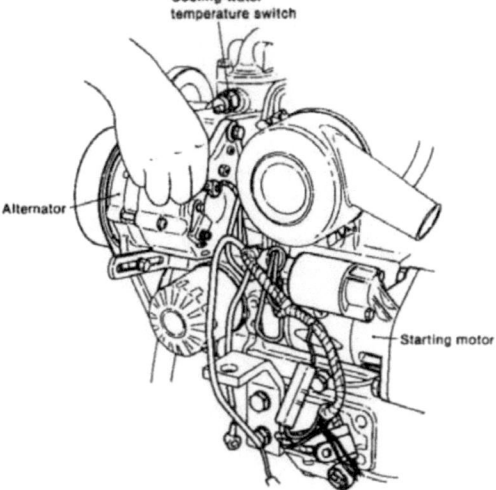

1-5. Remove the CSW hose

(1) Remove the CSW hose between the CSW pump and heat exchanger.

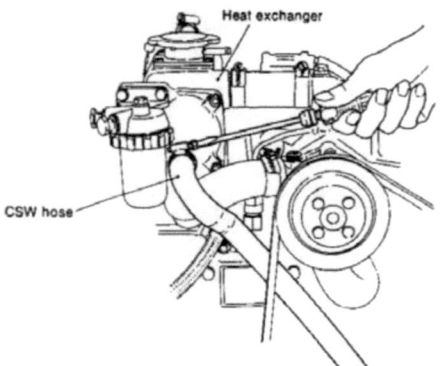

1-6. Disconnect the fuel piping

(1) Fuel tank to feed pump
(2) Feed pump to fuel filter
(3) Fuel filter to fuel injection pump

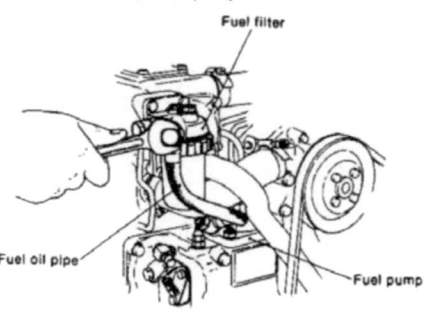

(2) Remove the CFW hose between the heat exchanger and CFW pump.

NOTE: CFW = Cooling Fresh Water

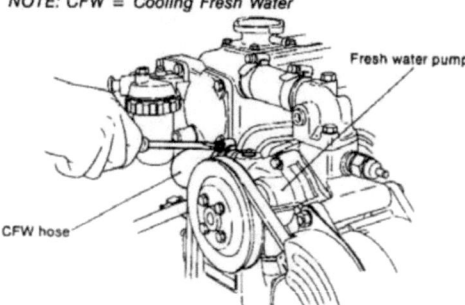

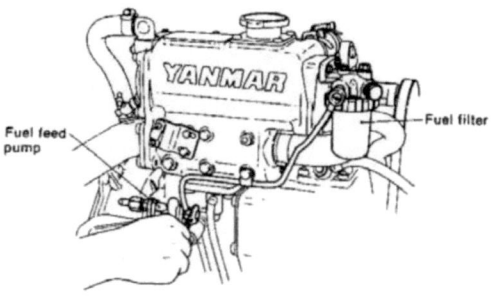

(3) Loosen the hose clamp on the CFW hose between the cylinder head and heat exchanger. The hose clamp at the heat exchanger side or the cooling water outlet connection side only need be loosened.

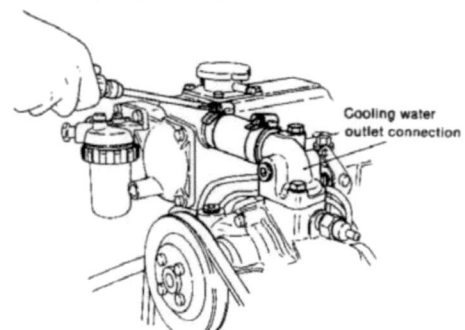

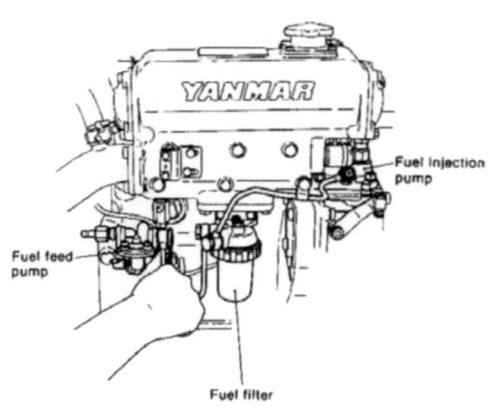

1-7. Removing the fuel filter

The fuel filter can be removed as assembled on the heat exchanger. However, to make removal of the heat exchanger easier, the filter should be removed separately.

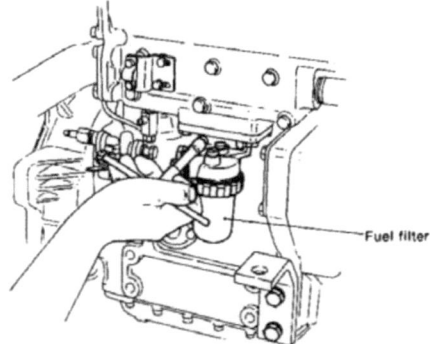
Fuel filter

1-8. Remove the fuel feed pump

In cases of model 3GMF, the heat exchanger drain plug may jam against the fuel feed pump pipe connecter. The fuel feed pump should be removed before removing the heat exchanger.
In cases of models 2GMF and 3HMF, the heat exchanger can be removed without removing the fuel feed pump.

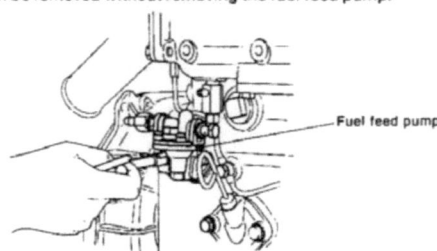
Fuel feed pump

1-9. Remove the remote control bracket

The heat exchanger fixing nut cannot be removed without first removing the remote control bracket.

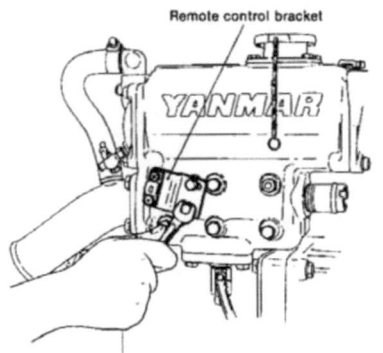
Remote control bracket

1-10. Remove the heat exchanger.

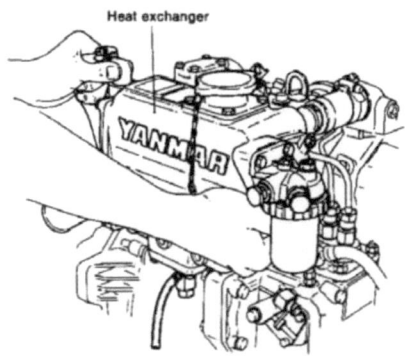
Heat exchanger

1-11. Remove the alternator

(1) Loosen the adjusting bolt and remove the V-belt
(2) Remove the alternator

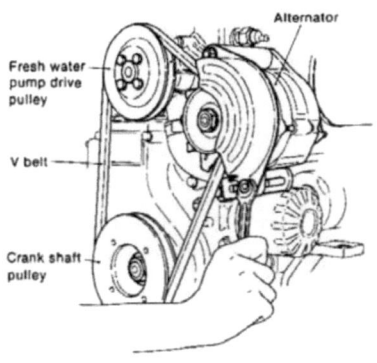
Alternator
Fresh water pump drive pulley
V belt
Crank shaft pulley

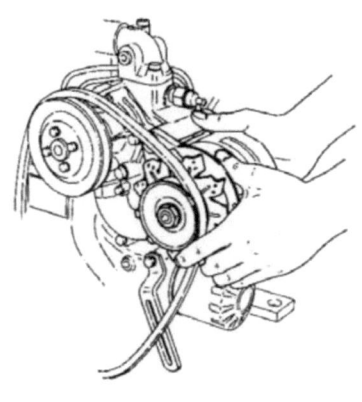

Chapter 15 Disassembly and Reassembly
1. Disassembly of Fresh Water-Cooled Engine

SM/GM(F)(C)·HM(F)(C)

1-12. Remove the CFW pump
Remove the CFW pump by loosening the hose clamp on the CFW hose between the CFW pump and cylinder block at the cylinder block end.

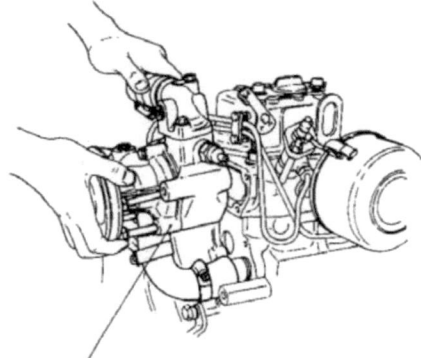
Fresh water pump

1-13. Remove the air intake silencer
Remove the intake silencer clip and the filter element. Then remove the set screw and the cover.

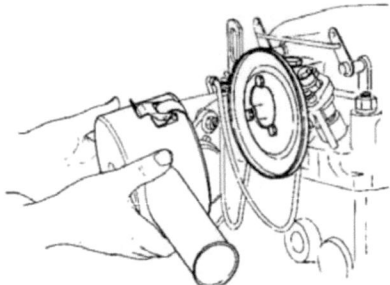

1-14. Remove the fuel high pressure pipe and fuel return pipe.

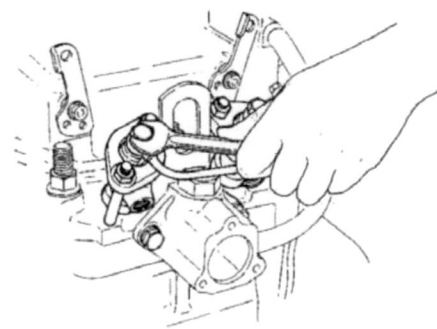

1-15. Remove the starter motor

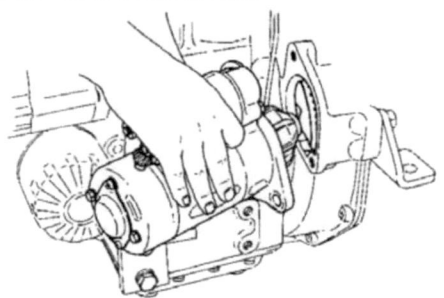

1-16. Remove the oil filter

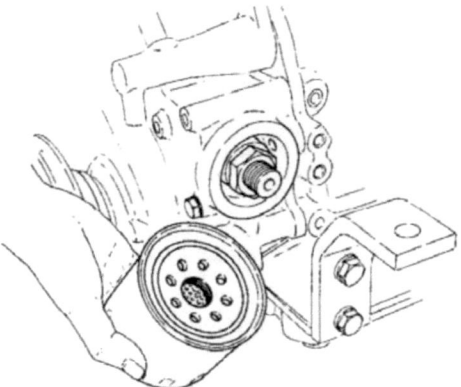

1-17. Remove the CSW pump
Loosen the water pump mounting bolts, remove the V-belt by sliding it toward the crankshaft side, and remove the sea water pump.

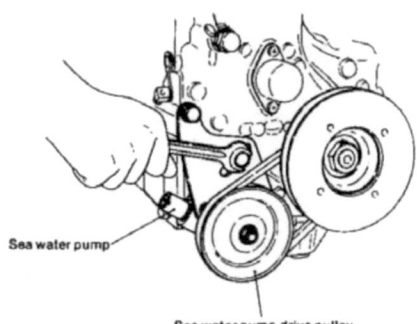
Sea water pump
Sea water pump drive pulley

Chapter 15 Disassembly and Reassembly
1. Disassembly of Fresh Water-Cooled Engine

SM/GM(F)(C)·HM(F)(C)

1-18. Remove the rocker arm chamber

(1) Remove the breather pipe at the side of the intake pipe
   [intake manifold for model 3GM30F and 3HM35F]
(2) Remove the rocker arm chamber

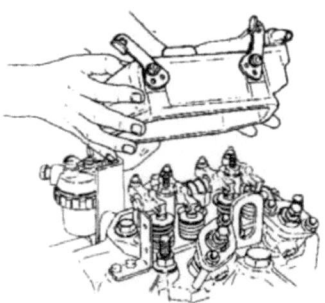

After this step disassembly is carried out in the same sequence as for the sea water-cooled engine.

The details are given in Section 4.3.14 "Remove the rocker arms "P14-23 of the sea water-cooled engine manual.

For reference

4-3.14 Remove the rocker arms

(1) Remove the mounting nut and remove the rocker arm shaft assembly.

(2) Pull the push rods.

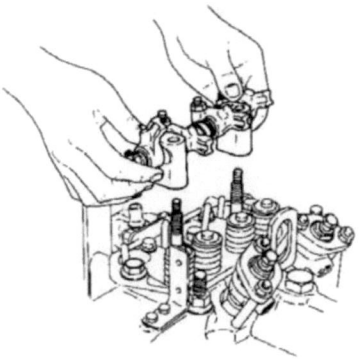

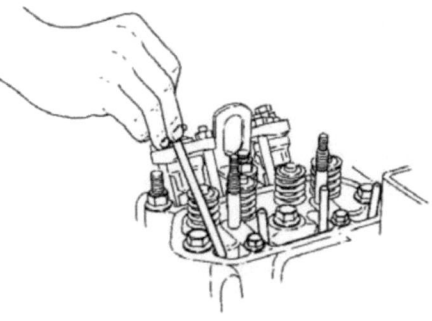

(3) Remove the cotter pins of the intake and exhaust valve springs.
NOTE: Arrange the parts by cylinder no., intake and exhaust.

Chapter 15 Disassembly and Reassembly
1. Disassembly of Fresh Water-Cooled Engine

4-3.15 Remove the cylinder head
(1) Disconnect the lubricating oil pipe.
(2) Remove the cylinder head nuts in the prescribed order, and remove the cylinder head.
(3) Remove the gasket packing.

4-3.17 Remove the injection pump
(1) Remove the injection pump nut.
(2) Remove the gear case side cover, move the governor lever 2, take out the fuel injection pump by matching the control rack with the cut-off part of the gear case.

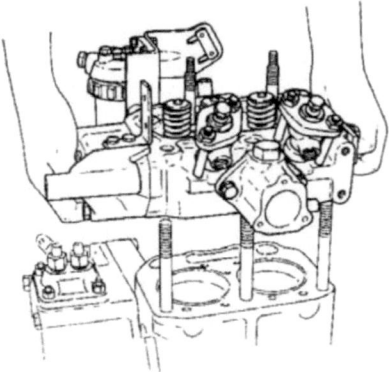

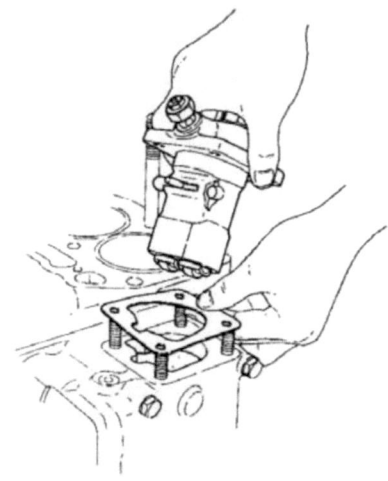

NOTE: Clearly identify the front and back of the gasket packing.

4-3.16 Remove the crankshaft pulley
Remove the crankshaft pulley end nut and remove the V-pulley and key.

(3) Remove the injection timing adjustment shims.
CAUTION: Note the number and total thickness of the timing adjustment shims.

4-3.18 Remove the timing gear case
(1) Remove the gear case

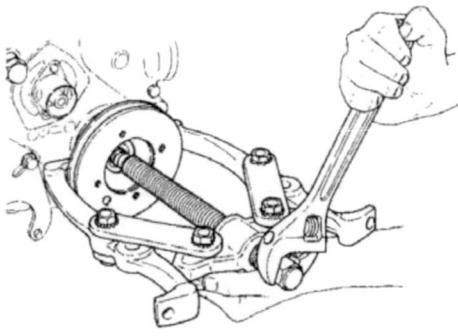

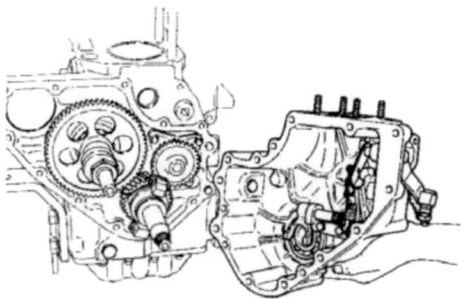

(2) Remove the thrust collar, thrust needle bearing, and governor sleeve.

4-3.19 Remove the clutch assembly

Loosen the mounting flange bolts and remove the clutch assembly.

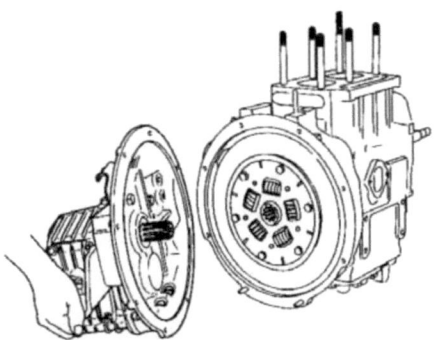

4-3.20 Remove the flywheel

(1) Remove the damper disk.

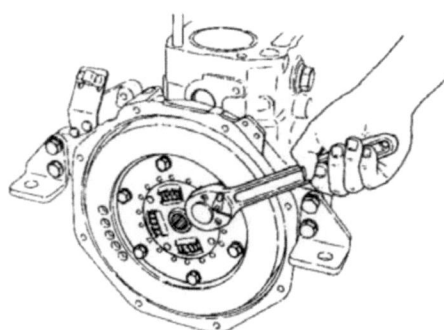

(2) Remove the flywheel.
Screw-in the two bolts to secure the clutch disk (slightly to the left and right of the flywheel) and remove it by pulling on the bolts.

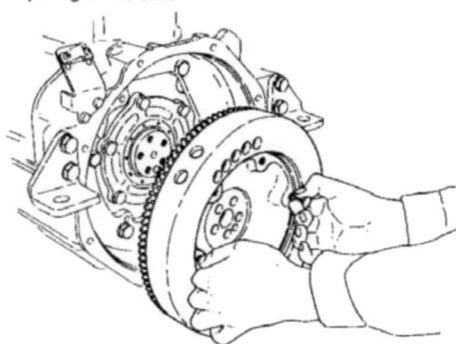

4-3.21 Remove the flywheel housing

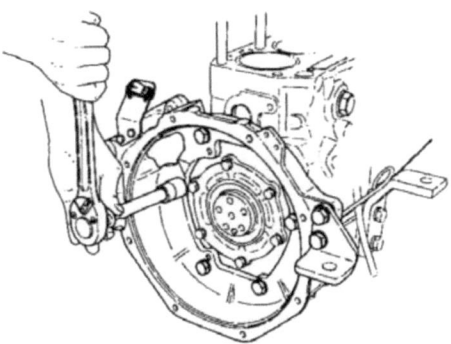

4-3.22 Remove the lubricating oil dipstick
4-3.23 Remove the feed pump

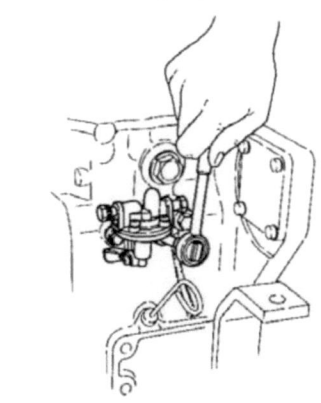

4-3.24 Remove the fuel cam

Remove the camshaft end nut and remove the fuel cam.

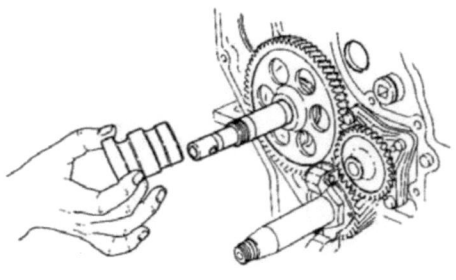

Chapter 15 Disassembly and Reassembly
1. Disassembly of Fresh Water-Cooled Engine

SM/GM(F)(C)-HM(F)(C)

4-3.25 Remove the governor weight assembly

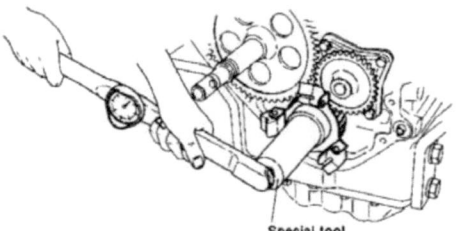

Special tool

Remove the crankshaft end nut and remove the governor weight assembly.

4-3.26 Remove the lubricating oil pump and driving gear assembly

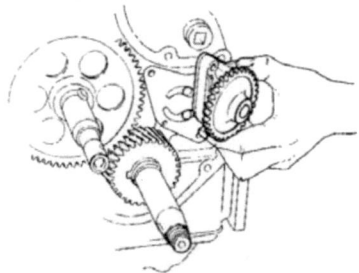

4-3.27 Remove the camshaft gear and the crankshaft gear

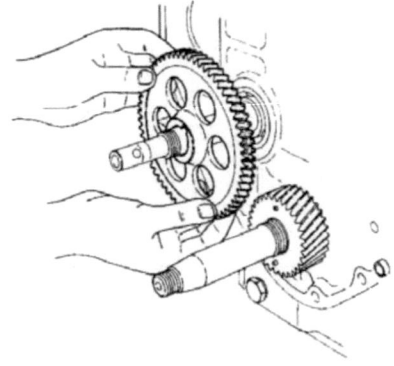

4-3.28 Turn the engine onto its side

(1) Remove the engine feet of the crankshaft side
(2) Turn the cylinder block over so that the crankshaft side is on the bottom.

4-3.29 Remove the oil pan and the oil intake pipe

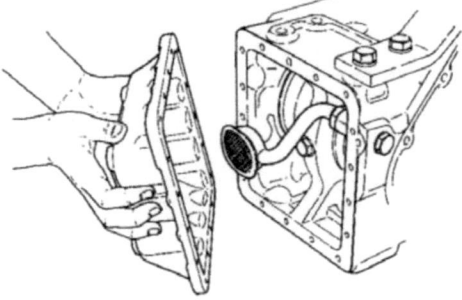

4-3.32 Remove the mounting bolt of the intermediate main bearing

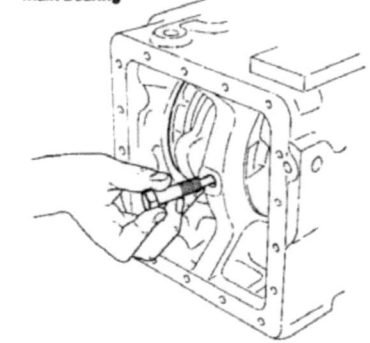

There are two intermediate main bearings, viz. No.1 and No.2, for engine model 3GM30F.

4-3.33 Pull the crankshaft

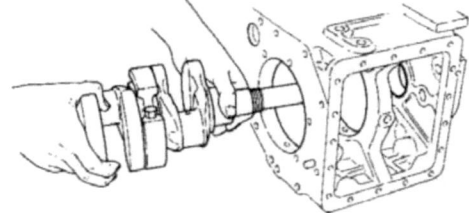

Chapter 15 Disassembly and Reassembly
1. Disassembly of Fresh Water-Cooled Engine

SM/GM(F)(C)·HM(F)(C)

4-3.34 Remove the camshaft

(1) Remove the camshaft bearing set screw.

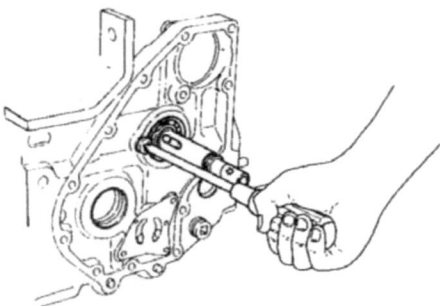

(2) Place the cylinder block upside down or raise the cylinder block by inserting a plate beneath it in order to prevent contact between the tappet and the cam.

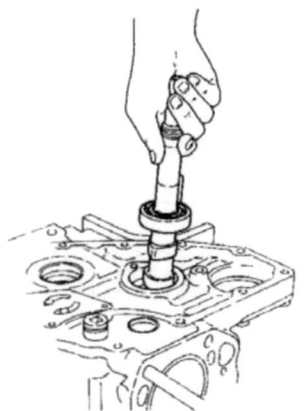

(3) Check that all the tappets are separated from the cam, and pull the camshaft out.

4-3.35 Remove the tappets

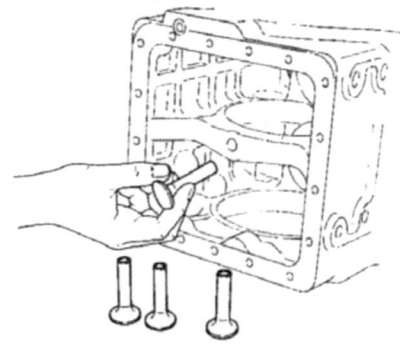

NOTE: Arrange the removed tappets by cylinder no. and intake and exhaust groups.

2. Reassembly of fresh water-cooled engine

In general, the reassembly of the fresh water-cooled engine is the same as that for a sea water-cooled engine, except for cooling system components such as the heat exchanger, fresh water pump, cooling water pipe and related parts.

For details of the reassembly sequence refer to chapter 14, Page 14-28~14-43 (Reassembly of Direct Sea-Water Cooling Engine)

For reference

5-3.1 Insert the tappets

(1) Turn the cylinder block over or turn it upside down.
(2) Coat the tappets with oil and insert into the tappet holes.

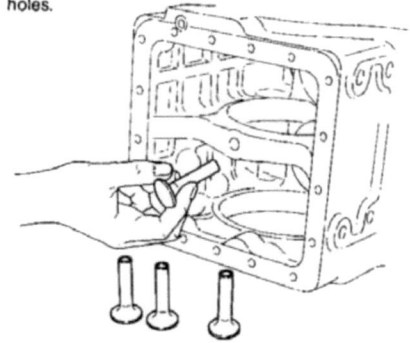

NOTE: Assemble the tappets in their original positions, paying careful attention to the cylinder numbers and intake and exhaust groupings.

5-3.2 Insert the camshaft

(1) Coat the camshaft bearing section with oil and insert the camshaft into the cylinder block by tapping the shaft end with a plastic hammer.

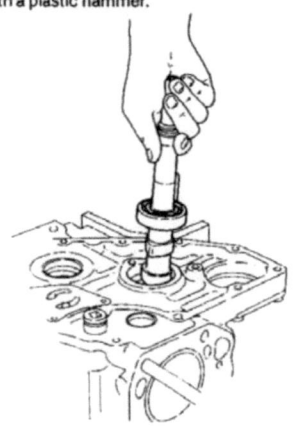

NOTE: Be careful not to damage the groove in the end of the shaft.

(2) After inserting the camshaft, check that it rotates smoothly before tightening the camshaft bearing set screw.

Tightening torque	2kgf-m (14.5ft-lb)

5-3.3 Install the crankshaft

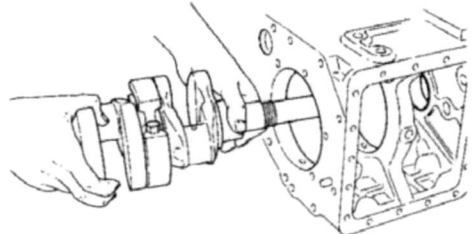

5-3.4 Tighten the set bolt of the intermediate main bearing

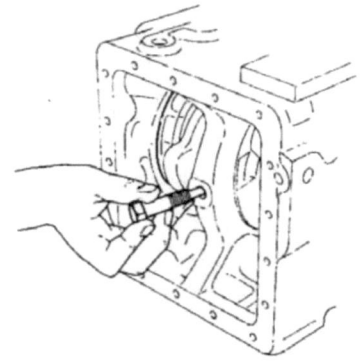

There are two intermediate main bearings, viz No.1 and No.2, for model 3GM30F

kgf-m(ft-lb)

	2GM20F, 3GM30F	3HM35F
Tighten torque	4.5~5.0 (32.5~36.2)	7.0~7.5 (50.6~54.2)

15-11

5-3.5 Install the main bearing housing

(1) Coat the oil seal section with oil.
(2) Insert the main bearing housing and tighten.

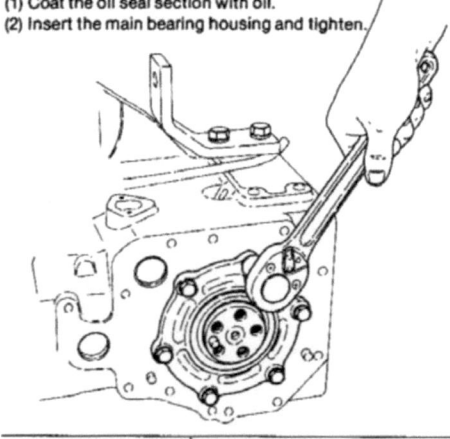

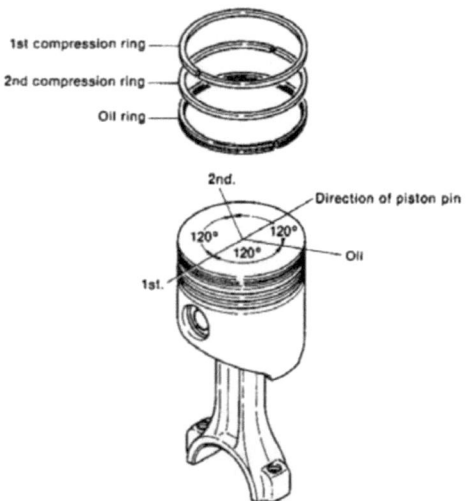

| Tightening torque | 2.5kgf-m (18ft-lb) |

(3) Check that the crankshaft rotates smoothly.
(4) Measure the crankshaft side gap, and adjust it to the prescribed value using the thickness of the packing.

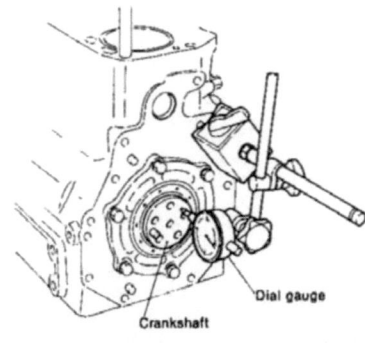

(4) Insert the piston connecting rod assembly so that the side of the connecting rod large end with the identification number is on the exhaust side.
Install the piston rings with a piston ring inserter.

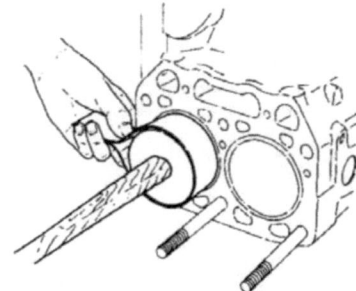

| Crankshaft side gap | 0.09 ~ 0.18mm (0.035 ~ 0.0071in.) |

5-3.6 Assemble the piston and connecting rod assembly

(1) Coat the crankpin section with oil and position it so that the insertion side crank is at the top.
(2) Coat the piston and crankpin bearing with oil.
(3) Position the piston rings so that the gaps are 120° apart; make sure that there is no gap at the side pressure section.

(5) After the connecting rod large end makes contact with the crankpin, push the piston crown down slowly to turn the crankshaft to bottom dead center.
(6) Align the connecting rod cap and connecting rod large end matching mark and tighten the connecting rod bolts.

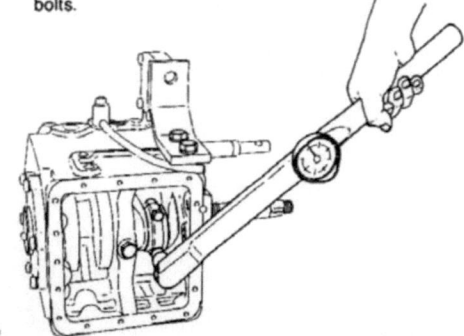

Chapter 15 Disassembly and Reassembly
2. Reassembly of Fresh Water-Cooling Engine

CAUTION: 1. Be careful to tighten the connecting rod bolts evenly.
2. Coat the bolt threads and washer face with oil.

	2GM20F, 3GM30F	3HM35F
Tightening torque	2.5(18.1)	4.5(32.5)

kgf-m(ft-lb)

(7) Measure the side clearance

Side clearance	0.2~0.4mm (0.0079~0.0157in.)

(8) Check that the crankshaft rotates smoothly.

5-3.7 Install the lubricating oil intake pipe
Coat the threads with "Screw Lock Super 203M", screw the pipe in and lock with the nut.

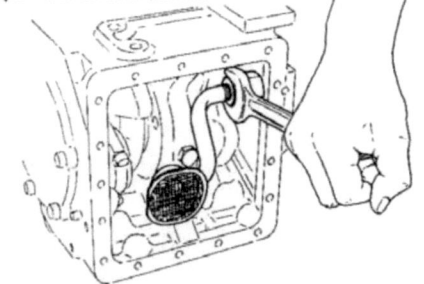

Screw-in distance	8 ~ 10mm (about 6 turns) (0.3149 ~ 0.3937in.)

5-3.8 Install the engine bottom cover (oil pan)
(1) Change the packing.
(2) Install the bottom cover.

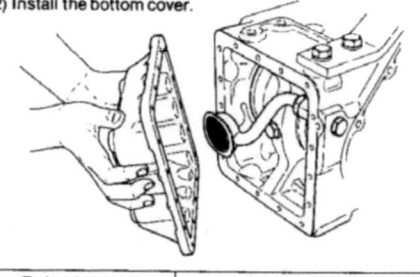

Tightening torque	0.9kgf-m(6.5ft-lb)

5-3.9 Install the mounting flange
(1) Set the engine upright.
(2) Align the positioning pins and tighten the flange.

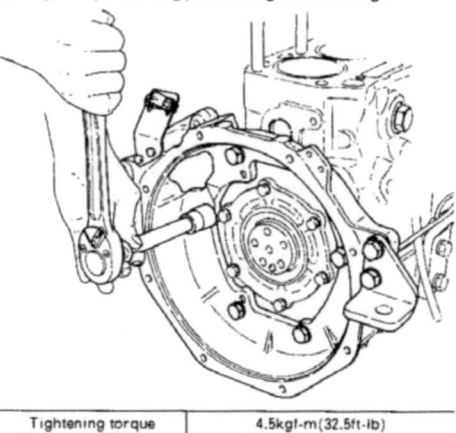

Tightening torque	4.5kgf-m(32.5ft-lb)

5-3.10 Install the flywheel
(1) Align the reference pins.
(2) Install the flywheel.

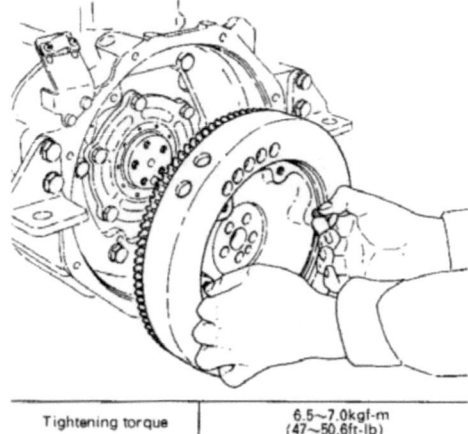

Tightening torque	6.5~7.0kgf-m (47~50.6ft-lb)

NOTE: After tightening, check the end run-out.

Chapter 15 Disassembly and Reassembly
2. Reassembly of Fresh Water-Cooling Engine

SM/GM(F)(C)·HM(F)(C)

5-3.11 Install the clutch assembly

(1) Install the clutch disc on the flywheel.

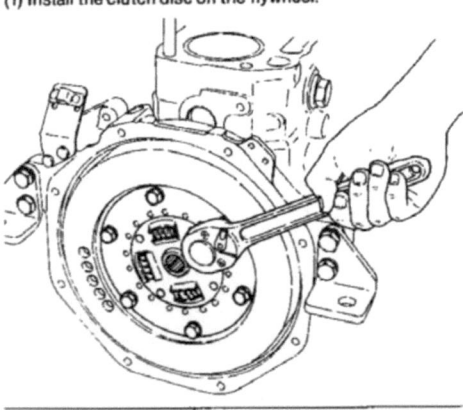

Tightening torque	2.5kgf-m(18ft-lb)

(2) Align the disc and input shaft spline, and install the clutch assembly on the mounting flange.

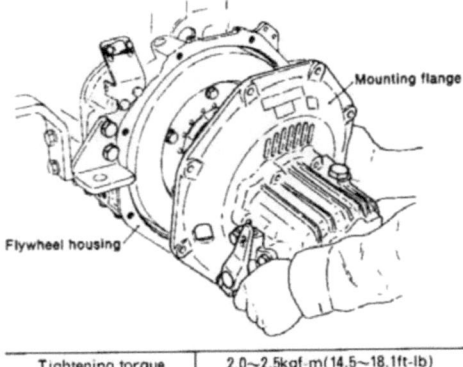

Tightening torque	2.0~2.5kgf-m(14.5~18.1ft-lb)

5-3.12 Install the engine feet and set the engine in position

(1) Dipstick flange and dipstick.
(2) Fuel pump.

5-3.13 Install the lubricating oil pump

Install the lubricating oil pump and driving gear assembly.

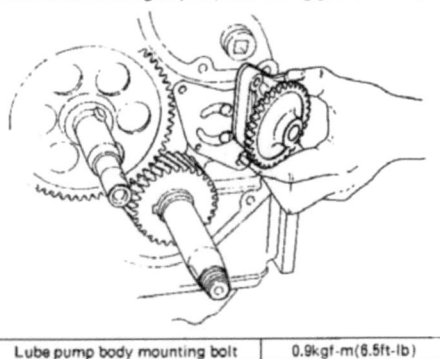

Lube pump body mounting bolt	0.9kgf-m(6.5ft-lb)

5-3.14 Assemble the camshaft gear and fuel cam

(1) Coat the shaft hole of the camshaft gear with oil and insert the gear.
(2) Coat the fuel cam with oil and insert the cam by aligning the "0" mark opposite the camshaft gear.

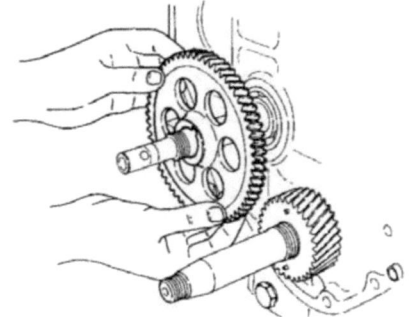

(3) Tighten the camshaft end nut.

Tightening torque	7~8kgf-m(50.6~57.9ft-lb)

5-3.15 Assemble the crankshaft gears

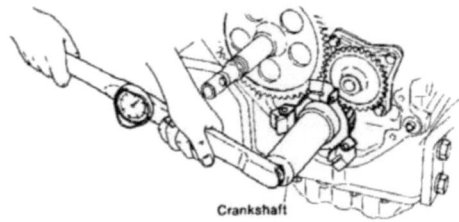

(1) Coat the crankshaft section and the inside of the gear with oil.
(2) Align the matching marks of the camshaft gear and the crankshaft gear and insert the crankshaft gear.

15-14

Printed in Japan
0000A0A1361

(3) After inserting the crankshaft gear, check the backlash.

Backlash	0.05 ~ 0.13mm (0.0020 ~ 0.0051 in.)

(4) Install the governor weight assembly and tighten the crankshaft end nut.

Tightening torque	8~10kgf-m (57.9~72.3ft-lb)

5-3.16 Install the governor sleeve

Install the governor sleeve, thrust needle bearing and thrust collar.

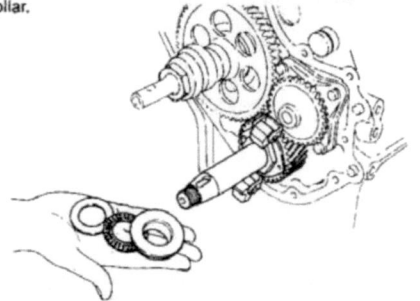

5-3.17 Install the timing gear case

(1) Coat both sides of the new packing with "Three Bond 3B8-005" and install.
(2) Install the timing gear case.

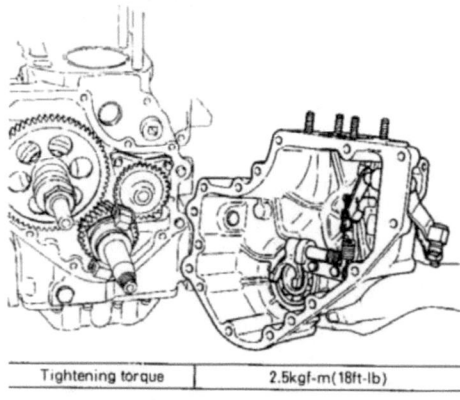

Tightening torque	2.5kgf-m (18ft-lb)

5-3.18 Install the crankshaft V-pulley

(1) Install the crankshaft key.
(2) Coat the crankshaft V-pulley and the inside of the oil seal with oil.

(3) Insert and tighten the V-pulley, making sure that the lip of the oil seal is not distorted.

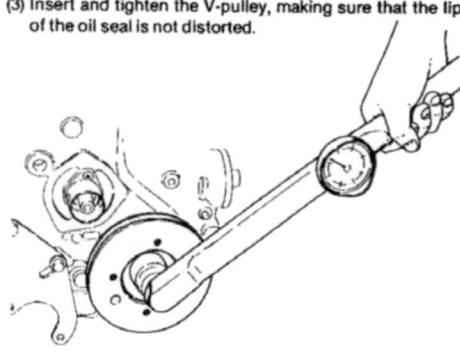

Tightening torque	10kgf-m (72.3ft-lb)

5-3.19 Install the water pump

Install the V-belt to the crankshaft V-pulley and install the water pump.

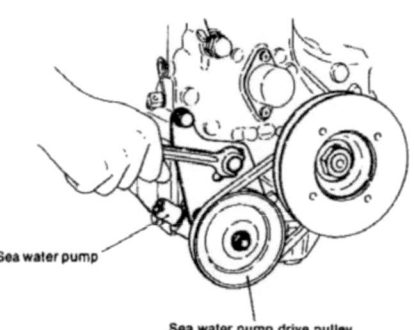

Sea water pump

Sea water pump drive pulley

(1) V-belt tension

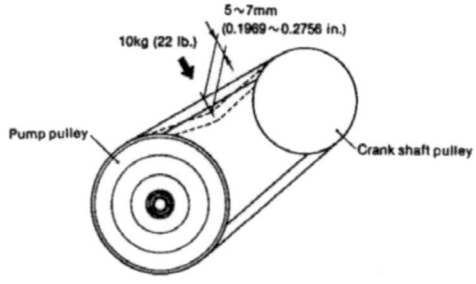

Pump pulley

Crank shaft pulley

V-belt tension Pushed with a force of 10kg (22 lb.)	5 ~ 7mm (0.1969 ~ 0.2756 in.)

Chapter 15 Disassembly and Reassembly
2. Reassembly of Fresh Water-Cooling Engine

(2) Tightening torque.

Tightening torque	2.5kgf-m(18ft-lb)

(3) Insert the pump by looking through the gear case side cover, and align the governor No.2 lever and rack connecting part.

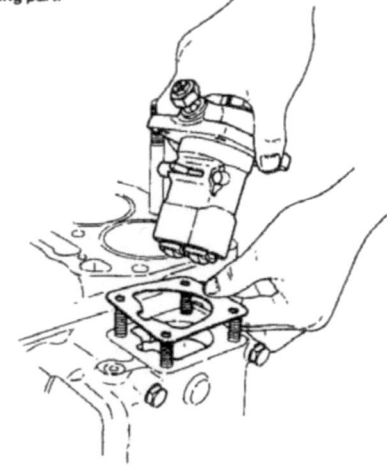

(4) Tighten the fuel pump

Tightening torque	2.5kgf-m(18ft-lb)

(5) Install the gear case side cover.

5-3.20 Install the cylinder head
(1) Install the gasket packing.

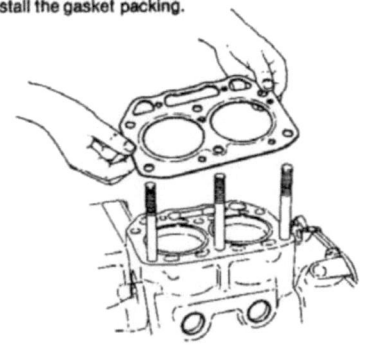

CAUTION: Take particular notice of the surfaces to be fitted.
Keep the TOP mark on the cylinder head side.

(2) Insert the cylinder head, being careful not to damage the threads of the tightening bolts, and tighten the nuts in the tightening sequence.

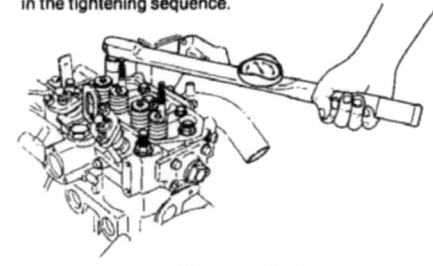

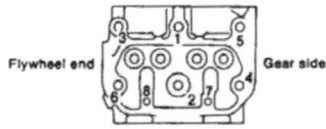

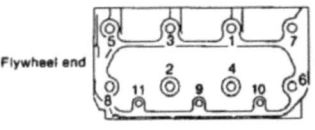

kgf-m(ft-lb)

Tightening torque		2GM20F, 3GM30F	3HM35F
	Main (M12)	12.0(86.8)	13 (94.0)
	Sub (M8)	3.0(21.7)	3 (21.7)

(3) Install the water pipe
(from the thermostat cover to the cylinder inlet joint).

5-3.21 Install the rocker arms
(1) Install the push rods on the tappets.

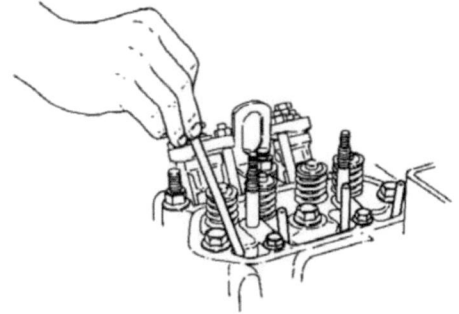

Chapter 15 Disassembly and Reassembly
2. Reassembly of Fresh Water-Cooling Engine

(2) Coat the inside of valve spring retainer with oil.
(3) Install the rocker arm shaft assembly and tighten the nut.

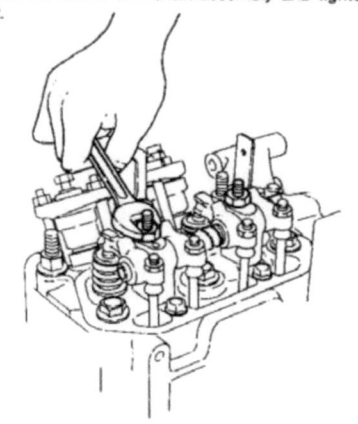

| Tightening torque | 3.7kgf-m(27ft-lb) |

CAUTION: 1. Loosen the valve head clearance adjusting screw in advance.
2. Check that the arm moves smoothly.

(4) Adjust the intake and exhaust valve head clearance and lock with the nut.

| Intake and exhaust valve head clearance (engine cold) | 0.2mm (0.008in.) |

5-3.22 Install the rocker arm cover

(1) Install the rocker arm cover.

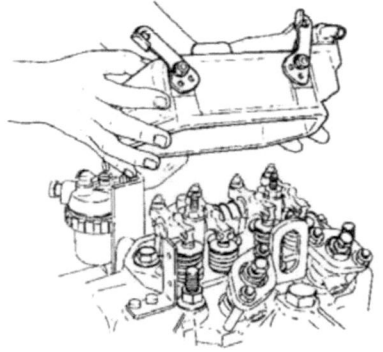

(2) Install the breather pipe to the air intake pipe (intake manifold ... 3GM30F).

The following sequence is different from that of a sea water-cooled engine.

2-1. Install the starter motor

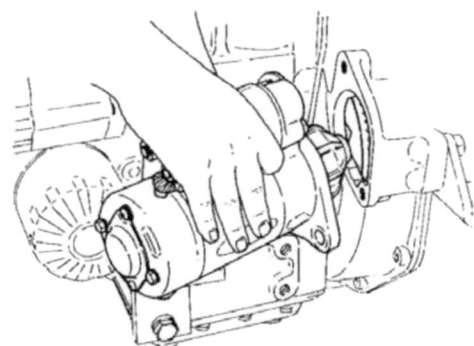

2-2. Install the oil filter

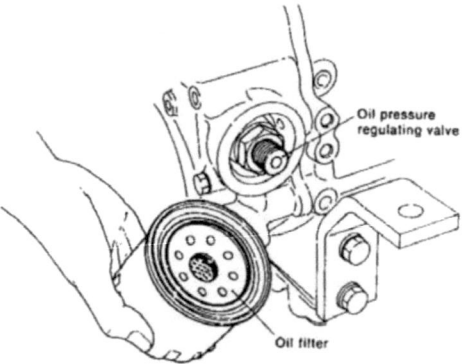

2-3. Assemble the high pressure fuel pipe and fuel return pipe

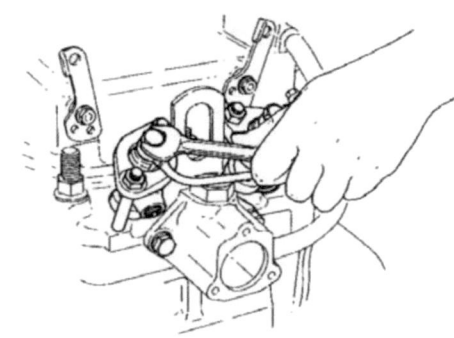

Chapter 15 Disassembly and Reassembly
2. Reassembly of Fresh Water-Cooling Engine

SM/GM(F)(C)·HM(F)(C)

2-4. Install the intake silencer

(1) Install the intake silencer cover to the air intake pipe.
[intake manifold . . . 3GM30F and 3HM35F]
(2) Install the intake silencer and tighten it with the clip.

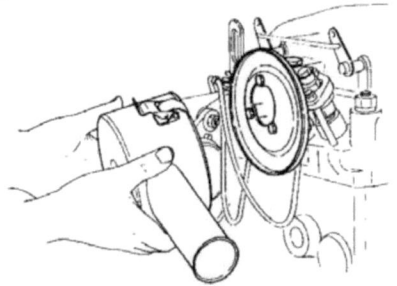

2-5. Assemble the CFW pump

Mount the CFW pump and replace the CFW hose between the CFW pump and cylinder block by connecting at the CFW pump and at the cylinder block.

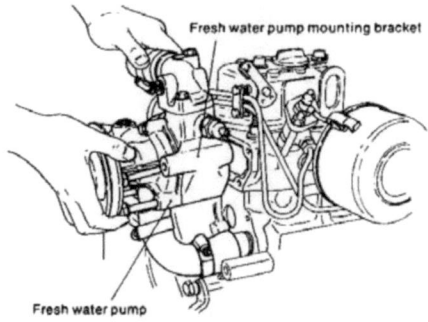

Fresh water pump

Tightening torque for CFW pump fixing bolt	2~2.5kgf-m (14.5~18ft-lb)

2-6. Install the alternator

(1) Install the alternator to the bracket.

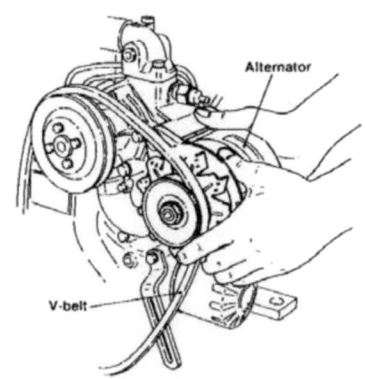

(2) Install the V-belt and tighten the adjusting bolt while adjusting the V-belt tension.

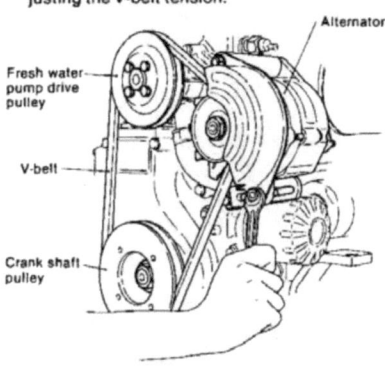

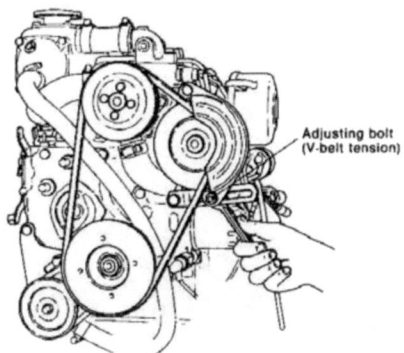

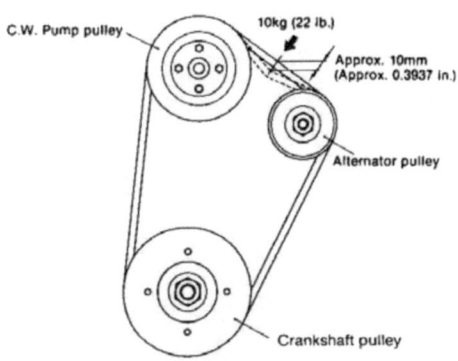

V-belt tension Pushed with a force of 10kg (22 lb.)	Approx. 10mm (Approx. 0.3937 in.)

15-18

Printed in Japan
0000A0A1361

Chapter 15 Disassembly and Reassembly
2. Reassembly of Fresh Water-Cooling Engine

SM/GM(F)(C)-HM(F)(C)

2-7. Assemble the heat exchanger
Mount the heat exchanger and replace the CFW hose at the thermostat cover side by connecting the hose to the heat exchanger. Tighten the hose clamp after the heat exchanger is assembled.

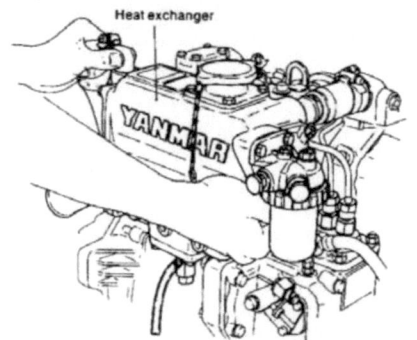
Heat exchanger

Tightening torque for heat exchanger fixing bolt	2~2.5kgf-m (14.5~18ft-lb)

2-8. Assemble the fuel feed pump (3GM30F)
For model 3GM30F engine assemble the fuel feed pump after the heat exchanger is assembled.

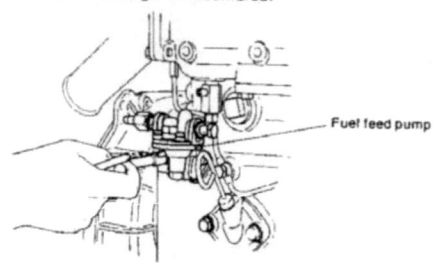
Fuel feed pump

2-9. Assemble the fuel oil pipe
(1) Fuel feed pump—fuel filter
(2) Fuel filter—fuel injection pump

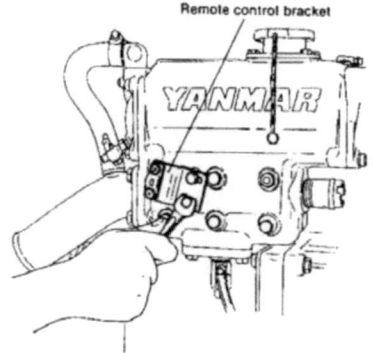
Remote control bracket

2-10. Assemble the cooling water pipe
(1) CFW hose (heat exchanger—CFW pump)

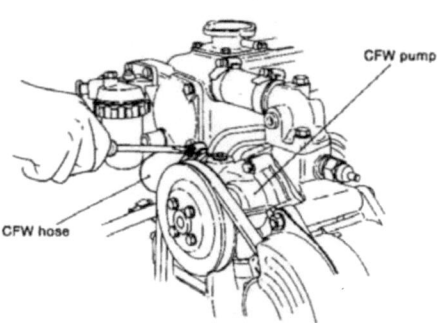
CFW pump
CFW hose

(2) CSW hose (CSW pump—heat exchanger)

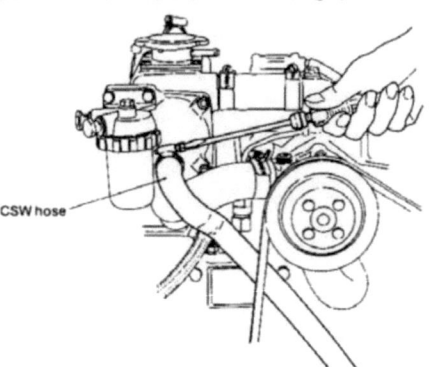
CSW hose

2-11. Connect the electrical wiring

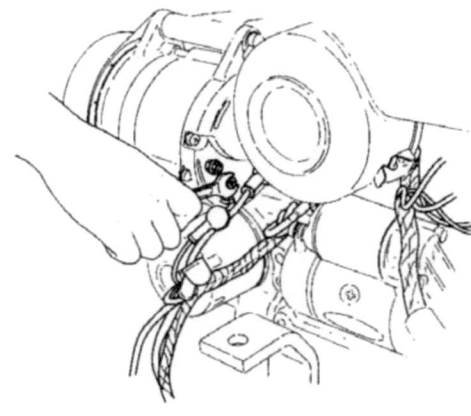

2-12. Install the remote control cables

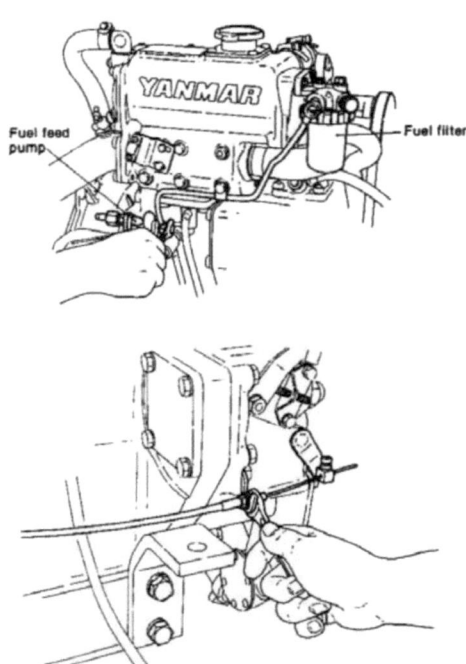

2-13. Connect the interior piping

Chapter 15 Disassembly and Reassembly
3. Tightening Torque

SM/GM(F)(C)-HM(F)(C)

3. Tightening Torque

The bolts and nuts used in this engine employ ISO general metric threads stipulated in JIS (Japanese Industrial Standards). Pay careful attention to the thread dimensions when replacing bolts and nuts.
Tighten the bolts and nuts to the tightening torque given in the table below.

3-1 Main bolt and nut tightening torque

Location	Bolt/nut		1GM10(C)	2GM20(F)(C)	3GM30(F)(C)	3HM35(F)(C)	Remarks
Cylinder head	Cylinder head tightening bolt and nut	Thread diameter	M10	M12	M12	M12	Nut and bolt
			—	M8	M8	M8	Aux. bolt
		Quantity	4	6	8	8	Nut and bolt
			—	2	3	3	Aux. bolt
		Tightening torque kgf-m(ft-lb)	7.5(54.248)	12.0(86.8)	12.0(86.8)	13(94.029)	Nut and bolt
			—	3.0(21.7)	3.0(21.7)	3(21.699)	Aux. bolt
	Rocker arm support nut	Thread diameter	M8	M10	M10	M10	
		Quantity	1	2	3	3	
		Tightening torque kgf-m(ft-lb)	3.7(26.762)				
	Exhaust manifold nuts	Thread diameter × pitch mm	M8 × 1.25				
		Quantity	2	3	6	6	
		Tightening torque kgf-m(ft-lb)	4.5(32.549)				
	Anticorrosion zinc	Thread diameter	—	M25	M25	M25	
		Quantity	—	1	1	1	
		Tightening torque kgf-m(ft-lb)	—	5~6(36.165~43.398)			
Timing gear	Timing gear case mounting bolt	Thread diameter × pitch mm	M6 × 1.0	M8 × 1.25			
		Quantity	12	12	12	12	
		Tightening torque kgf-m(ft-lb)	0.9(6.510)	2.5(18.083)			
	Camshaft end nut	Thread diameter × pitch mm	M20 × 1.5	M20 × 1.5	M20 × 1.5	M18 × 1.5	
		Quantity	1				
		Tightening torque kgf-m(ft-lb)	7~8(50.631~57.864)				
	Governor weight set nut	Thread diameter × pitch mm	M26 × 1.5				
		Quantity	1				
		Tightening torque kgf-m(ft-lb)	8~10(57.864~72.330)				

Chapter 15 Disassembly and Reassembly
3. Tightening Torque

SM/GM(F)(C)-HM(F)(C)

Location	Bolt/nut		1GM10(C)	2GM20(F)(C)	3GM30(F)(C)	3HM35(F)(C)	Remarks
Cylinder block	Mounting flange bolt	Thread diameter × pitch mm	M10 × 1.5				
		Quantity	6				
		Tightening torque kgf-m(ft-lb)	4.5(32.549)				
	Bottom cover bolt	Thread diameter × pitch mm	M6 × 1.0				
		Quantity	13	17	21	23	
		Tightening torque kgf-m(ft-lb)	0.9(6.510)				
	Oil pressure switch mounting	Thread diameter	PT 1/8				
		Quantity	1				
		Tightening torque kgf-m(ft-lb)	1.0(7.233)				
Crankshaft, pistons	Main bearing housing bolt	Thread diameter × pitch mm	M8 × 1.25				
		Quantity	6				
		Tightening torque kgf-m(ft-lb)	2.5(18.083)				
	Connecting rod bolt	Thread diameter × pitch mm	M7 × 1.0			M9 × 1.0	
		Quantity	1 × 2 = 2	2 × 2 = 4	3 × 2 = 6		
		Tightening torque kgf-m(ft-lb)	2.5(18.083)			4.5(0.6221)	
	Crankshaft V-pulley bolt	Thread diameter	M18				3HM35(F)(C) Counterclock- wise screw
		Quantity	1				
		Tightening torque kgf-m(ft-lb)	10(72.330)				
	Flywheel bolt	Thread diameter × pitch mm	M10 × 1.25				
		Quantity	5				
		Tightening torque kgf-m(ft-lb)	6.5~7.0(47.015~50.631)				
	Diameter disk bolt	Damper diameter × pitch mm	M8 × 1.25				
		Quantity	6			8	
		Tightening torque kgf-m(ft-lb)	2.5(18.083)				
	Intermediate main bearing housing bolt	Thread diameter × pitch mm	—	M8 × 1.25			
		Quantity	—	2 × 2 = 4	3 × 2 = 6		
		Tightening torque kgf-m(ft-lb)	—	3.0~3.5 (21.699~25.316)		4.5~5.0 (32.549~36.165)	
	Intermediate main bearing housing set bolt	Thread diameter × pitch mm	—	M10 × 1.25			
		Quantity	—	1		2	
		Tightening torque kgf-m(ft-lb)	—	4.5~5.0 (32.549~36.165)		7.0~7.5 (50.631~54.248)	
Cooling system	Water temperature sender bolt	Thread diameter	PT 3/8				
		Quantity	1				
		Tightening torque kgf-m(ft-lb)	1.0~1.5(7.2330~10.850)				

Chapter 15 Disassembly and Reassembly
3. Tightening Torque

SM/GM(F)(C)-HM(F)(C)

Location	Bolt/nut		1GM10(C)	2GM20(F)(C)	3GM30(F)(C)	3HM35(F)(C)	Remarks
Cooling system	Anticorrosion zinc mounting (Cylinder block)	Thread diameter × pitch mm					1GM10(C): Flange type 2GM20(C), 3GM30(C) and 3HM35(C): Plug type
		Quantity		1		2	
		Tightening torque kgf-m(ft-lb)		5~6(36.165~43.498)			
	Cooling water inlet joint	Thread diameter × pitch mm					
		Quantity			1		
		Tightening torque kgf-m(ft-lb)					
	Water pump body bolt	Thread diameter × pitch mm	M6×1.0		M8×1.25		
		Quantity	3		2		
		Tightening torque kgf-m(ft-lb)	0.9(6.5097)		2.5(18.083)		
Fuel system	Nozzle nut	Thread diameter × pitch mm		M20×1.5			
		Quantity	1	2		3	
		Tightening torque kgf-m(ft-lb)		10(72.330)			
	Delivery valve holder	Thread diameter		M18			
		Quantity	1	2		3	
		Tightening torque kgf-m(ft-lb)		4.0~4.5(28.932~32.549)			
	Fuel injection nozzle flange nut	Thread diameter × pitch mm		M8 × 1.25			
		Quantity	2×1=2	2×2=4		2×3=6	
		Tightening torque kgf-m(ft-lb)		2(14.466)			
Clutch system	Clutch housing nut	Thread diameter × pitch mm		M8×1.25			(*2) GM-series: M18 x 1.5 3HM35(F)(C): M24 (*3) a: 39.5 (1.5551) b: 32 (1.2598) c: 7 (0.2755)
		Quantity		8			
		Tightening torque kgf-m(ft-lb)		2.0~2.5(14.466~18.083)			
	Clutch mounting bolt	Thread diameter × pitch mm		M8×1.25			
		Quantity		8			
		Tightening torque kgf-m(ft-lb)		2.0~2.5(14.466~18.083)			
	Output shaft coupling tightening nut	Thread diameter × pitch mm		(*2)			
		Width B/C mm(in.)		30/34.6(1.1811/1.3622)		(*3)	
		Quantity					
		Tightening torque kgf-m(ft-lb)		10±1.5 (72.330~10.850)		9.5 (68.714)	
Electric system	Starter motor mounting top	Thread diameter × pitch mm		M10×1.5		M12	
		Quantity		2			
		Tightening torque kgf-m(ft-lb)		4.5~5.0 (32.549~36.165)		7.5~8.0 (54.248~57.864)	
	AC generator mounting bolt	Thread diameter × pitch mm		M8×1.25			
		Quantity		3			
		Tightening torque kgf-m(ft-lb)		2.2~2.7(15.913~19.530)			

3-2 General bolt and nut tightening torque

kgf-m(ft-lb)

Diameter of thread	General bolts 7T	Pipe joint bolts
M6	0.9±0.1 (5.9 ~ 7.2)	—
M8	2.5±0.2 (16.6 ~ 19.5)	1.2 ~ 1.7 (8.7 ~ 12.3)
M10	4.7±0.3 (31.8 ~ 36.2)	—
M12	8.0±0.5 (54.2 ~ 61.5)	2.5 ~ 3.5 (18.1 ~ 25.3)
M14	13.0±0.5 (90.4 ~ 97.6)	4.0 ~ 5.0 (28.9 ~ 36.2)
M16	20.5±0.5 (144.7 ~ 151.9)	5.0 ~ 6.0 (36.2 ~ 43.4)

4. Packing Supplement and Adhesive Application Points

The packing used in this engine is asbestos sheet sealed at both mating faces.
Be sure to use the correct supplement in accordance with the table below.

Location	Packing (coated)	Packing agent and adhesive
Cylinder head	Both sides of cylinder head side cover packing Cylinder head top and bottom casting sand hole plug Rocker arm chamber packing (rocker arm chamber side) Both sides of cylinder head gasket packing Intake and exhaust manifold bolt threads Exhaust manifold stud bolt thread Rocker arm support stud bolt Cooling water outlet joint threads	"Three Bond No. 4" "Three Bond No. 50" "Screw Lock Super 203M" "Screw Lock Super 203M"
Timing gear	Both sides of timing gear case packing Both sides of fuel injection timing adjustment shims Both sides of governor chamber packing Governor drive shaft bearing cover packing	"Three Bond 3B8-005" "Screw Lock Super 203M" "Three Bond 3B8-005"
Cylinder block	Both sides of oil pan packing Outside surface of cylinder liner Cooling water pipe joint threads Lubricating oil suction pipe threads Lubricating oil intake pipe blind plug threads Oil pressure regulator valve threads Oil pressure switch threads Cylinder head bolt stud Mounting flange face Lube oil pump face Both sides of bushing shell packing Both sides of dipstick flange packing Both sides of fuel pump packing	"Three Bond 3B8-005" White paint "Three Bond No. 20" "Screw Lock Super 203M" "Three Bond 3B8-005"
Crankshaft, piston	Crankshaft V-pulley key groove tightening section Connecting rod bolt threads	"Three Band 3B8-005"
Cooling system	Both sides of water pump packing Both sides of water pump packing Anticorrosion zinc flange threads Water temperature switch threads Water drain joint (cylinder, exhaust pipe)	"Three Bond No. 2" "Three Bond No. 4"
Clutch system	Mounting flange face Clutch housing face